ISBN 978-1-5278-0037-3
PIBN 10901051

1 MONTH OF
FREE
READING

at

www.ForgottenBooks.com

By purchasing this book you are
eligible for one month membership to
ForgottenBooks.com, giving you
unlimited access to our entire
collection of over 1,000,000 titles via
our web site and mobile apps.

To claim your free month visit:

www.forgottenbooks.com/free901051

English
Français
Deutsche
Italiano
Español
Português

www.forgottenbooks.com

Mythology Photography **Fiction**
Fishing Christianity **Art** Cooking
Essays Buddhism Freemasonry
Medicine **Biology** Music **Ancient
Egypt** Evolution Carpentry Physics
Dance Geology **Mathematics** Fitness
Shakespeare **Folklore** Yoga Marketing
Confidence Immortality Biographies
Poetry **Psychology** Witchcraft
Electronics Chemistry History **Law**
Accounting **Philosophy** Anthropology
Alchemy Drama Quantum Mechanics
Atheism Sexual Health **Ancient History**
Entrepreneurship Languages Sport
Paleontology Needlework Islam
Metaphysics Investment Archaeology
Parenting Statistics Criminology
Motivational

Journal of the
Society of Motion Picture Engineers

3℔

VOLUME 52 **JANUARY 1949** **NUMBER 1**

Subscription to nonmembers, $10.00 per annum; to members, $6.25 per annum, included in
their annual membership dues; single copies, $1.25. Order from the Society's General Office.
A discount of ten per cent is allowed to accredited agencies on orders for subscriptions and
single copies. Published monthly at Easton, Pa., by the Society of Motion Picture Engineers,
Inc. Publication Office, 20th & Northampton Sts., Easton, Pa. General and Editorial Office,
342 Madison Ave., New York 17, N. Y. Entered as second-class matter January 15, 1930,
at the Post Office at Easton, Pa., under the Act of March 3, 1879.

Society of
Motion Picture Engineers

342 MADISON AVENUE—NEW YORK 17, N. Y.—TEL. MU 2-2185

BOYCE NEMEC . . . EXECUTIVE SECRETARY

OFFICERS

1949–1950

PRESIDENT	PAST-PRESIDENT	SECRETARY
Earl I. Sponable	Loren L. Ryder	Robert M. Corbin
460 W. 54 St.	5451 Marathon St.	343 State St.
New York 19, N. Y.	Hollywood 38, Calif.	Rochester 4, N. Y.

EDITORIAL VICE-PRESIDENT	EXECUTIVE VICE-PRESIDENT	CONVENTION VICE-PRESIDENT
Clyde R. Keith	Peter Mole	William C. Kunzmann
120 Broadway	941 N. Sycamore Ave.	Box 6087
New York 5, N. Y.	Hollywood 38, Calif.	Cleveland 1, Ohio

1948–1949   1949

ENGINEERING VICE-PRESIDENT	TREASURER	FINANCIAL VICE-PRESIDENT
John A. Maurer	Ralph B. Austrian	David B. Joy
37-01 31 St.	25 W. 54 St.	30 E. 42 St.
Long Island City 1, N. Y.	New York 19, N. Y.	New York 17, N. Y.

Governors

1948–1949	1949	1949–1950
Alan W. Cook	James Frank, Jr.	Herbert Barnett
25 Thorpe St.	426 Luckie St., N. W.	Manville Lane
Binghamton, N. Y.	Atlanta, Ga.	Pleasantville, N Y.
Lloyd T. Goldsmith	William H. Rivers	Fred T. Bowditch
Warner Brothers	342 Madison Ave.	Box 6087
Burbank, Calif.	New York 17, N. Y.	Cleveland 1, Ohio
Paul J. Larsen	Sidney P. Solow	Kenneth F. Morgan
508 S. Tulane St.	959 Seward St.	6601 Romaine St.
Albuquerque, N. M.	Hollywood 38, Calif.	Los Angeles 38, Calif.
Gordon E. Sawyer	R. T. Van Niman	Norwood L. Simmons, Jr.
857 N. Martel Ave.	4431 W. Lake St.	6706 Santa Monica Blvd.
Hollywood 46, Calif.	Chicago 24, Ill.	Hollywood 38, Calif.

Report of the President*

THIS IS THE fourth semiannual report which I, as President, have had the honor of presenting to the Society of Motion Picture Engineers. It will be a review of the high lights of activity during the two-year period of my presidency.

The scope of Society activity has been defined to include all phases of pictorial rendition of action; whether it be from film, as in motion pictures, electronics, as in television, or other device. As will be apparent later in this report your Society is doing an able job in each of these fields.

Under the guidance of Paul J. Larsen the Society of Motion Picture Engineers was the initiating and only active body in gaining frequency allocations for theater television. This activity has been carried on through the years until now our efforts have gained recognition and individual companies in the industry are applying for and gaining frequency allocations which this Society made available to this industry.

The Society of Motion Picture Engineers stands out as a world leader in thinking and action in the field of theater television. The "firsts" which were brought about as a result of Society demonstrations are numerous and interesting. Some of them are listed below:

1. The use of television facilities as a visual public-address system.

2. The use of television facilities as a means of instruction—both to show close-ups of equipment and instructor.

3. The showing of members of the audience on the screen to the entire audience.

4. The filming, fast processing, and television projection of persons entering a meeting room.

5. The presentation of television on a twenty-foot screen.

6. The presentation of a television-broadcast pickup of a sports event and the immediate playing of a film transcription of the same event in order that the audience might again watch the high lights of the sport in action.

The papers on the art of television which are now published in the JOURNAL stand out as the one authentic record and statement of technical fact in regard to all phases of television as applied to motion pictures.

The standardization work of the Society holds an important position, as there are now more standards as applied to motion pictures than to any other industry. \This is important to this industry for it

* Presented October 26, 1948, at the SMPE Convention in Washington.

is through standardization that we gain the universal and world-wide market which we enjoy for our product.

The Society and the newly constituted Motion Picture Research Council have established definite co-operative procedures of handling technical problems of mutual interest. As of this writing, joint committees exist in the fields of sound, standards, and test films. There is now no overlap between the test-reel activity of the Society and that of the Research Council. The efforts of one organization augment and supplement without conflict that of the other, which is very gratifying.

The Society of Motion Picture Engineers having outgrown its previous quarters has now moved into a new suite of rooms in the Canadian Pacific Building, 342 Madison Avenue, New York 17. The entrance-room number is 912 and the telephone number is Murray Hill 2-2185. The staff of the Society includes Boyce Nemec, Executive Secretary, William H. Deacy, Jr., Staff Engineer, Helen M. Stote, JOURNAL Editor, Sigmund M. Muskat, Office Manager, and secretarial help, Helen Goodwyn, Dorothy Johnson, Thelma Klinow, Ethel Lewis, and Beatrice Melican.

We are proud of this staff and the work that they are doing. We want you to know that you can call upon them and your Society whenever the Society or our staff may assist you.

The JOURNAL of the Society has taken on a new format and a policy of giving to our readers in so far as possible the knowledge and information they are interested in reading. Our editorial staff has done an excellent job, and I am happy that the journals are now being issued on schedule which was not always possible during the war years.

Mr. John A. Maurer is continuing to do a colossal job as Engineering Vice-President. Both through committee and by personal effort his activity and accomplishments must have reached all of you.

The convention in Chicago emphasizing 16-mm and 8-mm motion pictures, the convention in New York with a theater engineering conference and exhibit, and the Hollywood convention with emphasis on motion picture production in television have all been most successful and a record of which we are justly proud. I know that the Washington convention will be remembered along with these other successes of our Society.

I want to thank you all for the support which I have received. It has been an honor and a pleasure to serve the Society, and as I change my status from that of President to that of Past-President, I shall carry on with you to aid and serve you in whatever way I may.

Res ectfull submitted,

Film-Collection Program[*]

By HOWARD LAMARR WALLS

ACADEMY OF MOTION PICTURE ARTS AND SCIENCES, HOLLYWOOD, CALIFORNIA

Summary—The Library of Congress attempted to restore to the screen the first twenty years of motion picture achievements by the optical printing of photographic paper rolls submitted for copyright registration from 1897 to 1917, but Congress refused to provide funds for the work. The Academy of Motion Picture Arts and Sciences, with the co-operation of The Library of Congress, now undertakes to do this work.

As CURATOR OF MOTION PICTURES in The Library of Congress, it was my privilege to appear before your Society in New York City on May 5, 1943, and to announce at that time the results of photographic tests that had been conducted with a view to restoring to the screen the first twenty years of motion picture achievements.[1] Sometime during the latter half of 1942, we at the Library had come to the conclusion that positive paper rolls of motion pictures submitted for copyright registration between 1894 and 1917 might be transferred to celluloid by some special photographic process. In following through with this thought, we approached John G. Bradley, then chief of the Division of Motion Pictures in the National Archives, for whatever assistance he could give us. Carl Louis Gregory, who was serving with the Archives at that time as a photographic engineer, was assigned to the problem. After a few tests at optical printing by reflected light, Mr. Gregory found that these opaque paper prints could successfully be copied.[2] With this discovery, it became evident that the foundations of our motion picture art-science, hitherto considered to be irretrievably lost, could now be resurrected through one major project.

It was the Library's intention to proceed at once with the copying of the nearly two and one half million feet of the paper films comprising the collection, but the exigencies of the war defeated it. The entire facilities of the motion picture division had to be given up wholly to the servicing of the seized films of enemy aliens to the various war agencies in behalf of the Alien Property Custodian. In the meantime, the paper rolls were carefully housed in a specially adapted

* Presented May 21, 1948, at the SMPE Convention in Santa Monica.

room pending the end of the war and a more opportune time to continue with the plan of restoration.

By the end of 1946, the Library was ready to resume its intention. It was estimated that the work would cover a period of approximately five years and would cost nearly a quarter of a million dollars. The method of procedure was set forth in the Library's budget for the fiscal year June, 1947, to June, 1948, but the Senate Appropriations Committee, after considering the Library's needs in this connection, refused to provide funds for the work. Moreover, it denied the Library's wish to assume the responsibility of establishing a much-needed national film collection—an activity in which the Library had been engaged, with the assent of previous Congresses, since 1942. The Committee's refusal was upheld in the votes of the House and Senate, and the Library's motion picture program was abruptly halted. The divisional staff, which had grown to thirteen persons, had to be summarily dismissed.

As one who had acquired a conversant knowledge of the collection of paper prints and their almost incredible value, I naturally became quite concerned about the future of the material. In a move to prevent further deterioration of the paper rolls and to get them copied to celluloid for their various social uses, I appealed to the Librarian of Congress to lend them to the Academy of Motion Picture Arts and Sciences if the Academy Foundation would undertake the task of copying them. He agreed to this, and I came on to Hollywood to present the case.

Both Jean Hersholt, president of the Academy, and Walter Wanger, chairman of the Academy Foundation, quickly expressed their interest in a cultural project of such importance. I am happy to be here today as curator of the Academy's Motion Picture Collection and to tell you that the beginnings of our screen history will be saved after all. I would like to pay a special public tribute to Mrs. Margaret Herrick, the executive secretary of the Academy, whose intense interest in the proposed project kept the idea alive and going during the few weeks that were necessary for the Academy's Board of Governors to arrive at the final and affirmative decision.

The Academy Foundation is pleased to have this opportunity to demonstrate briefly to the Society the results of the first efforts at converting the paper prints to celluloid. To us, it is a technological achievement that borders on the miraculous. We think that perhaps you will agree. The films were made by Film Effects of Hollywood on an optical printer designed and built by Douglas Heidanus. I would like to tell you briefly about our plans, about what the Academy Foundation hopes to accomplish with these films as they come off the optical printer.

Most of you are familiar with the workings of the Academy of Motion Picture Arts and Sciences, which had its inception in 1927. The Academy has striven, and is working today, to advance the arts and sciences of motion pictures and to foster co-operation among the creative leadership of the motion picture industry for cultural, educational, and technological progress. During the more than twenty years of its operation, it has recognized outstanding achievements by conferring annual Awards of Merit that serve as a constant incentive within the industry and focus wide public attention on the best in motion pictures. It has conducted co-operative technical research and stimulated the improvement of methods and equipment.

For some time, the Academy has wanted to extend its operations to include a motion picture service along closely defined social and historical lines. It feels that the motion picture medium should be preserved in a collection of films so brought together as to provide an important national source of research material for students of the cinema and other types of scholarly users. It believes that this is the only way by which the history of the motion picture can be written without conjecture; it wants to make it possible for the story of the motion picture to be effectively assembled for visual presentation in the classrooms of universities, colleges, and high schools as a part of their regular curricula.

The construction of such a collection to these ends is now under way with The Academy Foundation supporting the Academy's aim. The films that will be made from the photographic paper prints lent to us by The Library of Congress will, of course, serve as the foundations of the collection, while a carefully prepared plan of selection (designed to eliminate as far as possible mere personal judgment) will be put into operation to procure for preservation other important cinematic landmarks made up to current times.

The freedom of selection and the consequent magnitude of the general collection will depend on the availability of funds that can be obtained to carry on the work as we progress. Contrary to common belief, the Academy is not supported financially by wealthy film-producing companies; its routine expenses are paid from the membership dues of cinematographers, actors, writers, and others. The

Academy Foundation, now headed by Y. Frank Freeman, was established to receive tax-free donations and thereby support the broad cultural work of the Academy proper. It is through the support of this Foundation that the work on the paper prints will be accomplished, that additional social and historical aspects of the motion picture medium will be given expression.

In concluding my remarks, I must add that the Foundation does not yet have the quarter of a million dollars required to convert the paper prints. We have only a shoestring and a lot of ambition. We must have donations if we are to save the art-science of the motion picture and put it to universal use. We are open to all the suggestions you care to make regarding sources that can help us do the job.

REFERENCES

(1) H. L. Walls, "Motion picture incunabula in The Library of Congress," J. Soc. Mot. Pict. Eng., vol. 42, pp. 155–159; March, 1944.

(2) C. L. Gregory, "Resurrection of early motion pictures," J. Soc. Mot. Pict. Eng., vol. 42, pp. 159–170; March, 1944.

The Motion Picture Theater

PLANNING—UPKEEP

A new book carrying the above title and published by the Society of Motion Picture Engineers was announced in December, 1948. Motion picture exhibitors, purchasing agents, and theater architects will find a wealth of valuable information in its thirty-eight articles and 427 pages on the technical aspects of motion picture theater planning, construction, maintenance, modernization, and theater television.

Copies are available from the Society at $5.00, postage paid within the United States, except that copies sold in New York City are subject to an additional ten-cent city sales tax. Copies shipped outside the continental United States or its possessions are $5.50, postage paid.

THE MATERIAL included in this book consists, for the most part, of papers presented and discussions which took place at the 62nd Semiannual Convention of the Society of Motion Picture Engineers in New York City, October 20–24, 1947.

Recognizing at that time that theater owners all over the United States were planning for the construction of new theaters and the modernization of existing ones, it was felt that the Society was both obligated and in a position to furnish these theater owners with the latest scientific information on major phases of theater design and construction. Furthermore, it was evident that architects assisting in these plans had not had an opportunity, except through the trade papers, both to learn of trends in other parts of the country and to exchange information with other architects regarding novel features.

It was therefore decided early in 1947 that the major portion of the 62nd Semiannual Convention should be devoted to Theater Engineering. A preliminary survey conducted among theater organizations and circuits, and the enthusiastic response which it received, confirmed the Society's feeling that such a program was desirable. A special committee, consisting of leading theater architects and leading theater-circuit executives, responsible for construction and maintenance, was appointed to organize the conference. They determined that individual sessions would be devoted to specific principal subjects of interest, and that experts, most of them representing manufacturers of theater equipment, would be invited to present short technical papers written in a style that would be of interest to theater owners, purchasing agents, construction men, and theater managers. They further planned that a major portion of each session would be devoted to a discussion period permitting the various experts to question one another and also to permit those attending the Conference to offer opinions or ask questions from the floor. In order to accomplish this, a specially designed public-address system was obtained so that all in the room could hear the entire discussion easily.

To enhance the effectiveness of the Conference further, an exhibit of new and interesting theater equipment and materials was held in an adjoining room. The Society was successful in obtaining a number of outstanding exhibits of this nature, which proved to be of tremendous interest to those attending the Conference. Credit for the success of the Conference, which made publication of this book possible, is in no small part due to the Theater Conference Papers Committee, as well as the Editorial Vice-President, Convention Vice-President, and the Staff of the Society. The members of the Special Papers Committee and the sessions for which they were responsible are as follows:

Chairman
Leonard Satz
Century Circuit

Auditorium Design	**Lighting**
Martin F. Bennett	*Wallace W. Lozier*
Radio Corporation of America	*National Carbon Company*
Floor Coverings	**Acoustics**
Charles Bachman	*Charles S. Perkins*
Warner Brothers	*Altec Service Corporation*
Television	**Safety and Maintenance**
Donald E. Hyndman	*Henry Anderson*
Eastman Kodak Company	*Paramount Pictures*
Television Projection	**Ventilating and Air Conditioning;** **Promotional Display**
Paul J. Larsen	*Seymour Seider*
Consultant	*Empee Construction Corporation*

To ensure the success of the Conference, great care was taken in the selection of Session Chairmen, all of whom had attained a high standing in the industry.

Physical Construction	**Lighting**
Leonard Satz	*Lester B. Isaac*
Century Circuit	*Loew's, Incorporated*
Auditorium Design	**Safety and Maintenance**
John Eberson	*Henry Anderson*
Architect	*Paramount Pictures*

Floor Coverings
A. Griffin Ashcroft
Alexander Smith and Sons Carpet Company

Television Projection
Paul J. Larsen
Consultant

Television
Donald E. Hyndman
Eastman Kodak Company

Acoustics
Harvey B. Fletcher
Bell Telephone Laboratories

Ventilating and Air Conditioning;
Promotional Display
Seymour Seider
Empee Construction Corporation

One of the high lights of the Conference was a demonstration of large-screen television by the Radio Corporation of America, attended by over five hundred people.

The success of the Conference was indicated by record attendance, including a generous cross section of theater men and architects from all parts of the United States and twenty representatives from foreign countries. The Society, recognizing the tremendous interest in Theater Engineering, now offers the entire motion picture industry an opportunity to read or review all of the technical papers and the ensuing discussions as they took place at the conference.

James Frank, Jr.
Theater Conference Chairman

Effect of Television
on Motion Picture Attendance*

By RALPH B. AUSTRIAN

NEW YORK, NEW YORK

Summary—The purpose of this study is to obtain an indication of the effect of television upon motion picture attendance habits. Such a study could suggest the extent to which television will affect box-office receipts when set ownership has become more widespread than it is at present. Telephone numbers of 550 owners of home television sets were selected at random from the four major boroughs of New York City. Interviews were completed with 415 owners of sets presently in working order.

THE SAMPLE

FOR A STUDY of this type, telephone interviews seemed most appropriate, since they provided the double advantage of speed and economy. A recent report by The Pulse, Inc., shows that 87 per cent of the set owners in New York City have telephones, indicating that an adequate sample of set owners could be reached by telephone.

With a list of 10,000 set owners as a base, the telephone numbers of 550 were selected completely at random. The study was limited to the four major boroughs of New York City: Manhattan, Bronx, Brooklyn, and Queens.

Interviews were completed with 415 set owners. Interviews could not be completed with the remainder of our list for various reasons:

No longer had a set or temporarily out of order . . . .   13 per cent
Refused (too busy, ill, unable to speak English) . . . .   4 per cent
No answer . . . .   7 per cent

In a study of motion picture going conducted at home, it is important to keep to a minimum the number of people who are lost because they are not at home when called. In general, these are more active people—and likely to be frequent patrons of motion pictures. If this study were limited to a single attempt to reach each family, some bias would be introduced into the sample. For this reason, four and five "call backs" were made in every case where

* Presented May 17, 1948, at the SMPE Convention in Santa Monica.

there was a "no answer." The resulting 7 per cent "no answer" is considered satisfactory for the purpose of the present study.

The Questionnaire

The questionnaire used in this study is shown below.

Questionnaire—Television Telephone Survey

Good (Morning) (Afternoon) This is the General Research Bureau. We are conducting a survey among television-set owners.

1. Do you have a television set at home? Yes ( ) No ( )
2. Is it in working order? Yes ( ) No ( )
   (IF "NO" TO EITHER QUESTION 1 OR 2, TERMINATE INTERVIEW)
3. How long have you had it?_____
4. Do you think your television set has had any effect on either increasing or decreasing the number of evenings you spend at home?

   Increased ( )   Decreased ( )   Had No Effect ( )

   (IF "INCREASED")
   4(a) About how many evenings a week more do you spend at home?_____
   (IF "DECREASED")
   4(b) About how many evenings a week less do you spend at home?_____
5. Let's take the movies as an example. Since you got your television set, do you think you go to the movies more often or less often, or is there no difference?

   More often ( )   Less often ( )   No difference ( )

   (IF "MORE OFTEN" OR "LESS OFTEN")
   5(a) How often do you go to the movies now?
   _____times a week or _____ · times a month DK ( )
   5(b) About how often did you go to the movies before you got your television set?
   _____times a week or · _____ times a month DK ( )

COMMENTS:_____ _____

_____

PHONE NUMBER:_____ SEX: Male ( ) Female ( )

BOROUGH:_____

INTERVIEWER'S NAME:_____ · _____DATE:_____

Some thought was given to the possibility that the structure of this questionnaire may have biased the results of this survey. It might be argued that the introduction and the wording of the questions would lead television-set owners to feel obliged to report a change.

With this in mind, check interviews were made with a sample of people comparable to those included in the study itself. With these people, no mention was made of television. Rather, the same questionnaire was used with the substitution of "new radio" for "television set." Respondents in the check interview were asked to report if there had been any changes in their motion picture attendance since they bought their new radios.

These check interviews did not elicit any reports of gross changes in picture attendance. In fact, it was difficult for people to grasp why there should be a change.

It seems reasonable to conclude, therefore, that there is no bias inherent in the *structure* of the questionnaire or the *wording* of the questions. If there is any tendency which leads to exaggerated answers, it lies in the fact of television-set ownership itself. But there is no reason to believe that the answers found in the study give an incorrect picture of the trend.

NOTE OF CAUTION

Everything connected with television is changing rapidly from day to day. This study reflects present conditions; it is not presented as a prediction of future developments.

As television programs improve, the medium is likely to provide increasingly stiff competition for the motion picture producer. In interpreting the results of this study, the dynamic state of television should be kept in mind.

Although most of the sets owned by the people interviewed in this study were bought in the last year or two, many of them date back to before the war. It is perhaps too much to expect that a housewife who has had a television set for five or six years could give an absolutely accurate report of her motion picture attendance habits before she bought the set. What she tells us is not what *actually* happened (which is subject to errors of memory), but only what she *recalls* has happened.

Since the end of the war there has been a general decline in attendance at motion pictures. It is reasonable to assume that this has affected both set owners and nonowners. Interpretation of any reports of changes in attendance made to us in this survey must be tempered by knowledge of this general trend.

This survey is limited to home-set owners. Thus, any effects of television reported here may be an *underestimation* because no account is taken of the effects on nonowners who view television at the homes of friends, in bars, and other public places.

Summary of Findings

Television has had a definite social impact on the families interviewed. Three quarters of them report that they spend more evenings at home now that they have a set.

This effect has extended to motion pictures. Half of the set owners interviewed report that they go to the movies less often after buying a set.

Most of the people who are going to the movies less were formerly very heavy goers. The movies are losing some of their best customers.

Findings

Half of the people interviewed (51 per cent) report that they attend motion pictures less often since purchasing a television set.

The remainder state that television has had no effect on their movie-going habits (except for three people who state that they go more often now).

Table I shows a comparison of the findings obtained in this study with the results of a survey conducted in the fall of 1947 in the Los Angeles area by Television Research, Inc., of that city. Notice the similarity in results.

Table I.

	F. C. & B. New York	TV Research Los Angeles
	Per Cent	Per Cent
Attend less than before	51	46
Attend about the same	48	53
Attend more often	1	1

Attendance Reduction

Fifty-one per cent of the people interviewed in the present study reported a change in movie attendance. In what direction was this change made?

Table II shows a consistent shift to less-frequent movie attendance. The majority of persons who reported a change appear to have been very heavy movie-goers before they got their set. Their attendance has dropped from an average of "once every few days" to an average of somewhat less than once a week.

Table II

The Trend Away from Frequent Movie Attendance

As Reported by the 51 Per Cent Whose Habits Have Changed with Television
(211 People)

	Before TV	After TV
	Per Cent	Per Cent
Every few days	57	4
Once a week	33	39
Every two or three weeks	7	28
Infrequently	3	29
	100	100

Notice that scarcely any (only 3 per cent) of these set owners were infrequent movie-goers before buying a set; afterward, fully 29 per cent fell into this category. The behavior of those who report a change in their movie-going habits deserves more detailed study.

Table II shows that the majority of set owners had been very heavy movie-goers before buying their television sets.

Fully 90 per cent (that is, 57 plus 33 per cent) used to go once a week *or oftener*, according to their own statements. Now consider the habits of these heavy movie-goers after buying a telvision set.

Table III

Change in Movie-Going

Reported by Those Who Used to Attend Frequently

Movie-Going after Television	Every Few Days Per Cent	Once a Week Per Cent
Every few days	5	0
Once a week	68	0
Every two or three weeks	10	62
Once a month	6	24
Infrequently	11	14
	100	100

In Table III, the column showing the post-television habits of those who used to go "every few days" shows that most of these people have now dropped to only once a week in attendance. An additional 27 per cent (all of whom used to go every few days) now go even less than once a week.

As for those who used to go once a week, most of them now go only once every two or three weeks. The rest go even less frequently.

It is reasonable to assume that television would influence other forms of social behavior as well as motion picture attendance. In order to get a better understanding of the effect of television upon home set owners, all members of our sample were asked whether or not television had affected the number of evenings they spend at home. Three fourths of the people we talked to reported that they spend more evenings at home since buying their sets. The remainder said that television has had no effect upon their habits in this respect.

To what extent has the increase in "staying at home" affected movie-going?

Of the 75 per cent who said evenings at home have increased, *63 per cent said they attend the movies less often.*

It seems reasonable to assume that if these persons are staying home more and attending the movies less, television has had considerable impact on their social life. Motion pictures seem to have been hit hard by this increase in "stay-at-home" habits.

Some people, who tend to minimize the impact of television, have advanced the theory that although movie-going may fall off when a set is new, attendance will pick up again as the novelty wears off.

These assumptions were not borne out by the data accumulated in this study. The age of the set did not appear to have any relationship to reported changes in movie-going.

We must remember that television is still comparatively new, and it is still too early to judge the reaction of set owners as they become accustomed to this medium.

FURTHER EVIDENCE

As this study neared completion, a check study was conducted by Dr. Thomas Coffin of Hofstra College, Hempstead, New York, working in co-operation with Foote, Cone, and Belding.

The check study was conducted by personal interview. Interviews were completed with 270 families in and around Hempstead. Of this number, 135 owned home television sets; the other 135 families were similar in all respects except that they did not own television sets.

The interviewers asked set owners the same questions as did Foote, Cone, and Belding in the telephone survey reported here.

All families, as well, were asked for specific reports of entertainment activities such as attendance at movies, sports activities, radio listening, and so on, during the preceding week.

Results of the interviews provide support for Foote, Cone, and Belding's findings. In the personal interviews, 58 per cent report a decrease in movie attendance. (F. C. & B. reported 51 per cent).

The data on actual attendance at movies and other entertainments provide some validation of these findings. Most significant is the finding that set owners actually bought *20 per cent fewer* movie tickets during that week when compared with the *matched* sample of non-owners.

Navy Photography in the Antarctic*

By CHARLES CURTIS SHIRLEY

BUREAU OF AERONAUTICS, NAVY DEPARTMENT, WASHINGTON, D. C.

Summary—The primary purpose of Operation "Highjump," 1946–1947, was to train personnel, test equipment, and improve operational techniques in subzero temperatures. Every phase of the operation and the performance of equipment undergoing tests were photographed in color motion pictures with a view toward producing technical and training films for educational purposes.

The many difficulties inherent in photographic operations in subzero temperatures and polar regions require special techniques. These and the malfunctions of cameras, and causes and suggestions for improvements are treated. The Navy is developing cameras more suitable for use in frigid areas.

WHEN THE UNITED STATES NAVY Task Force SIXTY-EIGHT departed for the Antarctic on Operation HIGHJUMP in December, 1946, with it went sixty-eight Navy, Army, Marine Corps, and Coast Guard photographers. The primary objective of the Operation was to train personnel and test equipment in subzero temperatures. The secondary objective was to chart and photograph little known, or unknown areas. The duty of the photographers was to photograph every phase of the operation in motion pictures and still photographs both in color and black and white. The photographs and motion pictures were to be used for documentary and technical records, training films, and to teach personnel cold-weather techniques. The time permitted Operation HIGHJUMP in the Antarctic was very short. Ships could not remain in those icy waters longer than two months because of the danger of becoming icebound and being crushed by giant ice floes when winter began. In that short period over 241,000 feet of motion picture film were exposed, 109,327 aerial and still photographs taken and processed, and the necessary prints made.

There was no special attempt made to conduct tests on photographic equipment in the strict sense of the word. Many different types and models of cameras were used to record tests, experiments, and the over-all Operation as events occurred. There was no time to stage any action; it had to be photographed as it happened. Thus from a military standpoint the photographers received valuable training and learned many new techniques as they went along. No better operational cold-weather test could have been conceived for

* Presented May 21, 1948, at the SMPE Convention in Hollywood.

the cameras. Hourly records were kept of temperatures. Records were kept of camera malfunctions and there were many.

It is known that it gets colder in some parts of the United States than the temperature encountered by Task Force SIXTY-EIGHT Operation HIGHJUMP. Also some will recall that they have successfully made motion pictures in their own home community in temperatures colder than −50 degrees Fahrenheit without experiencing undue diffi-

Fig. 1—Navy photographer and camera suspended on a cargo platform from a crane of the USS *Burton Island* to obtain scenes of the Navy Icebreaker as she progresses through the ice pack in the Ross Sea.

culty with their cameras. The answer to this is that it takes many hours to chill a camera thoroughly. On an operation such as the South Pole area expedition, cameras remained exposed to the cold and, as a general rule, the ones intended for outside work are never brought in a warm building. There is a very good reason for this. After a camera has been thoroughly chilled in very cold temperatures and is brought into a warm place, every portion of it becomes thoroughly wet from condensation, even between the lens elements. Prolonged practice of this nature will eventually result in corrosion and rust if the camera is not completely dismantled and every part dried each time this occurs. It is very doubtful if a person in his right mind would stay out in subzero temperatures in his home environment and take pictures for very long at a time. The chances are he would prepare his cameras, then dash out and spend a few minutes taking pictures, and return to warmer quarters. This could not possibly be considered as a good cold-weather test for cameras. It must be remembered that a soldier, sailor, or marine who is fighting in subzero temperatures cannot be expected to find a warm building either to defrost his camera or himself.

Fig. 2—The author skiing to camera location with a heavy motion picture camera and storage battery. This illustrates the necessity for a more portable camera for cold-weather operations. It also indicates that all hand-held cameras had frozen up and failed.

PREPARATION OF CAMERAS

All cameras were completely delubricated and relubricated with cold-weather lubricants and tested in cold chambers. Close tolerances of working parts were made larger where possible. Yet failures occurred on all motion picture cameras from plus 15 degrees Fahrenheit to minus 27 degrees Fahrenheit.

It is quite apparent now that the reason our cameras passed cold tests in the laboratory and would not function properly in the field in much warmer temperatures, is that they were not completely chilled during laboratory tests. Cameras should remain in cold chambers, with film loaded until thoroughly chilled. As a general practice, most cold tests are only for a few hours. The total time to chill a camera thoroughly depends upon the size and mass of metal in its construction. It is important that the film be chilled too. Film becomes very brittle in subzero temperatures, and sometimes this causes malfunction. It is conceivable that in some instances it would require twenty-four or more hours to chill a camera thoroughly.

ELECTRIC-POWERED VERSUS SPRING-DRIVEN CAMERAS

Electric-powered cameras as a class are more reliable than spring-driven cameras for subzero temperatures. Standard 35-mm motion

Fig. 3—Navy photographers about to leave on a photographic assignment are given last-minute instructions.

picture cameras were used with two types of 24-volt electric motors, ball-bearing and sleeve-bearing type. The sleeve-bearing type slowed to half speed at +15 degrees Fahrenheit. The ball-bearing type functioned satisfactorily as low as −27 degrees Fahrenheit. The spring-driven cameras all failed on the plus side of the Fahrenheit scale after a few hours' exposure to low temperature.

Successful motion pictures were obtained from the air at temperatures of −40 degrees Fahrenheit with small 16-mm spring-driven cameras by keeping a stream of warm air sprayed on them. This would be impossible to do by a man operating a camera on the ground.

The large studio-type cameras were cranked by hand when the electric motors failed. Having been in the Antarctic previously with Rear Admiral Byrd, the officer in charge of all photography during Operation HIGHJUMP required all photographers to become proficient in hand-cranking cameras. It was expected that electric motors would fail. The average small, hand-held, spring-driven camera cannot be hand-cranked satisfactorily in extreme cold.

The reasons for failures of instruments in subzero temperatures are not new. It is well known that the differential of thermal expansion and contraction of moving parts of different metals is the cause. In

Fig. 4—Photographers hauling cameras on a sled at Little America.

cameras the usual causes are moving steel parts being frozen by the contraction of aluminum or magnesium alloys. The contraction of these light alloys is much greater than steel. Consequently, if moving parts housed in these alloys are not properly bushed with steel bushings and ball bearings and have proper tolerances, failures will occur at low temperatures. Tolerances can be increased on almost any camera so that it will operate at subzero temperatures, but when it is returned to moderate temperatures these tolerances usually become so great that the camera will be useless until it is reworked and the tolerances reduced. Navy specifications require that all cameras operate in any temperature likely to be encountered.

In the past the Navy has, except for special-purpose cameras, generally procured cameras that were basically engineered for the commercial market. A few changes were generally specified but these were usually minor. Cameras produced for the commercial market are not built to operate in all the temperatures required by the Navy. There are very good reasons for this. Such a camera, one which will operate in a temperature range of from −67 to +141 degrees Fahrenheit, probably can be produced, but the expense would be prohibitive for a commercial camera. All Navy cameras should not be expected to functon under these extreme conditions. In the

Fig. 5—A weasel equipped for photography at Little America. Large, heavy cameras were made portable in this manner.

event of another war, it is reasonable to believe that extreme temperatures will be encountered by some portion of the Navy almost every day, and cameras, like other instruments, must not fail. The contention is that a great many expensive cameras engineered for the above-mentioned temperature range may neve: be used in colder temperatures than freezing or higher temperatures than experienced in the tropics. A more reasonable solution, and a less-expensive one, would be to produce special subzero-weather cameras, required to operate from −70 degrees Fahrenheit up to freezing, and temperate-weather cameras for a range of about −10 up to 141 degrees Fahrenheit.

FILM

Film is a great source of trouble in subzero temperatures. It becomes very brittle and will break easily. Sharp bends must therefore be avoided. Research is required to find a more suitable plastic for film bases for subzero weather than is currently used. It may be that this research will prove that the present type of emulsion is causing most of the trouble. The differential of thermal expansion and contraction between the plastic base and the emulsion is no doubt the cause of some of the reactions of film in subzero temperatures. That is probably why film is inclined to curl very tightly toward the emulsion in cold temperatures. It is doubtful if industry

Fig. 6—Filming scenes of O. F. Bowe as he progresses through a dangerously crevassed area on the shelf ice west of Little America. The safety line around the author is attached to a weasel.

has ever been called upon to produce film that will be pliable in extremely low temperatures. There is very little commercial need for it.

Static is also a great source of trouble in extremely low temperatures. Research should be conducted to try to reduce or eliminate this.

TRIPODS

The friction-head variety is undesirable in subzero temperatures. Smooth operation even in mild climates requires that the head be packed with grease. The grease has to be replaced with a lighter lubricant of oil and graphite for cold-weather operations. With the heavy grease removed, the head wobbles and jerks when panned. Gyro-type or cranked-head tripods are probably satisfactory but the Task Force had none of these.

COLD-WEATHER TECHNIQUE

Much was learned from experience during Operation HIGHJUMP that may be helpful to anyone taking pictures in cold weather. Some of these techniques are listed below.

1. Never breathe on a lens; it will cause condensation which freezes instantly. The resulting ice is very difficult and sometimes impossible to remove unless the lens is taken in to a warm place.

Fig. 7—Photographer dressed in cold-weather gear on board the USS *Mount Olympus*, Flagship of Rear Admiral Richard H. Cruzen, Task Force Commander.

2. Never attempt to clean a lens with an ungloved hand. Body heat will be transmitted to the lens and cause it to ice over. Avoid breathing on the eyepieces and viewfinders for the same reason.

3. Keep the eye far enough from the eyepiece so that body heat will not be transmitted to it. There is danger of the eyelids' sticking to an eyepiece if they come in contact with unpainted metal.

4. Avoid touching unpainted metal surfaces with the bare skin. Painful injury will result, especially from touching unpainted steel. The skin will stick to it as if glued. A portion of skin is always lost if one is careless about this.

5. Do not take a thoroughly chilled camera from the cold to a warm place with the intention of using it immediately. The camera cannot be used until its temperature equals the surrounding warmer temperature. Several hours are required as a general rule for the camera and lenses to complete collecting moisture from condensation and thoroughly dry. If a camera is removed to the cold before being completely dry, icing will result. This can be serious if it occurs on the iris diaphragm or in some interior moving part of the camera. Keep cameras to be used in extreme low temperatures stored outside and those for interior work indoors in a warm temperature.

Fig. 8—Loading a motion picture camera barehanded at Little America, a painful task in subzero weather. Note the crank in place which indicates that the camera had to be hand-cranked because the electric motor had failed.

6. Never take a warm camera out in a blizzard with the expectation of getting good pictures. Drifting snow striking the lens will melt, and in a very short time the lens will be covered with ice.

RECOMMENDATIONS FOR A COLD-WEATHER CAMERA

A motion picture camera for subzero photography for combat use should embody the following features:

1. It should be light in weight and portable.

2. It should be as void of unpainted surfaces as possible, especially those surfaces which have to be touched or adjusted with bare hands.

3. It should be semiautomatic magazine load, regardless of whether it is 35-mm or 16-mm. One of the most painful things imaginable is threading a 35-mm studio-type camera barehanded in temperatues of −50 degrees Fahrenheit or even more moderate temperatures. It is possible to use skintight gloves which will prevent considerable loss of the skin, but it is impossible to thread such a camera with the hands adequately gloved for warmth in subzero temperatures. One of the photographers on Operation HIGHJUMP became very proficient at loading and threading a large studio-type camera with fully mittened hands by using a pencil to adjust the film.

4. All adjustments should be possible with the hands encased in three heavy woolen mittens plus an outer leather mitten shell.

Fig. 9—Photographic operations headquarters during Operation HIGHJUMP at Little America, Antarctica. Camp and living quarters may be seen in background.

5. Finders and eyepieces, if employing lenses (except for focusing, lenses are not recommended), should be well insulated with rubber and well ventilated with holes between the rubber and lens so that body heat will escape before it reaches the glass surface of the lens.

6. Electric power is preferred. The power source should be compact and capable of being strapped to the body in such a manner that it will not handicap movement.

7. Hand-held-type cameras are preferable and should be equipped with a shotgun-type stock and trigger.

8. A choice of three to four taking lenses and finder lens, mounted in a turret that can be rotated easily and positively locked.

9. It should function smoothly at −67 degrees Fahrenheit.

EXPOSURE

An entirely unexpected phenomenon encountered in the Antarctic is the abundance of reflected light. This is surprising to one who follows the practice of increasing exposure as the distance north or south of the equator is increased. In preparation for an expedition to the Antarctic, the average photographer would assume that a very fast emulsion would be required. As a general rule, motion picture film faster than Weston 50 is not usable without neutral density or other exposure-reducing filters.

Another surprising phenomenon in the Antarctic is the fact that there is more light on overcast days than on bright, clear, cloudless days. This is explained by the fact that practically all of the light striking the surface is reflected. Upon striking the bottom of the overcast it is reflected back to the ice surface. A continuation of this reflection back and forth causes a build-up of effective light. Thus, the Antarctic is probably the only place in the world that requires less exposure on overcast days, except that it is reasonable to believe a similar situation exists in the Arctic.

Because of the abundance of light in Antarctica it is necessary to mask off a portion of the photocell aperture on the average exposure meter to obtain a reasonably accurate reading. On two of the most popular type of meters the indicating needle registers beyond the highest calibrations. By masking off 50 or 75 per cent of the cell aperture and multiplying the reading obtained by the percentage mask off, fairly accurate exposure calculations may be obtained.

RULES FOR ICE PHOTOGRAPHY

Much can be written about ice photography. Probably the most authoritative work on ice photography is that by Herbert G. Ponting, who accompanied Captain Robert Falcon Scott of the British Royal Navy to the Antarctic as a photographer in the early part of the century.

1. For best detail on an ice surface, the picture should be taken against the light source.

2. Detail is not possible on a flat ice surface when the view is 180 degrees from the light source.

3. The angle relative to the light source can be determined by observations from several positions, and will vary in accordance with the results desired.

4. A good "Rule of Thumb" for exposure is to expose for detail in shadows.

ACKNOWLEDGMENT

Motion Picture Photography at Ten Million Frames per Second*

By BRIAN O'BRIEN AND GORDON MILNE

INSTITUTE OF OPTICS, UNIVERSITY OF ROCHESTER, ROCHESTER, NEW YORK

Summary—A new procedure is used in which the image of a rectangular picture is broken up and reassembled into a long narrow strip. After processing, the negative must be reconstructed into a rectangular motion picture frame by projection printing through an optical system similar to that which formed it.

To EXPOSE A conventional motion picture at a speed of several million frames per second would require a speed of film movement of the order of 200,000 feet per second for 16-mm film, a rate entirely beyond anything attainable at present. As an alternative the image may be swept across stationary film at speeds much higher than it is possible to move the film itself, but this procedure imposes a very serious limit upon the length of film and hence the number of frames which can be exposed.

The required film or image speed could be much reduced if the conventional shape of the motion picture frame were altered to make the dimension along the direction of film movement very small. If at the same time the dimension across the film were increased a corresponding amount to preserve the same picture area, the same total number of just-resolvable picture elements might be retained. In the camera to be described this change of image shape is accomplished by a stationary optical system through which the motion picture negative is exposed. After processing, the negative film is printed by projection through a similar optical system which reconstructs the image back to the shape of the original object and thus to the approximate proportions of the normal rectangular motion picture frame. In this manner motion pictures in excess of ten million frames per second have been produced with very moderate film velocities, the arrangement permitting the photography of a very large number of motion picture frames in a single sequence. By the use of an automatic printer the final positive is presented as a standard 16-mm motion picture print.

* Presented October 21, 1947, at the SMPE Convention in New York.

The optical system changes the shape of the picture by cutting the original image of the object into a series of narrow strips, redisposing these strips end to end, and reimaging them as a single long strip extending across the motion picture negative film. We have termed this process image dissection, and the optical system which accomplishes it an image dissector. The basic idea of cutting an image into strips is not new, and methods for accomplishing this have been described by Walton.[1] However, to meet the requirements of ultraspeed photography it has been necessary for us to devise a new type of optical system.

The principle of the image dissector is shown in Fig. 1. Consider a number of small identical objective lenses L, as shown in Fig. 1, with their optic axes perpendicular to the plane of the paper. For convenience these lenses are cut with flat edges and blocked together as shown. If the line joining their centers is slightly inclined to the horizontal, and a distant event is viewed by these lenses, each will form a separate but identical image of the distant event, and these images will appear as a flight of steps as shown in the upper part of Fig. 1.

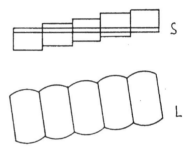

Fig. 1—A multilens unit L and its images on a slit S, illustrating the method of image dissection.

If these images are received on a metal plate containing a narrow slit S, which is not inclined but truly horizontal as shown in the figure, it is evident that this slit will pick just the top of the picture from the first image, a strip of the picture next below from the second image, next below that for the third image, and so on. Since all the images are alike, it will be evident that the various parts of the slit from left to right carry the equivalent of one complete picture of the event. This we have referred to as image dissection, and in the example shown in Fig. 1, the rectangular image is dissected into five narrow strips which are assembled end to end to form one long narrow strip which passes through the slit. If this narrow strip is imaged on photographic negative film which is moving in a direction from top to bottom referred to in Fig. 1, it will be seen that vertical streaks will be formed on the film, their position corresponding to the light and dark portions of the original picture. If the slit is made sufficiently narrow, and the number of lenses sufficiently great so that the whole picture is represented entirely in this narrow slit, then the blurring of the picture which results from the motion of the film cannot be greater than the width of the slit, and this in turn can be made as small as the finest detail which the photographic negative is capable of resolving. While such a streak negative could be analyzed and the necessary information secured from it for many scientific purposes, it is far more convenient to print the negative back on to positive film by projection through the very optical system which formed it. In this manner the picture is rectified from streak images back to an ordinary frame of motion picture film. Each position of the slit across the negative film gives a new frame of the final motion picture.

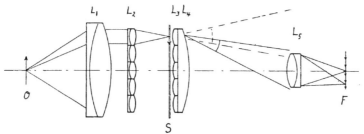

Fig. 2—Optical arrangement of a 5-element image dissector as used for near events.

A schematic diagram of the complete optical system is shown in Fig. 2, including the provision for photographing an object which is not at infinity. Light from the object at O is collimated by the first objective lens L_1, and then received by the multiple objective lens system L_2. These lenses form the multiple images in the plane of the slit S. The image of the slit S is formed on the film at F by the main lens L_5, which must be a very well-corrected photographic objective. In order that the pupils of the multiple optical system shall fall on the final objective L_5, two sets of specially designed field lenses, L_3 and L_4, are mounted very close to the plane of the slit. By this arrangement the effective photographic speed of the combined optical system is equal to the photographic speed of the lens L_5, subject only to surface reflection losses of the lenses L_1 to L_4, inclusive. These reflection losses can be made quite small by suitable nonreflection treatment. In the present camera the final lens L_5 is a photographic objective operating at $f/2.0$.*

In our Model I camera there are fifteen small objectives L_2 instead of the five shown in Fig. 2. When used with a single slit at S this produces a final picture having a total of fifteen elements only. To increase the number of elements multiple slits may be used. Referring again to Fig. 1, if the inclination of the line joining the centers of the lenses L be reduced to one half, and if the slit width be reduced to one half, then it is evident that only the upper half of the original picture will be covered by the five slit elements. Suppose now that a second slit be placed parallel to S but spaced below it by just five times the width of either slit. Under these conditions with a single set of five lenses L, the number of elements of the final picture will be

Fig. 3—Fifteen-element image dissector. The slit jaws may be seen as a faint dark streak through the multicondenser unit. Above this is a device for placing fiducial marks on the negative. The multilens unit is largely concealed, but its focusing adjustment shows prominently (large knurled head in foreground).

Fig. 4—Fifteen-element image dissector viewed from object side. The end of the multilens unit may be seen near the middle of the picture, and below it the thumbscrew for adjusting tilt.

doubled, the first slit taking care of the upper half of the final picture and the second slit, the lower half. Photographs of the complete image dissector with single slit are shown in Figs. 3 and 4, without the final photographic objective L_5. A photograph of the double-slit system is shown in Fig. 5. Although the two slits are very close together, their images on the final negative film must be widely spaced to avoid overlap of the streak image. This is accomplished by the reflecting prisms

---

* This lens is a Kodak Ektar of 45 mm focal length, designed for object at infinity. It is here required to work at 5 to 1 conjugates. F. E. Altman, of the Eastman Kodak Company, very kindly arranged for a supplementary lens system which is used in front of the Ektar to permit the latter to operate at its designed conjugates.

Fig. 5—Image dissector with double slit in place.

shown in the top and bottom right of Fig. 5, which are provided with small micrometer screws to adjust their position. With the arrangement shown in Fig. 5, a 30-element picture is formed.*

In Fig. 6 is shown the manner of handling the negative film which was described some years ago.[2] A shallow drum, open at the top, is mounted on the

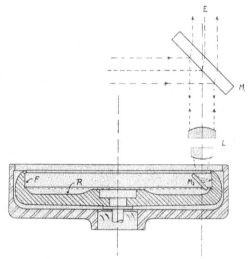

Fig. 6—Cross section of high-speed rotor illustrating method of imaging slit on the moving film.

shaft of a vertical high-speed motor. The inner circumference of the drum is machined to accept one turn of 16-mm film cut with a gauge 24 inches long so that it just fits within the drum with a negligible gap. Light from the image dissector reaches the mirror M_1, is reflected down through the main photographic objective L, and reflected once more by the lower mirror M_2, to form the

* The small objective L_2, although simple cemented achromats, must be of good quality. L. V. Foster, of the Bausch and Lomb Optical Company, very kindly arranged for the procurement of the optical elements of the Bausch and Lomb 40-mm microscope objective which proved very satisfactory.

image on the rotating film. At top rate of speed this film is driven past the image at the rate of 400 feet per second. By proper attention to mechanical and optical detail it is possible to resolve approximately 100 lines per millimeter on this film. At a resolution of 80 lines per

Fig. 7—The complete camera with image dissector in place.

millimeter with the film traveling 400 feet per second, the individual exposures are of one-ten-millionth-second duration, and in effect ten million separate motion picture frames per second are photographed.

The fully assembled camera is shown in Fig. 7. The diameter of the rotor case is about 10 inches and the over-all height about 18 inches. The present complete camera weighs about 60 pounds, much of this weight being in the cast-iron base. The driving motor, rated at **12,000** revolutions per minute, is shown immediately below the rotor case. Below the motor, directly connected to its shaft is a miniature inductor type of alternating-current generator. The frequency of the alternating-current output of this generator is a very accurate measurement of rotor speed, a matter of some importance in certain scientific studies with the camera. The main objective lens L_5 is in the vertical

Fig. 8—The printer arranged for automatic rectification of the streak negative (placed in the precision carriage at left) into a finished 16-mm projection print.

column above the rotor, and the image dissector appears on the upper arm, the housing being removed to show it more clearly.

When using an image dissector with a single slit, the 24-inch length of film will carry more than 60,000 slit images. Since each complete slit image constitutes a motion picture frame, it is evident that, if necessary, a single scene of more than 60,000 frames' duration can be photographed. In using a double-slit system giving a 30-element picture, the number of frames which may be photographed without overlap is much reduced. In the present model this is limited to continuous runs of 1600 frames, which is still sufficient for photographing a great variety of events.

Since at a speed of ten million frames per second the exposure time is limited to one-ten-millionth second, the problem of getting sufficient light for proper photographic exposure can be very difficult. If one calculates the illumination required upon ordinary opaque objects of average reflectance, the value turns out to be between twenty and one hundred million foot-candles. To obtain such an illumination upon even one square foot would require several thousand kilowatts of the most efficient light sources such as carbon or mercury arcs, and such power requirements are obviously out of the

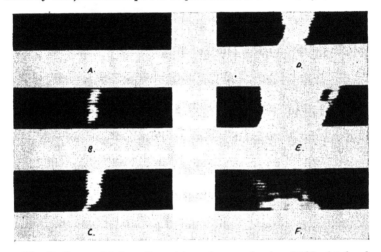

Fig. 9—Reassembled motion picture frames showing 0.125-microfarad condenser charged to 30 kilovolts discharged through 0.003-inch vertical iron wire. Magnification on print approximately six times, approximate intervals: A, 0 second; B, 0.05 microsecond; C, 0.1 microsecond; D, 0.4 microsecond; E, 1 microsecond; F, 3 microseconds. Note the initial expansion rate of over 10 kilometers per second. This expansion rate is actually too fast for proper resolution.

question except in specially equipped laboratories. Fortunately modern electrical-discharge flash lamps provide an ideal solution because the duration of the high-intensity illumination required is quite short. Suppose a scene sequence of 2000 frames is required at ten million frames per second. This means that the illumination must continue for only one-five-thousandth second. Flash lamps of the types described by Edgerton and his associates can easily produce a total flux of 10^9 lumens for times greater than 10^{-4} second. For example, a General Electric type FT 524 lamp operated from a capacitor of 100-

microfarad capacitance charged to 4 kilovolts will produce the order of 10^9 lumens for about 2×10^{-4} second, an operating discharge which is entirely practical in this lamp.   A capacitor and power supply weigh only about 75 pounds, providing a practical and at the same time portable light source.   A timing mechanism is, of course, necessary to

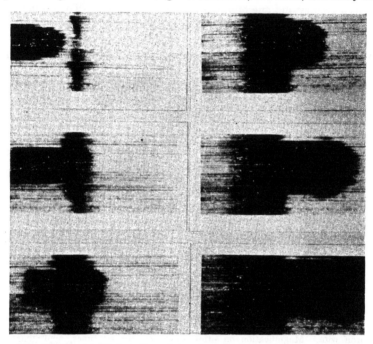

Fig. 10—Reassembled motion picture frames of a 0.22-caliber rifle bullet passing through 2-mm glass plate. Photographed at 5,000,000 frames per second, prints reproduced at intervals of approximately 40 frames or 8 microseconds. The actual bullet is about $^1/_2$ inch long. Note as the bullet strikes the plate a fracture wave travels vertically at approximately three times bullet velocity. After the bullet has passed a cloud of glass fragments remains "suspended" in mid-air, the forces acting on these particles now being relatively small.

discharge the lamp at the appropriate moment required by the object to be photographed, but in most applications this is comparatively simple. Numerous timing methods have already been described by Edgerton.

After the negative film is processed it is necessary to print it back through a system similar to that of the image dissector, in order to reconstitute the original picture. This is done with a simple automatic printer shown in Fig. 8. The negative film is mounted on a micrometer slide which is advanced automatically the required interval by means of a solenoid-operated ratchet. The reconstructed image is received on standard 16-mm positive film carried in the small motor-driven camera provided with a microswitch on the single frame shaft. This switch closes the solenoid circuit momentarily during the pull-down interval of the camera, advancing the negative by any desired amounts from one to twenty frames. Thus if the original photograph has been made at a higher speed than is required to show the motion it is only necessary to print every second, fifth, or even every twentieth frame in making the final 16-mm film.

In Fig. 9 is shown a sequence of the explosion of a metallic wire subjected to a heavy capacitor discharge. This is a very bright self-luminous event and presents no illumination problem. In Fig. 10 is shown a photograph of a rifle bullet initiating a transverse fracture wave in a vertical glass plate. In Fig. 11 is shown an enlarged reproduction of the negative from which Fig. 10 was made. It will be noted that the upper half of the photographs of Fig. 10 have been supplied by the

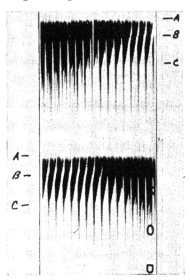

Fig. 11—Negative streak image of rifle bullet breaking glass panel. The bullet is entering the field at A, the fracture wave is commencing at B, and the bullet leaves the field approximately at the position C.

upper portion of Fig. 11, while the lower halves of each picture have been provided by the lower half of Fig. 11. A single frame of the final motion picture is represented by a very narrow strip extending from left to right all the way across the upper portion of Fig. 11, plus a similar narrow strip extending across the lower portion of the figure. With the film speeds used at present this strip on the original negative is only one-eightieth millimeter high with the camera running at ten million frames per second.

The most conspicuous feature of the present result, other than high speed, is the very poor image quality. Although the camera is useful in its present form for the analysis of certain types of fast events, the poor image quality is a very serious practical limitation. This is fully recognized, and this paper is in the nature of a progress report. Another form of the camera giving similar speed but much better image quality is now under construction, and it is hoped that a report upon this may be made at an early date.

REFERENCES

(1) U. S. Patents Nos. 2,021,162 (1935); 2,061,016 (1936); 2,088,732 (1937); 2,088,626 (1937); 2,089,155 (1937); 2,112,002 (1938); assigned to Scophony Ltd., London.

(2) "High-speed running film camera for photographic photometry," *Phys. Rev.*, vol. 50, p. 400; 1936.

Visual Test Film

THE 35-MM VISUAL TEST FILM first announced jointly by the Society of Motion Picture Engineers and the Motion Picture Research Council in January, 1947, is now available on safety base only. The price per roll is $22.50 postage paid in the United States, except in New York City where 45 cents must be added for city sales tax. When shipped outside the United States or possessions, the postage-paid price is $25.00.

Recent changes in the method of printing have resulted in improved steadiness, making this film a reliable performance test for 35-mm projectors. Vertical unsteadiness is measured in per cent of picture height while horizontal unsteadiness is measured in per cent of picture width. Unsteadiness values as low as one fourth of one per cent may be determined readily.

Screen masking may be adjusted and projector alignment checked with the use of the Focus and Alignment target. A Travel Ghost target provides a sensitive indication of shutter-timing errors and a series of vertical and horizontal lines indicate lack of sharp focus or curvature field.

Complete instructions for use are supplied with this film.

Comparison of Lead-Sulfide Photoconductive Cells with Photoemissive Tubes*

By NORMAN ANDERSON and SERGE PAKSWER

CONTINENTAL ELECTRIC COMPANY, GENEVA, ILLINOIS

Summary—A comparison is given for lead-sulfide photoconductive cells versus photoemissive tubes with *S1* and *S4* response with respect to signal (expressed as voltage sensitivity dV/df) at different color temperatures of the exciting light source. Spectral response, linearity, uniformity, and frequency response of lead-sulfide cells are also discussed.

At present most of the phototubes used in sound reproduction are caesium-oxide gas-filled tubes. Of late, however, following a publication by R. J. Cashman[1] considerable interest has been shown by the motion picture industry in the lead-sulfide cell. In the past year some additional insight has been gained as to the problems arising with the application of these cells to sound reproduction and it seems that the motion picture industry should be apprised of some of these problems in order to design their future equipment. In this paper we shall endeavor to discuss the lead-sulfide cells as compared to photoemissive tubes and to point out some modifications of the present practice which have to be followed in order to get optimum performance from lead-sulfide cells.

SPECTRAL RESPONSE

As SHOWN PREVIOUSLY, the lead-sulfide cells have a much higher response to infrared between 1 and 3 microns than the presently used photoemissive-type tubes. The spectral-response curve enters into consideration in several factors of the sound-reproduction system.

A. Spectral response of the tube itself.

B. Spectral distribution of the exciting-lamp source.

C. Spectral distribution of the transmission and refraction characteristics of the optical glasses used in the optical system.

D. Opacity to radiation of the sound tracks in different regions of the spectrum.

The spectral response of the lead-sulfide surface is a variable and to some extent controllable feature in the production of such cells. It has been shown[2] that the spectral response depends greatly on the relative amounts of oxygen and sulfur in the sensitive layer. In

---

* Presented February 12, 1948, at the Midwest Section Meeting in Chicago.

Fig. 1, curve A shows the response of a cell containing very little lead oxide. Curve C shows the spectral response of a cell with a high content of lead oxide, and curve B gives the spectral response of an intermediate-type layer. Another factor influencing the spectral response seems to be the thickness of the coating. It can be expected that the final shape of the spectral-response curve for cells used by the motion picture industry can to some extent be adapted to values required by the three other factors.

In Fig. 2 are represented spectral-distribution curves of a black body at different true temperatures in degrees centigrade. These figures are relative figures corrected to the same watt output and it is shown that the maximum shifts toward longer wavelengths at lower temperatures. For use with a lead-sulfide surface, a source with a maximum between 1 and 2 microns seems proper. For this reason it has been suggested to use indirectly heated exciter lamps. Work is being done at present on such lamps.

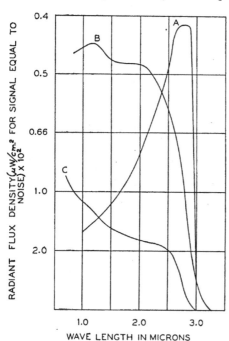

Fig. 1—From L. Sosnowski, et al., reference 2. Reprinted by permission of *Nature* Magazine. Curve A—Small amount of lead oxide. Curve B—Intermediate. Curve C—High content of lead oxide.

The final spectral-distribution curve of the exciter lamp will then include the emissivity factor of the radiating material and the color temperature at which these lamps can be run without decomposition or deterioration of the radiator.

There are few data available on the infrared transmission of crown and flint glasses. Some published data[3] show that for Jena glasses the coefficient of transmission of crown glass starts to decrease considerably between 1.7 and 2.5 microns depending on the type of glass. Flint glasses seem to be transparent to somewhat longer wavelengths. This transmission as well as the refraction characteristics leading to the possibility of designing chromatically corrected systems requires further study.

The same consideration is true for the opacity of different film materials in the infrared.

Summing up, it should be pointed out that for practical applications in the future a certain standardization on the type of spectral response of the lead-sulfide cell would be of interest.

SIGNAL SENSITIVITY OF LEAD-SULFIDE TUBES AS COMPARED TO PHOTOEMISSIVE-TYPE TUBES

A paper was published by A. Cramwinckel[4] in which different types of photoemissive tubes were compared, with relation to their sensitivity, to photovoltaic selenium cells. It seemed to us of interest to extend the comparison to lead-sulfide tubes. Because of the difference in mechanism of photoconductive and photoemissive types a common basis of comparison had to be found. As such the "voltage sensitivity" as shown in the book of Zworykin and Wilson[5] seems to give the most comprehensive results. "Voltage sensitivity" is the voltage developed across the series resistor per unit flux and its value is determined by the following equations:

EQUATIONS FOR VOLTAGE SENSITIVITIES

PHOTOEMISSIVE TUBES	PHOTOCONDUCTIVE TUBES

$$\frac{dV}{df} = \frac{Rs}{1 + Rf(ds/de)} \qquad \frac{dV}{df} = \frac{ER \cdot dr}{(R + r)^2 2df}$$

$$\cong Rs \text{ (vacuum tubes)} \qquad \qquad = \frac{Edr}{4rdf} \text{ (optimum value when } R = r)$$
$$\cong RsA \text{ (gas tubes } R < 1 \text{ megohm)}$$

V = voltage developed across load resistor R
R = load resistor
f = flux in lumens or watts
s = luminous sensitivity of photoemissive tube in microamperes per lumen at some specified color temperature
e = voltage developed across the photoemissive tube
ds/de = change in s per unit change in voltage across photoemissive tube
A = gas amplification of gas photoemissive tubes
E = voltage supply in circuit
r = resistance of photoconductive cell when exposed to a radiant flux f

We have plotted the curves of voltage sensitivity using these equations (Fig. 3). Voltage sensitivity in volts per microwatts is plotted as a function of the color temperature of a calibrated tungsten lamp.

Corrections have been made so that equal radiant energies are obtained for each color temperature. These corrections are tabulated in various books. The corrections used here were taken from Moon.[6] The correction factor in going from 2870 to 1700 degrees Kelvin is about 12.3. For the 200-watt tungsten lamp used in obtaining these data, the difference between true temperature and color temperature is very small and hence may be neglected when using a color-temperature scale. If a lamp is calibrated in volts at a fixed color temperature, then the voltage that must be applied to obtain any other color temperature can be calculated by a formula given by Moon.[7] This was the method used here to obtain the range of color temperature given in the graph.

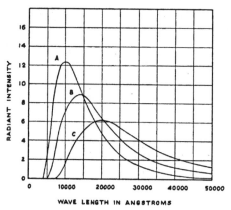

Fig. 2—Spectral distribution of the radiation from a black body. The wattage output in this case was kept constant.
Curve A—2970 degrees Kelvin. Curve B—2150 degrees Kelvin. Curve C—1500 degrees Kelvin.

Curves 1 and 2 give the voltage sensitivities of the S1 and S4 photoemissive vacuum tubes with a load resistor of 1 megohm. Curves 3 and 4 give the voltage sensitivities of the S1 and S4 gas tubes with a load resistor of 1 megohm. The voltage sensitivity of the gas tubes is limited by the factor $R\,(ds/de)$ which becomes quite appreciable as the load resistor is increased. However, if R is chosen as 1 megohm or less the voltage sensitivity of the gas tube becomes as many times greater as the gas-amplification factor A. For these data the gas amplification for the blue tube is taken as four and that of the red tube as eight. It must be noted that curves 3 and 4 give the maximum voltage sensitivity that may be expected by a load resistor of 1 megohm; in practice it will be lower than that indicated in the graph. Curves 5 and 6 give the theoretical voltage sensitivity of the S1 and S4 vacuum tubes with a load resistor of 100 megohms. In practice a high load resistor of this order can be used only when weak light levels are being detected.

The rated current sensitivities of the photoemissive tubes used here are as follows:

S4 vacuum tube—40 microamperes per lumen

S1 vacuum tube—20 microamperes per lumen

S4 gas tube—160 microamperes per lumen (gas amplification—4:1)

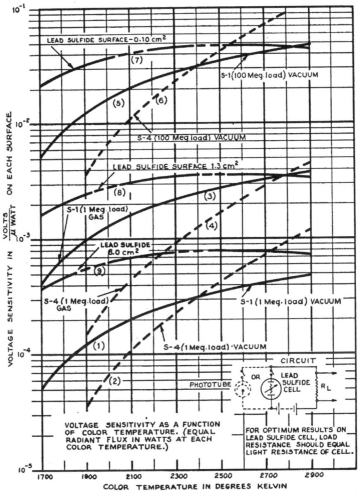

Fig. 3—Voltage sensitivity of different photoelectric surfaces as a function of the color temperature

S1 gas tube—160 microamperes per lumen (gas amplification—8:1)

The voltage sensitivity of the photoemissive tube is independent of area for a constant flux. However, on the lead-sulfide cell, the voltage sensitivity varies inversely with the area, again for constant flux. For this reason, various curves are plotted in which the area of the lead-sulfide surface is increased in the range 0.1 to 6.0 square centimeters. Data were obtained by taking an average of a number of cells having an area of 0.1 square centimeter. The curve showing the lead-sulfide surface of 1.3 square centimeters was calculated in order to make

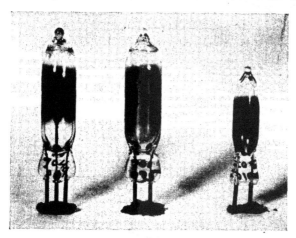

Fig. 4—Lead-sulfide cells.

comparisons with the photoemissive gas tubes which are being used now in sound reproduction. The curve plotted for the 6-square-centimeter lead-sulfide surface was plotted for comparison with the photoemissive surfaces used in taking these data as the areas of these cells were 6 square centimeters.

For comparison of these surfaces as to color temperature let us use curves *8, 3,* and *4.* It is evident from these curves that the blue tube is increasing in voltage sensitivity as the color temperature increases toward 3000 degrees Kelvin. The red tube is increasing in voltage sensitivity also but at a much slower rate; however, the lead-sulfide surface is fairly flat in voltage sensitivity in this region of color temperature having a light peak near 2500 degrees Kelvin. At low color temperature of the order of 1900 degrees Kelvin the voltage sensitivity of the lead-sulfide cell is still fairly flat while that of the blue tube is decreasing very rapidly being down a factor of about 17 from that of the lead-sulfide cell and that of the red tube being down by a factor of 2.5. At lower color temperatures the voltage sensitivity of the lead-sulfide cell falls off quite rapidly and this would have to be taken into account in designing an indirectly heated exciter lamp as a source for the lead-sulfide cell.

From these considerations it can be readily seen that the best advantage of the lead-sulfide cell will be had when the optics are such as to permit the use of small-area cells of the order of 0.1 square centimeter ($^1/_8 \times ^1/_8$ inch). If a cell of this area is used the advantage in voltage sensitivity over the photoemissive type cell will be about a factor of 10. These curves for lead sulfide were calculated on a basis of a matching load resistor for each value of light resistance. In practice this is not feasible and some loss is to be expected in voltage sensitivity for this reason.

DIMENSIONAL REQUIREMENTS

It follows from the discussions of the last paragraph that an adequate lead-sulfide cell should have a small area which would permit small over-all size of the tubes and would require particular care in the exact positioning of the light spot to the contacting pins or other exterior components of the tube. On Fig. 4 are given photographs of lead-sulfide cells which satisfy these requirements. These cells offer an advantage of uniformity in mass production, close tolerances of the sensitive area with respect to conductive pins, and direct insertion into a socket in which they can be held rigidly in place without the need of a base. This development is in line with the trend of the radio-tube industry to miniature and subminiature sizes.

LINEARITY WITH LIGHT AND VOLTAGE

Photoemissive vacuum tubes are linear with light levels up to the order of 5 lumens and the response is independent of voltage if the voltage is maintained above that needed for saturation. Certain cells may be limited in linearity due to the existence of electrical leakage, fatigue, or an accumulation of space charge. Photoemissive gas tubes are linear with a light flux up to 1 lumen on the surface but not linear with voltage above 25 volts. Lead-sulfide cells of average sensitivity are linear with light levels up to 0.01 lumen on the surface. The linearity breaks off at lower light levels than this for very sensitive

cells and continues beyond 0.01 lumen for cells of lower sensitivity.
The linearity with voltage for constant light illumination is very good;
for instance, one very sensitive tube with a grid area of 1 mm wide ×
5 mm long had a drop of only 10 per cent in voltage sensitivity from
the expected linearity relationship when the voltage changed from 22
to 90 volts. This was with a light level beyond the linearity region of
this cell. For lower light levels the linearity with voltage should be
even better.

FREQUENCY RESPONSE

Photoemissive vacuum tubes are flat in frequency response to very
high frequencies. Photoemissive gas tubes are fairly flat in frequency
response up to 10,000 cycles but drop appreciably at higher frequencies. Lead-sulfide cells have a fairly flat frequency response to 10,000
cycles but then drop quite rapidly above 10,000 cycles. Variations of
frequency response between individual tubes require additional study.

UNIFORMITY OF SIGNAL AND FREQUENCY RESPONSE

It has been observed that sensitivity and frequency response vary
somewhat at different points of the sensitive surface. This characteristic of lead-sulfide cells also requires additional study in conjunction with the optical systems to be developed for their use.

Summing up the contents of this paper it seems that the lead-sulfide
cell is showing considerable advantage in sound reproduction. Its
application will require a modification of the present mechanical,
optical, and electrical design of motion picture projectors and the
optimum operation of these cells will follow an accurate balancing of
all the factors involved.

REFERENCES

(1)   R. J. Cashman, "Lead-sulfide photoconductive cells for sound reproduction," J. Soc. Mot. Pict. Eng., vol. 49, pp. 342–348; October, 1947.

(2)   L. Sosnowski, J. Starkiewicz, and O. Simpson, "PbS photoconductive
cells," Nature, vol. 159, pp. 818–819; 1947.

(3)   Smithsonian Physical Tables, 1934, p. 386, Table 438.

(4)   A. Cramwinckel, "The sensitivity of various phototubes as a function of
the color temperature of the light source," J. Soc. Mot. Pict. Eng., vol. 49, pp.
523–530; December, 1947.

(5)   V. K. Zworykin and E. D. Wilson, "Photocells and Their Applications,"
John Wiley and Sons, New York, 2nd ed., 1934, p. 187.

(6)   P. Moon, "The Scientific Basis of Illuminating Engineering," McGraw-
Hill Book Company, New York and London, 1st ed., 1936, p. 146.

(7)   Page 162 of reference 6.

Volume Compressors for Sound Recording[*]

By W. K. GRIMWOOD

KODAK RESEARCH LABORATORIES, ROCHESTER, NEW YORK

Summary—This paper deals in a general way with volume compressors of the type used in sound recording. The subject matter is divided into six sections: the desirability of volume compression, compressor characteristics, problems arising from the use of compressors, classification of the types of compressors with the advantages and disadvantages of each type, compressor design, and the measurement of compressor performance.

MANY DEVICES[1,2] have been developed by the communications industry for the automatic control of volume in telephonic transmission. Of these devices, the limited-range compressor and the peak limiter have come into general use for 35-mm sound-on-film recording, and two others, the volume-operated gain-adjusting device[3] and the limited-range expander, may be useful for 16-mm sound recording. This paper will be concerned primarily with the problems of design and use of compressors and limiters for sound recording

DESIRABILITY OF AUTOMATIC CONTROL

The range of sound levels to which the ear is sensitive is much greater than the range which can be linearly accommodated by any known method of sound recording. Fortunately, sound intensities which reach the upper threshold of hearing are rare, and very faint sounds are usually submerged in the ambient-noise level. Thus, the range of sound levels encountered in recording is not greatly in excess of the capabilities of 35-mm sound-on-film recording, although the range is considerably more than is now practical on 16-mm film. Photographic recording is limited at high amplitudes by 100 per cent modulation of the exposing light and at low amplitudes by the granular structure of the developed image. By manual adjustment of the amplification the sound levels to be recorded may be brought to lie within these two limits, low levels being brought up and high levels suppressed, a process which requires skill and experience on the part

[*] Presented May 17, 1948, at the SMPE Convention in Santa Monica.

of the operator. Since overmodulation of the exposing light results in noticeable and unpleasant distortion, a factor of safety must be allowed so that unexpectedly high peak sound levels will not cause serious distortion, manual control being too slow to react to sudden changes in level. It is evident that if an amplifier could rapidly, and without nonlinear distortion, change its gain so that its output level were limited to correspond with the overload level of the light modulator, then the operator could record at a higher average level without danger of objectionable distortion of the peak levels. Furthermore, if an amplifier could, without distortion, control its own gain so that the range of input levels were divided by a factor to result in a lesser range of output levels, then a wide range of sound levels could be compressed into the limited range of the recording medium. If the amount of such compression were no greater than that normally done by manual control, expansion upon reproduction would not be necessary.

Compression of the sound volume range may be desirable for reasons other than the characteristics of the recording medium. For example, experimental home recordings on 16-mm film often sound as though the volume range reproduced were greater than the volume range of the original sound. This effect is probably due partly to the acoustics of the rooms in which the recordings were made and partly due to the monaural character of the recording process. It is reasonable to expect that the naturalness of recordings made in the home (where rooms are more reverberant than are sound studios) would be improved by the use of compressors in recording.

The volume range that can be reproduced satisfactorily is equally as important as the range that can be recorded. The volume range that can be used in a motion picture theater is surprisingly narrow, much shorter than the range that can be recorded.[4] This is because audiences are intolerant of loud sound reproduction and the audience noise level is high; some form of compression must, therefore, be used in making recordings to be reproduced where the noise level is high and the maximum level limited by listener preferences.

COMPRESSOR CHARACTERISTICS

The static input-output relations of a compressor or limiter amplifier differ from those of an ordinary amplifier in a manner which may best be understood by referring to Fig. 1. Curve O-O' represents the input-output relations of a normal amplifier, O-a-A represents these same relations in a compressor amplifier, and *O-b-B* applies to a limiter amplifier. Points *a* and *b* are known as thresholds. The input-output relation of the amplifiers is defined in terms of the slope (decibel scale) of the curve above the threshold and the input range in decibels between the threshold level and the maximum useful output level. For applications of sound recording, the maximum useful output level would be taken as the output level corresponding to 100 per cent light-modulation. In Fig. 1 an output level of zero decibels (point *x*) has been taken as the output at which the light-modulator will be fully modulated. The specification of the com-

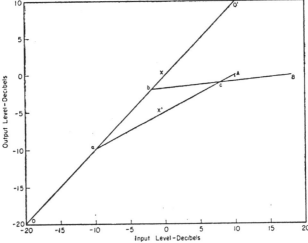

Fig. 1—Idealized input–output characteristics of volume compressor.

pressor curve would therefore be: slope $\frac{1}{2}$, range 20 decibels; for the limiter curve, the specification is: slope $\frac{1}{10}$, range 20 decibels. Alternatively, these two characteristics may be specified more simply as a compression of 20 into 10 decibels and as a compression of 20 into 2 decibels, respectively. When operating with a fixed slope, compression is sometimes expressed as the decibel difference between point *x* on the linear curve and the corresponding ordinate of the compressor curve (*x'*). This method of expressing the amount of compression is not very useful in setting up operating conditions, particularly when the slope is very low, as it is for limiter operation, but it is very useful in the practical use of a compressor, a meter sometimes being calibrated to give a direct indication of the amount of compression. Experimental curves are very similar to those of Fig. 1, except that points a and b are not sharply defined. The threshold should, then, be defined as the point of intersection of the extrapolations of the linear portions of the curve. Since this involves, plotting the curves, it is more convenient in practice to define the threshold arbitrarily as, for example, the point at which the output is compressed by $1/2$ decibel. So long as the method of definition is specified, there need be no confusion.

The values of compression and of limiting used here for illustration are representative of those generally used in sound recording. The compresssions usually used range from 20 into 10 to 30 into 15, and the limiter characteristics are in the region of 10 into 1 to 20 into 1.

The input-output relations just discussed were defined as static; the dynamic input-output relaton must be linear or else the nonlinear distortion will be intolerable. The difference between static non-linearity and dynamic linearity is one of operating time. If the change in gain takes place at a rate so slow that individual cycles of the lowest audio frequency to be transmitted are not measurably altered in shape, then there will be no nonlinear distortion. Such slow operation would defeat the primary purpose of compressors but it is also possible to change the gain very rapidly upon application of the audio signal and to release the gain to normal very slowly upon cessation of the signal. In this case, there will be nonlinear distortion only during the short period during which the gain is changing rapidly and, if this period can be made sufficiently short, the ear will not detect the distortion. Compressor amplifiers, therefore, are designed to decrease their gain very rapidly upon the sudden application of a signal and to increase the gain to normal very slowly upon sudden removal of the signal. This timing is controlled by the charge and discharge time constants of a simple resistance-capacitance network. The time required for a specified percentage completion of gain change from the normal (uncompressed) value to the compressed equilibrium value upon the instantaneous application of a signal is known as the operating time. The time required for the same percentage of gain change in the return of the compressed gain to the uncompressed equilibrium value upon the instantaneous cessation of the input signal is known as the release time. The percentage change in gain for which the time is given is usually either 63 per cent (the value resulting when the time equals the resistance-capacitance product) or

99 per cent. Both figures are open to objection: the 99 per cent figure because it requires an unduly high accuracy of measurement, the 63 per cent figure because it gives a misleadingly short operating time. The operating time is determined by the charging of a capacitor through a rectifier and the discharge time by the discharge of the capacitor through a fixed resistor. The resistance of a rectifier is a function of the voltage drop across the rectifier, increasing as the voltage drop decreases. Thus, when a signal is suddenly applied, the capacitor will start to charge according to the exponential charging law, but, as the voltage across the capacitor builds up, the voltage across the rectifier decreases, the rectifier resistance increases, and the capacitor charge builds up to its ultimate value much more slowly than would be predicted from the exponential law. An amplitude change of 90 per cent of the ultimate change in amplitude is recommended as a satisfactory compromise in the specification of the action times.

The operating time of commercially available compressor amplifiers is usually from 1 to 2 milliseconds and the release time from 100 to 400 milliseconds. Experimental evidence[5] indicates that performance is improved as the operating time is decreased, the lower limit being one of practical design.

The steady-state distortion characteristics of a compressor amplifier are those of a normal amplifier except at very low audio frequencies. If a long release time is used, there will be no increase of low-frequency distortion, but if a short release time is desired, the timing capacitor will discharge sufficiently between cycles of the audio input to affect the wave form so that some compromise must be made between discharge time and distortion. It is possible to devise circuits in which the release time would be a function of the duration of the applied signal but such refinements have not as yet come into use for sound recording.

Some transient distortions are present in compressors. One of these, previously mentioned, is the distortion inherent in changing amplification at a rate comparable with the instantaneous rate of change of the signal. The gain may, however, be changed in a period of time shorter than that required to produce an impression on the ear so that this type of distortion is of no practical importance. Another, and frequently serious, distortion is the generation of pulses known as "thump" from the fast operating time. If a high-frequency audio signal is suddenly applied to a compressor amplifier, the static potentials in the variable transmission circuit suddenly change to new levels and remain at these levels until the amplitude of the input signal is changed. Since the rate of change from one level to the other is of the same order of magnitude as the instantaneous rate of change of a high-frequency audio signal, the change in potentials will be transmitted by the audio amplifier; furthermore, the change from one potential level to another will alter the charge on coupling capacitors between the variable trnsmission circuit and the output audio amplifier, and the change in potential on these capacitors as they discharge is transmitted by the audio amplifier. Thus, unless some means are used to prevent the control signal from appearing in the audio output, a sudden increase in compression will result in a pulse appearing in the audio output, this pulse taking the form of a sharp rise in potential followed by an exponential decay, the duration of the pulse being determined by the time constants of the amplifier coupling circuits.

Because of the slow decay time, the audible effect of the pulse is such that it is usually called "thump." The term "thump" is often given to any audible effect of similar nature; it is not restricted to pulses directly due to the control signal. Thump is not necessarily fundamental to compression but is usually present to some degree, just as nonlinear distortion is present to some degree in a normal amplifier. One of the most serious problems in the design of a compressor for 16-mm recording is the development of a circuit in which the thump component can be held to tolerable levels without frequent inspection and maintenance. Thump can be serious out of all proportion to its direct audible effect. It may, particularly when operating on a limiter characteristic, modulate the audio signal and cause the output amplitude to exceed that shown by the steady-state characteristic for periods of time long enough for the resulting overload of the light-modulator to be plainly audible in the finished record. Tolerable levels of thump components will not be given here, but will be discussed more fully in one of the following sections.

The frequency response of a compressor or limiter does not, in general, differ from that of a normal amplifier. Because of the compression action, there are certain requirements for the frequency characteristics of the control signal path but they do not necessarily affect the audio signal transmission path. The use of a compressor does have some effect upon the recording channel as a whole, in that the compressor alters and restricts the preferred location of frequency equalization in the channel. Both of these effects will be considered in some detail in the next section.

The noise level of a compressor should not differ significantly from that of an ordinary amplifier. This means that the bulk of the system noise should arise in sources located ahead of the point at which compression action takes place. If it arises beyond this point, the signal-to-noise ratio will be reduced as the compressor acts.

COMPRESSOR PROBLEMS

The solution of one problem usually gives rise to several others. Certainly this is true of the use of compressor amplifiers. One of the most serious of these new problems is what has been termed "spectral-energy distortion."[6] Consider the case of a speech sound which starts with a consonant followed by a vowel. Consonants are usually composed of high-frequency components of low amplitude, while vowels are predominantly low-frequency components of high amplitude. Therefore, it is probable that the compressor or limiter will not be actuated by the opening consonant but will be operated by the following vowel. The result (sometimes called "essiness") is over-accentuation of some sibilant sounds which, because of long pauses or widely fluctuating speech levels, find the compressor in the passive condition. If we place in the control path of the compressor an equalizer (or "de-esser") which boosts the high frequencies so that the compressor responds to lower levels of high-frequency components than to those of low-frequency components, then this accentuation of sibilants will largely disappear. The effect is entirely eliminated only when the compressor threshold follows a frequency-versus-level curve which matches the frequency spectrum of the sound source.

The spectral-energy distribution of the sound to be recorded is, however, a function of acoustic conditions, type of source (speech, music, or noise), and frequency response of the recording channel. The energy distribution of speech alone varies with the individual and with the effort level of the speaker. Thus, the exact compensation of spectral-energy distortion becomes a very complex problem. Fortunately, satisfactory results are obtained in practice with a single "de-esser" equalizer whose characteristics are based upon the average spectral-energy distribution of speech. Some further refinement probably is desirable in studio recording and can be obtained readily by having several equalizer characteristics available to the operator, say, one for each of three effort levels of speech and a fourth for music. The frequency characteristic of the "de-esser" equalizer is not, in general, the inverse of the spectral-energy distribution of the source, but should take into account the average of all frequency discriminations, whether acoustical, electromechanical, or electrical in nature, between the sound source and the equalizer itself. The correction of spectral-energy distortion, while desirable, is less necessary in a limiter amplifier than in a compressor amplifier because the limiter is operated above the threshold less frequently than is the compressor.

A closely related problem arising from the use of compressor amplifiers is that of the location of the compressor in the recording channel.[5] It is obvious from the input-output relations of a compressor or limiter that for levels above the threshold the effect of any frequency discrimination ahead of the compressor will be reduced by the compression ratio. Since such discrimination is usually intentional, this effect is undesirable. Elimination of this effect calls for a rearrangement of the recording channel such that all equalizing is placed after the compressor, but this solution is not entirely satisfactory since the mixer operator must have some equalizing under his control in order to adjust for the set acoustics and for differences in the source material. Much of the equalizing done by the mixer is of such a nature that it automatically corrects for variables that otherwise would distort the normal spectral-energy distribution of the source. Thus, a satisfactory solution is to place the bulk of the equalizing beyond the compressor and to leave a bare minimun of variable equalizing ahead of the compressor. If more than one microphone is used simultaneously, the spectral-energy distribution of the sources may be different and the equalizing required may be different so that ideally a compressor should be used in each input circuit. This condition is even more likely to exist in re-recording than in original recording. When a limiter amplifier is used, it is not permissible to put most of the equalizing beyond the limiter. If the limiter is to protect against overmodulation of the light-modulator, any equalizing between the limiter and the modulator must be restricted to the type which decreases the level of some frequency components. Because the limiter acts infrequently and on the highest levels, the effect of equalizing ahead of the limiter is not serious. This restriction applies only to equalizing in the transmission path beyond the point at which the control-signal path branches off from the main path. When the control circuit branches from the main path beyond the point at which the actual gain-changing takes place, both paths are equally affected by equalizing inserted between these two points. Hence, in a limiter, the modulator will still be protected against overload and in a compressor, the equalizing can be considered as being beyond the compressor.

Monitoring the recording level is more of a problem in a channel using a compressor or a limiter than in a channel using only normal amplifiers. The object of level monitoring is to know, at all times, the recording level in terms of the overload point of the light-modulator. If a volume indicator is placed ahead of the compressor, its indication may be correlated with light-modulation at any one frequency, but, unless the frequency characteristic of the indicator is matched to that of the recording channel between the point of connection of the indicator and the light-modulator, the indications will be of little value. The volume indicator may be placed beyond the compressor and the equalizer. In this case, the accuracy of reading the indicator must be multiplied by the inverse of the compression slope if the precision of the indication is to be held to the same value as in a channel without a compressor, the input volume range indicated by the meter being increased by the amount of the compression. This is not necessarily a disadvantage; in fact, a good case can be made for the increased volume range shown by the indicator. When a limiter is used, a volume indicator beyond the limiter is of little use, since a wide range of input levels is compressed into a very small range of output levels. If the volume indicator can be given the same freqency response as the limiter, a location ahead of the limiter is satisfactory. A meter reading the amount of compression or of limiting may be used to supplement the volume indicator, but it is not a satisfactory substitute because a compression indicator gives no indication of levels below the threshold level.

The release timing of a compressor presents some minor problems in that the release time should depend partly on the type of material being recorded. When the average sound level fluctuates fairly rapidly, as in speech, a short release time is desirable, but when the level may change slowly, as in music, a longer release time is preferred. Although the release time can be made to change automatically with the duration of the sound, there seems to be little justification for the complexity of an automatic control so long as a change in equalization (as between speech and music), which is not readily made automatic, must be made by the operator.

COMPRESSOR CIRCUITS

A compressor consists basically of a circuit whose transmission can be varied by a control signal and a second circuit which derives this control signal from the audio signal. Because the control signal may be derived from either the audio input or the audio output, compressors may be grouped into one of two classes: the forward-acting type in which the control signal is derived from the audio input (Fig. 2 (a)), and the backward-acting type in which the control signal is derived from the audio output (Fig. 2 (b)). These two classes have quite different input-output relations; the forward-acting type usually has a compression slope which decreases as the input level increases. Thus, the input-output curve approaches a maximum and it may have a negative slope beyond the maximum. The exact form of the curve depends upon the characteristics of the control circuit and of the

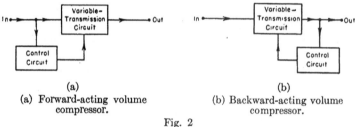

(a)

(a) Forward-acting volume compressor.

(b)

(b) Backward-acting volume compressor.

Fig. 2

variable transmission circuit and will change with any shift in the characteristics of either of these circuits.

The compression ratio of the backward-acting circuit is nearly constant over a wide range of input levels and the input-output curve is not greatly changed by the characteristics of the variable transmission circuit. The backward-acting compressor has one serious disadvantage which is not present in the forward-acting type. Since it is a form of feedback circuit, self-oscillation is possible and careful circuit design is necessary to avoid instability of this type. Compressors and limiters at present used in 35-mm sound recording are universally of the backward-acting type. Inasmuch as the purpose in using a compressor is to reduce the range of signal levels applied to the modulator without noticeably altering the original volume relations, an input-output curve having a constant slope of less than unity (above the threshold) is to be preferred to a curve in which the slope decreases gradually from unity to zero, thus completely destroying the syllabic volume relations of high-level input signals. While this latter curve is satisfactory for limiter operation, the limiter function of preventing overload of the light-modulator makes control of the output level by the output level preferable to control by the input level. Because of the general acceptance of the backward-acting compressor as the more desirable type, the remainder of this paper will be devoted to this type, though for the greater part of the text it will not be necessary to distinguish between the two types.

Compressors may be further classified in terms of the form of control of the variable transmission circuit. We may term "one-dimensional" all circuits in which the electrical transmission is controlled by an electrical signal. Those compressor circuits in which the electrical transmission of the input signal is controlled by another form of energy (such as mechanical) will be termed "two-dimensional." Included in this class are those circuits in which the input signal is nonelectrical and the control signal is electrical. No commercial fast-acting compressors of the two-dimensional type are known to the writer, but this type has some very worth-while advantages over the purely electrical type if the major problem of slow-action time can be overcome, and there are some interesting possibilities. One slow-acting compressor of this type uses a thermistor to convert the control signal energy into heat which, in turn, controls the transmission through the audio signal path.[7] There exists the possibility of compressing by using the control signal to vary the field strength of the magnetic field of a dynamic or of a velocity microphone.[*] Another possibility is a step-by-step compressor which would use high-speed relays operated by the control signal to insert attenuators in the audio-transmission path.[**] Another possibility is the adaption of the carbon-pile regulator to the high-speed, low-level operation required in a compressor.

These possibilities are sufficient to disclose the two main advantages of a two-dimensional system: first, the absence of an electrical connection between the variable-transmission circuit and the control circuit permits the transmission to be altered without the generation of transients in the audio path arising from the reaction of the one circuit on the other; second, the dynamic linearity of the audio path is not affected by the control signal. These two factors are fundamental limitations of the one-dimensional system. If the electrical

* J. G. Streiffert, of the Kodak Research Laboratories, private communication.
** T. G. Veal, of the Kodak Research Laboratories, private communication.

transmission of one path is to be under the direct control of another electrical signal, some special means must be employed to prevent the control signal from appearing in the first path, and the transmission of the desired signal can only be altered by a nonlinear element which must simultaneously cause nonlinear distortion of the desired signal.

The means taken to prevent control-signal components from appearing in the audio output may be made the basis of further classification of one-dimensional compressors. Three methods have been used: the carrier method, the compensation method, and the balance method.

The carrier type of compressor[8] uses an oscillator and a balanced modulator to shift the audio spectrum up into the carrier-frequency range and a demodulator to step back down to audio frequencies. The variable-transmission circuit is placed in the carrier link. This circuit may be of either the variable-mu or the variable-impedance type, although the former ordinarily would be preferred for its simplicity. The shift to carrier frequencies makes it possible to separate the control signal from the audio signal on a frequency basis and also to filter out nonlinear distortion products on a frequency basis. The disadvantages of this scheme are that an oscillator, a demodulator, and an accurately balanced modulator are necessary.

The compensator type of compressor[9,10] uses a variable-mu type of tube to control the audio gain, changes in the plate current of this tube due to the control signal being compensated by an opposite change in the screen current of a second variable-mu tube. This circuit has the advantage of not requiring push-pull operation but has the disadvantage of requiring the changes of plate current of one tube to be exactly matched by changes in the opposite sense in the screen current of another tube.

Practically all compressors in actual use at the present time separate the control signal from the audio signal by balancing the two circuits with respect to each other. If the audio signal is applied out-of-phase to the two inputs of a push-pull amplifier and the control signal is applied in-phase to the same inputs, the control signal and the audio may be separated on a phase basis. This method has two disadvantages: the variable-mu transmission circuit must be push-pull (which requires two accurately matched nonlinear elements), and some means must be used to cancel the in-phase components appearing in the output of this circuit. Both requirements can be met without great difficulty or circuit complexity.

Audio-frequency compressor amplifiers may be divided into variable-mu types and variable-impedance types. The variable-mu type is used almost exclusively for sound-on-film recording and for radio broadcasting. Two versions of the variable-mu compressor [11-13] are currently used: compressors designed primarily for limiter operation use mixer-type tubes such as the 6L7, those designed primarily for operation over a characteristic curve having a slope of the order of one half use remote cutoff pentodes such as the 6K7. The variable-mu type of compressor amplifier has one main advantage over other types of compressors: simplicity. This circuit simplicity is the result of two properties of vacuum tubes; first, the gain of a vacuum tube is controllable by the potential of a grid whose impedance is so high that for practical purposes there is no power drawn from the source of the control signal; second, the vacuum tube can be so used that when the amount of compression is a maximum the variable-transmission circuit may still have a net gain in signal level, and hence there is no net loss in the variable-transmission circuit to be made up by the addition of amplifier stages.

The variable-mu compressor has also a major disadvantage which results from the use of tubes to control gain. In order to control the circuit gain and at the same time to balance the audio-transmission path with respect to the control signal, it is necessary to have two matched nonlinear characteristics. The characteristics of vacuum tubes, however, vary considerably from tube to tube and drift with aging of any individual tube. It is possible, by the use of a negative feedback, to make circuits using vacuum tubes linear to any assignable degree. No similar technique is known whereby a circuit may be forced to have a given degree of nonlinearity. Hence, when nearly identical nonlinear characteristics are necessary in two vacuum tubes, they can be obtained only by a process of selection from a group of aged tubes and the matching of the tubes should be checked at frequent intervals, preferably each time the apparatus is used and at least twice during each day of continuous use. This disadvantage of the variable-mu compressor is not particularly serious in professional sound-recording work because the scale of operations is such that routine maintenance and checking of all equipment is standard procedure.

Compressors and limiters of the variable-impedance type comprise a very extensive group, not only because a variable impedance may be used in many ways to control the transmission of a circuit but also because any nonlinear element has potential applications in compressor design. Many variable-impedance compressor circuits have been published in which the plate impedance of a vacuum tube is the variable element.[3,14-19]   Plate impedance may be used in several ways: as one portion of a variable voltage divider, as one of the feedback impedances in a feedback amplifier, as a means of effectively changing the connections of two transformers from series aiding to series opposing; other configurations can be devised. Circuits using tubes as variable-impedance elements are subject to the same disadvantage as that cited against the variable-mu circuits. There may be some difference in degree since simpler tube types may be used in the variable-impedance circuits but matched characteristics are still necessary. These circuits have, in general, the characteristic in common with other variable-impedance circuits that they are variable-loss circuits. This loss must be made up by amplification in some other part of the circuit. As in the variable-mu compressors, there is the advantage that no appreciable power is drawn from the control-signal source.

Some further mention should be made of those circuits in which the variable impedance is part of the beta circuit of a feedback amplifier. Such circuits are usually accompanied by a claim of superior merit because they are feedback circuits, especially in respect to harmonic distortion. What merit these circuits possess cannot be attributed to negative feedback. As in all other one-dimensional compressor circuits, the variable-transmission elements cannot of themselves distinguish between the audio signal and the control signal; both signals operate on the same nonlinear characteristic. Distortion is, then, determined by the curvature of the nonlinear characteristic over the maximum range of amplitudes of the audio signal. Strictly speaking, the usual feedback equations are not applicable to compressor circuits; the derivation of these equations assumes linear transfer characteristics.

A second group of variable-impedance compressors[20,21] uses passive nonlinear elements, known as varistors, in the variable-transmission circuit. Copper-oxide rectifiers are the most commonly used varistors, silicon carbide (Thyrite) has been used in these Laboratories in an experimental compressor, and germanium-crystal rectifiers have been tried experimentally. The advantage of varistors over vacuum tubes is their stability. Two or four units may be selected for matched-impedance characteristics and they will remain matched over long periods of time. The varistor compressor, in common with other variable-impedance types, compresses by introducing a loss of energy into the circuit. In fact, the variable-transmission portion of the compressor is often called the "variolosser." Varistors have one disadvantage not present in vacuum-tube variolossers in that power is required to control their impedance. The timing capacitor of the control circuit cannot supply sufficient power without either increasing the action time or shortening the discharge time, so that it is necessary to insert a direct-coupled impedance-changer tube between the timing capacitor and the variolosser. The varistor is a relatively low impedance device. If it were made sufficiently high in impedance to be negligible current drain on the timing condenser, an impractically high voltage would be needed for control and the impedance would be too high for use in audio-frequency circuits. Also, the discharge time would be a function of the amount of compression.

COMPRESSOR DESIGN

The purpose of this section is to point out some of the principles to be followed in the design of compressor amplifiers; no specific circuits will be presented. While much of the discussion will be of general applicability, it is intended to apply specifically to those circuits which may be represented by the block diagram of Fig. 3.

The choice of the type of compressor will naturally be arrived at by weighing the advantages and disadvantages of the various types in relation to the requirements of the particular application. In sound-on-film recording, the use of compressor amplifiers may be divided among 35-mm and 16-mm apparatus and studio and portable equipments, the four combinations having different requirements. These four uses put different emphasis on such factors as size, weight, cost, distortion, flexibility of operation, stability of operating characteristics, and routine maintenance requirements. This latter factor is of the utmost importance in equipment designed for 16-mm amateur sound recording and is probably more important in 16-mm professional work than in 35-mm usage. Cost, size, and weight are all important factors in the 16-mm field, even in studio recording, and gains can be made in these respects at the expense of flexibility of operation. The one factor that cannot be sacrificed, if good 16-mm recordings are to be made, is the factor which is usually (and rather vaguely) called quality. More exactly, quality means freedom from distortion, permanence of electrical characteristics, and physical durability.

The desired input-output curve may affect the choice of the type of compressor to be used and will greatly influence circuit details. In 35-mm recording, some studios use a compression of, roughly, 20 into 10 decibels, while others use a limiting characteristic of about 10 into 1 decibel. Thus, compressor amplifiers may be designed primarily for one or the other of these two types of characteristic. For 16-mm recording, on the other hand, it now seems probable that the highest average quality will be obtained by combining the two characteristics. That is, a compression of 20 into 10 decibels should break at a second threshold into a compression of 10 into 1 decibel. Thus, the circuit will not be the same as would be used for either characteristic alone.

The choice of a circuit will also be influenced by the operating level. In order to reduce thump to a miinmum, it is necessary to work the variable-transmission circuit at the maximum audio level consistent

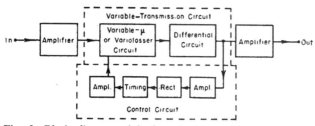

Fig. 3—Block diagram of backward-acting volume compressor.

with nonlinear distortion requirements. Hence, if this part of the compressor can handle a signal level of 10 volts, to operate it at a level of 1 volt would, in effect, increase the thump level by a factor of 10.

The control circuit, especially that part of it which is associated with the rectifier, is very important to the proper functioning of the compressor. The ratio of discharge to charge time is approximately 1000. Thus, if 5 megohms is taken as a practical limit to the value of the discharge resistor, the charging impedance should not exceed 5000 ohms throughout the frequency range to be handled by the compressor. The rectifier, which for minimum operating time should be used in a full-wave circuit, must be fed from a low-impedance source. Voltage feedback offers a practical means of obtaining very low source impedance; if the rectifier is fed from a transformer, voltage feedback from the primary of the transformer will not be effective unless a high-quality transformer is used. The rectifier should be chosen for low plate resistance and should be operated with as high a signal level as is practical since the plate resistance decreases as the voltage drop across the rectifier increases. A high operating level also reduces the effect of variations in the contact potential on the threshold level. Barrier-layer-type rectifiers do not usually have a sufficiently high ratio of back-to-forward resistance to be useful, but the germanium-crystal diode may be satisfactory if a sufficient number are used in series.

Spectral-energy distortion has been mentioned earlier in this paper. Its correction will usually require some boosting of the high frequencies and perhaps of the low frequencies. If it is desired to do this boosting in the rectifier amplifier, it may readily be done in the feedback path. If done in this manner, the amount of feedback remaining at the frequency of maximum boost should be enough to hold the internal impedance of the amplifier at this frequency to a value sufficiently low to have little effect on action time. The bulk of the charging impedance should be the internal resistance of the rectifier. The low-frequency response of the control-signal circuit must be handled with care in a backward-acting compressor. Backward-acting circuits become unstable when the phase shift around the loop formed by the variable-transmission circuit and the control-signal circuit reaches 90 degrees and the signal level exceeds the threshold level. Phase shift introduces a time delay between the point at which the gain is increased or decreased and the rectifier so that the gain changes overshoot the correct value on both decreases and increases. The remedy for this effect is to keep the phase shift around the loop low, down to frequencies below which no signal of amplitude exceeding the threshold level can reach the gain-changing portion of the circuit.

It will be apparent from the block diagram of Fig. 3 that the output amplifier and the rectifier amplifier can be combined. If the output amplifier is made push-pull, the rectifiers may be connected through blocking capacitors to the plates of the output stage. A simpler circuit is thus obtained though at the expense of flexibility: voltage feedback should be used to provide a low source impedance for the rectifiers, the power-handling capacity of the amplifier must be sufficient to supply the peak charging current required by the timing-capacitor-charging circuit, the gain of the amplifier must be adequate to provide the proper operating level for the rectifiers and must remain fixed, and the frequency response should be such as to give the desired degree of compensation for spectral-energy distortion. If the amplifier is to supply power directly to a light-valve or recording galvanometer, the only one of these requirements likely to be trouble-some is that of frequency response. The equalizing needed to com-pensate for 16-mm film losses and that needed to reduce spectral-energy distortion are sufficiently similar to make the combining of the output and rectifier amplifiers a practical possibility.

It was mentioned earlier that a combination of compression and limiting is desirable for 16-mm recording. This may be done with-out very much complication by providing two full-wave rectifiers in which the threshold biases and the loop gains of the two rectifier cir-cuits have been adjusted to give the desired threshold levels and com-pression ratios. There are several means of adjusting the loop gains: the compressor rectifier may be fed from cathode followers to provide a low-impedance source and the gain set by voltage dividers in the grid circuit of the cathode followers; if the limiter rectifier is fed from a transformer, the compressor rectifier may be supplied by another winding or by taps on the same winding; if a circuit is used which requires a direct-current stage following the timing capacitor, this stage may be made a double-triode inverter stage so that two timing capacitors may be used and the gain set by adjustment of a tap on the discharge resistor of the capacitor controlling the com-pressor characteristic. In this case, the rectifiers may be fed from the same source but must be connected to give control signals of opposite polarities.

In Fig. 3, there is a block labeled "differential circuit." The func-tion of this circuit is to remove control-signal pulses before they can overload the amplifier. In purely electrical compressor amplifiers, the change in gain is always accompanied by a change in the static oper-ating point of the variable-transmission circuit, and in the balanced type of circuit this change takes the form of in-phase pulses whose magnitude may be from ten to one hundred times the amplitude of the desired out-of-phase signal. With a bridge configuration of the vari-able-transmission circuit controlled by a truly push-pull control sig-nal, these pulses would not appear at the audio terminals of the bridge but, except in the doubly balanced type, they do appear in the audio circuits and usually of such amplitude that they will overload the output amplifier unless removed by a differential circuit. Perhaps the best-known differential circuits are the transformer and the push-pull choke. An in-phase signal applied to either produces no net flux so that a push-pull transformer or choke is effectively a short circuit for in-phase components of the signal. Differential response may be obtained with vacuum-tube circuits.[22] Three variants of a basic differential tube circuit have been used in these Laboratories in this application. By using some negative feedback these circuits are easily made so stable that after the initial adjustment their characteristics remain unchanged by tube aging or replacement. Either single-sided or push-pull output may be used and moderate amplification can be realized.

Because the static operating point of the variable-transmission circuit changes with the amount of compression, reactances associated with this circuit may be a source of trouble. For example, the plate and screen currents of variable-mu tubes change with the amount of compression. These circuits should be returned directly to a plate supply of good regulation, not decoupled by a resistance-capacitance filter. If a resistance-capacitance decoupling circuit is used, it may be responsible for either an overshooting or a slow creep of the compressed signal, depending upon the time constant of the resistance-capacitance circuit.

Although variable-mu tubes present a serious problem in the matching and maintenance of their characteristics, they are widely used for the variable-transmission circuit of commercially available compressors. The tube types most used are the 6L7 with the audio input to No. 1 grid and the control signal to grids No. 1 and No. 3, and the 6K7 with both signals applied to No. 1 grid. The stability of the operating characteristic may be improved by a very large resistor common to the cathodes of the push-pull variable-mu tubes. If another tube is used as a cathode resistor, a high impedance is obtained without an excessive voltage drop, and by applying the control signal to the grid of this tube, the sensitivity is greatly increased. The high common cathode impedance also makes the variable-mu stage self-inverting so that a push-pull input stage is not essential. With this type of compressor, a balance control is needed to adjust the circuit for minimum thump from time to time. This control may be a potentiometer of roughly 200 ohms connected between the cathodes of the variable-mu tubes, with the slider wired to the plate of the control tube. A triode-connected variable-mu pentode seems to be the best choice for the control tube, the internal impedance of the tube and the plate-operating voltage of the compressor tubes being determined largely by the choice of cathode resistor for the control tube. The combination of a push-pull stage with a large common cathode resistor results in nearly complete cancellation of even-order harmonic distortion but, at high audio levels, a form of thump may be present which can be traced, not to compression action, but to the changes in average plate current which accompany even-order distortion, these changes being not necessarily best balanced when the balance for compression thump is optimum.

Both Thyrite and copper-oxide varistors have been used in these Laboratories in experimental compressors of the variable-impedance type. Thyrite is a high-resistivity material in which the current is proportional to a power of the voltage, the power being about 4 for disks made to operate in the range of 1 to 20 volts across the disk. The nature of the material limits the use of Thyrite to relatively high-level operation, the smallest disks made by the General Electric Company handling an audio level of 1 volt with less than 1 per cent distortion. The characteristics of Thyrite are satisfactorily stable, the main problem in its use in a compressor being that of obtaining precisely matched disks. This problem can be overcome by selecting disks from a large group or, perhaps more economically, by cutting the disks from a rod form and grinding the pieces to the exact size.

Copper-oxide rectifier disks of the type used for modulators are very satisfactory for use in compressors and are obtainable in matched groups, matching of the disks by a process of selection being accepted practice in making balanced modulators. A single copper-oxide disk of this type will handle between 10 and 15 millivolts of audio signal with not more than 1 per cent distortion. By using a number of disks in series, higher levels may be handled.

Either type of varistor may be used as one arm of a voltage divider in the grid circuit of a vacuum tube. The varistor should, however, be used as the shunt arm since this portion of the divider has across it the lower audio voltage. From a circuit point of view, it is more convenient to use four varistors in a bridge network, each half of the bridge being the shunt portion of a voltage divider. Since the varistor network is carrying direct current, it will ordinarily be isolated from the preceding amplifier and the following differential amplifier by coupling capacitors. If the varistors are not well matched, the charging current of these capacitors will exaggerate the thump, so a thump level which is not audible may cause circuit difficulties. Similarly, the circuit preceding the varistor network should be of high impedance so that the two sides of the push-pull input circuit need not be well matched. If the input circuit is a phase inverter, it should not be the phase-splitter type which has low internal impedance for one phase and high impedance for the other. As a matter of more theoretical than practical interest, it might be mentioned that if the source impedance is sufficiently high, the varistor network can be fed from a single-sided source without increasing thump, this type of feed being impractical because there is no longer cancellation of even-order distortion.

When the amount of compression is large, the audio level across the varistor network will be high at the instant of application of the signal. For example, if there is a 20-decibel compression, the audio signal appears across the varistor at ten times its ultimate amplitude. Under these conditions, there are likely to be thump pulses appearing in the output which are not consistently repeatable either in amplitude or in polarity and the action of the compressor in response to a suddenly applied audio input will, when examined with an oscilloscope, at times appear perfect and at other times will show evidence of severe overcompression. This effect may be reduced to negligible proportions by using biased diodes across the varistor network to limit the instantaneous amplitude to a value slightly above the steady-state level. Another method, which unfortunately would add considerably to the circuit complexity, is to combine some of the advantages of the forward-acting compressor with those of the backward-acting type. It is practical in a forward-acting compressor to delay the audio signal sufficiently to prevent the instantaneous audio level from ever becoming excessively high by inserting a delay network in the audio path beyond the point of connection of the control-signal path.[17] It should then be possible to feed the output of the control-signal rectifier into the same timing circuit as is used for the backward-acting portion of the compressor and, by proper adjustment of gain and threshold, to obtain a condition in which the instantaneous audio amplitude is limited by the forward-acting circuit while the equilibrium input-output characteristic is determined by the backward-acting circuit.

Because varistors change their impedance in response to the current through them, a direct-current amplifier stage is necessary between the timing capacitor, which cannot supply appreciable current, and the varistor network. Of a number of circuits which might be used in this position, the one, known as the Voltohmyst circuit, seems to have the most advantages. The transconductance of the circuit can be controlled by the relative values of the individual and the common cathode resistors, the use of fairly large individual cathode resistors makes the circuit adequately independent of tube changes, and there is partial phase inversion by virtue of the common cathode resistor. This latter point is of importance because the in-phase signal at the audio terminals of the varistor bridge decreases as the phase inversion of the control becomes more perfect.   The circuit also has the advantages that it is adaptable to control signals of either polarity and the normally unused grid is available for use with auxiliary circuits.   It is desirable for two reasons to operate the control stage in a slightly unbalanced condition: a very slight shift from a perfectly balanced condition may result in a reversal in the direction of current flow through the varistors at the start of compression, which will have a serious effect on the compression action.   Also, the varistors should be biased to give an insertion loss of about 6 decibels, or their impedance will have to change greatly before they can cause any appreciable compression.

MEASUREMENT AND SPECIFICATION OF COMPRESSOR PERFORMANCE

Thorough testing of audio amplifiers requires a formidable array of equipment, such as an audio oscillator, a calibrated attenuator, an audio-frequency voltmeter, a cathode-ray oscilloscope, distortion-measuring equipment, and a square-wave generator.   All these items are useful in measuring the performance of compressor amplifiers, and additional specialized equipment is desirable for measurements of timing and thump.

The steady-state characteristics of a compressor amplifier may be measured with the equipment and techniques ordinarily used in testing audio amplifiers and need not be discussed here except in so far as the tests are modified by the peculiarities of compressors.   The input-output relation should be measured and plotted on decibel scales. While such a graph is needed for a complete specification of the input-output relation, most backward-acting compressors may be specified satisfactorily in this respect by numerical values for the threshold level and the range and slope of the compression region.   The range and slope are adequately specified by expressing the useful working range above the threshold in decibel-input range and decibel-output range.   For example, 20:10 decibels means that the output level increases 10 decibels above the threshold level for a 20-decibel increase in input level.   The threshold level may be defined as the output level which is $1/2$ decibel below the uncompressed output level. This definition of threshold level is recommended because most compressor amplifiers have provision for turning off the compression. Hence, under actual operating conditions, this definition allows the threshold level to be checked very easily.

Frequency response, noise level, and distortion should be measured in both the compressed and the uncompressed condition and should be specified for the maximum operating output level in the compressed state. Noise level and distortion should also be given for the same output level in the uncompressed condition, and frequency response should preferably be given for both low- and high-output levels. Distortion should be measured at a number of frequencies, particularly in the compressed state and at low frequencies. A compressor amplifier should also be checked for instability by making a frequency run at high-input level covering the lower audio range down to about 5 cycles per second. Instability is most likely to occur in the frequency range of 10 to 20 cycles per second and then only at input levels high enough to cause compression.

The tests so far described differ from those ordinarily used only in extent and thoroughness and, while they adequately specify the steady-state performance of a compressor, they give no information as to what happens during and immediately following a change in gain. Meeting steady-state performance requirements is not difficult. It is the manner in which a compressor makes the transition from the uncompressed to the compressed condition, or vice versa, that will usually determine the merit of a compressor amplifier. The transition performance of a compressor can be specified fairly well from measurements of the action times and the thump level. Action times were defined earlier, and it was pointed out that there is no generally accepted definition. Consequently, a figure for operating time or release time is meaningless unless accompanied by a definition of these times. For the reasons given earlier, the writer prefers to define action time as the time interval between the start of compression or decompression (in response to an instantaneous change of level) and 90 per cent completion of the change in gain. Both times should be specified for a stated amount of compression because the change in gain is not necessarily linear with the voltage on the timing capacitor, so that the rate of change of gain will depend not only on the rate of change of the timing-capacitor voltage but also on the voltage level, and in addition, the charging rate of the capacitor is not exponential because the rectifier impedance is a function of the charging current.

The measurement of action times is difficult and requires equipment which is not commonly available. The times have been defined in terms of the amplifier gain and gain can be measured only when there is an input signal. The measurement of action times therefore resolves itself into a measurement of the growth and decay times of the envelope of an audio-frequency signal. The release time may be measured by switching a high audio-frequency input signal from a level which gives a chosen amount of compression to a lower level which gives an output below the threshold level. An oscillogram of the output signal can then be measured to determine the release time. A variable-width sound recorder can be used as a recording oscillograph but is not satisfactory for the measurement of release time because the narrow track width and high rate of film travel gives a trace which has a small amplitude and a long base line. A cathode-ray oscilloscope having a long persistence screen, a single sweep, and calibrated time markers could probably be used with fair accuracy without the necessity of making a photographic record.

The measurement of operating time is more difficult than the measurement of release time because the action is so short, preferably less than 1 millisecond, and hence the operating time becomes comparable to the period of the highest audio frequency which can be transmitted by the amplifier. Also, the presence of thump tends to obscure the amplitude changes due to the operating time. Fortunately, the actual value of the operating time becomes less and less important as the time becomes shorter and shorter; the object of a very short operating time is largely to prevent overload of the light-modulator. Consequently, visual examination of recordings made on a variable-width sound recorder of the highest audio frequency that can be recorded will give all the information regarding operating time that is really essential.

If a pulse generator and a single-sweep oscilloscope (preferably with a long persistence screen) are available, the operating time may be estimated with fair precision without the delay involved in making a recording. When a single pulse of known duration, say, 1 millisecond, is applied to the input of a compressor, the pulse amplitude being set to equal the peak amplitude of the input level for which the operating time is to be determined, the pulse appearing at the output of the compressor will have a sharp leading edge of high amplitude which decays exponentially into a flat top, the flat top persisting for the remainder of the duration of the pulse. The fraction of the pulse width at which the exponential portion of the output pulse has dropped to 10 per cent of its initial amplitude (the flat portion being taken as zero) is the operating time in milliseconds.

An oscilloscope is invaluable for examining the action of a compressor for defects. If a signal of 5000 cycles per second is suddenly applied to the compressor-input terminals, the output signal as viewed on the screen of a scope with the sweep set to a low frequency, say, 20 cycles per second, should show a pattern of constant amplitude except for a pulse of duration of 1 cycle or less of the input signal, this pulse appearing at the leading edge of the pattern. At such low sweep rates, the individual cycles of the pattern will not be resolved but the envelope of the output signal will be clearly visible. If the pattern shows a rounded envelope joining the first pulse to the flat portion of the pattern, indicating that a number of cycles have amplitudes exceeding the final amplitude, the charging time is too long, probably because the rectifier has too high an internal impedance or is being operated at too low a voltage, so that the timing capacitor is partially charged very quickly but the remainder of the charge is accumulated slowly through the increasing impedance of the rectifier. When the input switch is closed, the output pattern should appear instantly to the eye and remain steady without bounce or creep, these effects being most probably due to storage of energy by a reactance in some part of the circuit where the current drain is a function of the amount of compression. Bounce is likely to be caused by a storage circuit of short time constant and creep by a circuit of long time constant.

Since thump is the visible evidence of the chief problem in compressor design, methods by which it may be measured and specified are of great importance. Routine checking of compressor performance is sometimes done by measuring the degree of in-phase balance. The in-phase balance check is made simply and quickly without the use of special equipment but has two disadvantages: First, the test, as used, is specific for each model of compressor amplifier and is not easily generalized to be equally useful on any and all compressors. Second, optimum in-phase balance and minimum thump do not necessarily represent identical operating conditions for all compressor designs. A method of directly measuring thump, which has been used by one of the motion picture studios, is illustrated[12] in Fig. 4. A 7000-cycle carrier is switched on and off by the 2-cycle oscillator and an electronic switch, switching disturbances are filtered out by a high-pass filter, and the remaining signal is applied to the compressor to be tested. The 7000-cycle carrier is removed from the compressor output by a low-pass filter, leaving the low-frequency thump components which are measured with a volume indicator. The thump components should measure at least 55 decibels below the unfiltered output. This method is applicable to the measurement of any type or design of compressor and it measures thump relative to the signal amplitude. For experimental work on compressors, it is best to make some modifications in this method of testing. The switching rate should be variable and the volume indicator should be replaced by a cathode-ray oscilloscope. The switching rate should be adjustable because thump sometimes is a maximum at a particular rate,

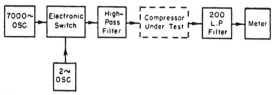

Fig. 4—Block diagram of circuit for testing volume-compressor action.

and the lowest rate should allow the period during which the signal is off to be about twice the release time of the compressor in order to allow a return to normal gain during the off period. The highest rate need not be greater than ten interruptions a second. A scope is recommended in place of a meter because the indication of a meter for pulses of short duration depends upon such factors as the duration and repetition rate of the pulses and the dynamic response of the meter, and even though the dynamic characteristics of the meter were standardized (as they are for the volume indicator), the meter reading and the permissible thump level would have to be correlated for every interruption rate to be used in testing. The use of an oscilloscope permits the actual peak amplitude of thump components to be measured relative to the peak amplitude of the signal and also allows qualitative examination of the wave form of the thump.

There remains the question as to what level of thump is permissible and to this question no entirely satisfactory answer is known to the writer. Listening tests made on direct-speech-pickup and music reproduced from vertical-cut transcriptions have indicated that a thump level (measured with a scope as described in the preceding paragraph) of 5 per cent is not perceptible. These tests were made with a compression slope of one half and the amount of compression reached about 15 decibels on signal peaks. The results should not be taken as conclusive; operation on a slope of $1/10$ (limiter) may require a lower thump level and thump may be more serious in sound recording than it is in direct monitoring. A thump level of 5 per cent is rather high, compressor operation is fairly satisfactory, but limiter operation with this level of thump is not. The thump adds to the signal amplitude so that the output level, instead of being controlled by the signal amplitude, is controlled by the amplitude of signal plus thump. When the loop gain is high, overcompression may result and recovery from this condition is delayed by the release time. In working with compressors using the varistor bridge, it has been found that when thump was reduced from 5 per cent to 1 or 2 per cent the balance and stability of the circuits associated with the varistor circuit became much less critical factors. Perhaps the most satisfying answer to the question of permissible thump level is that if the operation of a compressor amplifier, when examined by the interrupted-signal method, appears clean and consistent under all operating conditions, then the thump level is unimportant.

CONCLUSION

The need for compressors in sound-on-film recording and the problems inherent in their use have been discussed, followed by an analysis of compressor characteristics, the types of compressors available, and the methods of evaluating their performance. Although improvements and refinements can be expected, the volume compressor has reached a state of technical development which makes it a necessary part of sound-recording equipment. The technique of measuring and specifying performance has not kept pace with this technical development. There is a need for standardization of nomenclature and test procedure that will permit exact specification of volume-compressor performance.

REFERENCES

(1) S. B. Wright, "Amplitude range control," *Bell Sys. Tech. J.*, vol. 17, pp. 520–538; October, 1938.

(2) A. C. Norwine, "Devices for controlling amplitude characteristics of telephonic signals," *Bell Sys. Tech. J.*, vol. 17, pp. 539–554; October, 1938.

(3) S. B. Wright, S. Doba, and A. C. Dickieson, "A Vogad for radiotelephone circuits," *Proc. I.R.E.*, vol. 27, pp. 254–258; April, 1939.

(4) W. A. Mueller, "Audience noise as a limitation to the permissible volume range of dialogue in sound motion pictures," *J. Soc. Mot. Pict. Eng.*, vol. 35, pp. 48–59; July, 1940.

(5) M. Rettinger and K. Singer, "Factors governing the frequency response of a variable-area recording channel," *J. Soc. Mot. Pict. Eng.*, vol. 47, pp. 299–327; October, 1946.

(6) B. F. Miller, "Elimination of spectral-energy distortion in electronic compressors," *J. Soc. Mot. Pict. Eng.*, vol. 39, pp. 317–324; November, 1942.

(7) J. A. Becker, C. B. Green, and G. L. Pearson, "Properties and uses of thermistors—thermally sensitive resistors," *Bell Sys. Tech. J.*, vol. 26, pp. 170–212; January, 1947.

(8) U. S. Patent No. 2,379,484, Robert L. Haynes, assigned to RCA (1946).

(9) W. H. Stevens, "Variable slope with constant current," *Wireless Eng.* (London), vol. 21, pp. 10–12; January, 1944; *Electronic Ind.*, vol. 3, p. 176; March, 1944.

(10) A. N. Butz, Jr., "Surgeless volume expander," *Electronics*, vol. 19, pp. 140–142; September, 1946.

(11) W. L. Black and N. C. Norman, "Program-operated level-governing amplifier," *Proc. I.R.E.*, vol. 29, pp. 573–578; November, 1941.

(12) J. K. Hilliard, "The variable-density film-recording system used at MGM studios," *J. Soc. Mot. Pict. Eng.*, vol. 40, pp. 143–176; March, 1943.

(13) J. P. Taylor, "Limiting amplifiers," *Communications*, vol. 17, pp. 7–10, 39–40; December, 1937.

(14) R. C. Mathes and S. B. Wright "The Compandor—an aid against static in radio telephony," *Bell Sys. Tech. J.*, vol. 13, pp. 315–332; July, 1934.

(15) G. W. Cowley, "Volume limiter circuits," *Bell Labs. Rec.*, vol. 15, pp. 311–315; June, 1937.

(16) G. Q. Herrick, "Volume compressor for radio stations," *Electronics*, vol. 16, pp. 135 and 323; December, 1943.

(17) D. E. Maxwell, "CBS automatic gain-adjusting amplifier," *Tele-Tech*, vol. 6, pp. 34–36, 128; February, 1947.

(18) L. B. Hallman, Jr., "Practical volume compression," *Electronics*, vol. 9, pp. 15–17, 42; June, 1936.

(19) H. H. Stewart and H. S. Pollock, "Compression with feedback," *Electronics*, vol. 13, pp. 19–21; February, 1940.

(20) S. Doba, Jr., "Higher volumes without overloading," *Bell Labs. Rec.*, vol. 16, pp. 174–178; January, 1938.

(21) O. M. Hovgaard, "A volume-limiting amplifier," *Bell Labs. Rec.*, vol. 16, pp. 179–184; January, 1938.

(22) J. F. Toennies, "Differential amplifier," *Rev. Sci. Instr.*, vol. 9, pp. 95–97; March, 1938.

Some Distinctive Properties of Magnetic-Recording Media*

By R. HERR, B. F. MURPHEY, AND W. W. WETZEL

MINNESOTA MINING AND MANUFACTURING COMPANY, ST. PAUL, MINNESOTA

Summary—Information is presented relative to the adjustment of bias current in magnetic recordings and the various effects of bias changes on distortion, frequency response, overload characteristics, and permanency are discussed. Other factors which influence frequency response are outlined briefly and it is shown that the inherent frequency response of the medium is difficult to divorce from effects due to the recording system. The problem of noise is presented in general terms and the nature and level of the noise from a direct-current saturated medium is advanced as an important criterion of quality.

INTRODUCTION

IN THE COURSE OF the authors' research in connection with the development of magnetic-recording tapes, it has become necessary to develop techniques for rapid evaluation of tapes of very widely different properties. These techniques are exactly those required by a user who wishes to get the best possible performance from a magnetic tape, and thus are of some general interest. The remarks are, for the most part, quite generally true of such other magnetic media as wires, sheets, disks, and cylinders. Many of the important characteristics of a magnetic material may be best obtained by measurements of its magnetic properties, but only data of the sort obtained by conventional erase, record, and reproduce heads are discussed here. The data were taken on $1/4$-inch magnetic tape on a loop tester employing a modified Ranger erasing head, a modified Brush recording head, and an unmodified but selected Brush reproducing head. Various speeds have been used from 5 to 36 inches per second; most data were taken at 9.2 inches per second.

HIGH-FREQUENCY BIAS

It has been recognized for several years that to obtain good reproduction from magnetic media it is desirable to add to the audio current in the recording head a certain amount of high-frequency bias

* Presented May 18, 1948, at the SMPE Convention in Santa Monica.

current.[1] From the point of view of distortion, the frequency of the
bias current is not critical so long as it is high enough not to beat with
any appreciable harmonics of audio current; also, it will be shown
later that from the point of view of noise it is desirable to have the
bias frequency high.   However, the current value must be selected
with care.   It would seem simple enough to vary the bias current
until best results occurred, but evidently it is easy to err since many
conflicting systems have been used for setting bias. Yet the bias
value is most important in obtaining maximum output with minimum

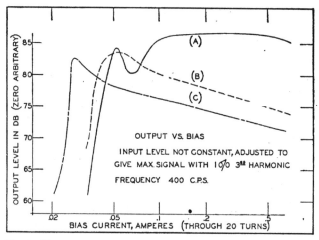

Fig. 1—Effect of bias current on output obtainable with 1 per cent
third-harmonic distortion, at 400 cycles per second.

distortion, and we believe that it can be selected systematically, at
least for media of desirable characteristics.

Most frequently one sees the effect of bias current depicted by
curves showing how the output (playback) level of a tape signal and
the distortion of this signal vary with bias current for a constant audio
recording current.   These curves are easy to obtain experimentally,
but difficult to interpret.   However, one is not usually concerned
with the input current required.   The curves of Fig. 1 were plotted
with this in mind.   Here the input is not held constant; the curves
show the maximum level obtainable at 400 cycles per second with 1
per cent third-harmonic distortion, as a function of bias current. We
measure third-harmonic distortion in preference to total distortion

since the latter may be affected by noise level, whereas the former may be related theoretically with the magnetic properties of the medium. Even harmonic distortion is negligible in magnetic recording if the recording is done on a magnetically neutral tape by alternating-current bias of good wave form. Even harmonic distortion will result from magnetized heads or the use of a direct-current (permanent-magnet) erase. The curves are for three "types" of oxide

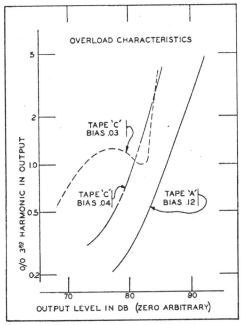

Fig. 2—Variation in distortion with output level, near the overload point, 400 cycles per second.

tapes into which we find the hundreds of oxides tested may be divided roughly, although no two will be exactly alike. Unless other considerations may be shown to enter, we would start by choosing the bias to give a maximum in such a curve as shown in Fig. 1.

Regarding the three shapes of the curves, one may say that it is preferable to have a broad maximum rather than a sharp one simply from a design and control point of view, but it is also desirable quite generally. The sharp maxima are usually associated with bad overload characteristics and also with an output sharply varying with

small bias changes for constant input. This will apply to sharp
maxima as in curve (*C*) or the first sharp maximum in the curve (*A*).
For example, Fig. 2 shows some overload characteristics where third-
harmonic distortion is plotted on a log scale against output in decibels
and it may be seen that the sharp maximum in curve (*C*) of Fig. 1
which corresponds to the irregular overload characteristic of Fig. 2 is
not a desirable bias setting, and the bias should be increased perhaps
20 per cent to a value which gives a smooth overload characteristic.
On the other hand, overload characteristics are good for bias set for
the maximum of curves (*A*) and (*B*) in Fig. 1; the data for tape *A*

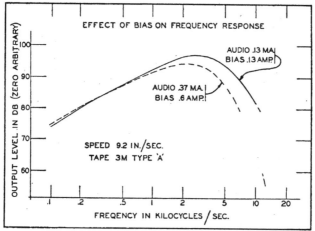

Fig. 3—Frequency response for constant-current recording at 9.2
inches per second, as affected by bias current. Brush heads.

are shown in Fig. 2   In general, if the bias setting seems very critical,
it is an unsatisfactory tape or an incorrect bias setting.

   The discussion above is based on data at one frequency and there
may be some question about whether the best bias for one frequency
is the best for another.   There has been a tendency to standardize
tests at 400 cycles per second which we believe should be continued
until there is evidence for some better test frequency or frequencies.
One definite result noted with all media is that a higher bias current
leads to poorer high-frequency response.   It is believed that this
effect is caused by the partial erasing action of stray bias field and thus
is somewhat a function of the design of the recording head, but whatever the cause it leads one to use the lowest value of bias possible if

the maximum in Fig. 1 is so broad as to allow a variation in bias without sacrifice of signal level. For example, Fig. 3 shows two unequalized frequency-response curves (constant-current recording) made on the tape of Fig. 1, curve (A), one with a high (0.6-ampere) and the other with a low (0.13-ampere) bias. These values of current are those used in a separate 20-turn bias winding on the Brush recording head. It can be seen that this increase of bias leads to a very large loss of highs. Ordinarily for high-quality work, the bias will be set without regard to frequency response and the speed or equalization adjusted to give the required high-frequency. response.

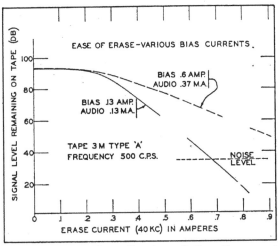

Fig. 4—Stability of recordings in small alternating-current fields and erase currents required to obliterate, as affected by bias current used in recording.

There are other slight frequency dependent effects of bias which are not well understood, but which are relatively minor for a good head and tape design. These are connected with the nonuniformity of the bias field through the thickness of the medium at the recording head and the relative effectiveness in playback of surface and subsurface layers of the magnetic medium as a function of recorded wavelength and probably with other factors which we do not understand.

Another effect of bias current which is of interest and possible importance is its effect on ease of erase. Fig. 1 shows that it is possible to record with two different bias currents and correspondingly different audio currents and get the same level recording with the same distortion. Two such recordings will not differ in playback, but are very different in permanency as Fig. 4 shows. In this figure is plotted the remaining level of signal as successively higher erase currents are used in the erase head. It is seen that the signal recorded with high bias is harder to erase and much less affected by weak alternating-current fields, which may be an advantage or disadvantage depending upon the use to be made of the recording.

FREQUENCY RESPONSE

The subject of frequency response of a medium is one which has been treated rather thoroughly in previous papers in most respects. It is impossible to specify the frequency response of the medium as such, since, at least at the present state of the art, it is always affected to some degree by the heads used in the recorder. The effect of gap width in the playback head is discussed by Holmes and Clark,[1] and of gap design in the record head by Clark and Merrill.[2] The bias used affects the frequency response probably in accordance with the record-head design. The problem of azimuthal alignment of record and playback heads always enters in some slight degree for the shortest wavelengths that may be recorded. Demagnetization in a tape is probably an important factor, but not so important as it was once thought to be[3] or as it apparently is in wire recording. The part which coercive force plays in the effect of demagnetization may be considerable, and thus gives a large variation in frequency response so far as the medium is concerned, but we have found that this effect may be negligible in practice. The reason is that with the higher coercive force a larger bias field is normally required, and the effect of the larger bias field in reducing high-frequency response may be of the same order as the increase in high-frequency response resulting from the higher coercive force. Thus with our particular recording head and operating with bias adjusted as discussed above, we have found the frequency response not to be simply related to coercive force. Conceivably, with a more ideal recording head, a direct relationship could be observed, and with a very poor head an inverse relationship is possible. In any event, we have come to regard frequency response as a less important characteristic of a tape than we once did. It appears that in an economical "home" recorder, where it is desired to get optimum performance at a low tape speed (7.5 inches per second), the quality is more limited by distortion, hum, and inexpensive components than by frequency response, while in a high-quality machine, operating at 18 or 30 inches per second tape speed, the frequency response is usually adequate with any of a wide variety of tapes.

Two factors affecting the frequency response of a tape other than its magnetic properties, are its thickness and its smoothness. Regarding thickness, it appears that the signal from a given tape at very short wavelengths is obtained almost solely from a very thin surface layer, so that changes in thickness above a very thin minimum do not affect the signal at high frequencies. On the other hand, a signal

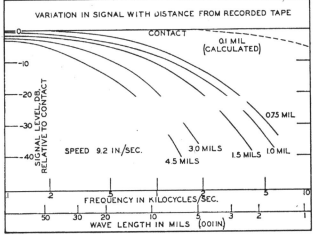

Fig. 5—Attenuation in signal strength caused by various separations between recorded tape and playback head, as a function of frequency (at 9.2 inches per second tape speed) or wavelength (at any speed).

of long wavelength utilizes the full thickness of the medium up to any practicable dimension, and thus is affected by changes in thickness. For a given recording current and bias, the signal obtained will be somewhat (though not directly) proportional to thickness at low frequencies, and roughly independent of thickness at high frequencies. One qualifying restriction, however, to an unlimited increase in thickness for better bass response comes, once again, from head design. For the usual ring-type recording head, the fields, both audio and bias, drop off rapidly with distance from the head, and there is a limit to the tape thickness which may be used without suffering distortion from the gradient existing in the bias field. To this extent, the optimum tape thickness is highly dependent upon the nature of the field of the recording head.

The surface of the tape (and the heads) must be as smooth as possible for good high-frequency response. From simple dipole theory, one may expect the field from a very short wavelength record to fall off very rapidly with distance from the tape, and this is indeed the case. Fig. 5 shows the results of an experiment in which various frequencies were recorded on a tape which was then played back with the playback head in contact and also separated from the tape by various thin paper shims. The playback signal level relative to that obtained in contact is plotted against frequency so that the attenuation introduced by various separations is shown as a function of frequency (or wavelength). The dotted curve is an extrapolation to a very small separation of 0.0001 inch (2.5 microns) and it can be seen that even this minute separation causes a noticeable attenuation at very short wavelengths. Thus, while it may be practicable so far as equalization is concerned to record wavelengths of the order of 0.001 inch, more reliable and reproducible results will be obtained if higher speeds are utilized to keep the minimum recorded wavelength longer. The use of a longer minimum wavelength also simplifies other problems such as head azimuth adjustment.

Noise

The subject of noise is the one in which there is probably least agreement among various investigators. A treatment of erased noise and noise as a function of direct-current magnetization is given by Holmes,[4] based on measurements on wire. The noise on a tape may theoretically be reduced to that produced by randomly oriented domains by an erasing head which truly demagnetizes the tape, and probably the dominant noise such a tape should produce is the slight microphonic noise in the playback head due to tape friction. Our experiments show that this condition may be approached, based on tapes demagnetized by relatively complete methods, although we have never seen a practical ring-type erase head achieve this noise level. The same remarks apply to such noise as may be generated by the bias field in the recording head, with one exception discussed below. We have found that any measurement of the erased noise level of a tape was in reality more nearly an evaluation of the erase and record heads and their electrical supplies.

An erase head or record head which has a small permanent mag-netization will cause the tape to be noisy, and this· same effect can be caused by a small direct-current component or by an unsymmetrical wave form in erase or bias currents. If there are direct-current com-ponents in the erase field or if, in the extreme case, the tape is erased by a permanent magnet, the noise is increased; any irregularities in the tape cause irregular magnetization and the head, which is sensi-tive to the time rate of change of flux through it, picks up these ir-regularities. For this reason we have used a direct-current erase extensively in evaluating tapes since any shortcomings of the tape,

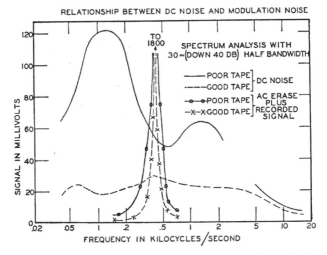

Fig. 6—Effect of tape irregularities of long wavelength in producing sidebands near a recorded signal.

whether they be imperfect dispersion of the oxide in the binder or roughness of the front or back surface of the magnetic layer, are detected by this method. Furthermore, analysis of the noise into a frequency spectrum frequently indicates the source of noise and suggests the cure for it. The noise in direct-current wipe thus is a valuable tool in research but also is a measure, in some degree, of the noise which will result in practice from inevitable slight imper-fections in the recording system. It also has value as an easily re-produced condition which may be made the basis of a reference level in comparisons between various laboratories using different recording equipment.

The direct-current noise is also of significance in indicating the "modulation" noise to be expected. This is the noise which occurs in the presence of a signal and which is absent in an erased tape. Chapin[5] first showed that the modulation noise near the frequency of a recorded note was composed of sidebands formed by the major components of direct-current noise. Fig. 6 shows an example of two tapes identical except for their direct-current noise. These tapes have·equal signal levels at the same harmonic distortion on our system. Fig. 6 shows the spectra of their direct-current noises, taken with a Hewlett Packard wave analyzer with 30-cycle half bandwidth, following an unequalized amplifier. The tape with a poor oxide dispersion has excessive low-frequency direct-current noise. It also shows the spectra resulting from recording a 400-cycle signal. It may be seen that the tape with low direct-current noise produces a purer tone and less background noise. The bandwidth plus certain wow in our equipment make the data somewhat inaccurate, particularly for the better tape, but the difference between the tapes is clear.

This circumstance of noise resulting in the form of sidebands from the recording of a signal is the source of a possible noise caused by bias current, noted as an exception in the first paragraph on noise. While an ideal erasing head will subject the tape to gradually decreasing fields and thus demagnetize it, a recording head must be built to have the transition from gap field to no field as abrupt as possible. Thus the bias frequency is normally recorded to some degree upon the tape, and irregularities in the tape will modulate the bias signal. Irregularities of wavelengths near that of the recorded bias signal may produce sidebands in the audio range. This effect in most recorders is less serious than effects of direct-current components or poor wave form, but is a form of limiting noise which can scarcely be avoided. It may, however, be reduced to insignificance by using such a high bias frequency that an insignificant level is recorded. The frequency required will vary with tape, head, and speed, being proportional to speed.

There are other forms of noise which are difficult to measure qualitatively but annoying to listen to, such as clicks and pops and such intermittent pulses which have been present in many different kinds of magnetic tape and which, likewise, are due to faults in construction of tape. They are not a necessary part of magnetic recording and may be eliminated by proper coating of a good dispersion upon an adequate backing.

A final, rather special, form of noise is the signal irregularity caused by nonuniformity of the coating. This may be tested by recording a uniform signal and playing back through a suitable graphic recorder. In the past year, the uniformity of tapes, both within a single roll and from various production runs, has been much improved. Uniformity within ± 0.3 decibel in a roll and ± 0.6 decibel among many different rolls may now be expected.

Acknowledgment

The authors wish to thank the management of the Minnesota Mining and Manufacturing Company for permission to publish the data contained in this paper.

References

(1) L. C. Holmes and D. L. Clark, "Supersonic bias for magnetic recording," *Electronics*, vol. 18, p. 126; July, 1945.

(2) D. L. Clark and L. L. Merrill, "Field measurements on magnetic recording heads," *Proc. I.R.E.*, vol. 35, pp. 1575–1580; December, 1947.

(3) W. W. Wetzel, "Review of the present status of magnetic-recording theory," *Audio Eng.*, vol. 32, p. 28; January, 1948.

(4) L. C. Holmes, "Some factors influencing the choice of magnetic medium," *J. Acous. Soc. Amer.*, vol. 19, pp. 395–403; May, 1947.

(5) D. M. Chapin, "Measurement and calculation of under-signal noise in magnetic recording," Program 33rd Meeting Acoustical Society of America, May, 1947.

Discussion

Mr. Lewin: We made some tests at the Signal Corps using a Brush and regular tape, I think the type called No. 110, and we have experienced considerable trouble with what we call an echo. You referred to it as a print-through effect, an echo effect, 'from one layer to another, and so far, we have not been able to find what causes it. It is not always consistent. Sometimes we do not get it. Other times we get it at a fairly low recording level.

Mr. R. Herr: There will always be a certain transfer effect from one layer to another, but the fact that it does not occur all the time indicates that it is not necessarily associated with magnetic recording. It does vary with level, of course. The fact that it occurs only sometimes indicates that the roll on which it occurs has been handled in some way differently from the roll in which it does not occur. At least two factors operate to increase that effect. One of those is the presence of an alternating magnetic field. If you have an alternating field in the vicinity of a recorded tape, you will increase the transfer, the echo effect, enormously. Such fields are not likely to be encountered—I am sure you have not deliberately placed the recording in a strong field, but even a weak field can have an appreciable effect. Another factor is heat. A roll stored at high temperature will show the effect to a larger degree than a roll stored at low temperature. That is a sort of idealization process of an unrecorded layer in a field of recorded layers, and it is like a piece of iron in a weak field being hit with a hammer—it will acquire more magnetization than without the mechanical strain.

MR. LEWIN: We did try to expose a roll, before recording, to a strong field, and we used the voltage regulators which ordinarily gives quite a strong field, and did not notice any effect at all.

MR. HERR: Before recording, you should not expect any effect from that, but once it is recorded, you have a modulated layer adjacent to another layer in the field of the recorded layer, a very weak field, then you apply an alternating field to this entire region, and the previously unrecorded layer is strongly magnetized, even by the weak field of the recorded layer. What you have done before you record it has no bearing on this.

QUESTION: Sometimes you get it—not from one roll to another, but in one part of a roll, and then you will not hear it again for the rest of the roll.

MR. HERR: That I cannot explain.

QUESTION: We have gotten it fairly consistently, with two different recorders and two different recording heads.

MR. HERR: You say it is not related to the level? If it were related to the level, then there would be a fairly obvious explanation because the characteristic for such a field, in the absence of bias, is highly nonlinear, but except for that, I cannot explain it; the tape should be the same unless it was wound differently or treated differently in one place or another.

QUESTION: An interesting effect we noticed was that if we did not rewind the reel after recording it, the print-through appears as an echo, as it followed the modulation. If we did rewind it, then it appeared as an anticipation.

MR. HERR: I have noticed that effect myself.

Wide-Track Optics for Variable-Area Recorders*

By L. T. SACHTLEBEN

RCA Victor Division, Camden, New Jersey

Summary—Wide variable-area sound tracks that meet Academy Research Council dimensional specification[1] RC-5001 have the advantage that they yield an improvement in ratio of signal to noise. The optical system for the new RCA Type PR-31 de luxe recording machine is equipped to make sound tracks to this standard. RCA recorders of earlier types can also be converted to record tracks to this standard.

The optics that provide this feature are confined to the slit and objective lens assembly and involve no extensive changes in the optical system. They consist primarily of an objective lens of new design that provides additional magnification in the direction of the slit length to meet the RC-5001 standard. When systems of earlier types are converted, a new slit and objective-lens barrel are installed, but no other optical changes are required. The converted system will produce all types of area tracks that could be made before conversion, and may be equipped for phototube monitoring.

THE SIGNAL-TO-NOISE ratio of a sound track is improved by increasing the width of the sound track. If the noise originates primarily from a grainy structure in either the dense or clear portions of the sound track, this structure being characterized by a very large number of very small particles that are randomly distributed, it can be shown that the amplitude of reproduced noise will vary as the square root of the effective width of the sound track. At the same time the amplitude of the reproduced signal will vary directly as the effective width of the track. A net gain in signal-to-noise ratio of 3 decibels will therefore result from doubling the effective width of the sound track. The noise that is due to fog or a very dirty condition in the clear portions of a variable-area print will approximately respond in this manner to an increase in effective track width. Also the noise or hiss that accompanies the reproduction of a variable-density print responds in the same way.

Another type of noise is found in sound tracks. It is of a different nature and is objectionable primarily when signal levels are very low or when no signal is present. Under this condition the transparent areas of the printed track are reduced to a minimum, or the density is

* Presented May 20, 1948, at the SMPE Convention in Santa Monica.

January, 1949   Journal of the SMPE   Volume 52   89

raised to a maximum, by the action of the noise-reduction system.
Noise caused by film grain or dirt is therefore very low, but occasional
minute pinholes, or small transparent areas in the print, due to dirt or
other imperfections in the negative, give rise to discrete sounds variously known as "ticks," "pops," or "crackles."　　These sounds are
individually recognizable because they occur relatively infrequently,
as compared to the sounds that occur frequently enough to produce
the continuous effect of hiss.　For this reason they constitute an objectionable residue of noise under conditions that ideally should be
noiseless.

The effect of widening the sound track is to reduce the relative
amplitude of these discrete "pops" even more than hiss.　Doubling
the track width doubles the number of discrete sounds per second but

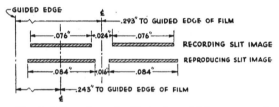

Fig. 1—Dimensions of wide-track variable-area recording slit image and wide-track reproducing slit image according to Academy Research Council Specification[1]
RC-5001.

has no effect on their absolute amplitude as long as they occur separately.　The amplitude of the desired signal is, however, doubled.
The ratio of signal-to-discrete-noise level is therefore increased 6 decibels.　The fact that twice as many "pops" occur, each of a level that
is relatively 6 decibels lower than before, is of no consequence as long
as their probability of simultaneous occurence is still substantially
zero.　Sound tracks that are wide enough to double the amplitude of
the useful signal effectively are therefore especially effective in suppressing residual surface noise of this nature at very low recorded
levels.

Fig. 1 illustrates the present 35-mm industry standard dimensions of
the wide variable-area sound track.　The negative track is recorded
in two 0.076-inch-wide halves, with a separation of 0.024 inch between
them to facilitate isolation of the halves in push-pull reproduction.
The print is scanned to a total width of 0.184 inch, in two 0.084-inch
illuminated segments separated by a 0.016-inch septum.　As compared

with the 35-mm standard narrow track, the recorded signal will be greater in the ratio 0.152/0.076, the reproduced hiss from the transparent part of the track will be greater in the ratio $\sqrt{0.152/0.076}$, and the level of the "pops" will be unchanged. The hiss from the dense parts of the print at densities in the range 1.3 to 1.5 is so low as to be negligible for tracks of both widths.

Special optics have been designed that make any RCA variable-area studio recording optical system, of the PR-23 and later types, convertible for recording the wide sound tracks. The optical system for

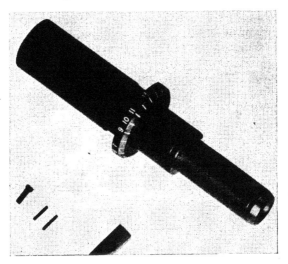

Fig. 2—Wide-track conversion assembly ready to install in narrow-track optical system.

the new PR-31 de luxe recorder is also available equipped with the special optics for wide-track recording. These optics are contained within the barrel or assembly that houses the slit and final objective of the system. Fig. 2 shows the appearance of a conversion assembly ready to install.

In addition to the optical changes, the distance from the slit to the film must be increased about 1/8 inch by relocating the optical system, and the system must also be moved laterally 0.050 inch to permit one half of the wide track to be located in the standard position, with its center 0.243 inch from the edge of the film. This permits one half of the track to be reproduced in any standard soundhead.

Inasmuch as the wide-track technique is intended primarily to improve the quality of original sound tracks which will later be re-recorded to standard release negatives, it is advisable that wide tracks be of the push-pull type. Variable-area push-pull sound tracks in common use are of three types, known as Class A, Class B, and Class AB. Each has its own peculiar characteristics and advantages from the operational standpoint, but all have quality advantages over any of the tracks of the nonpush-pull type.

The special wide-track optics were first designed with the conversion of existing optical systems in view. To this end details were worked out in such way that any of the push-pull apertures that had been designed for the PR-23, or later studio recording optical sytems,

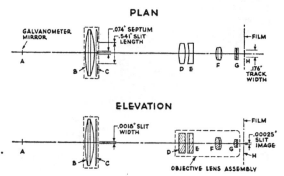

Fig. 3—Schematic of the modulating branch of the optical
system showing wide-track optics.

could be used without change in the production of wide tracks. As a result the wide tracks are fully modulated at a galvanometer amplitude that is about 1 decibel below the corresponding amplitude for narrow-track push-pull recording. In the case of the Type PR-31 recorder optical system, apertures have been especially designed for wide-track push-pull work, and this optical system fully modulates all types of sound tracks at a uniform galvanometer amplitude.

Fig. 3 shows two schematic views of the modulating branch of a variable-area recording optical system, that extends from the galvanometer mirror to the film, and illustrates the nature of the wide-track optics. The optics in all other parts of the optical system are the same for either wide-track or narrow-track work. In the figure, A is the galvanometer mirror, B is the slit condenser, and C, the slit. H is the film. Parts D, E, F, and G constitute the special wide-track optics which are combined to form a well-corrected anamorphote objective.

In the transverse plane (plan view of the figure), lenses D and E constitute an air-spaced cylindrical lens which co-operates with the cemented spherical doublet F to image the slit upon the film. Cylindrical lens G has no power in the transverse plane. In the longitudinal plane (elevation view of the figure), cemented spherical doublet F co-operates with cemented cylindrical doublet G to image the slit upon the film. Cylindrical lenses D and E have no power in the longitudinal plane. The reduced image of the slit at the film is formed at different magnifications in the two planes, so that the ratio of its length to its width is much greater than the corresponding ratio for the slit itself. This ratio is about 2.44 times as great for the image as for the slit, and it is by this inequality of the two ratios that the slit image is made long enough to record a track that is 0.176 inch wide. The anamorphote objective is corrected for all the usual aberrations in both planes. The air-spaced cylinders D and E are designed to compensate the curvature introduced by lens F, and the result is that the wide-track optical system exhibits no more image curvature than a narrow-track system.

Anamorphote-lens combinations require critical adjustment of the azimuth of the cylindrical components in order to perform properly. Special fixtures and procedures are employed to align the lenses by optical test, and the mountings are especially designed to hold the lenses securely in the position so determined, after they are assembled into the optical system.

Since the increase in track width has been obtained without any increase in the area of the slit, this obviously results in exposure of the wide-track negative falling to about 40 per cent of the exposure for a narrow-track negative. Compensation for this exposure loss is made by employing fine-grain stocks similar to Eastman 1372 and exposing them with white light. Adequate negative densities are obtainable, and the quality of the sound-track image is substantially equivalent to that obtained with ultraviolet exposure and emulsions similiar to Eastman 1357.

Fig. 4 illustrates the action of the modulator and noise-reduction shutter, in the Type PR-31 recorder optical system equipped for recording Class A push-pull wide track. A drawing that illustrates the character of the negative track produced is also shown. The slit S and aperture images A and B, which are in the plane of the slit, are

shown as they would be seen to appear from the position of the galvanometer mirror. The dotted line V is the outline of the noise-reduction shutter vane located just beyond the plane of the slit, but not
visible. An opaque portion or septum SP occupies the center of the
slit. The slit is thus divided into two active segments, one being
illuminated by aperture image A and the other by B. The cutting
edges of these images cross their respective segments of the slit at
points midway between their centers and ends, when no modulation is

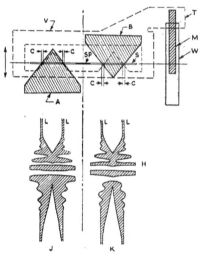

Fig. 4—Action of the Class A push-pull
aperture and slit in the Type PR-31 recorder wide-track optical system.

present, and the central half of each segment of the slit thus is fully
illuminated. Two edges of the opening in the shutter vane also
cross each active segment of the slit and block all of the light entering
the slit from A and B with the exception of four small equal portions
C. In the accompanying drawing of the negative sound track, the
four strips or bias lines L are produced by the portions C of the slit
that are illuminated by A and B, but are not covered by the shutter
vane.

As modulation currents are applied to move images A and B up and
down together, a motor controlled by the modulation currents causes
the vane V to move downward to uncover sufficient additional portions of the active segments of the slit to accommodate the excursions

of A and B. When modulation currents cease, the condition shown in the illustration is resumed. An auxiliary image M moves with images A and B. This image lies in a window W in the plane of the slit, and its top portion is limited by the tab T on the shutter vane. When the image M is reimaged on the monitor screen, it provided a means for observing the displacements of images A and B, and also of the shutter vane V.

The Class A push-pull sound track produced by this arrangement is in two parts, J and K. Each of these is a normal duplex track with noise reduction, and can be played by itself as a complete record. The two tracks J and K are in 180-degree phase relation to each other. The push-pull track has the advantages that the noise-reduction envelopes of the portions J and K are in phase and cancel in push-pull reproduction, and that even-harmonic distortion of photographic origin also cancels. The use of push-pull is thus essential when mak-

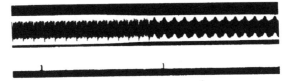

Fig. 5—Typical wide variable-area sound track—Class A push-pull print (recorded on converted optical system).

ing original negatives of the highest possible quality. Portion H illustrates overshooting.

Fig. 5 shows a section of a typical Class A wide-track push-pull print of recorded speech. This was made with a narrow-track optical system that has been converted to record wide sound tracks.

Studio operational experience with optical systems converted to wide track indicates their frequency-response performance is comparable with narrow-track systems. No special compensation was required by the conversions. Over-all distortion measured from playback negatives, including distortion of signal generator and recording and reproducing channels, is as follows:

Level Referred to 100 Per Cent Modulation	Distortion
−2 decibels	2.4 per cent
−4 decibels	2.0 per cent
−8 decibels	1.8 per cent

The reduction in relative level of the "ticks" and "pops," which are the most serious noise at very low levels or in silent parts of variable-area records, has been very satisfactory.

Exposure characteristics follow:

Lamp Current—7.2 amperes in 7.8-ampere lamp
Negative Density—2.7
Print Density—1.3 to 1.5
Negative Stock—Eastman 1372
Processing—Commerical laboratory

ACKNOWLEDGMENT

Acknowledgment is due the help of many people who participated in developing the wide-track optics.   Credit is particularly due Mr. G. L. Dimmick, Mr. J. L. Pettus, Mrs. Mary Smuck, who did the computing, and Mr. E. S. Leslie, who responsibly accepted and solved many of the practical problems of assembling the first units that were built.   The requirement for cylindrical lenses of adequate quality for this design was met by the Herron Optical Company of Los Angeles.

REFERENCE

· (1)   Academy Research Council Specification RC-5001 is now incorporated in American Standards Z22.69-1948 and Z22.70-1948 (Universal Decimal Classification*UDC 778.534.4), J. Soc. Mot. Pict. Eng., vol. 51, pp. 547–548; November, 1948.

Trend Control in
Variable-Area Processing*

By F. P. HERRNFELD

Ansco, Los Angeles, California

Summary—This paper compares two alternate methods of obtaining a signal for cross-modulation testing. It also describes a meter which will give from a single print density the information formerly given only by a series of prints of different densities. This meter will be of help in predicting trends in processing variations due to changes in developing.

INTRODUCTION

As POINTED out by Baker and Robinson,[1] a cross-modulation test affords an extremely accurate means of determining correct negative and print densities for given conditions of laboratory processing for a variable-area sound track. This type of test is now in universal use. The modulated test tone consists usually of a 9000-cycle carrier, amplitude-modulated by 400 cycles The peak amplitude of this modulated wave is adjusted to about 90 per cent of a fully modulated track. A 1000-cycle tone is recorded as part of the test to be used as a reference level, its amplitude being the same as the peak amplitude of the modulated wave. The exposure on the sound negative is held to such a value that, when the negative is printed onto a positive film and developed normally, minimum cross-modulation products will be present. In order to determine the proper printing exposure (the developing time is held to a certain value fixed by the picture requirements) several prints of different densities have to be made. The cross-modulation products are measured through a 400-cycle band-pass filter and are plotted, in reference to the 1000-cycle tone, against print densities. From the curve one selects the proper printing point, which is optimum cancellation. This optimum, which usually measures below the noise level of the system, can never be reached unless the oscillator generating the modulated wave has an output which is free of the 400-cycle, the modulating frequency. Few laboratories have such an oscillator available, and fewer are equipped to build one.

---

* Presented May 18, 1948, at the SMPE Convention in Santa Monica.

From the mathematics of an amplitude-modulated carrier wave, generated in a push-pull modulator, one can see that the alternating-current output wave consists of a carrier and two sidebands. Assumed in this statement is that the cancellation of the modulation frequency is complete, i.e, that the circuit is balanced and that either the galvanometer and/or the film will work as a low-pass filter capable of suppressing other components generated in the process of modulation. Fig. 1 shows a microphotograph of a cross-modulation test made with a Radio Corporation of America cross-modulation oscillator. The film used was Eastman 1372 exposed with white light and developed in high contrast sound-track negative developer.

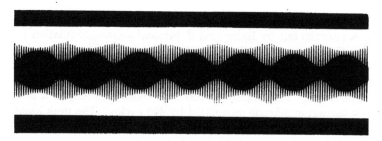

Fig. 1

SINGLE-SIDEBAND RECORDING SYSTEM

The cross-modulation test in itself does not require a signal which must contain a carrier and two sidebands.[2] The test can be made with identical results by using a single-sideband modulated carrier. In this case, instead of modulating a high frequency by 400 cycles, it is easier to mix two high-frequency signals in a linear network (resistances) which have a different frequency equal to the modulating frequency. Two signals of equal amplitude will give a 100 per cent modulated wave, i.e., they will add to zero when opposite in phase and to twice the peak amplitude when in phase. The single-sideband frequency can be either the carrier frequency plus or minus the modulating frequency, i.e., in the case of 9000; 8600 or 9400 cycles. As present-day standards call for only 80 per cent modulation, the sideband should have only 80 per cent amplitude in reference to the carrier. Fig. 2 shows a microphotograph of a single-sideband modulated carrier signal. Again Eastman 1372 film was exposed with white light and developed in high-contrast sound-track negative developer.

With phase-shift oscillators, such as manufactured by General Radio or Hewlitt-Packard, one will have no trouble in keeping the difference in frequency sufficiently constant. As the signal is applied at a very low level it is advisable to build the two oscillators out through high-loss pads, before combining the two signals to eliminate interaction. If the harmonic contents of both oscillators are high it is advisable to feed one of the signals through a low-pass filter. Usually a single-section, constant-K filter,[3] with a cutoff frequency equal to the oscillator frequency is sufficient. Under these conditions, the writer has measured more than a 70-decibel (the limit of the test equipment) differential between the wanted and unwanted signal.

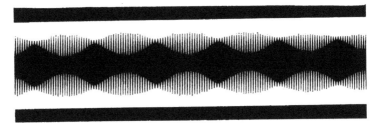

Fig. 2

PHASE METER FOR MEASUREMENTS

Under certain conditions, as in print production, it is impossible to make a series of print densities. From the electrical measurement, one can only determine if the print falls within the limits given, usually −30 decibels relative to 1000 cycles on either side of the minimum. If the cross-modulation products, for some unforeseen reason, should exceed the limits, visual means are at present the only recourse for determining if the print is too dark or light. Trends within the limits, which, if known, may prevent future excessive cross modulation can hardly be determined by the visual method.

In Fig. 3 there is shown a meter built for a different purpose that may be of help in determining trends in processing a variable-area sound track. The signal is picked up and amplified by usual means. The input to the meter is adjusted by P_1 so that the carrier amplitude measured by M_1 is +10.0 decibels relative to 1 milliwatt across 600 ohms (0 dbm*). Part of this signal is fed through a unity gain amplifier into a ring modulator.

* Decibels with respect to 0.001 watt.

Part of it through a 15-decibel, 600/600-ohm isolation pad, a single-section 400-cycle (f/fm = 1.2) constant-K band-pass filter[4] to an amplifier having 42 decibels gain. In setting up the equipment, the gain from the input terminals of the isolation pad P_4 to the output of the amplifier should be adjusted to 25 decibels. That is, so M_3 can be

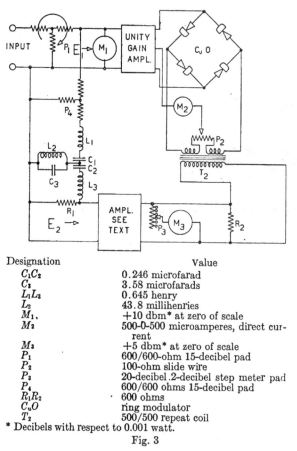

Designation	Value
C_1C_2	0.246 microfarad
C_3	3.58 microfarads
L_1L_3	0.645 henry
L_2	43.8 millihenries
M_1,	+10 dbm* at zero of scale
M_2	500-0-500 microamperes, direct current
M_3	+5 dbm* at zero of scale
P_1	600/600-ohm 15-decibel pad
P_2	100-ohm slide wire
P_3	20-decibel .2-decibel step meter pad
P_4	600/600 ohms 15-decibel pad
R_1R_2	600 ohms
C_uO	ring modulator
T_2	500/500 repeat coil

* Decibels with respect to 0.001 watt.

Fig. 3

calibrated to measure the cross-modulation products; i.e., the 400-cycle tone contained in the original signal, directly without applying a correction factor. M_3 has a sensitivity of +5.0 dbm at zero on its scale and is calibrated to −10.0 relative to its own zero. An auxiliary potentiometer in series with M_3 decreases the sensitivity of the meter

by 20 decibels in 2-decibel steps. Thus M_3 will read, if M_1 is adjusted
to "0" at the 1000-cycle reference signal, cross-modulation products
of either a single- or a double-sideband 400-cycle modulated signal
from -10.0 to -40.0 directly.

The 400-cycle tone is then fed across the other two terminals of the
ring modulator. To balance the ring modulator after construction of
the meter, a 9000-cycle tone at a level of $+10$ dbm is sent into the
instrument and P_2 (a slide wire) is adjusted until M_2, the phase meter,
does not give an indication; i.e., the ring modulator is balanced. This
adjustment does not have to be reset unless the ring modulator becomes defective and a new one has to be installed.

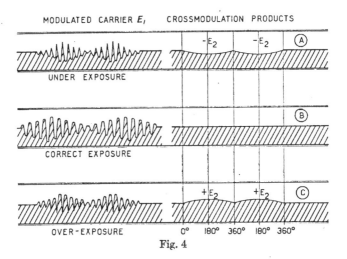

Fig. 4

A modulated signal, which does not contain cross-modulation products, will behave similar to a single tone; that is, the ring modulator
stays balanced and therefore M_2 reads zero. If the modulated signal,
hereafter called E_1 (Fig. 3) contains cross-modulation products, the
400-cycle tone will pass through the 400-cycle band-pass filter unim-
peded, is amplified and will appear as E_2 across meter M_3, and is subsequently fed into the ring modulator. E_2 will unbalance the ring modulator one way or the other, depending on the phase of the voltage E_2
with respect to voltage E_1. This unbalance will result in a positive or
negative indication of meter M_2, the magnitude of which will depend
on the amplitude of E_2 as E_1 is fixed. As the rectifiers or diodes are

connected in full wave, the detector operates over the full cycle. The output is nearly linear for variation of E_2.

Fig. 4 shows the resulting wave forms of a cross-modulation test when under-, correctly, or overexposed. It also makes it evident that E_2 is always either in phase or 180 degrees out of phase with E_1. If E_1 is held at a constant amplitude and E_2 is either in phase (Fig. 4-C) or 180 degrees out of phase (Fig. 4-A) with E_1, I_{M2} (Fig. 3) can be calibrated for a given laboratory condition directly in under- or overexposure in terms of "lights" or density. After calibration, the phase output of the reproducer or amplifier is of no account as long as the

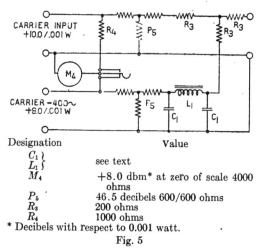

Designation	Value
C_1 } L_1 }	see text
M_4	+8.0 dbm* at zero of scale 4000 ohms
P_5	46.5 decibels 600/600 ohms
R_3	200 ohms
R_4	1000 ohms

* Decibels with respect to 0.001 watt.

Fig. 5

phase relationship of E_1 to E_2 inside the meter stays the same. The phase sensitivity of the meter can be altered to suit particular conditions by inserting an attenuator between M_3 and T_2.

Fig. 5 shows the circuit diagram and values for the cross-modulation single-sideband generator.

REFERENCES

(1) J. O. Baker and D. H. Robinson, "Modulated high-frequency recording as a means of determining conditions for optimal processing," J. Soc. Mot. Pict. Eng., vol. 30, pp. 3–18; January, 1938.

(2) A. Narath, "Entstehung und Beseitigung des Donnereffektes bei Zackcentonfilmen," Zeit. für Tech. Phys., vol. 5, pp. 121–130; May, 1937.

(3) T. E. Shea, "Transmission Networks and Wave Filters," D. Van Nostrand, New York, N. Y., October, 1929, p. 225.

(4) See p. 230 of reference 3.

LOREN L. RYDER, PAST-PRESIDENT OF THE SOCIETY, AND EARL I. SPONABLE, INCOMING PRESIDENT

Sixty-Fourth
Semiannual Convention

THE 64TH SEMIANNUAL CONVENTION of the Society of Motion Picture Engineers was held at the Hotel Statler in Washington, D. C., October 25–29, 1948. There were 400 members and guests who registered for the seven technical sessions and the two symposia on high-speed photography. The Banquet was attended by 265, and there were 209 at the Luncheon.

Eric Johnston, president of the Motion Picture Association was the Guest Speaker at the Luncheon. John Russell Young, president of the Board of Commissioners, Washington, District of Columbia, also spoke.

At the Banquet, Mr. Ryder presented the Progress Medal Award to Peter Mole, and the Samuel L. Warner Award to Nathan Levinson. Sixteen Active Members were elevated to the grade of Fellow. Mr. Ryder also introduced Mr. Sponable, the new president of the Society.

On Thursday evening, a special session was held at the Naval Photographic Center. This included a tour of the Center, and was followed by a technical session held there.

Mrs. Nathan D. Golden, hostess for the Ladies' Committee, prepared an interesting program of sightseeing for the women guests.

PROGRESS MEDAL AWARD

ON OCTOBER 28, 1948, Mr. Peter Mole was presented with the 1948 Progress Medal Award, given for outstanding achievement in motion picture technology. Mr. Mole was chairman of the Pacific Coast Section of the Society shortly after it was formed in Hollywood. He has also been a member of the Board of Governors and is Executive Vice-President of the Society for 1949–1950.

Born in Termini, Sicily, Italy, he was brought to the United States when he was six years old. After being educated in various technical schools he joined the engineering staff of the General Electric Company at Schenectady, New York, where he was active in the development of the General Electric searchlight and a high-intensity rotating carbon-arc theater projection lamp.

In 1923 he left the General Electric Company and moved to California where he became interested in motion picture studio lighting, first with the Metro-Goldwin-Mayer Studios in the electrical department. After receiving his groundwork training in actual production, he went to work for a motion picture studio lighting equipment manufacturer.

With his technical background plus experience in the studios he was ideally suited to enter the field of the manufacture of specialized equipment for an industry that was growing so fast its requirements changed almost from month to month.

It was not long before he joined forces with Elmer C. Richardson, another design engineer, and Fielding C. Coates, a studio chief electrician, and formed the Mole-Richardson Company.

Mr. Mole and other members of his organization have been the authors of numerous papers on motion picture studio lighting which have been published in the JOURNAL, as well as serving on many

PROGRESS MEDAL AWARDED FOR ACHIEVEMENT IN MOTION PICTURE TECHNOLOGY

Society committees. His technical contributions and those of his organization are outstanding.

Mr. Mole's success in his chosen field is not due entirely to his ability to organize and operate an engineering and manufacturing organization to meet the needs of a unique industry. He has an unusual insight into the intangibles created by the art form in motion picture production. He knows that engineering perfection must not transcend utilization in an industry where dramatic effect is the end result; yet he has been able to design and produce highly specialized lighting tools which satisfy both the artist and the engineer. It is for these reasons that,

"It is the unanimous recommendation of the Progress Award Committee that the Progress Medal Award of the Society of Motion Picture Engineers be awarded this year to Mr. Peter Mole for his many and continuing contributions to Motion Picture Lighting technique and equipment.

"Mr. Mole and his organization have pioneered in the development of lighting equipment during each of the successive stages in the advance of motion picture lighting practice for more than twenty years. The wide use of their products manifests the success of their efforts and achievement. Future lighting developments which are currently being studied undoubtedly will reflect the benefits of Mr. Mole's experience as new and improved lighting becomes available for studio use.

"It is fitting to note that Mr. Mole's organization has been the recipient of four Certificate Awards from the Academy of Motion Picture Arts and Sciences, together with recognition by the United Nations Conference on International Organization."

●　●　●

SMPE—SAMUEL L. WARNER
MEMORIAL AWARD FOR 1948

THE WARNER BROTHERS established the Samuel L. Warner Memorial Award, to be given each year by the Society of Motion Picture Engineers at its Fall Convention to an engineer selected by the Society, who has done the most outstanding work in the field of sound motion picture engineering. The 1948 Award was presented to Nathan Levinson who has had a long and successful career in radio communications as well as in sound motion pictures. He started his radio work as an engineer for Marconi, prior to the First World War, and served in the United States Army Signal Corps during that war, rising to the rank of major. After the war he joined the Bell System as a commercial engineer in the radio broadcast field.

Shortly after this, when sound was proposed as an adjunct to motion pictures, various attempts were made to interest the studios in sound for motion pictures, but only Samuel L. Warner was convinced of its commercial possibilities. The history of the growth of sound in motion pictures is a matter of record and need not be repeated here.

During World War II the Navy asked the Warner Brothers to take over the manufacture of a special combat camera, after others had failed. Mr. Levinson undertook this responsibility in addition to directing Warner Brothers' Sound Department and successfully produced and delivered these cameras.

He pioneered such ideas as the intercutting of variable-area and variable-density sound tracks for increased volume range, the commercial use of control track for increased volume range, and one of the first camera blimps, permitting the cameraman to come out of the soundproof camera booth imposed by the advent of sound recording.

The use of 16-mm motion pictures with high-speed development, while not an original idea with Mr. Levinson, was, under his guidance, commercialized for recording race-track events. Currently, Mr. Levinson is playing an important role in the development of television for theater use and as a tool for the production of sound motion pictures. He is also active in the commercial development of a new color system for motion pictures.

The industry is proud of Mr. Levinson and owes to him a debt of gratitude for his many technical contributions to the advancement of the art.

SAMUEL L. WARNER MEMORIAL AWARD PRESENTED ANNUALLY FOR MOST OUTSTANDING WORK IN SOUND MOTION PICTURE ENGINEERING

JOURNAL AWARD

The Journal Award Committee recommended that the Award for the year 1948 be given to Messrs. J. S. Chandler, D. F. Lyman, and L. R. Martin for their paper, "Proposals for 16-mm and 8-Mm Sprocket Standards," published in the June, 1947, issue of the JOURNAL. This deals with the design of that fundamental motion picture mechanism, the sprocket wheel, and the variables that affect the interaction of the film and the sprocket.

They have also proposed standard sprocket-design formulas which allow the engineer to accommodate the variables that apply to his particular problem, being adaptable to any application regardless of the size and function of the sprocket or the path and shrinkage of the film. If changes are made in the physical properties of the film or if research discovers conditions of improved operation, the formulas can be adjusted.

● ● ●

Dr. J. S. Chandler was born in Nebraska and attended the Georgia School of Technology, obtaining a B.S. degree in Mechanical Engineering in 1934 and an M.S. degree in 1936. He became a Research Fellow at the Pennsylvania State College the following year and received his Ph.D degree in Mechanical Engineering in 1938. Since that time he has been employed in the Sound Recording Section of the Physics Department in the Kodak Research Laboratories, where his work has included the studies of the mechanical components of cameras, printers, and other equipment in connection with sound film. He is the author of several previous papers, including one published in the JOURNAL of the Society of Motion Picture Engineers dealing with design considerations of film sprockets.

● ● ●

Mr. Donald Franklin Lyman was born in Winsted, Connecticut. He was graduated from Massachusetts Institute of Technology in 1921 with the B.S. degree in Mechanical Engineering. After employment with the Western Electric Company for three years, he was engaged in 1924 as special engineer in the Development Department at the Camera Works division of the Eastman Kodak Company.

He has worked on fire-control apparatus for airplanes, has done a considerable amount of work for the American Standards Association's Committee on Motion Pictures, and at present, his work involves the development and engineering inspection of amateur motion picture apparatus.

• • •

Mr. Lawrence Randall Martin is assistant to the manager of the Camera Works division of the Eastman Kodak Company and has been associated with the company since July 6, 1931.

Mr. Martin was first employed as a draftsman; two years later he became a product designer, and subsequently chief engineer on motion picture cameras.

In January, 1940, he was named production technician, representing the company in contracts with the Sperry Gyroscope Corporation, the Ford Motor Company, and the United States Army Ordnance Division. He later joined the manager's office to act as liaison between the company and the Armed Services.

At the present, Mr. Martin is responsible for co-ordination of efforts of all departments in new products programs.

He was born in Punxsutawney, Pennsylvania, and received the Mechanical Engineering Degree from Cornell University in 1931.

He has served on several committees of the Society and is particularly interested in the industrial applications of motion pictures.

Fellow Awards—1948

AT THE BANQUET held on October 27, 1948, President Ryder presented sixteen Active Members of the Society with the Fellow Award. The names of the recipients are as follows:

FRED T. ALBIN
Radio Corporation of America

PAUL ARNOLD
Ansco Corporation

GEORGE W. COLBURN
George W. Colburn Laboratories

G. RICHARD CRANE
Western Electric Company

GLENN L. DIMMICK
Radio Corporation of America

HAROLD E. EDGERTON
Massachusetts Institute of Technology

THOMAS T. GOLDSMITH
Allen B. DuMont Laboratories, Inc.

HAROLD C. HARSH
Ansco Corporation

MATTHEW T. JONES
National Carbon Company

DONALD F. LYMAN
Eastman Kodak Company

PIERRE MERTZ
Bell Telephone Laboratories

OSCAR F. NEU
Neumade Products Inc.

RAY R. SCOVILLE
Western Electric Company

NORWOOD L. SIMMONS
Eastman Kodak Company

CHARLES O. SLYFIELD
Walt Disney Studios

H. EDWARD WHITE
Eastman Kodak Company

Section Meetings

Central

The October 22, 1948, meeting of the Central Section of the SMPE was held in the Auditorium of the Engineering Building in Chicago, in joint session with the Chicago Section of the Institute of Radio Engineers.

Kenneth Jarvis of the IRE called the meeting to order and introduced R. T. Van Niman who, in turn, introduced the first speaker, Ernest F. Zatorsky, director of sound recording, The Jam Handy Organization, Detroit.

His paper titled "Microphone Placement Techniques as Applied to Motion Picture Sound Recording" outlined problems involved in securing good quality sound pickup without having the microphone appear in the picture. He advocates use of one microphone directly above the camera line and in front of the subject.

The next paper was "Synchrolite for Television Film Projectors" by L. C. Downes, Television Engineering Section, General Electric Company, Syracuse. This paper described a pulse light source for a standard projector operating at 24 frames per second with the shutter removed, the pulse rate being 30 per second to synchronize with tube scanning. The lamp is Krypton filled. The flash points are a tungston alloy and arc at about 70 to 80 amperes. Operation of the unit consumes 400 to 500 watts, and the life of the lamp is rated at 50 hours. The temperature at the film gate is very low, and the light delivered to the television pickup is 50 foot-candles. The pulse circuits were described in detail, and open discussion followed this presentation.

The November 12, 1948, meeting of the Central Section was called to order by R. T. Van Niman in the rooms of the Western Society of Engineers. About 110 members and guests were present. Short reports on the Washington convention and the election of national officers were given.

"Carbon-Arc Projection," a technicolor film produced by National Carbon Company was presented first. Preliminary comments were given by C. E. Heppberger of this Company. The film was projected with a 16-mm carbon-arc machine, and problems in the making of the film and techniques used were explained.

"A Discussion of High-Quality Sound Reproduction" was given by John K. Hilliard, of the Altec Lansing Corporation.

Mr. Hilliard outlined the requirements for high-quality sound reproduction from both the objective and subjective points of view, pointing out that an overall flat system frequency response, the supposed ideal from the objective point of view, does not necessarily produce a satisfactory illusion of reality in the reproduced sound. He discussed many of the factors which are believed to be responsible for this situation such as level differences between original and reproduction, various types of distortions introduced acoustically, electrically, mechanically, or photographically, and psychological conditioning of listeners. The considerations upon which the motion picture industry's optimum electrical-response curves for theater reproducing equipment are based were outlined, and Mr. Hilliard stated that similar curves for 16-mm reproducing equipment are being plotted.

Section Meetings

Acoustical-response measurements on typical good-quality theater equipment were presented, and dynamic power-level curves were shown which indicate that all components of sound systems must be capable of handling with low-distortion transients of extended frequency range and peak powers many times the so-called "normal" system ratings if the reproduced sound is to more than superficially·resemble the original.

Mr. Hilliard's formal presentation was followed by a discussion period of almost equal length participated in by many members of the Chicago audio group who attended the meeting. Spirited arguments developed as to the precise meaning of some of the terms used in discussing "high-quality" sound reproduction, and many dissenting opinions were expressed regarding such conclusions as have so far been drawn in the audio field. The one point of general agreement appeared to be that only a beginning has been made in reproducing sound, and that a great deal of further research and development is needed, particularly with respect to the psychological aspects of the general problem.

Pacific Coast

An audience of approximately 150 members, admitted by membership card only, filled the Western Electric Review Room to witness a program consisting of a symposium of papers on "The Problem of Sound Reproduction on 16-Mm Kodachrome," at the October 12, 1948, meeting of the Pacific Coast Section. R. G. Hufford of the Eastman Kodak Company presented a paper on the "Sensitometric Characteristics of the Kodachrome Sound Track," which was followed by a short sound demonstration film. Robert V. McKie of the RCA Victor Division gave a short talk on the establishment and maintenance of commercial processing tolerances for making variable-area Kodachrome sound track. J. G. Frayne of the Electrical Research Products Division of Western Electric outlines the methods of making variable-density sound track on Kodachrome and introduced a new system of "electrical printing" whereby the sound track is re-recorded onto each Kodachrome release print. Dr. Frayne gave a demonstration of the quality obtained by this latter method.

The November 9, 1948, meeting of the West Coast Section was held at the recently completed broadcast studio of Radio Station KHJ and the Mutual-Don Lee network. Approximately 350 members and their wives attended this very interesting and informative evening's entertainment and tour through the studio.

The studio management arranged to permit the audience to see a live-talent television broadcast as the opening phase of the evening's program. A portion of the audience saw the program reproduced on a large-screen television receiver in an adjacent auditorium in the studio. Following this television activity, W. Carruthers, chief engineer of Station KHJ, and F. L. Hopper of Western Electric, discussed some of the outstanding architectural, acoustical, and electrical features of this studio, and Harry Lubcke, director of television at Station KHJ, discussed some of the interesting aspects of the television experiences of this organization.

The remainder of the evening was devoted to a conducted tour in small groups through the entire studio, with engineering experts available for questions from the audience concerning the engineering features of the equipment in the studio.

110

Meetings of Other Societies

Inter-Society Color Council Meeting

The seventeenth annual meeting of the Inter-Society Color Council will be held on Wednesday, March 9, 1949, Hotel Statler, New York City. The meeting will consist of a Discussion Session at which committee chairmen will report on the following problems:

2—Color Names (Revision of), Deane B. Judd

6—Color Terms, Sidney M. Newhall

7—Color Specifications, Walter C. Granville

12—Studies of Illuminating and Viewing Conditions in the Colorimetry of Reflecting Materials, Deane B. Judd

14—A Study of Transparent Standards Using Single-Number Specification, Robert N. Osborn

A Business Session will conclude the afternoon meeting. Anyone interested is invited to attend. Hotel reservations should be made directly to the hotel at least ten days prior to the meeting, indicating that you are attending the ISCC meeting.

Meetings of the Optical Society of America are scheduled for the same hotel, March 10–12. It is usual for one or more O.S.A. sessions to be devoted to color, and it is therefore suggested that all ISCC members who are interested plan to remain for these meetings.

PSA Convention

The high light of the recent convention of the Photographic Society of America, held in Cincinnati, Ohio, November 3–7, 1948, with more than 700 photographers and technicians in attendance, was an all-day clinic of the Technical Division on "Photography in Industry."

Among the outstanding technical papers presented were those by J. I. Crabtree, "Rapid Processing of Films and Papers"; Edward H. Loessel, "Making Duplicates of Color Transparencies"; H. G. Morse, "High-Speed Flash Photography in Black and White and Color"; H. A. Miller, "Direct Positive Transparencies by Chemical Reversal"; Harvey P. Rockwell, "Light Measurement in Photography"; and Allen Stimson, "Exposure and Light Measurement."

The Motion Picture Division of PSA presented a number of interesting amateur and commercial films in six well-attended sessions. Among the papers dealing with motion pictures were: "Title Backgrounds by the Experts," by Dennis R. Anderson; "A Challenge to Your Talents," by Mrs. Warner Seely; "Electrical Remote-Control Unit for Movie Cameras," by Belgrave F. Gostin; "Home Movies in Agricultural Education," by George F. Johnson; "How to Make a Movie," by Charles H. Coles; "Making Movies of Football," by Harris B. Tuttle; "Direct 16-Mm Productions," by Larry Sherwood; and "Photometric Calibration of Motion Picture Camera Lenses," by M. G. Townsley.

Book Reviews

Informational Film Year Book, 1948

Published (1948) by the Albyn Press, 42 Frederick St., Edinburgh, 2, Scotland. 200 pages. 21 figures. $5^3/_4 \times 8^3/_4$ inches. Price, 12s. 6d. net.

This second volume of the *Informational Film Year Book* follows the same pattern as the initial volume published last year. However, many additional data have been included making the book of even more value than before to producers and consumers of nontheatrical motion pictures. Eight feature articles appear on various phases of documentary films, filmstrips, and equipment. The remainder of the book consists of listings and directories covering films of the year, who's who in documentary, various organizations and societies the world over, film producers, special service firms, and equipment suppliers. The pictorial section, although not everything to be desired, adds flavor to the book and prevents it from becoming a mere compilation of articles and other information. Because of the international character of the publication, it would be helpful if a few notes were given concerning the contributors since it is not generally the case that authors are well known outside their respective countries.

LLOYD E. VARDEN
Pavellé Color, Inc.
533 W. 47 St., New York, N. Y.

The High-Current Carbon Arc, by Wolfgang Finkelnburg

Published (1947) by the Office of Military Government for Germany (U. S.) Field Information Agency, Technical, Final Report No. 1052, through the Office of Technical Services, U. S. Department of Commerce (Publication Board No. 81644). Paper covers, photo offset from typewritten manuscript. 219 pages + x pages. 129 figures. 8 tables. 90 references. $7^1/_2 \times 10$ inches. Price, $5.00.

This book, in the first German edition of 1944, was prepared as a confidential text for the guidance of scientists in Germany working on carbon-arc searchlight development. Since the Allies came to rely on radar communication while the enemy was still expanding the size and intensity of his antiaircraft searchlights, these latter reached a much higher state of development than here in the United States. For instance, 450- and 1000-ampere carbon-arc searchlights were in active combat use and an advanced stage of development, respectively, in Germany, as compared with the 195-ampere maximum employed by United States forces. Since searchlight arcs differ in no important theoretical respect from those employed in motion picture photography and projection, these developments are of particular interest to technicians in the motion picture industry. The book presents "the whole knowledge" in Germany of the physical properties, theory, and application of carbon-arc light sources. In addition to a treatment of previously published material, a large amount of hitherto unpublished information from the author's own laboratory as well as from other German workers and firms is included.

112

Book Reviews

The author classifies carbon arcs as of two fundamental types, low-current and high-current, depending upon the current density at the anode, and independent of the composition. He finds, for instance, that the plain "low-intensity" carbon arc, as we term it, behaves much as a high-intensity arc if it is operated at comparable current densities. Of course, the unsteadiness and the noise of this overload low-intensity arc are too great for most commercial uses, but the physical processes in the arc itself, particularly at the anode, are found to be identical with those which govern high-intensity arc operation. The first part of the book is devoted to a detailed description of the operating characteristics which distinguish the two fundamental types of arcs.

The next chapters deal with the operating properties of high-current carbon arcs, including not only the radiant output, but studies of arc temperature, carbon consumption, the transport of material through the arc, arc behavior in pure gases and at different pressures, magnetic properties of several types of arcs from the standpoint of stabilization, and chemical processes within the arc. This prepares the reader for the following theoretical section of the book, where a theory is presented explaining the fundamental differences between low- and high-current carbon arcs. In particular, the increase in arc voltage with increasing current which distinguishes the behavior of the high-current carbon arc from the negative characteristic of the low-current type, and the observed relations between current, brilliancy, and core composition are explained in terms of the author's theory.

The concluding chapter of the book deals briefly with applications of the high-current carbon arc, the section on searchlights being of particular interest on account of the very high current units employed in Germany. The German practices in the motion picture studio and projection fields and medical therapeutics are interestingly described. A concluding section, treating the carbon arc as a tool in high-temperature chemistry studies, well illustrates the author's hope that the book serve primarily as a basis and incentive for further research on the theory and application of the high-current carbon arc. To anyone interested in this field, the book provides a very worth-while background.

F. T. Bowditch
National Carbon Company
Cleveland 1, Ohio

George Mitchell Receives ASC Award

The American Society of Cinematographers presented George Mitchell, Associate member of the SMPE, with a certificate of appreciation in recognition of his ceaseless pioneering in the field of motion picture photographic equipment, and his immeasurable contribution to the advancement of cinematography as an art and as a science.

The award, the first given by the Society in its thirty years of existence, was made in a surprise presentation to Mr. Mitchell, September 11, 1948. Similar awards will be made by the Society from time to time, to others whose contribution to cinematography is considered equally noteworthy.

Current Literature

THE EDITORS present for convenient reference a list of articles dealing with subjects cognate to motion picture engineering published in a number of selected journals. Photostatic or microfilm copies of articles in magazines that are available may be obtained from The Library of Congress, Washington, D. C., or from the New York Public Library, New York, N. Y., at prevailing rates.

American Cinematographer
29, 11, November, 1948
Cinecolor Moves Ahead (p. 373) N. KEANE
Filming the Olympic Games (p. 374) F. FOSTER
Exponent of the Moving Camera (p. 376) H. A. LIGHTMAN
Seven New Lenses for 16-Mm Cameras (p. 384) A. ROWAN

Audio Engineering
32, 11, November, 1948
Sound on Film (p. 24) J. A. MAURER
32, 12, December, 1948
Suggested Wiring Standards for Motion Picture Recording Equipment (p. 16) G. RUDOLF

British Kinematography
13, 3, September, 1948
The Origin and Development of the Matte Shot Process (p. 74) W. P. DAY
Power Supplies for Motion Picture Studios (p. 83) F. S. HAWKINS
13, 4, October, 1948
The Processing of Colour Films (p. 109) J. H. JACOBS
Safety Regulations in Kinemas and Theatres (p. 121) A. F. STEEL
Light Production From the Carbon Arc (p. 128) H. P. WOODS

Communications
28, 11, November, 1948
High Fidelity Tape Recording (p. 16) R. BARUCH

International Photographer
20, 10, October, 1948
Thomascolor (p. 7) A. WYCKOFF
Production in Germany (p. 12) W. B. KELLY

International Projectionist
23, 10, October, 1948
Causes and Prevention of Damage to 35-Mm Theatre Release Prints (p. 5)
Television: How It Works (p. 12) W. BOUIE
23, 11, November, 1948
Safety Film: Projection Factors (p. 9) H. B. SELLWOOD
Television: How it Works. Pt. 5 (p. 17) W. BOUIE

Proceedings of the I.R.E.
36, 10, October, 1948
The Chemistry of High-Speed Electrolytic Facsimile Recording (p. 1224) H. G. GREIG

Radio and Television News
40, 5, November, 1948
The Recording and Reproduction of Sound. Pt. 21 (p. 50) O. READ

⌁ New Products ⌁

Further information concerning the material described below can be obtained by writing direct to the manufacturers. As in the case of technical papers, publication of these news items does not constitute endorsement of the manufacturer's statements nor of his products.

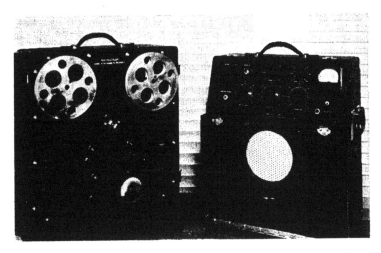

Magnagram

Recogram Recorders Company, 11338 Burbank Boulevard, North Hollywood, California, recently announced the "Magnagram" M-116. This is a completely portable location recorder which is contained in two separate cases.

Contingent upon film speed, the unit is capable of recording up to thirty minutes. With 400-foot reels the mechanical unit can be blimped for silent operation. The transparent door leaves the working parts readily visible.

The magnetic film is caused to pass over an independent floating drum connected with a dynamically balanced flywheel in the retreat from the recording-reproducing heads. This motion, coupled with dampening arms confines flutter to approximately $1/10$ of one per cent.

For ease in editing magnetic film the M-116 is designed to pass the track backwards as well as forwards over the heads without altering the normal threading.

All components of the magnagram are mounted on 19-inch Western Electric relay-rack panels. The drive assembly employs a hysteresis-synchronous motor with prelapped nylon gears.

～ New Products ～

Further information concerning the material described below can be obtained by writing direct to the manufacturers. As in the case of technical papers, publication of these news items does not constitute endorsement of the manufacturer's statements nor of his products.

Photocell Cable

Transradio, Ltd., 138A Cromwell Road, London, S.W.7, England, has announced a special photocell cable, type PC1-T. It is claimed that its losses are but a fraction of those common to the usual type of low-loss cable. It is vibrationproof, nonmicrophonic, highly flexible, monaging, and unaffected by oils.

Fig. 1—Photocell cable.

This cable is designed to afford that constancy of electrical characteristics required for cable connections to portable equipment. It finds particular application to all kinds of photoelectric-cell work, sound-film projection, and microphones, where stability of electrical values is necessary.

Movette Camera

American Cinefoto Corporation, 1560 Broadway, New York 19, N. Y., recently introduced the Movette, a process for taking animated portraits.

A high-intensity stroboscopic light combined with a multiple camera operates at 180 frames per minute and uses 4- × 5-inch cut film.

The camera is without shutters. A built-in commutator arrangement fires the flash lamp in rapid sequences during which period the negative holder inside the camera completes its travel across the optical unit. The motor and flash cease to operate after the last picture has been taken.

The Movette plugs in any 110-volt, 60-cycle alternating-current light socket; the charging voltage is 2000 volts and the effective flash $1/10,000$ of a second.

A series of 42 pictures are produced in about 15 seconds. Developing the film and making an 11- × 14-inch enlargement are done in the conventional manner.

EMPLOYMENT SERVICE

POSITIONS WANTED

Situation wanted in Television Broadcasting Industry. Possess 7 years' administrative, 5 years' motion picture engineering, and 4 years' electronic teaching experience. For details write "Anonymous," 190 Hutchison Blvd., Mount Vernon, N. Y.

Mechanical and Electrical Design Engineer with experience on sound reproducers, amplifiers, and photoelectric devices. Kjeld Bogedam, c/o Nielsen, 244 Fisher Ave., Tottenville, S. I., N. Y.

Journal of the Society of Motion Picture Engineers

| VOLUME 52 | FEBRUARY 1949 | NUMBER 2 |

Subscription to nonmembers, $10.00 per annum; to members, $6.25 per annum, included in
their annual membership dues; single copies, $1.25. Order from the Society's General Office.
A discount of ten per cent is allowed to accredited agencies on orders for subscriptions and
single copies. Published monthly at Easton, Pa., by the Society of Motion Picture Engineers,
Inc. Publication Office, 20th & Northampton Sts., Easton, Pa. General and Editorial Office,
342 Madison Ave., New York 17, N. Y. Entered as second-class matter January 15, 1930,
at the Post Office at Easton, Pa., under the Act of March 3, 1879.

Society of Motion Picture Engineers

342 MADISON AVENUE—NEW YORK 17, N. Y.—TEL. MU 2-2185
BOYCE NEMEC . . . EXECUTIVE SECRETARY

OFFICERS

1949–1950

PRESIDENT	PAST-PRESIDENT	SECRETARY
Earl I. Sponable	Loren L. Ryder	Robert M. Corbin
460 W. 54 St.	5451 Marathon St.	343 State St.
New York 19, N. Y.	Hollywood 38, Calif.	Rochester 4, N. Y.

EDITORIAL VICE-PRESIDENT	EXECUTIVE VICE-PRESIDENT	CONVENTION VICE-PRESIDENT
Clyde R. Keith	Peter Mole	William C. Kunzmann
120 Broadway	941 N. Sycamore Ave.	Box 6087
New York 5, N. Y.	Hollywood 38, Calif.	Cleveland 1, Ohio

1948–1949 & 1949

ENGINEERING VICE-PRESIDENT	TREASURER	FINANCIAL VICE-PRESIDENT
John A. Maurer	Ralph B. Austrian	David B. Joy
37-01 31 St.	1270 Avenue of The Americas	30 E. 42 St.
Long Island City 1, N. Y.	New York 20, N. Y.	New York 17, N. Y.

Governors

1948–1949	1949	1949–1950
Alan W. Cook	James Frank, Jr.	Herbert Barnett
25 Thorpe St.	426 Luckie St., N. W.	Manville Lane
Binghamton, N. Y.	Atlanta, Ga.	Pleasantville, N Y.
Lloyd T. Goldsmith	William H. Rivers	Fred T. Bowditch
Warner Brothers	342 Madison Ave.	Box 6087
Burbank, Calif.	New York 17, N. Y.	Cleveland 1, Ohio
Paul J. Larsen	Sidney P. Solow	Kenneth F. Morgan
508 S. Tulane St.	959 Seward St.	6601 Romaine St.
Albuquerque, N. M.	Hollywood 38, Calif.	Los Angeles 38, Calif.
Gordon E. Sawyer	R. T. Van Niman	Norwood L. Simmons, Jr.
857 N. Martel Ave.	4431 W. Lake St.	6706 Santa Monica Blvd.
Hollywood 46, Calif.	Chicago 24, Ill.	Hollywood 38, Calif.

Three-Color Subtractive Photography*

By W. T. HANSON, JR., AND F. A. RICHEY

KODAK RESEARCH LABORATORIES, ROCHESTER 4, NEW YORK

Summary—The color-vision characteristics of the eye are discussed and the rules which are followed are used to show the requirements for the "perfect" additive and subtractive three-color photographic processes. Since these requirements are not achieved in practice, a theoretical study of a practical color process may not always give an adequate analysis of its usefulness. However, such an analysis may point out some of the pitfalls which occur in practice. For example, many subjects may appear to be the same color to the normal eye and yet give different results when photographed. Also, any given color may be reproduced incorrectly by any process in use.

The effects on picture quality of changes in contrast, balance, and a variety of other variables are shown. The restrictions which some of these factors place on the use of color films are mentioned.

COLOR PHOTOGRAPHY is now an accepted reality and from all appearances is here to stay. During the past twenty years the motion picture industry has witnessed the slow but steady growth of color, both in the professional and amateur fields. Many problems have been encountered and many, but not all of them, have been overcome. However, the average motion picture goer will lay his money down for a color picture in preference to one in black and white.

How did color photography get here? Contrary to popular conception, color photography is not an invention and even the individual color systems and color processes are something more than inventions.

The basic concepts of modern color photography are almost one hundred years old, and it is now a very complex and abundant field. A good many physicists, chemists, psychologists, physiologists, and artists have contributed. Until recently the terminology has varied with the profession as well as the eccentricity of the individual, so that much of the literature is not easy reading.

Practical color processes do not emerge from theoretical discoveries alone. The successful existing color processes are here because of the parallel evolution of dye chemistry and photographic chemistry and,

* Presented May 18, 1948, at the SMPE Convention in Santa Monica.

most important, because of countless trial-and-error experiments, and the ingenious solution of thousands of small problems.

Color photography must always start and end with the mechanism of color perception by the human eye, that is, the color process must first "see" the scene in a manner *approximating* that of the human eye. It must then reproduce that scene in such a manner that it seems *plausible* to the eye.

It has long been known that all colors could be matched by mixing amounts of three so-called primary colors. With a given set of three primaries taken from the spectrum, each of the other colors of the spectrum can be duplicated by a mixture of certain intensities of the original three. There is an important reservation in this generaliza-

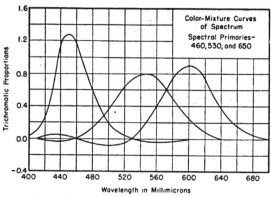

Fig. 1—Color-mixture curves of the spectrum.

tion, for it will be seen from Fig. 1 that in certain regions of the spectrum negative amounts of the three primaries must be permitted. Since negative quantities of the primaries cannot exist, the equivalent is obtained in practice by adding the third primary to the colors which cannot be matched.

If another set of three wavelengths had been chosen as the primaries, a similar but different set of curves would have resulted.

It is frequently desired to express the color-mixture data obtained with one set of primaries in the equivalent amounts of a different set of primaries. This process is illustrated in Fig. 2.

The orange-yellow color at the top can be matched by the primaries, R, G, and B. The squares show the unit amounts of these primaries and the rectangular arrangement at the right shows the amounts of the three required to match the color. If we wish to express this color in terms of another set of primaries, R', G', and B', we can first find the amounts of R', G', and B' which are required to match exactly the original unit amounts of R, G, and B. By using R', G', and B' in these ratios, the total amounts of R, G, and B shown in the upper right can be matched. Then the sum of the three values of R', the sum of the three values of G', and the sum of the three values of B' will match the original color.

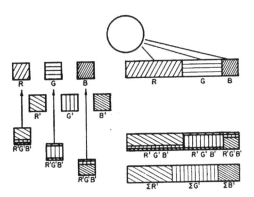

Fig. 2—Graphical transformation of primaries.

Transformation of Primaries

$$C = rR + gG + bB$$
$$R = a_{11}R' + a_{12}G' + a_{13}B'$$
$$G = a_{21}R' + a_{22}G' + a_{23}B'$$
$$B = a_{31}R' + a_{32}G' + a_{33}B'$$
$$C = r(a_{11}R' + a_{12}G' + a_{13}B') + g(a_{21}R' + a_{22}G' + a_{23}B') + b(a_{31}R' + a_{32}G' + a_{33}B')$$
$$C = (ra_{11} + ga_{21} + ba_{31})R' + (ra_{12} + ga_{22} + ba_{32})G' + (ra_{13} + ga_{23} + ba_{33})B'$$

This operation is usually carried out mathematically. The equations which are involved are shown below. There a color, C, is matched by an amount, r, of the primary, R, an amount, g, of the primary, G, and an amount, b, of the primary, B. These primaries can, in turn, be defined in terms of the amounts (a_{11}, a_{12}, a_{13}, a_{21}, etc.) of a second set of primaries, R', G', and B'. If the unit amounts of R, G, B, and R', G', B' are defined by the amounts required to produce a white of a certain brightness, then the substitutions and factoring give the last form of the equation.

In a similar manner, the amounts of the primaries, R', G', and B', which are required to match the various *spectrum* colors can be calculated from the color-mixture curves of the primaries, *R*, *G*, and *B*. This then gives us a new set of color-mixture curves. The change from one set to the other is a linear transformation. There are an *infinite* number of primaries and corresponding color-mixture curves which describe the color-vision characteristics of the human eye, and all of them tell exactly the same story. These of course include as primaries purely hypothetical colors which cannot exist in practice. Proper choice of hypothetical primaries can lead to mixture curves with no negative regions.

Another important characteristic of color vision is the relative brightness of different colors. Equal energies of different colors are

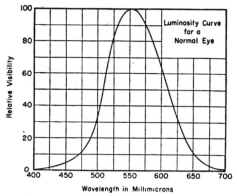

Fig. 3—Luminosity curve for a normal eye.

not of equal brightness or luminosity. If the relative luminosity of spectrum colors of an equal energy is measured, the curve in Fig. 3 results.

For many years there was a good deal of confusion in the field of color measurement because a variety of workers used different primaries in determining the color-mixture curves and different methods of measuring the luminosity of the different spectrum colors. Different types of equipment led to slightly different results, because of certain inaccuracies and also because of the fact that the individual observers vary in their characteristics. For this reason a standard system of color specification became necessary for all the various workers in the field of color.

This standard system was set up by the International Commission on Illumination and is called the ICI system. This group selected the previously adopted luminosity curve as a standard and defined its three primaries to meet certain requirements. First, all real colors should be matched with positive amounts. Second, one primary should be such that one of the mixture curves would be identical to the luminosity function. By using the best available color-mixture data, the primaries, X, Y, Z, and the corresponding mixture curves were established to define the "standard observer." Obviously the primaries do not represent real colors. These standard color-mixture curves are shown in Fig. 4.

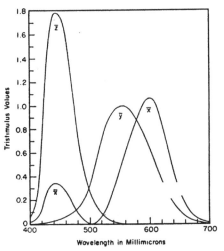

Fig. 4.—Standard ICI color-mixture curves.

Maxwell, in 1855, suggested that positives made from black-and-white negatives which, in turn, were made through red, green, and violet filters, could be used to control the amounts of red, green, and violet light transmitted by filters and that these, when superimposed, would give a color reproduction of the original scene.

All additive systems of color photography are modifications or applications of this invention made more than 90 years ago.

This system of Maxwell's was an additive method. Shortly afterwards this principle was extended by du Hauron, who showed that images made through the red, green, and blue filters and printed in superposition in cyan, magenta, and yellow dyes or pigments would also give a fair reproduction of the original scene. This extension of Maxwell's system is the basis for all the subtractive color systems. However, it was many, many years before the sensitizers, dyes, and photographic materials in general were available for the application of these simple principles.

For many years there were heated arguments as to the exact requirements for the sensitivity distributions of the three emulsion-filter combinations to be used in obtaining the three records for color photography. Certain workers in the field felt that narrow bands of sensitivity in the red, green, and blue regions of the spectrum gave the most satisfactory results. Others felt that the sensitivity distributions of the three emulsions had to match the sensation curves of the eye. Others attempted to match the sensitivity-distribution curves of the emulsions with the absorption curves of the dyes or pigments being used in making subtractive prints. Although these earlier efforts did not lead to a resolution of these theoretical problems, enough practical experience was gained so that when improved sensitizers, emulsions, and techniques of making colored photographic images were developed it was possible to work out empirically methods of making quite satisfactory color photographs. Continued progress in the "techniques" of making colored images and superimposing them have brought color photography to the present state.

Finally, knowledge as to the theoretical requirements for "exact" photographic color reproduction followed close on the heels of better data describing the color-vision characteristics of the eye. Hardy and Wurzburg[1] applied the principles of colorimetry and the characteristics of the human eye to the problem of establishing the theoretical requirements for the perfect additive three-color photographic process They showed that the sensitivity distributions of the three emulsions used in obtaining the three images must correspond with the color-mixture curves determined with the three primaries which were used in showing the additive color picture. These distribution curves would be some linear transformation of the color-mixture curves obtained by using other primaries, including the standards selected by the ICI. Of course, for any primaries which could be used in practice, that is, real colors, even spectrum colors, these film sensitivities would require negative proportions of certain regions of the spectrum. Although a number of suggestions have been made as to how such negative sensitivities might be achieved, and some methods have been patented, to date no satisfactory practical solution has been found.

Later, Yule[2] and MacAdam[3] extended these principles to the problem of subtractive color photography. Although it was not possible to establish the so-called ideal dyes for use in subtractive photography, MacAdam showed that for one set of dyes actually being used in practice, it was possible to establish so-called additive primaries which would describe the behavior of subtractive mixtures of these dyes and was able to show that by the use of six masks it should be possible by photographic means to obtain a very close approximation to "exact" color reproduction. The basic principles were those developed by Hardy and Wurzburg, and the sensitivity requirements of the three emulsions in this process were the color-mixture curves derived from the primaries. These, of course, contained negative portions at certain regions in the spectrum.

This can be summarized by stating that the theoretical requirements which a subtractive color process must fulfill in order to give "exact" color reproduction have not been established completely. They do indicate a need for negative sensitivities and for the use of six masks. The first of these needs cannot be fulfilled at all and the second is entirely impractical. So, the color processes have to struggle along without fulfilling these requirements, and they do give satisfactory results.

However, even though present-day color processes do give satisfactory results, there are certain deficiencies which must exist because of the failure to meet the requirement that the film-sensitivity distribution be a linear transformation of the color-mixture data of the eye. For example, there are an infinite number of energy distributions of light which appear the same to the eye. A color film will not necessarily see such colors as being alike.

A pair of dye combinations which produce a very close visual match is shown in Fig. 5. These spectrophotometric measurements show the densities of the two combinations to light of the various wavelengths in the visible range. As is often the case, these colors which appear to be identical have very different absorption characteristics. When photographed with one of the commercially successful color films, the resulting photographs are also very different.

At first thought, one may say, "This doesn't make too much difference because we shall never encounter two colors of this sort side by side." However, the fact that the two colors do not match indicates that *at least* one of them is not properly reproduced. In fact, *any* color might be improperly reproduced by any of the present color processes which in normal practice give excellent results. Fortunately most of the colors which we normally encounter have more or less continuous light-absorption ·bands and are reproduced fairly accurately. It may appear that an undue amount of emphasis is being put on this type of problem. The important point is that in dealing with flowers, new types of fabrics, or with new color situations in general, it is wise to make a test with a given photographic process to see that it will reproduce adequately the specific colors which are important rather than to assume that the process is perfect and start shooting.

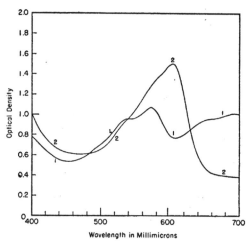

Fig. 5—Spectrophotometric curves of matching colors.

For most practical applications of color photography, the reproduction of colors need not be theoretically perfect. Even with very pleasing color pictures, an analysis of the individual colors will reveal considerable departure from the hue, saturation, and brightness of the original colors of the scene. However, when combined in a picture of familiar and pleasing composition, the color reproduction is plausible enough to give the impression of correct reproduction.

Let us now look at some of the requirements which can and must be fulfilled in obtaining satisfactory color photographs.

The first of these requirements is color balance. This is usually best observed in the accuracy with which grays of various brightnesses are reproduced. One might consider this the minimum requirement of a color process. However, the errors encountered in matching grays to the original subject are present in about the same degree in the reproduction of all colors.   In the case of pastels and other colors of low saturation, this error in balance may become a serious distortion.

Color balance is measured by reading a scale of grays with a color densitometer and plotting the densities of the dyes against the logarithm of the exposure.   By definition,[4, 5] the equivalent neutral densities (END) of the dyes of a given color process are those which, in superposition, will appear gray under the viewing conditions for which the color film is designed.   A correctly balanced color process

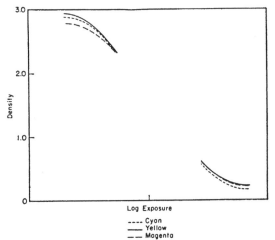

Fig. 6—Correctly balanced neutral scale.

would have a gray scale in which all three dye curves were superimposed.   Slight deviations from this ideal are usually encountered at very low densities and also in the region of maximum density.

If, however, the color balance is uniformly high in any of the dyes, magenta in the case of Fig. 7, the picture through such a process will also show a decided shift to a magenta balance.   This is noticed not only in the reproduction of grays but in a change in all the colors of the pictures.   Thus, blues become more purple, yellows become more orange, and greens become darker.   This type of distortion results also from any change in color temperature of the exposing light from that for which the color film has been balanced.

This uniform shift in color balance of a color film may at first

appear to be very objectionable but when a picture is viewed by projection in a darkened room the distortion appears to become less objectionable with continued viewing. This accommodation of the
visual process, or color adaptation, does tend to make an off-balance
picture appear more nearly satisfactory by projection. If, however,
as in motion picture projection, the color balance shifts from scene to
scene the color change is very noticeable.

In Figs. 6 and 7 the gammas of all the dye scales were equal.
If the gammas of the three dye images are not equal, Fig. 8, the color
photograph varies in color balance from one density level to another

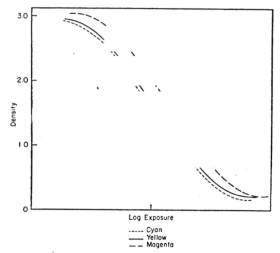

Fig. 7—Magenta-balanced neutral scale.

and the distortion in color reproduction varies depending on the color
and its brightness. In a picture through the process represented
in Fig. 8, the light densities would be much too green and the darker
portions of the pictures much too magenta. This is a very undesirable type of distortion, for the eye cannot become adapted to both
errors. Such a picture continues to be objectionable, no matter how
long we look at it. From the shapes of the curves it is easy to see why
such a distortion has been termed a kink.

Assuming that the first requirement is fulfilled and that correct
color balance and matched relative gammas of the three dye images
can be achieved, and these are no small assumptions, we are still faced

with a difficult decision: "What gamma or contrast level is most desirable for a specific color process?"   In black-and-white photography, the question is completely answered by the requirements for pleasing tone reproduction.   In color photography, the problem is complicated by the fact that color saturation varies with the gamma. At low gamma, we can have all the advantages of pleasing tone rendition and greater latitude but we must pay for these advantages by sacrificing color saturation.   At high gamma, color saturation is satisfactory but latitude must be low.

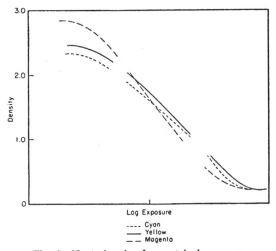

Fig. 8—Neutral scale of unmatched gammas.

The relationship between gamma and color saturation can be explained quite easily by expressing the amounts of the three image dyes present in a given area of a color photograph in terms of equivalent neutral density.   For example, suppose that a given color is reproduced by a color process having a gamma of 1.0 by the amounts of the dyes: cyan, 0.8; magenta, 0.1; and yellow, 0.6.   If the same process were operated at a gamma of 1.5, the reproduction densities would be: cyan, 1.2; magenta, 0.15; and yellow, 0.9.   It is possible to determine the color densities (in terms of END) over and above the gray content (in terms of END) merely by subtracting the density of the dye occurring in the lowest amount.   The results of this subtraction from the two groups of densities above are given in Table I.

TABLE I

THE EFFECT OF GAMMA ON COLOR SATURATION

	Cyan	Magenta	Yellow	Cyan-Magenta	Yellow-Magenta
At $\gamma = 1.0$	0.8	0.1	0.6	0.7	0.5
At $\gamma = 1.5$	1.2	0.15	0.9	1.05	0.75

The much greater density differences at the gamma of 1.5 represent a significant increase in color saturation.*

Experience has shown that the color saturation obtained at higher contrast is quite desirable so that in practice, processes are usually operated at a relatively high contrast. The desired tone reproduction must be obtained by much flatter lighting than normally would be used in black-and-white photography. It may be noted in passing that the opposite approach is not satisfactory, that is, a low-contrast process with contrasty lighting. The maximum color saturation possible is limited by the density range and gamma of the color process and is only slightly affected by lighting variations.

The decision between usable contrast and acceptable color saturation again arises in duplicating a color film. The process of duplicating a color film results in a loss in color saturation, providing the contrast is reproduced at the same level as in the original picture. This loss in saturation is a result of the properties of the dyes available for use in color photography. The only way this color-saturation loss can be improved is by increasing the contrast of the reproduction. This again results in a very definite compromise in the over-all quality of the reproduction. A more complete discussion of this problem is given in a separate paper by Miller.[6]

After considering some of the theoretical problems involved in obtaining more nearly perfect color reproduction and some of the more elementary variables in color processes, we should like to stress that what is necessary and what is desired in a color process depend very largely on the manner in which the process is to be used. The requirements for a good motion picture print in color are different in many respects from the requirements for a good reflection print process.

* This method of expressing colors in terms of END does not express a quantitative value for hue and saturation such as a Munsell notation or the like, but it is a useful technique in the field of color photography.

Evans[7] has directed attention to the many psychological effects which complicate any orderly analysis of color vision and color photography. Brightness constancy, color and brightness adaptation, and simultaneous contrast have a profound influence in all color systems.

These phenomena result from the fact that the visual process does not function merely as a physical instrument for measuring the stimuli from different areas. On the contrary, the appearance of an object is always affected by the spatial relation of the object to other objects and to lighting conditions. The eye always sees things as the observer thinks they really are rather than as they happen to appear at the moment. A simple example is that of a white object in a shadow near a black object in full sunlight. Although the luminance of the white object may be much less than that of the black object under these conditions, the eye immediately recognizes the true brightness relationship of the two objects. This brightness-constancy effect may not be shown to the same degree in a photograph as in the original scene since the viewing conditions are in most cases entirely different.

The eye varies in sensitivity to light over a considerable range depending on the intensity level under which it is used. Something similar to this brightness adaptation causes the eye to become adapted to colored light so that it tends to accept that color as white. The maximum effect, of course, is realized when all of the light reaching the eye is of the same color.

Simultaneous contrast has been related to the adaptation of the eye to local areas of a picture. That may be simplified by stating that areas of complementary or contrasting colors appear to be increased in saturation by their proximity within the picture. In addition to selecting colors which produce a desired hue in the finished color photograph, simultaneous contrast can be used to striking advantage in obtaining pleasing color pictures.

These effects are distinctly beneficial in the projection of color transparencies in a darkened room and are therefore very important in the success of many motion picture processes. Furthermore, these psychological effects explain in part why the data obtained in an isolated field of a colorimeter and the mathematical derivations from such data may have very little correlation with the infinite variety of conditions which can be encountered in photography and in the presentation of the resulting pictures of everyday objects.

You will recall the emphasis on the word *approximation,* in comparing the color photographic process to the visual process. We think it is still safe to state that the perfect color process has not yet been realized.   There are, however, many successful color processes which can give very pleasing results in spite of the many compromises which must be present in each system.   This means that to obtain satisfactory results the user must learn quite a lot about the color process with which he is working.   As he learns what a particular color process will, and equally important, what it will not do, his success with that process will become more consistent.

Close co-operation between the user of color photographic materials and the manufacturer of them has been and will continue to be very important in obtaining satisfactory results with what we now have and know.   This sort of co-operation is also necessary for the introduction of improved color photographic materials and techniques which will give better results.

REFERENCES

(1) A. C. Hardy and F. L. Wurzburg, Jr., "The theory of three-color reproductions," J. Opt. Soc. Amer., vol. 27, pp. 227–240; July, 1937.

(2) J. A. C. Yule, "The theory of subtractive color photography," J. Opt. Soc. Amer., vol., 30, pp. 322–331; August, 1940.

(3) D. L. MacAdam, "Subtractive color mixture and color reproduction," J. Opt. Soc. Amer., vol. 28, pp. 466–480; December, 1938.

(4) R. M. Evans, "A color densitometer for subtractive processes," J. Soc. Mot. Pict. Eng., vol. 31, pp. 194–202; August, 1938.

(5) M. H. Sweet, "A precision direct-reading densitometer," J. Soc. Mot. Pict. Eng., vol. 38, pp. 148–173; February, 1942.

(6) T. H. Miller, "Masking: A technique for improving the quality of color reproductions," J. Soc. Mot. Pict. Eng., this issue, pp. 133–155.

(7) R. M. Evans, "Visual processes and color photography," J. Opt. Soc. Amer., vol. 33, pp. 579–614; November, 1943.

Masking: A Technique for Improving the Quality of Color Reproductions*

By T. H. MILLER

EASTMAN KODAK COMPANY, ROCHESTER 4, NEW YORK

Summary—Currently available subtractive color-photography processes provide pleasing pictures of most natural objects. However, when an original color photograph is the subject, as in the cases of duplicating and copying, the resulting reproduction is usually not satisfactory when compared with the original. Differences between the original and the reproduction are primarily due to the high photographic contrast and the optical characteristics of the dyes in the original. Masking to improve the quality of color reproductions involves making an auxiliary image, generally by a photographic method, and registering it with the original color transparency. Reproductions are made from the combined transparency and mask.

I. INTRODUCTION

BOTH THE SCIENCE and art of color photography have developed rapidly since the introduction of Kodachrome film in 1935. Color motion pictures and still-picture transparencies of excellent quality are now a reality for both amateur and professional photographers using only conventional cameras and processing equipment.

In the early years of Kodachrome the original color pictures were used only for projection for the enjoyment of the family and friends. In manufacturing and processing, therefore, the films were balanced for optimum quality when shown by projection in darkened rooms. Even now most color pictures are used only for projection and the processes are still balanced for best results for this condition.

Interest in color and color pictures spread rapidly and with the introduction of Kodachrome and Ektachrome sheet films, professional and commercial color photography gained further momentum. Color added interest and appeal to educational, entertainment, industrial, and other motion pictures previously made only in black and white. Color slidefilms, color prints, and color reproductions of photographs in magazines and on billboards are now becoming the rule rather than the exception.

---

* Presented May 18, 1948, at the SMPE Convention in Santa Monica.

Contrary to the amateur and early professional practice, a single original transparency or motion picture seldom satisfies the requirements of the present-day professional color photographer. The professional generally must supply many reproductions from his original: duplicate motion pictures or slidefilms for simultaneous showings in all parts of the country, color prints for publicity releases, advertising folders, display ads, and many other forms of color reproductions.

It is more difficult to make good color reproductions than it is to make good color originals. The nature of the problem (and some of the reasons) can be stated briefly as follows:

(1) The original pictures made by three-color subtractive processes, such as Kodachrome, for example, provide very acceptable, though perhaps not exact, reproductions when used to photograph most natural objects. There are some unusual objects with sharp absorption bands that do not photograph well because of the relationships between the absorption bands of the object and the sensitivities of each of the three film-emulsion layers. A color photograph is one of these objects. The spectral characteristics of a color photograph and the subject are frequently quite different and while the two may appear similar they may photograph differently.

(2) The photographic contrast of color films is relatively high. With the dyes that can be used in a practical color process this is necessary to provide sufficient color saturation. The contrast, though high, is not beyond the acceptable limit for making color originals. However, when the high-contrast original is reproduced on a second high-contrast material, the contrast is increased and generally, then, exceeds the acceptable limit.

(3) Color originals are seldom judged by comparing them with the subject. The quality of duplicates is almost always judged on the basis of side-by-side inspection or projection with the original color photograph.

(4) The dyes that can be used in color processes are not perfect. Their hues and relative brightnesses are often far from ideal, though close enough to provide pleasing color originals. When a reproduction is made the optical characteristics of the dyes cause the exaggeration of the relative brightnesses of some colors and shifts in the hues of others, often to the extent that the reproduction is not acceptable.

In order to use films of the Kodachrome type in the production of theater release prints it would be necessary to make a first reproduction to work in the special effects. The release prints would then be reproductions of reproductions and the deficiencies would be still more serious.

Masking is a technique, used in making color reproductions, either to correct fully or to minimize the reproduction deficiencies resulting from the high photographic contrast or the optical characteristics of the dyes in the original photograph or both.

A *mask* is an auxiliary image (generally a photographic image) used in register with an original color picture to modify the characteristics of the original for purposes of reproduction. Depending on the nature of the mask it may be used to modify contrast, to change the relative brightnesses of some colors, or to shift the hue or saturation of some colors.

Masks are generally made by contact-printing the original color photograph onto a light-sensitive film, processing the film, and then visually or mechanically registering the color original and mask. The reproduction is made from this combination.

Masks may be negative (if made from a positive), or positive (if made from a negative). Masks may be black and white (as in the case of silver masks) or colored (as in the case of single or multi-colored-dye masks).

The combination of mask characteristics and the contrast required depends upon the nature of the original, the character of the unmasked reproduction, and the modifications desired. Since it is the purpose of this paper to describe masking for the improvement of color reproductions, the following will be considered in order:

(1) the characteristics of the color original;
(2) the characteristics of the unmasked color reproduction;
(3) the modifications of the color original which are necessary if the reproduction is to match the original.

Reproductions of color transparencies may be made on films incorporating dye systems which are either the same or different from the dye system of the original film. To differentiate between these two techniques it is convenient to refer to the former as *duplicating* and to the latter as *copying*.

In descriptions of color reproductions made on the same type of films as the original photographs the technique will be referred to as *duplicating* and the result, a *duplicate*. An example is the printing of a Kodachrome original onto a Kodachrome film. When the reproduction is made by using a different process from the original, the technique will be referred to as *copying* and the result a *copy*. The masking problem differs for these two reproduction techniques.

II. The Color Original

Since Kodachrome film is used for making more original color pictures than any other process, the characteristics of such originals will be discussed first. Other processes are essentially the same and the specific and important differences will be noted later. There are several different Kodachrome films with different dye characteristics; so the examples which follow are not to be regarded as representing any specific Kodachrome film, but simply as representative of a re-

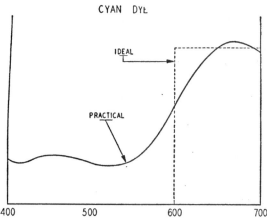

Fig. 1—Spectrophotometric curves of ideal and practical cyan dyes.

versal color process where the couplers are added in the processing rather than incorporated in the film at the time of manufacture. Of the two reversal processes, Kodachrome is an example of the former type and Ektachrome is an example of the coupler-included type.

Tungsten lamps used in projectors and illuminators radiate energy at all wavelengths and when viewed produce a sensation of white light. When the spectral distribution of energy of this light source is altered by the selective absorption of a dye, the sensation of white is no longer produced and the light appears to be colored.

Both Kodachrome and Ektachrome pictures consist of three superimposed dye images, one cyan, one magenta, and one yellow. The dye images in combination absorb the complementary colors of those seen in the color picture. The cyan dye absorbs its complementary, red, the magenta dye absorbs green, and the yellow dye absorbs blue. Absorption characteristics of dyes are shown by spectrophotometric

MAGENTA DYE

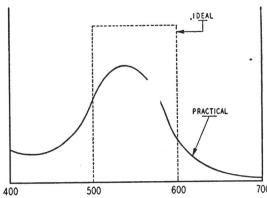

Fig. 2—Spectrophotometric curves of ideal and practical magenta dyes.

YELLOW DYE

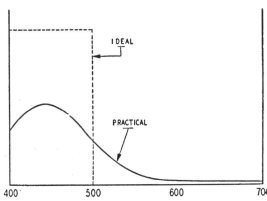

Fig. 3—Spectrophotometric curves of ideal and practical yellow dyes.

curves, the extent of the absorption being indicated by the height of the curve at each wavelength.

An ideal cyan dye would absorb red only as indicated by the dotted line in Fig. 1. The characteristic of the Kodachrome-type cyan, which absorbs an appreciable amount of green and blue, is shown by the solid line. Since the green and blue absorptions are about equal, the dye by itself appears relatively darker than the ideal cyan indicated.

An ideal magenta dye would absorb green only, transmitting all of the red and blue (Fig. 2). The Kodachrome-type magenta does not absorb all of the green and does absorb some blue and red. If the magenta absorbed blue and red equally, it would simply be darker than an ideal dye, but since it absorbs much more blue than red, the hue is different from the ideal magenta—it is shifted toward red.

An ideal yellow would absorb blue only (Fig. 3), transmitting all the green and red. The Kodachrome-type yellow does not absorb all blue, does absorb some green, and a very slight amount of red.

To simplify the following descriptions, schematic spectrophotometric curves will be used to illustrate dye characteristics. These are shown in Fig. 4 (a), (b), and (c). The red absorptions of the magenta and yellow dyes are neglected because they are unimportant in the problem of reproduction as summarized in Table I.

Table I

Kodachrome-Type Dye Characteristics (Schematic)

| | Gamma Values* | | |
Dyes	Blue	Green	Red
Cyan	0.45	0.45	1.5
Magenta	0.45	0.90	...
Yellow	0.60	0.15	...

* Gamma values of cyan-, magenta-, and yellow-dye images resulting from exposure to white light and measurement with blue, green, and red light. These values and those given in subsequent tables are measured with respect to the original scene.

Combined in equal quantities the Kodachrome-type dyes described form a neutral scale because the sums of the absorptions of red, green, and blue light by the three dyes are equal. The combined dye diagrams (Fig. 5) show the contributions of each dye to the neutral.

III. The Unmasked Duplicate

By means of a single, simple formula it is possible to calculate the reproductions of the principal colors and their neutral scale in the duplicate. The principal colors referred to are the pure cyan, magenta, and yellow dyes of the original and their combinations in pairs to form red, green, and blue. For color duplicating, as for all other photographic reproductions, the contrast (or gamma) of the original multiplied by the gamma of the duplicating film gives the contrast of the duplicate picture.

$$\gamma(\text{original}) \times \gamma(\text{duplicating film}) = \gamma(\text{duplicate})$$

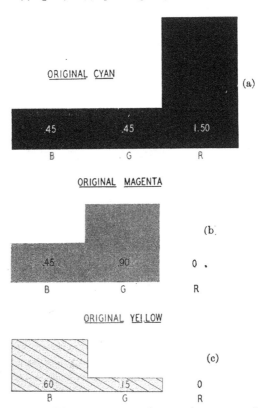

Fig. 4—Schematic spectrophotometric curves of (a) cyan, (b) magenta, and (c) yellow dyes.

When an original of this type is printed onto a film of the same kind, all the blue-light contrast is recorded by the blue-sensitive layer of the duplicating film and reproduced as yellow dye with a gamma of 0.6. All green-light contrast is reproduced in the green-sensitive

layer as magenta dye (gamma 0.9), and all red-light contrast is reproduced in the red-sensitive layer as cyan dye (gamma 1.5).

The reproduction of the original cyan—Applying the formula above

gamma of original = 1.5 (cyan-dye gamma to red light)

gamma of duplicating film = 1.5 (cyan-dye gamma to red light)

gamma of duplicate = 2.25 (cyan-duplicate gamma to red light).

Since the dye in the duplicate has the same chemical structure as
the original cyan, its blue, green, and red absorptions have the same
relationship as those of the original. Therefore, the cyan in the duplicate with red-light contrast of 2.25 has green-light contrast of 0.68
and blue-light contrast of 0.68.

NEUTRAL

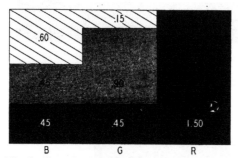

Fig. 5—Contributions of cyan, magenta, and yellow dyes to neutral scale.

Because the cyan dye in the original absorbs green (gamma = 0.45)
duplication of a cyan image will result in a magenta image coincident
with the cyan duplicate. The green-light contrast of the original
cyan is 0.45. This is reproduced as magenta at a gamma of 0.9, hence
the magenta addition has a green-light contrast of 0.45 × 0.9 = 0.41.
The magenta given in the example also absorbs blue and the blue contrast is 50 per cent of the green contrast. The blue-light contrast of
this magenta addition is thus 0.21.

The original cyan also absorbs blue (0.45), and this is reproduced
as yellow dye at a gamma of 0.60. There is, therefore, a yellow-dye
addition to the cyan image in the duplicate, the contrast of the yellow
addition being 0.45 × 0.60 = 0.27. This yellow dye absorbs some
green, 25 per cent of the blue absorption or 0.07.

The result of duplicating the cyan image of the original is thus:

Blue	Gammas to Green	Red	
0.68	0.68	2.25	cyan dye
0.21	0.41	0.0	magenta-dye addition
0.27	0.07	0.0	yellow-dye addition
1.16	1.16	2.25	

A comparison of the original cyan and the duplicate from it (Fig. 6) shows the duplicate to be higher in contrast and density than the

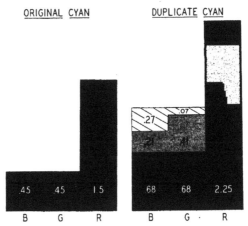

Fig. 6—Schematic spectrophotometric curves comparing an original cyan scale with the duplicate made from it.

original. It is darker but it has not changed hue because the green and blue absorptions are equal in the duplicate as they are in the original. While the high cyan contrast is due to the high contrast of the original and the duplicating film, the magenta- and yellow-dye additions are due to the green and blue contrast of the original cyan.

The reproduction of the original magenta

gamma of original = 0.9 (magenta-dye gamma to green light)

gamma of duplicating film = 0.9 (magenta-dye gamma to green light)

gamma of duplicate = 0.81 (magenta-duplicate gamma to green light).

In the example chosen the magenta dye has blue-light contrast equal to 50 per cent of its green-light contrast, and since the green-light gamma of the magenta dye in the duplicate is 0.81 the blue-light contrast is 0.40. These are the optical characteristics of the middle layer of the duplicate.

Because of the blue-light absorption of the magenta dye in the original a yellow-dye image is added to the magenta duplicate. The blue-light contrast of the original magenta is 0.45 and this is reproduced as yellow dye at a gamma of 0.6, giving a yellow addition with a blue-light contrast of 0.27 and a green-light contrast of 25 per cent of this or 0.07.

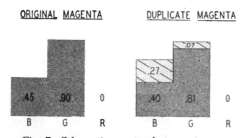

Fig. 7—Schematic spectrophotometric curves comparing an original magenta scale with the duplicate made from it.

The result of duplicating the original magenta image is thus:

	Gammas to		
Blue	Green	Red	
0.40	0.81	0.0	magenta dye
0.27	0.07	0.0	yellow-dye addition
0.67	0.88	0.0	

The blue absorption of the original magenta was 50 per cent of the green. This blue absorption resulted in the addition of yellow to the magenta areas in the duplicate raising the blue absorption to over 70 per cent of the green (Fig. 7). This shifts the hue of the magenta toward the red, meaning that original magenta areas are more red in the duplicate. The green-light contrast is lower in the duplicate than in the original. This indicates a loss in the saturation of magenta areas in addition to the hue shift.

The reproduction of the original yellow

gamma of original = 0.6 (yellow-dye gamma to blue light)
gamma of duplicating film = 0.6 (yellow-dye gamma to blue light)
gamma of duplicate = 0.36 (yellow-duplicate gamma to blue light).

The original yellow is relatively very low contrast. When duplicated the yellow has a gamma of only 0.36 to blue light and 25 per cent
as much or 0.09 to green light.

Since the original yellow dye absorbs some green light, magenta
dye is added to the yellow areas in the duplicate. The green-light
contrast of the original yellow is 0.15, and it is reproduced as magenta
with a green-light gamma of 0.9 making a magenta image with green-
light contrast of 0.14. This additional magenta absorbs blue with a
contrast of 0.07 (50 per cent of 0.14).

Results of duplicating the yellow are thus:

Blue	Gammas to Green	Red	
0.36	0.09	0.0	yellow dye
0.07	0.14	0.0	magenta-dye addition
0.43	0.23	0.0	

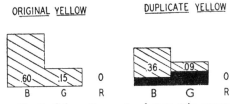

ORIGINAL YELLOW   DUPLICATE YELLOW

Fig. 8—Schematic spectrophotometric curves
comparing an original yellow scale with the
duplicate made from it.

A comparison of the yellow original and duplicate (Fig. 8) shows the
contrast, hence the color saturation, of the duplicate considerably
lower. Actually, it is so low that there is almost no yellow in the
duplicate. Whereas the original yellow exhibited green absorption of
only 25 per cent of the blue, the duplicated yellow has a green absorption of over 50 per cent of the blue. This causes a shift in hue of the
yellow toward orange.

Red in an original transparency is produced by the combination of
magenta and yellow dyes. Therefore, the reproduction of a red area
can be calculated by adding the characteristics of the magenta and
yellow reproductions. The same procedure can be applied to determine the nature of a duplicate green (cyan plus yellow) and a duplicate blue (cyan plus magenta).

Red in Original				Red in Duplicate			
B	G	R		B	G	R	
0.45	0.90	0.0	magenta	0.67	0.88	0.0	from magenta
0.60	0.15	0.0	yellow	0.43	0.23	0.0	from yellow
1.05	1.05	0.0		1.10	1.11	0.0	

Green in Original				Green in Duplicate			
B	G	R		B	G	R	
0.45	0.45	1.5	cyan	1.16	1.16	2.25	from cyan
0.60	0.15	0.0	yellow	0.43	0.23	0.0	from yellow
1.05	0.60	1.5		1.59	1.39	2.25	

Blue in Original				Blue in Duplicate			
B	G	R		B	G	R	
0.45	0.45	1.5	cyan	1.16	1.16	2.25	from cyan
0.45	0.90	0.0	magenta	0.67	0.88	0.0	from magenta
0.90	1.35	1.5		1.83	2.04	2.25	

From all of the above calculations the effect of duplication on the principal colors can be tabulated as follows:

cyan—much darker
magenta—slightly desaturated, hue shifted toward red
yellow—greatly desaturated, hue shifted toward red
red—duplicate closely matches original
green—much darker, slight shift in hue toward blue
blue—much darker, shift in hue toward green

Since the original neutral scale is a combination of all three dyes, its reproduction is the sum of the reproductions of cyan, magenta, and yellow.

	Gammas to		
Blue	Green	Red	
1.16	1.16	2.25	(duplicate from orginal cyan)
0.67	0.88	0.0	(duplicate from original magenta)
0.43	0.23	0.0	(duplicate from original yellow)
2.26	2.27	2.25	

This reproduction is neutral and its contrast is 2.25. This figure checks with that obtained in another way of calculating the neutral gamma. The original contrast is 1.5 and the duplicating film has contrast equal to 1.5; then the duplicate has contrast 2.25.

In summarizing the characteristics of the unmasked reproduction, it is important to repeat that the reproduction deficiencies are produced by the optical characteristics of the dyes in the original color transparency. In understanding color-reproduction problems, it is important to remember that these are characteristics of the dyes themselves and are not the results of improper film manufacturing or processing.

It is difficult to predict whether or not original color pictures would appear better than those afforded by present processes if "ideal" dyes were available. The real advantage of the "ideal" dyes would be realized in duplicating as there would be no "additions" to the duplicate of each layer as now encountered because dyes absorb in regions other than those for which they are primarily intended.

IV. Effect of Masking

It is impossible to single out one of the four undesirable dye absorptions described as the most serious of all, for each unwanted absorption is effective twice: once directly and once indirectly. The direct effects, additions of unwanted dye to each image, have been described. In addition, the presence of an unwanted absorption in one dye means that the dye primarily intended to absorb that color must be lower in contrast to maintain a satisfactory neutral scale. The color of the primary absorber in this relationship is thus desaturated in the original picture and even to a greater extent in the duplicate. The yellow dye of the example illustrates this effect. Since the cyan and magenta each absorb an appreciable amount of blue, the yellow must be quite low in contrast.

In a reproduction of a skyscape or a seascape, the direct effects of the blue and green absorptions of the cyan would certainly spoil the duplicate. In a reproduction of a yellow dress or flowers, the green absorption of the yellow dye would be the only direct source of trouble, though the indirect effect of the cyan-dye characteristics on the yellow saturation must not be forgotten as they may be just as important.

In transparencies containing all colors the most objectionable reproduction errors are frequently the excessive darkening of cyans, blues, and greens caused by the characteristics of the original cyan dye.

The green and blue absorptions of the original cyan of the example have gammas of 0.45 which are positive since the image is positive. A negative silver mask of the cyan layer alone, developed to a contrast of 0.45 and bound with the original cyan, would subtract 0.45 from the contrast of the cyan to blue, green, and red light as follows:

Blue	Gammas to Green	Red	
0.45	0.45	1.5	unmasked cyan-dye image
−0.45	−0.45	−0.45	mask image
0.0	0.0	1.05	

In practice such a mask can be made easily by printing the Koda-chrome or similar color transparency onto black-and-white film using *red* light. The cyan dye is the only dye that absorbs red, and thus the black-and-white picture is a record of the cyan layer only. The red-light contrast of the cyan dye in the original is 1.5, and the contrast necessary in the mask for canceling the contrast to green and blue is 0.45. Therefore, 1.5 × gamma to which the mask must be developed = 0.45. The mask development is thus = 0.3.

This is a *red-light mask*. It is a 30 per cent mask as far as the red-light contrast of the cyan is concerned but a whole or 100 per cent mask as far as the blue and green contrasts are concerned. The mask, of course, covers the neutral scale as well as the picture, reducing the neutral-scale contrast from 1.5 to 1.05.

While the red-light mask cancels the blue- and green-light contrast of the original cyan, lowers the red-light contrast of the cyan, and lowers the neutral-scale contrast, it does not alter the characteristics of the magenta or yellow images. Magenta and yellow areas in the original transmit all the red light used to expose the mask, and the contrasts of these two layers are unchanged by the mask. Only the density of the magenta and yellow areas is increased.

With the mask in register with the original transparency, the combi-nation is characterized as follows:

	Blue	Gammas to Green	Red
cyan dye plus mask	0.0	0.0	1.05
magenta dye	0.45	0.90	. . .
yellow dye	0.60	0.15	. . .
	1.05	1.05	1.05

The Duplicate from the Masked Original

The red-light contrast of the cyan dye with the mask over it is 1.05. The gamma of the duplicating cyan to red light is 1.50. Thus the duplicate from the masked cyan has red-light gamma = 1.575 or ap-proximately 1.50. As in the original, the cyan dye in the duplicate absorbs blue and green light with contrasts of 0.45 each.

There are *no additions of magenta or yellow* since the masked cyan has no green or blue contrast. The reproduction of the masked cyan, therefore, almost exactly matches the original cyan. The magenta and yellow reproduce exactly as they did when no mask was used, for the mask does not influence the contrast of either of these original images.

When the duplicates from the unmasked and masked cyans are compared, the duplicate from the masked cyan shows improved brightness but no change in hue (Fig. 9). Since blue and green in the original contain much cyan, the red-light mask results in improved

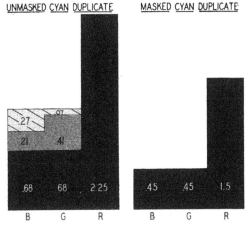

UNMASKED CYAN DUPLICATE   MASKED CYAN DUPLICATE

Fig. 9—Schematic spectrophotometric curves comparing duplicates from unmasked and masked cyan scales.

brightness notably in cyans but also in blue and green reproductions. In practice the use of a red-light mask also improves the reds. Without a mask, since the cyan is darker in the reproduction, the tendency is to print the duplicate somewhat lighter over-all to prevent excessively dark skies, grass, and other combinations of which cyan is a part. Then the reds are too light. When the mask is used and the duplicate properly printed, the reds reproduce relatively darker than in the unmasked duplicate.

The red-light mask does not compensate for any hue shifts encountered during duplicating. It only improves the relative brightnesses of cyans, blues, and greens compared with reds, yellows, etc., and is often referred to as a *relative-brightness mask*.

The masked neutral scale has a contrast of 1.05 and the duplicating film has a gamma of 1.5; so the contrast of the neutral scale in the reproduction is approximately 1.5, the same as in the original.

V. FEATURES OF THE SINGLE SILVER MASK

The use of a single silver mask as described above represents the most practical masking technique in duplicating, and recognition of the following features of the single silver mask is important to its successful use.

(1) With a single silver mask in register with a color transparency for duplication, the neutral scale of the transparency remains neutral but is lower in contrast.. The color balance of the original is not changed but the *brightness* of some colors with respect to other colors is changed.

(2) When colored light is used to expose the mask, the mask is dense in areas corresponding to the parts of the transparency transmitting that color, and the mask is transparent in areas corresponding to parts of the transparency absorbing the exposing light. In the masked transparency and in the duplicate made from it, colors transmitting the exposing light are made relatively darker and colors containing the complement of the exposing light are made relatively brighter. The use of red light for exposing the mask brightens the cyans; thus blues and greens appear brighter because they contain much cyan. The use of green light to expose the mask would lighten the magenta, and the use of yellow (red and green) light would lighten the blue by lightening both the cyan and magenta.

(3) No hue shifts are corrected by a single neutral mask. To correct for a hue shift it would be necessary to use the mask for exposing only one layer of the duplicating film. This would change the contrast of one component of the neutral scale, and it would then cease to be neutral. To restore the scale to a neutral balance would require a change in the contrast of one layer of the duplicating film. This cannot be done with the present materials and process.

VI. REQUIREMENTS FOR COMPLETE CORRECTION IN DUPLICATING

In order to compensate for all of the optical characteristics of the Kodachrome-type dyes that lead to duplicating errors it would be necessary to print the duplicate by means of three successive exposures of red, green, and blue light. Four separate masks would be required for the nullification of the four undesirable dye absorptions as follows:

(1) Mask of the cyan layer with contrast 0.45 which is capable of compensating for the green-light contrast of the cyan image. This can be made with red light.

(2) Mask of the cyan layer with contrast 0.45 which is capable of compensating for the blue-light contrast of the cyan image. This can be made with red light, and in the example can be the same mask as (1) because the blue-light contrast is the same as the green-light contrast.

(3) Mask of the magenta layer with contrast of 0.45. This can be used to compensate for the blue-light contrast of the magenta image. This mask can be made with green light from the transparency with the mask (1) in place. Mask (1) will prevent the green-light contrast of the cyan dye from influencing the green-light mask. While the green-light contrast of the yellow dye will interfere somewhat, the extent is insufficient to cause any significant trouble.

(4) Mask of the yellow layer with contrast 0.15. This can be used to compensate for the green-light contrast of the yellow dye. Such a mask should be exposed with blue light with masks (2) and (3) in place to prevent the blue-light contrast of the cyan and magenta dyes from influencing the mask being made with blue light.

When making the red-light duplicate printing exposure no mask is required since only the cyan dye absorbs red, and there are no dye additions in the duplicate due to absorptions in other layers. When making the green exposure, masks (1) and (4) should be registered with the transparency and for the blue exposure masks (2) and (3).

The net effect of these above masks leaves the neutral scale with

$$\gamma_B = 0.60 \qquad \gamma_G = 0.90 \qquad \gamma_R = 1.50$$

and the scale is no longer neutral. To make it neutral the processing contrast of both the yellow and magenta layers in the duplicate would need to be increased to bring them to the same value as the cyan image or still further masks would be required to neutralize the neutral scale. Full correction by these methods is obviously impractical.

In a summary of masking for duplicating the transparency onto the same material it can be stated that

(1) A single silver mask provides relative brightness correction, and it is a relatively simple technique.

(2) No hue-shift correction can be accomplished unless the gammas of the film layers are changed. This cannot be done in current processing procedures.

(3) Complete correction for unwanted dye absorptions is hypothetically possible but generally impractical, uneconomical and, in fact, usually unnecessary.

VII. THE PROBLEM OF COPYING

The problem of copying is somewhat more complicated than that of duplicating. In making a duplicate the principal objective is to reproduce the dye concentrations, and when this is accomplished the duplicate will match the original transparency.

An example of copying is the reproduction of Kodachrome-type transparencies on Ektachrome-type film, and the complications which arise become apparent on comparison of the dye systems of the Kodachrome and Ektachrome processes.

The dye system of Ektachrome-type film can be approximately described by Table II.

TABLE II

EKTACHROME-TYPE DYE CHARACTERISTICS

Dyes	Blue	Gammas to Green	·Red
Cyan	0.20	0.45	1.5
Magenta	0.45	0.85	...
Yellow	0.85	0.20	....

A red-light mask in combination with the Kodachrome-type original described above nullifies the green- and blue-light contrasts of the original cyan. The original cyan, therefore, copies on film of the Ektachrome-type as cyan only, no magenta or yellow additions; but the copy cyan has different optical characteristics from the original, and the two will not match visually.

Since the magenta and yellow dyes of the original and copy film also differ in their optical characteristics, any suitable masking technique becomes a matter of multiple masks calculated by much more detailed computations than are within the province of this paper.

Making separation negatives and dye-transfer prints from either type original is a third example of copying. In this case it is somewhat more practical to attempt complete correction of the unwanted absorptions: Multiple masks can be used and the contrast of each color separation controlled by the development time of the individual negatives.

By using the masks suggested above for the complete correction of the unwanted dye absorptions, separation negatives can be made from Kodachrome-type transparencies. Each negative is then a record of only one of the three images of the transparency. Since the blue-, green-, and red-light contrasts of the neutral scale are reduced to 0.60, 0.90, and 1.00, respectively, the separation negative development is adjusted to bring the contrasts of all negatives to the same value. Assuming that a gamma of unity is required, the negatives could be developed to gammas of 1.66, 1.11, and 1.0, respectively.

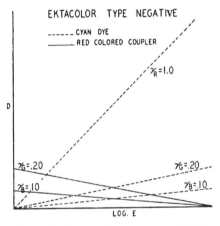

Fig. 10—Characteristic curves of Ekta-color-type cyan-dye scale (broken lines) and remaining red-colored cyan coupler (solid lines).

Neutral-Scale Gammas in Masked Kodachrome Original		Separation Negative Development Gamma		Contrasts of Separation Negative
B	0.60	×	1.66	=	1.0
G	0.90	×	1.11	=	1.0
R	1.00	×	1.00	=	1.0

VIII. COLORED COUPLER MASKS

During 1947 the Eastman Kodak Company announced a new color-negative sheet film taking material, called Ektacolor. Color prints can be made from Ektacolor negatives by the dye-transfer process through the use of three separation positive reliefs. Unique in the Ektacolor film is the automatic formation, during processing, of colored masks within two of the film's layers, to compensate for un-wanted absorptions of the cyan and magenta dyes. This represents the most recent advance in the science of masking. While Ektacolor, as announced, will be supplied only as a portrait film, the possibility of the basic principle's being extended to other color materials merits its description here.

Ektacolor is a three-layer film with the emulsions sensitive to blue, green, and red. Incorporated in the respective layers are yellow-, magenta-, and cyan-forming couplers. In the first film-processing

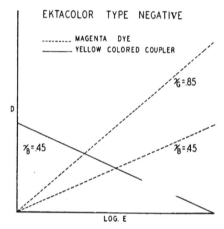

Fig. 11—Characteristic curves of Ektacolor-type magenta-dye scale (broken lines) and re-maining yellow-colored magenta coupler (solid lines).

step three negative silver images are simultaneously developed in the layers. The developer is oxidized in proportion to the silver de-veloped, and the developer reaction products combine with the couplers to produce negative dye images.

After the silver and unused silver halide are removed from the film, only the three negative dye images and the unused coupler remain. The coupler that produces cyan dye is reddish in color and the coupler that produces magenta is colored yellow. Since the original color of the couplers is destroyed as the new image dyes are formed, the remaining couplers comprise positive colored images. These are the masks.

The dye images formed from colored couplers in the Ektacolor-type film exhibit optical characteristics similar to those formed from colorless couplers in Ektachrome-type films. For photographic reasons, however, the over-all contrast of the Ektacolor-type is kept lower than the Ektachrome-type.

If the colorless coupler-dye gammas were adjusted, for example, as follows, a neutral scale with a gamma of 1.0 would result. The table, therefore, indicates the magnitudes of the corrections which must be accomplished by the colored coupler masks.

Dyes	Blue	Gammas to Green	Red
cyan	0.10	0.20	1.00
magenta	0.25	0.70	...
yellow	0.65	0.10	...
	1.00	1.00	1.00

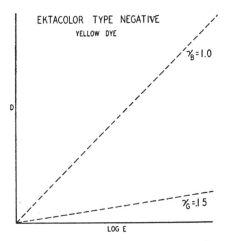

Fig. 12—Characteristic curves of Ektacolor-type yellow-dye scale.

The reddish-colored positive image formed by the unused cyan coupler cancels the negative blue- and green-light contrast of the cyan dye, and the yellow-colored positive image formed by the unused magenta coupler cancels the negative blue-light contrast of the magenta dye. At present there are no entirely suitable magenta-colored yellow couplers so the green-light contrast of the yellow dye is not masked.

Cancellation of the unwanted absorption values in the cyan and magenta layers will leave the total red-, green-, and blue-light absorptions unbalanced. The manufacturing and film processing are, therefore,. adjusted to give a balanced neutral scale when the unwanted cyan and magenta absorptions are masked. With the increased contrast in two of the layers, the dye absorptions in the present example are, therefore, approximately:

Dyes	Blue	Gammas to Green	Red
cyan	0.10	0.20	1.0
magenta	0.30	0.85	...
yellow	1.00	0.15	

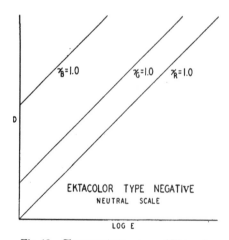

Fig. 13—Characteristic curves of Ektacolor-type neutral scale measured by blue, green, and red light.

In order to be completely effective as a mask for the cyan layer, the remaining reddish coupler must have blue-light contrast of 0.10 and green-light contrast of 0.20 with maximum densities to these two colors equal to the maximum densities of the respective absorptions of the cyan dye. These characteristics are provided and controlled during the making of the couplers (Fig. 10).

To be effective as a mask of the magenta layer the remaining yellow coupler must have a gamma of 0.30 to blue light and a maximum density to blue equal to the maximum density of the magenta dye to blue light (Fig. 11).

There is no correction in the yellow layer (Fig. 12).

With the corrections afforded through the use of colored couplers, the color negative closely resembles the ideal dye image characteristics (Fig. 13). Exceptions are the very slightly low magenta contrast and the equally slight green absorption of the yellow dye. From such a corrected negative separation, positives and color prints can be made substantially free of errors due to the unwanted dye absorptions.

IX. Conclusion

Fortunately it is easier to make a mask and use it in duplicating and copying (except for motion pictures) than it is to understand the exact reasons for its use. The manufacturers of color films issue instructions for masking in some detail, and these generally provide the best possible practical method for obtaining adequate color reproduction. It is important, though, for the color technician to understand the processes with which he is working, and to know their possibilities and limitations; only then can he make the best use of the instructions and materials which he has at his command.

Acknowledgment

The presentation of this paper at the Convention was illustrated with 73 color slides and 14 large color transparencies. The author acknowledges, with thanks, the contributions of those from various departments of the Eastman Kodak Company who provided data and laboratory assistance to make the paper and illustrations possible.

Foreword

The following three papers were presented at the 63rd Semiannual Convention of the Society of Motion Picture Engineers at Hollywood in May, 1948, at a joint meeting with the Inter-Society Color Council. Two other papers presented at the same sessions were most instructive, but consisted of physical demonstrations and color illustrations to such an extent that reproduction in the JOURNAL is impractical.*

To many of our readers the Inter-Society Color Council, which so generously prepared the program for the joint meeting, needs no introduction. However, for the benefit of others not acquainted with the ISCC, it may be noted that it is not just another technical society. Its members are mainly technical societies or associations having color problems, such as the American Pharmaceutical Association, Illuminating Engineering Society, and eleven others. The SMPE is a member, being represented by a committee of which Ralph Evans is chairman.

The purpose of the ISCC is "to stimulate and co-ordinate the work being done by various societies . . . leading to standardization, description, and specification of color, and to promote the practical application of these results . . . ." In accordance with this purpose, the SMPE will make the three following papers available.

C. R. KEITH

* "Color Phenomena," by Isay Balinkin, and "Seeing Light and Color," by Ralph M. Evans.

Spectral Characteristics of Light Sources*

By NORMAN MACBETH

Macbeth Corporation, New York 11, New York

AND

DOROTHY NICKERSON

United States Department of Agriculture, Washington, D. C.

Summary—New light sources are being developed regularly. No longer can industry depend upon carbon-arc lamps and incandescent lamps alone. A brief description of all the important light sources will be made with special emphasis on their spectral characteristics and their effect on colored objects.

Introduction

EFFICIENCY AND SIZE of light sources have an important bearing on the selection of lamps for motion picture photography, yet more and more, where color effects are desired, their color and spectral characteristics become important. Natural daylight, the carbon-arc, and incandescent tungsten-filament lamps provide most of the illumination today for the motion picture industry. While they remain the most important sources, there have been other developments during the past ten years that for one purpose or another have a special bearing or interest for the motion picture engineer. It is our purpose in this report to assemble as much typical information as is available on the spectral characteristics of both the standard sources and on the more newly developed fluorescent, mercury-cadmium, and concentrated zirconium-arc sources.

First, however, attention should be called to the fact that the shorthand method of specifying the color of illuminants in terms of color temperature is a practice that often obscures the differences between color and spectral composition. For that reason it seems important to indicate that the Planckian locus represents the locus of chromaticities of a black-body radiator at various temperatures, Fig. 1, and that it is only necessary—according to American practice—

---

* Presented May 20, 1948, at the SMPE Convention in Santa Monica.

in order to apply the term color temperature to a source, that it match the color of a black body at a given temperature. It is not necessary that it should match the black body in spectral distribution, Fig. 2.

The specification of the color of sources in terms of color temperature without an understanding of the limited meaning of the term has caused much confusion in color thinking as it concerns the illuminant.

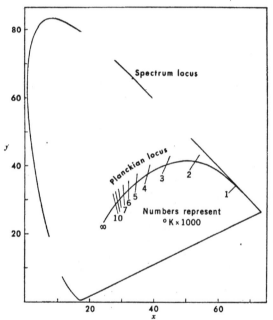

Fig. 1—The Planckian locus on a chromaticity diagram consists of points that represent chromaticities of a black-body radiator at various temperatures.

It should therefore be pointed out that the Optical Society of America, the American Psychological Association, and the Illuminating Engineering Society definitions[1] of color temperature refer to chromaticity only.* The British, on the other hand, in a definition[1] proposed by

* A note attached to the IES definition states that the term color temperature is "usually assignable only for sources which have a spectral distribution of energy not greatly different from that of a black body." In practice the IES does not hold to this restriction.

the Colour Group of the Physical Society, require that a source be of substantially the same spectral distribution in the visible region as a full radiator of the same color.   Thus, the Americans and British differ fundamentally in their concept of the term.

The meaning and use of the term color temperature as a short-cut method for describing the chromaticity of an illuminant should be

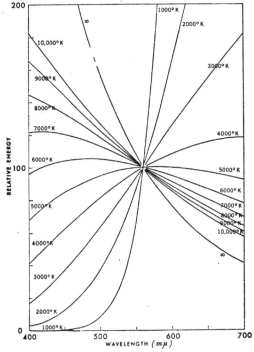

Fig. 2—Spectral-energy distribution in the visible region of the spectrum for a black body radiating at various temperatures.

clearly kept in mind for what it is.   For any real understanding of color processes, whether visual or photographic, it is necessary to take into consideration the more exacting specification of spectral distribution.   Thus, while illuminants in this report are often referred to in terms of the color-temperature scale, it should be remembered that it is not their color but only their spectral characteristics that will tell whether they are suitable for use with a given film, or to produce a specified result.

Another point that should be emphasized if we are to continue to use the term color temperature is that reciprocal temperature provides a better scale for expressing chromaticity differences than does temperature itself.   A chromaticity difference of 100 degrees Kelvin at 3000 degrees Kelvin is a very important color difference but at 6500 degrees Kelvin it is hardly significant.   In 1933 I. G. Priest[2] proposed that reciprocal temperature be adopted as the conventional parameter for specifying the chromaticity of incandescent illuminants and various

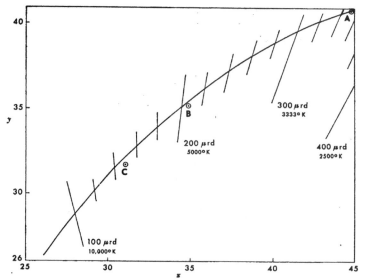

Fig. 3—ICI (x, y) diagram showing a portion of the Planckian locus in the range 2848 degrees Kelvin (A illuminant), through B and C illuminants to above 10,000 degrees Kelvin.   Intervals shown are 20 mireds (μrds) from 400 to 100, with equivalent color temperatures indicated.

phases of daylight and that the microreciprocal degree absolute centigrade (abbreviated "mired," or "μrd") be adopted as the most convenient unit.   There are several reasons for this proposal, all valid, as anyone will find who works with color difference specifications that relate to illuminants.[3]   On Fig. 3 the intervals from the color temperature of illuminants from 3000 to 10,000 degrees Kelvin are 20 mireds; compare this with the widely varying size of intervals when color temperature is used, as in the 1000-degree-Kelvin intervals on Fig. 1. This would be quite as evident on a uniform chromaticity diagram.

Regarding spectral distribution of light sources there are a number of reports that may be referred to for information of general interest in the motion picture field. The 1943 report of Linderman, Handley, and Rodgers[4] is well illustrated to show various lighting techniques. It discusses lamp requirements for color quality of light and for spectral distribution of film and includes a detailed listing and description of 22 types of carbon-arc and incandescent lamps used in motion picture studios. The SMPE Studio Lighting Committee Report of 1945[5] deals with operation and maintenance of carbon-arc and incandescent lamps, the section on incandescent lighting including a discussion of color temperature as one of the variables. A great deal of quantitative data for computational use in studying the color of illuminants and their effect on object colors is contained in the report on Quantitative Data and Methods for Colorimetry published in 1944 by the Committee on Colorimetry of the Optical Society of America.[6] Both this and the Linderman report contain a number of good references. Tables containing spectral radiancy and spectral-distribution data for computing ICI colorimetric values for black-body radiators at temperatures 2800 to 3800 degrees Kelvin for every 100 degrees Kelvin have been published by Adams and Forsythe,[7] but perhaps the most extensive and available tables of colorimetric computational data for Planckian radiators and a variety of actual sources, for use either by the weighted or selected ordinate method of calculation, are contained in the report of the OSA Committee on Colorimetry.[6] *

References devoted to specific illuminants are included in the appropriate section of the discussion that follows.

Since the authors have prepared this paper as part of a technical session on color arranged by the Inter-Society Color Council, there are other references that should be included at this point. Such references may not belong in a paper strictly devoted to a discussion of spectral characteristics of light sources, but the questions they raise

---

* Since the Santa Monica meeting at which this paper was given, a new book by Forsythe and Adams[8] has come to the authors' attention. While one might infer from the title that the book refers solely to fluorescent and other gaseous-discharge lamps, attention should be called to the fact that it includes a chapter on arcs of various types, one on light sources of short duration, one on delayed phosphorescence, and one on fluorescence and television. Any book on lamps or lighting by these authors is welcome, and to find one that brings the subject up to date in 1948 is particularly welcome, and should prove useful to any motion picture engineer interested in the spectral and other characteristics of current light sources.

must be considered before a motion picture or any other engineer can understand and control the direction and size of color change with change of illumination, and his attention should therefore be called to them. These questions involve studies of surround, adaptation, constancy—studies that fall more in the province of the psychologist than the physicist. Yet such questions must be studied in the future with quite as much care and in as much scientific detail as has been devoted in the past two decades to problems of color measurement and specification.

There is, for example, a book by Katz,[9] written in 1911 and translated from the German in 1935, that discusses various factors involved in studies of color constancy. More recently we have in this country the work of Helson, Judd, and Evans. The work of Helson and his students and associates at Bryn Mawr has resulted in a series of five papers already published,[10-14] one paper in press[15] and two Ph.D. dissertations to be published. The last two papers concern change in hue and saturation of aperture colors as a function of the composition and luminance of the surrounding field (R. V. Higbee) and the effect of general and local shadows on hue, lightness, and saturation (J. deB. Brugger). A study to find a formula for predicting the changes found in the Helson experiments has been made by Judd,[16] who expects to continue the work until a formula is found to fit all the requirements. Still more recently we have the new Evans book[17] which calls attention to many general illumination problems, just as he has been calling the attention of the SMPE and other groups to them during the past ten years in a series of well-illustrated lectures. Color needs to be seen if its effects are to be intelligently studied, and Ralph Evans' illustrated lectures have enabled more people to gain a picture of some of the many color problems not yet understood, certainly not yet answered, than has been possible by any other method of presentation. His book, which is well illustrated in color, should reach a wide audience. It is to be hoped that it may awaken many a young student of color to the possibilities that lie ahead in this field of color research.

While these few publications do not provide an answer to the entire problem, they do call attention to phases of it that are often omitted from consideration during technical color training. It is only when one comes up against practical problems that he finds there are these other matters that at present must be handled by experience or on a basis of trial and error. Actually, if results already published by Helson, Judd, and Evans, and in Holland by Bouma and Kruithof,[18,19] were more widely understood, a great many practical answers could be worked out today.   These references are therefore included so that the motion picture group may have its attention particularly called to this type of study.   They relate to color and illumination problems encountered every day in the motion picture industry.

Natural Illumination

Scientists have been trying for years to equal the color of natural illumination with reasonable efficiency. The spectral-energy distribution curves of daylight and sunlight are enough like those of a theoretical black-body radiator so that black-body-distribution curves are often used as standards for comparison.   For a discussion of

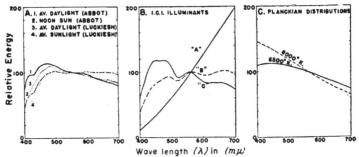

Fig. 4—Reference curves for energy distribution in the visible region for three types of standards used in studies of natural daylighting: (A) curves of natural sunlight and daylight; (B) ICI standard illuminants for colorimetric work; (C) black-body distributions in the 6500- to 8000-degree-Kelvin range.

spectral emittances of complete radiators, and a list of references to tables and charts specifying relative values for various temperatures, see[20] section 18.   There are certain exceptions to this resemblance, however, for absorption of energy in the earth's atmosphere, due chiefly to the presence of moisture, causes some differences between these curves as shown in measurements by Abbot,[21] and in proposals of Moon[22] for standard curves for engineering use.   As standards for colorimetric work the International Commission on Illumination in 1931 adopted three illuminants, known as ICI Illuminants[23,24] A, B, and C.   Illuminant A represents a tungsten-filament lamp that operates at a color temperature of 2848 degrees Kelvin with an energy distribution very like that of a black body; the other two are intended to represent two phases of natural daylight, B that of noon sunlight,

C that of average daylight. They are produced by using specified liquid filters with lamp *A*. For comparison these three types of standards used in studies of natural daylighting are shown in Fig. 4— those for natural daylight, for ICI standards, and for Planckian or black-body distributions in a range of color temperatures 6500 to 8000 degrees Kelvin.

It is difficult and time-consuming to make energy measurements of natural daylight; therefore, few data are available. Yet the color of

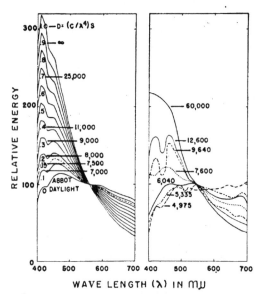

Fig. 5—Curves approximating spectral-energy distribution of daylight; On the left, as calculated by Gibson for a range of Abbot daylight to a limit blue sky; on the right, as measured by Taylor in 1939.

daylight may change through a wide gamut from sunrise to sunset, and from sunlight to blue sky, or to daylight provided by cloudy skies. K. S. Gibson of the National Bureau of Standards some years ago suggested a formula[25] by which the spectral energy of sun-outside-the-atmosphere (he used Abbot data) may be combined with the Rayleigh scattering equation to provide an approximate spectral-energy distribution for any given color temperature of skylight. Fig. 5 shows curves of this sort for color temperature of Abbot's daylight-outside-the-atmosphere (about 6000 degrees Kelvin), up to a limit blue sky (at ∝).   Paired with this diagram is one showing measurements made by Taylor and Kerr[26] in 1939 of daylight ranging from 4975 up to 60,000 degrees Kelvin.

Measured color temperatures of natural daylight are reported up to 60,000 degrees Kelvin.   On the other hand late-afternoon sunlight may measure as low as 2000 degrees Kelvin,[27] actually lower than the color temperature of most incandescent lamps which for ordinary use usually range from 2800 to 3000 degrees Kelvin (depending on size),

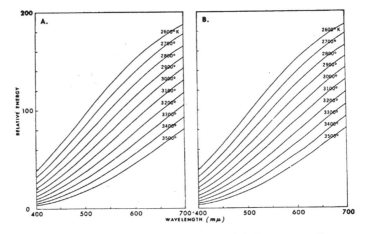

Fig. 6—Spectral-energy-distribution curves: (A) For tungsten filaments operating at equal voltage but different temperatures (from Linderman, Handley, Rodgers report;[4] (B) For Planck's formula, adjusted at 560 to the tungsten-filament values shown for (A). (The actual temperature of the tungsten filament is lower than that of a true black body when they are a color match.)

and for photographic work up to as much as 3400 degrees Kelvin for lamps of fairly short life.

Local atmospheric conditions may cause the color and intensity of natural light to vary considerably especially in cities where smoke and fumes may serve to filter the daylight, and result sometimes in major deviations from expected energy distributions of natural daylight.

Most manufacturers of color film have developed their emulsions so that they will be exposed either in a high color temperature of incandescent tungsten (3200 degrees Kelvin) or in an overcast daylight of about 6500 degrees Kelvin. For daylight film the best results are obtained when the daylight most closely resembles the color and spectral distribution for which the film was designed.* Practically no combination of cloud effects and daylight can produce the disastrous resultant color effects on color film that are produced by sharp, bright, spectral lines, such as those of the ordinary mercury lamp.

INCANDESCENT TUNGSTEN FILAMENT

The oldest sources, yet those most regularly used by motion picture engineers, are incandescent tungsten filaments and carbon arcs. Great progress has been made in both.

In the case of incandescent tungsten filaments, whose characteristics have been thoroughly described by Forsythe and Adams,[28] the early low-wattage lamps are now replaced by 5000- and 10,000-watt lamps that have increased both in color temperature and in lumen-per-watt efficiencies.

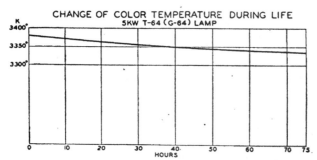

Fig. 7—Color-temperature characteristics of 5-kilowatt studio lamps during life applicable to lamps that are cleaned at least every 20 hours of burning by swirling tungsten cleaning powder about in the bulb. (From SMPE Studio Lighting Committee Report.[5])

This source is admirably suited for motion picture work because of dependability and low labor maintenance. Especially attractive is the close approximation of the spectral-energy distribution of the incandescent lamps to that of a black body at equivalent color temperature. Spectral-energy-distribution curves for tungsten filaments operating at equal wattage but different temperatures are shown in Fig. 6A. Paired with the distribution curves in Fig. 6A is a diagram, Fig. 6B, showing a series of distributions according to Planck's formula, adjusted at λ 560 to match the corresponding data for the

---

* Spectral sensitivities for various types of color film must be known before accurate colorimetric calculations can be made to select the illuminant of most suitable spectral distribution for use with a particular color film.

tungsten-filament lamps.   As may be seen, the relative distributions for incandescent tungsten filaments and for black-body radiators at equivalent color temperature are similar.   The actual temperature of the tungsten filament is lower than that of a true black body when they are a color match, about 100 degrees at 3000 degrees of color temperature.

The change in color temperature of lamps operated at constant voltage has been discussed by Judd[29] for certain lamps.   The change

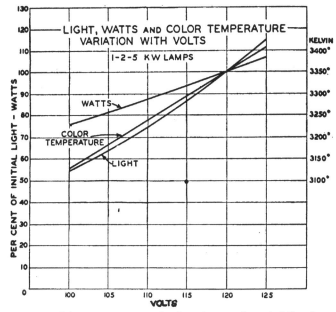

Fig. 8—Color temperature and other changes characteristic of gas-filled tungsten lamps when operated at other than rated voltage.   (From SMPE Studio Lighting Committee Report.[5])

in 5-kilowatt studio lamps, applicable to lamps cleaned every 20 hours of burning by swirling tungsten cleaning powder in the bulb, was reported by the SMPE Studio Lighting Committee in 1945.[5]   Fig. 7 is taken from their report.   Fig. 8, also taken from their report, indicates the color-temperature change that may be expected of lamps operated at other than rated voltage.

Lamps for black-and-white photography vary in color temperature from 2900 degrees Kelvin for 75-watt lamps to 3350 degrees Kelvin for 10,000-watt lamps and are used chiefly for other characteristics

than color temperature; for color photography the color temperatures
are important, and usually fall into two classes, either 3200 degrees
Kelvin intended for use with color films such as Eastman's Type B
or Ansco's Tungsten Type or 3200-degree-Kelvin film, or 3350-degree-
Kelvin lamps, often called "CP" lamps.

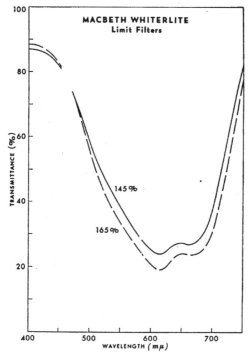

Fig. 9—Spectral transmissions of Macbeth Whiterlite filters used with high-temperature incandescent lamps to provide a daylight distribution suitable for use with Technicolor film. The two curves
represent tolerance limits.

Films such as Technicolor are made for daylight only. For such
films the incandescent lamp can still be used, usually the 3350-degree-
Kelvin lamp, in that case with a color-correcting filter suitable for
maintaining a spectral distribution of illumination that is required
for use with such film.* A film such as Technicolor requires some-

* If films were sensitive to the same wavelength distribution as the average human
eye, then daylight itself, or as good an artificial daylight as possible, should be used

what more transmission in the red than would be supplied by daylight filters designed for visual work. This excess in the red, while keeping the rest of the curve close to that of daylight, is supplied by Macbeth Whiterlite filters. Spectral transmissions of these filters are shown in Fig. 9. The two curves are those for tolerance filters supplied in

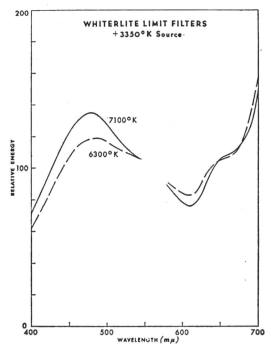

Fig. 10—Relative spectral-energy-distribution curves of 3350-degree-Kelvin tungsten-plus-Whiterlite limit filters. These distributions provide limit color temperatures of 6300 to 7100 degrees Kelvin.

for daylight film. The closest filtered approximation to daylight employs tungsten-filament lamps with Macbeth Daylite filters (Corning 5900).[3, 30-32] The thickness of a given filter controls the color temperature through a range from about 5000 degrees Kelvin to 8500 degrees Kelvin when used with lamps of required wattage. Such filters as these are supplied for visual tasks where carefully standardized artificial daylight is a requirement.[33, 34] Films, however, are not sensitive to the same distribution as the eye, and may require adjustments in transmission for certain portions of the spectrum in order to compensate for this difference. We note this diffence in order that there may be no confusion between the Daylite filters used for visual work and the Whiterlite filters supplied for use in color photography.

accordance with specification, graded according to transmission. One of the filters, 145 per cent (American Association of Railroads), produces a calculated color temperature of 7100 degrees Kelvin when used with a 3350-degree-Kelvin lamp.   The second filter, 165 per cent (American Association of Railroads), produces 6300 degrees

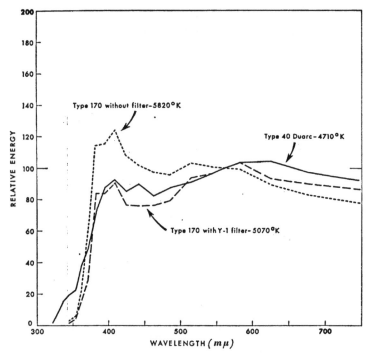

Fig. 11—Energy distribution of light from carbon-arc lamps used in studio lighting.   These curves represent the light emerging from the optical system: Type 40 Duarc at 4710 degrees Kelvin; Type 170 without filter at 5820 degrees Kelvin, with filter at 5070 degrees Kelvin.

Kelvin with the same lamp.   Relative-energy-distribution curves for a lamp-and-filter combination are shown in Fig. 10.

This combination of lamp and filter has the low labor maintenance factor of incandescent.   However, the efficiency in lumens per watt of the incandescent lamp is reduced nearly two thirds by filter absorption, which is a subtractive method of producing color.   The resulting energy distribution is, however, well suited for Technicolor. Filter holders designed for use with spot-type illuminators using incandescent sources have been available for some time.   A new development is a broad light fitted with filters.

CARBON ARC

Because the high-temperature carbon arc provides a great quantity of illumination from a single source and also provides a sufficiently close approximation to the spectral-energy distribution of sunlight, it is the most generally used source of illumination in the motion picture studios for color work.   It has the advantage of being a radiating source which, by use of various chemical combinations, can supply good spectral characteristics that resemble the continuous curve of a black-body radiator.   Color temperatures varying from 3500 to 3900 degrees Kelvin for low-intensity carbon arcs, and from 4500 to 6500 degrees Kelvin, high-intensity carbon arcs can be produced.

Spectral-energy-distribution curves of several types of carbon arcs are illustrated in the Linderman, Handley, and Rodgers report.[4] These are characteristic of the arc sources as directly viewed without the modification which occurs as a result of use with an optical system. In Figs. 11 and 12 the curves shown are characteristic of the radiation in the light beams as they emerge from the optical system.   Fig. 11 is a modification of Fig. 8 of a paper by Bowditch, Null, and Zavesky.[35] This shows the energy distribution of the most popular units employed in studio lighting, the Type 40 Duarc at 4710 degrees Kelvin which seems quite satisfactory either alone or mixed with sunlight for providing a proper balance for use with Technicolor, and the Type 170 lamp at 5820 degrees Kelvin, which when used with a filter gives 5070 degrees Kelvin, and produces a suitable spectral distribution.*

Fig. 12 shows the spectral-energy distribution of the light on the theater projection screen with two high- and one low- intensity projection arcs;   low-intensity carbon arc at 3870 degrees Kelvin,

* The fact that Technicolor film is adequately served by tungsten-plus-Whiterlite filters within color temperature limits of 6300 to 7100 degrees, and by carbon arc reduced from 5820 to 5070 degrees Kelvin by use of a filter, is a clear demonstration of the fact, discussed in the introduction to this paper, that color temperature may often prove to be a very unsatisfactory measure of the color characteristics of light sources, particularly when light sources depart substantially from the spectral distribution of a full radiator of the same color.   Experience shows that spectral distributions of either the tungsten-plus-Whiterlite filter or indicated filtered carbon arc provide satisfactory results with Technicolor, yet the color temperature of one averages 6700 degrees Kelvin, the other just over 5000 degrees Kelvin.

"Suprex" carbon arc at 5380 degrees Kelvin, and a 13.6-mm High-Intensity Projector carbon arc at 5600 degrees Kelvin.

When filters are used with carbon arcs, it has usually been supposed that they are necessary to take care of the peak energy at about 390 millimicrons considered characteristic of this source. However, as Bowditch points out, the 390-millimicron peak is apparently a char-

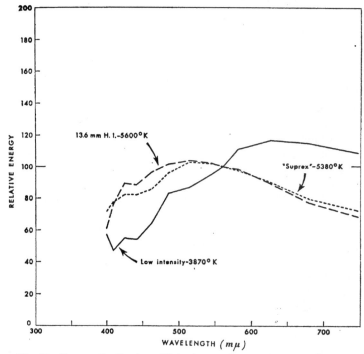

Fig. 12—Energy distribution of light from carbon-arc lamps used for projection. One low- and two high-intensity lamps are illustrated, low intensity at 3870 degrees Kelvin, Suprex at 5380 degrees Kelvin, and a 13.6-mm high-intensity at 5600 degrees Kelvin.

acteristic of the diffuse arc flame which is not effectively projected by the optical system. The filters are used to reduce the intensity in the far blue and near ultraviolet so as to provide a better balance with color processes designed for daylight. A gelatin-straw filter may be used, also a new glass coating which can be applied even to Fresnel lenses to supply color correction without the fading which characterizes the use of gelatin.

All of its advantages, which include high efficiency at temperatures much above those of tungsten, make the carbon arc a most important source in the motion picture industry. Its high brightness, plus its color and efficiency, makes it the standard source for theater projection.

Bowditch and his associates, who have discussed the spectral characteristics of carbon arcs before this Society at various times, are authorities in this field and their reports may be referred to for further information.[35-39]

FLOURESCENT SOURCES

The oldest of the new and radical deviations from standard light sources such as have been described is the flourescent lamp. It is ten years[8, 40] since this new source was announced. It offered, and still offers, exciting adventures in color through the use of fluorescent powders. Between 80 and 90 per cent of the total lumen output of a fluorescent lamp is derived from fluorescence; the balance is mercury light transmitted through the phosphors coated on the inside of the glass tube. To date, fluorescent lamps have been made available in a variety of colors—blue, green, pink, red, gold—but since our interest is primarily in sources that provide colors reasonably close to those of black-body radiators, this discussion will be confined to those fluorescent lamps that are called white or daylight in color and are usually assigned a color temperature by the manufacturers.

In Fig. 13 relative spectral-energy-distribution curves of three standard fluorescent lamps are shown: 3500 degrees Kelvin, 4500 degrees Kelvin white, and 6500 degrees Kelvin "daylight." If the spectral-energy distributions of these lamps are compared with natural daylight, or of black-body radiators at similar color temperatures, it will be found that these curves fall short in three respects. First, they are low in energy in the red end of the spectrum because, until recently, no phosphor has been available which could be activated sufficiently in the red region of the spectrum between 650 and 700 millimicrons. Second, they are low in the blue end because no blue phosphor of sufficient strength in the 400- to 420-millimicron region is available. Third, about 10 to 15 per cent of the radiation of these lamps is in sharp, bright mercury lines which distort color, color film in particular.

Recently, because of the very great importance to the photographic industry of correcting these deficiencies, particularly in the red, a new

phosphor has been developed for use in the red end.   This has been
studied by Froelich and described in a paper by Buck and Froelich.[41]
It is a double-activated calcium phosphor with a two-peak emission,
one at 650 millimicrons with 50 per cent intensity at 700 millimicrons,
another at 360 millimicrons.   The best initial efficiency obtained with
this phosphor in a 40-watt lamp was about 12.5 lumens per watt,

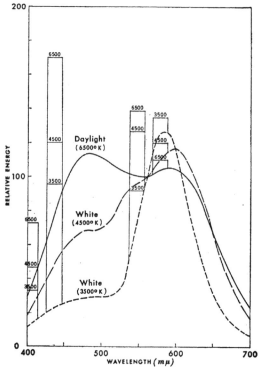

Fig. 13—Spectral-energy-distribution   curves   for
three standard fluorescent lamps: 3500 degrees Kelvin
white, 4500 degrees Kelvin white, and daylight 6500
degrees Kelvin.

which is low as compared with the 65 to 70 lumens per watt with the
phosphor for the red now used in the white lamps.   Unfortunately,
only one manufacturer has announced a fluorescent lamp employing
this new phosphor in lamps for general use, although others provide
it for restricted use, as in photocolor fluorescent lamp.   This is due to
the fact that while this red-corrected fluorescent lamp provides a much

closer approximation to the spectral-energy requirements for film and visual applications, the addition of this red phosphor lowers the over-all efficiency of the fluorescent tube enough so that manufacturers at the present time are convinced that maintaining present efficiencies is more important than the color improvement which they consider to be minor. There is no reason to believe that a new blue phosphor cannot be developed.

It may, however, take some time before phosphors of sufficient sensitivity are developed which will permit a heavy enough coating on the tube to absorb 100 per cent of the visible mercury radiation and still permit the ultraviolet radiation to excite the phosphors that lie on the glass surface, and thus produce sufficient visible radiation.

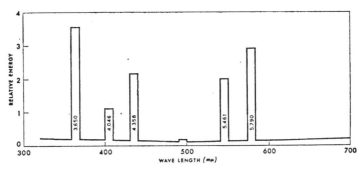

Fig. 14—Spectral-energy-distribution curve of 5000-watt "compact-source" mercury lamp. (Adapted from curve supplied by the British-Thomson-Houston Company.)

The size and large numbers of lamps needed to provide sufficient illumination for motion picture work, and the difficulty of applying optical controls, would continue to restrict their use for many purposes even if the spectral characteristics of these lamps were much improved.

MERCURY-CADMIUM LAMPS

For many years, lamp manufacturers have been developing higher-pressure mercury lamps. Research has, quite properly, been directed toward this end, since, especially for studio lighting, a high-brightness source is required and mercury lamps provide this feature.[42-44] Because of its spectral distribution, however, a straight mercury source has been poor even for black-and-white photography. The spectral-

energy distribution of a 5000-watt compact-source mercury lamp is
shown in Fig. 14.   The objectionable features include the bright mercury lines, and the great gaps in the visible region where the virtual
lack of energy makes certain sharp cutoff colors reproduce as black or
gray, while colors having a dominant wavelength on one of the mercury lines reproduce as overly bright.

As reported by Carlson[45] at a recent SMPE meeting, during the war
considerable work was done in England by the British-Thomson-
Houston Company on the combination of mercury and cadmium.   As
may be noted in Fig. 15, almost double the number of lines or bands
are produced by this combination as for mercury alone.   These lines

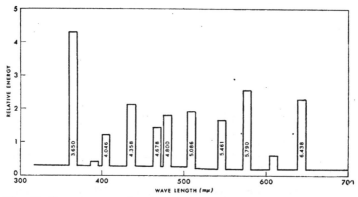

Fig. 15—Spectral-energy distribution of mercury-cadmium lamp. (Adapted from curve supplied by the British-Thomson-Houston Company.)

are well distributed over the visible region.   However, these lines or
bands still leave large gaps where little radiation is present, and thus
still produce unnatural effects on both colored objects and color film.

Although this source may prove good for black and white providing
a reasonable cost per unit can be established, for color photography
there seems to be a conflict of opinion.   Theoretically, unless new
metals can be found which will create new bands of approximately
equal brightness over the whole visible spectrum so as to simulate a
continuous black-body radiator, satisfactory results are not be be expected.

While visiting England last year, the senior author of this report
was told of questionable color results obtained by users who tried this
source for color photography.

CONCENTRATED ZIRCONIUM ARC

One of the latest lamp developments is the concentrated zirconium-arc lamp developed by the Western Union Telegraph Company and reported two years ago to the Optical Society of America and to the Society of Motion Picture Engineers by Buckingham and Deibert.[46,47]

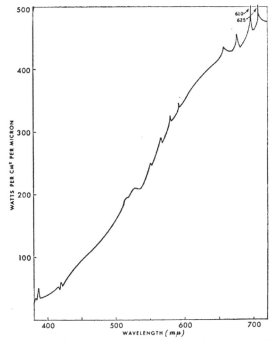

Fig. 16—Spectral-energy distribution of concentrated zirconium arc lamp. (Adapted from curves shown by Buckingham and Deibert.[46, 47])

Because the concentrated arc is a hot body, it is actually a continuous radiator over the entire visible spectrum. As may be noted from Fig. 16, its relative-energy distribution is smooth throughout most of the visible spectrum, providing a color temperature of about 3200 degrees Kelvin. Below 400 millimicrons and above 695 millimicrons there are several bands of energy caused by gases, but these are not evident in the visible portion of the spectrum, and should not interfere in color photography except to film sensitive in those regions.

While this concentrated arc is new and not yet fully developed so as to be placed in low-cost manufacturing, it has been used already in optical equipment because of the fine definition provided by the small point-type source of the lower-wattage lamps.

There is no reason why this 3200-degree-Kelvin source cannot be raised in color temperature by use of filters, just as incandescent is used with filters for Technicolor.

An important feature of the concentrated arc, both spectrally and photographically, is that the change of color temperature caused by voltage fluctuations is very small, and shows no greater change during life. Over a very wide voltage fluctuation (60 to 140 volts) for the 300-watt lamp, the color-temperature range is no more than 200 degrees Kelvin, and for normal voltage fluctuation of 5 or 10 volts, is very small indeed.

Although originally produced in sizes ranging from 2 to 100 watts, further developments during the past two years have been directed to the production of lamps larger than 100 watts. A number of major problems have had to be solved in their design and construction. A 300-watt lamp is now available commercially, a 1000-watt concentrated arc lamp is developed but not yet in commercial production, and work is proceeding on the design of a 5-kilowatt lamp.

The 300-watt lamp is provided with two anodes of molybdenum rod, $1/4$ inch in diameter by 4 inches long. In normal operation the brightness is around 50 candles per square millimeter, the lamp showing an efficiency of 1.8 candles per watt, about twice as efficient as the 100-watt concentrated arc lamp. One satisfactory design of the 1000-watt size shows the lamp contained in a 6-inch diameter bulb, with the cathode spot about $2^3/4$ inches from the bulb wall. As reported by W. D. Buckingham,* lamps of this type have had lives of several thousand hours in the laboratory. A few lamps made with a 4-inch diameter bulb appear also to have a reasonably long life. The candle power of these lamps, when operating at their normal current (50 amperes), is about 1600, the brightness centers around a value of 50 candles per square millimeter, the area of the spot is normally about 35 square millimeters, and their efficiency is about 1.3 candles per watt which is somewhat less than that of the 300-watt lamp. The 1000-watt lamp is expected to find application in the

---

* Reported by W. D. Buckingham at a conference on High-Intensity Light Sources, Northwestern University, June, 1948.

field of 16-mm motion picture projection, for the image of a rectangular cathode spot can be placed directly at the film gate of the projector with little loss of light. The design being followed in the development of 5-kilowatt lamps is similar to that of the 1000-watt size but the electrodes are larger and the anode made with four radiating fins. The luminous spot will probably be slightly over one-half inch in diameter.

Progress is also reported on the development of brighter lamps. Lamps twice as bright, but considerably less efficient, have been produced by use of hafnium oxide as a cathode-filling material. The

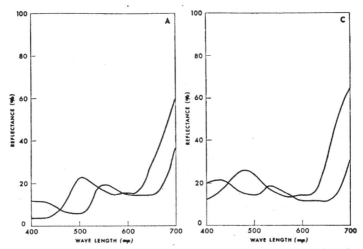

Fig. 17—Samples represented by the curves in A are a color match under ICI illuminant A. Those represented in C are a color match under ICI illuminant C. The colors of these pairs will differ greatly when one illuminant is substituted for another.

spectra produced by the hafnium and zirconium lamps are quite similar. The real advantage in the hafnium-type lamp, by comparison of the 1000-watt size, is that its brightness varies between 90 and 110 candles per square millimeter, while the brightness of a zirconium lamp usually lies between 45 and 55 candles per square millimeter. The efficiency of the 1000-watt hafnium lamp is 0.9 compared with 1.3 for the zirconium lamp. The cost and shortage of hafnium oxide is delaying commercial production of these newer type lamps, although experimental lamps are available.

DEMONSTRATIONS

In place of demonstrations of the effect of illumination that were a part of this paper as presented at the Santa Monica meeting, spectrophotometric curves of two pairs of samples are shown in Fig. 17. The importance of illumination in effecting color changes is indicated by the fact that the samples represented by two curves in Fig. 17A are a color match under tungsten light, 2848 degrees Kelvin, while the samples represented by the curves in Fig. 17C are a color match in daylight, about 6700 degrees Kelvin. These curves were supplied by F. T. Simon of Sidney Blumenthal and Company. Perhaps the best demonstration of a metameric pair that is known to the authors is one described by Dexter and Stearns.[48] Dr. Stearns was kind enough to provide samples of this pair which were shown under a series of five commonly used illuminants when this paper was presented at the meeting in Santa Monica. Also shown was the Macbeth Chromacritic which is widely used in the graphic-arts industry for viewing colored transparencies. In the Chromacritic the color and intensity of the illumination may be changed on a screen to a predetermined setting.

ACKNOWLEDGMENTS

The authors wish to acknowledge the co-operation of Ralph Farnum, Frank Carlson, B. T. Barnes, E. S. Steeb, Jr., of the General Electric Company; F. T. Bowditch and W. W. Lozier of the National Carbon Company; the British-Thomson-Houston Company; and W. D. Buckingham of the Western Union Telegraph Company, who supplied information regarding the various lamps described. They also wish to thank F. T. Simon of the Sidney Blumenthal Company, and E. I. Stearns of Calco Chemical Division, American Cyanamid Company, for the metameric samples used at the meeting to demonstrate color change of samples with change of light source.

REFERENCES

(1) S. M. Newhall and J. G. Brennan, "The ISCC comparative list of color terms (1948)," Inter-Society Color Council, 1949.

(2) I. G. Priest, "A proposed scale for use in specifying the chromaticity of incandescent illuminants and various phases of daylight," J. Opt. Soc. Amer., vol. 23, pp. 41–45; February, 1933.

(3) H. P. Gage and N. Macbeth, "Filters for artificial daylighting, their grading and use," Trans. Illum. Eng. Soc., vol. 31, pp. 995–1020; December, 1936

(4) R. G. Linderman, C. W. Handley, and A. Rodgers, "Illuminants in motion picture production," J. Soc. Mot. Pict. Eng., vol. 40, pp. 333–367; June, 1943.

(5) "SMPE Studio Lighting Committee Report," J. Soc. Mot. Pict. Eng., vol. 45, pp. 249–260; October, 1945.

(6) OSA Colorimetry Committee, "Quantitative data and methods for colorimetry," J. Opt. Soc. Amer., vol. 34, pp. 633–688; November, 1944.

(7) E. Q. Adams and W. E. Forsythe, "Radiometric and colorimetric characteristics of the black body between 2800 °K and 3800 °K," Denison Univ. Bul., J. Sci. Labs., vol. 38, pp. 52–68; December, 1943.

(8) W. E. Forsythe and E. Q. Adams, "Fluorescent and Other Gaseous Discharge Lamps," Murray Hill Books, New York, N. Y., 1948.

(9) D. Katz, "The World of Colour," Translated from the German by R. B. Macleod and C. W. Fox, Kegan Paul, Trench, Trubner and Company, London, England, 1935.

(10) H. Helson, "Fundamental problems in color vision. 1. The principle governing changes in hue, saturation, and lightness of non-selective samples in chromatic illumination," J. Exper. Psychol., vol. 23, pp. 439–476; November, 1938.

(11) H. Helson, "Color tolerances as affected by changes in composition and intensity of illumination and reflectance of background," Amer. J. Psychol., vol. 52, pp. 406–412; July, 1939.

(12) H. Helson, "Some factors and implications of color constancy," J. Opt. Soc. Amer., vol. 33, pp. 555–567; October, 1943.

(13) H. Helson and J. Grove, "Changes in hue, lightness, and saturation of surface colors in passing from daylight to incandescent-lamp light," J. Opt. Soc. Amer., vol. 37, pp. 387–395; May, 1947.

(14) H. Helson and V. B. Jeffers, "Fundamental problems in color vision. II. Hue, lightness, and saturation of selective samples in chromatic illumination," J. Exper. Psychol., vol. 26, pp. 1–27; January, 1940.

(15) H. Helson and W. C. Michels, "The effect of chromatic adaptation on achromaticity," J. Opt. Soc. Amer., vol. 38, pp. 1025–1032; December, 1938.

(16) D. B. Judd, "Hue, saturation and lightness of surface colors with chromatic illumination," J. Opt. Soc. Amer., vol. 30, pp. 2–32; January, 1940.

(17) R. M. Evans, "Introduction to Color," John Wiley and Sons, New York, N. Y., 1948.

(18) P. J. Bouma and A. A. Kruithof, "Chromatic adaptation of the eye," Philips Tech. Rev., vol. 9, no. 9, pp. 257–267; 1947–1948.

(19) A. A. Kruithof and P. J. Bouma, "Hue estimation of surface colors as influenced by the colours of surroundings," Physica, vol. 9, pp. 957–966; December, 1942.

(20) OSA Colorimetry Committee, "Physical concepts: Radiant energy and its measurement," J. Opt. Soc. Amer., vol. 34, pp. 183–218; April, 1944.

(21) C. G. Abbot, F. E. Fowle, and L. B. Aldrich, "The distribution of energy in the spectra of the sun and the stars," Smithsonian Misc. Coll., vol. 74, p. 15; 1923.

(22) P. Moon, "Proposed standard solar-radiation curves for engineering use," J. Frank. Inst., vol. 230, pp. 583–617; November, 1940.

(23) Commission Internationale de L'Eclairage, Proceedings of the Eighth Session, Published by University Press, Cambridge, England, 1931, pp. 219–220.

(24) D. B. Judd, "The 1931 ICI standard observer and co-ordinate system for colorimetry," J. Opt. Soc. Amer., vol. 23, pp. 359–374; October, 1933.

(25) K. S. Gibson, "Approximate spectral energy distribution of skylight," J. Opt. Soc. Amer., vol. 30, p. 88 (A), February, 1940.

(26) A. H. Taylor and G. P. Kerr, "The distribution of energy in the visible spectrum of daylight," J. Opt. Soc. Amer., vol. 31, pp. 3–8; January, 1941.

(27) N. Macbeth, Sr., "Color temperature classification of natural and artificial illuminants," Trans. Illum. Eng. Soc., vol. 23, pp. 302–324; March, 1928.

(28) W. E. Forsythe and E. Q. Adams, "The tungsten filament incandescent lamp," Denison Univ. Bul., J. Sci. Labs., vol. 32, pp. 70–131; April, 1937.

(29) D. B. Judd, "Changes in color temperature of tungsten filament lamps at constant voltage," J. Opt. Soc. Amer., vol. 26, pp. 409–420; November, 1936.

(30) D. Nickerson, "Artificial daylighting for color grading of agricultural products," J. Opt. Soc. Amer., vol. 29, pp. 1–9; January, 1939.

(31) D. Nickerson, "Artificial daylighting studies," Trans. Illum. Eng. Soc., vol. 34, pp. 1233–1253; December, 1939.

(32) D. Nickerson, "The illuminant in color matching and discrimination," Illum. Eng., vol. 36, pp. 373–391; March, 1941.

(33) D. Nickerson, "Color measurement and its application to the grading of agricultural products," U. S. Dept. Agr. Misc. Publ. 580, United States Department of Agriculture, Washington, D. C., 1946.

(34) D. Nickerson, "The illuminant in textile color matching, An illuminant to satisfy preferred conditions of daylight-match" (a report on Inter-Society Color Council Problem No. 13), Illum. Eng., vol. 43, pp. 416–467; April, 1948; summary report, J. Opt. Soc. Amer., vol. 38, pp. 458–466; May, 1948.

(35) F. T. Bowditch, M. R. Null, and R. J. Zavesky, "Carbon arcs for motion picture and television studio lighting," J. Soc. Mot. Pict. Eng., vol. 46, pp. 441–453; June, 1946.

(36) F. T. Bowditch and A. C. Downes, "The photographic effectiveness of carbon arc studio light-sources," J. Soc. Mot. Pict. Eng., vol. 25, pp. 375–382; November, 1935.

(37) F. T. Bowditch and A. C. Downes, "Spectral distribution and color-temperatures of the radiant energy from carbon arcs used in the motion picture industry," J. Soc. Mot. Pict. Eng., vol. 30, pp. 400–407, April, 1938.

(38) M. R. Null, W. W. Lozier, and D. B. Joy, "The color of light on the projection screen," J. Soc. Mot. Pict. Eng., vol. 38, pp. 219–228; March, 1942.

(39) R. J. Zavesky, M. R. Null, and W. W. Lozier, "Study of radiant energy at motion picture film aperture," J. Soc. Mot. Pict. Eng., vol. 45, pp. 102–108; August, 1945.

(40) G. E. Inman, "Characteristics of fluorescent lamps," Trans. Illum. Eng. Soc., vol. 34, pp. 65–78; January, 1939.

(41) G. B. Buck, II, and H. C. Froelich, "Color characteristics of human complexions," Illum. Eng., vol. 43, pp. 27–49; January, 1948.

(42) H. A. Breeding, "Mercury lighting for television studios," Proc. I.R.E., vol. 31, pp. 106–112; March, 1943.

(43) R. E. Farnham, "An appraisal of illuminants for television studio lighting," *J. Soc. Mot. Pict. Eng.*, vol. 46, pp. 431–440; June, 1946.

(44) W. E. Forsythe, E. Q. Adams, and B. T. Barnes, "Mercury vapor lamps," *Denison Univ. Bul.*, J. *Scientific Labs.*, vol. 37, pp. 107–132; August, 1942.

(45) F. E. Carlson, "New developments in mercury lamps for studio lighting," *J. Soc. Mot. Pict. Eng.*, vol. 50, pp. 122–138; February, 1948.

(46) W. D. Buckingham and C. R. Deibert, "The concentrated-arc lamp," *J. Opt. Soc. Amer.*, vol. 36, pp. 245–250; May, 1946.

(47) W. D. Buckingham and C. R. Deibert, "Characteristics and applications of concentrated-arc lamps," *J. Soc. Mot. Pict. Eng.*, vol. 47, pp. 376–399; April, 1947.

(48) F. C. Dexter and E. I. Stearns, "Example of metamerism," Letter to the editor, *J. Opt. Soc. Amer.*, vol. 38, pp. 816–817; September, 1948.

DISCUSSION

MR. CARL FREUND: Seeing these different light sources, I think the studio make-up department should pay attention to these. I have seen many make-up departments where the people are made up under fluorescent light and then set under incandescent light. When a Technicolor picture is shown people should be made up under proper color lighting.

MR. LONG: As you increase the pressure on cadmium-mercury lamps, the lines consolidate considerably, and in ranges of 20 to 30 atmospheres you have almost continuous lines. The cadmium does not come up quite so rapidly, but fills in greatly on the lines and appears more like a continuous picture.

MR. FREUND: I would like to point out one other very important fact in black-and-white photography. Usually we make tests of a star before we start to shoot a picture—usually on a test stage or some place, and measure the foot-candles correctly and establish a level for the lighting level. Later in the rushes, and when we go into production, usually different lights are used, or newer lamps, and then we are supposed to make the close-ups all over again and then everyone says to make it just like the test; so we have a chart on what diffusion we use, but we find that the skin texture never comes out right. It was different when we shot the test than when we actually shot the production—there was an older lamp or newer lamp in one or the other. But we have a different skin texture, even with the proper lamp.

MR. BELL: Has a photographic test been made and are the effects either more pronounced or less than they are here?

MISS DOROTHY NICKERSON: You get other effects than those just from the illumination, but I think Mr. Evans can answer it better, because I am sure they must have tried it.

MR. RALPH EVANS: In color photography, we are dealing with the three receptive systems somewhat similar to that of the eye, but it isn't expected that a pair of colors which match under any illumination should match when photographed by any of the known processes. The reason for the match is rather difficult to go into briefly, but it is not to be expected that the same colors that match under any illumination will match in color photography at the present time. The requirement would be that the three receptors have exactly the same properties as the television mechanism of the eye, and that at the present time is impossible.

Color-Order Systems[*]

By CARL E. FOSS

Color Consultant, Princeton, N. J.

Summary—The straightforward approach to any color problem requires consideration of the properties of illumination, the object, and the receptor. Ordinarily the receptor is the eye but in motion picture photography an intermediate step, the translation of the scene in front of the camera into a scene for presentation to the eye, is required. Since a successful final result from a color standpoint is a complex combination of many factors, color-reference points in the various stages of production are very welcome.

Many kinds of color-reference material are available and a careful appraisal will develop three points of difference.

One group includes collections of samples which illustrate the color gamut of colorants in prescribed mixtures, another group includes samples illustrating systems derived from color-mixture data, and a third group includes samples that illustrate systems that deal with various aspects of visual color space.

THE FUNDAMENTAL DEFINITION of color can be given as an equation:

$$\text{Radiant Energy} \times \text{Visual Process} = \text{Color}.$$

Color is the result of the evaluation of a particular kind of radian energy by the visual and related processes of an observer. There ar many distinct situations or conditions that give rise to color, bu these are all extensions or amplifications of this simple fact. In mos cases, the radiant energy has undergone some modification before a observer evaluates it. In fact, many modifications of the radian energy from a source or illuminant quite often take place before it i finally evaluated by the observer. It is therefore very helpful to think of the source of radiant energy as distinct and separate from th great variety of modifiers encountered. Thus we may expand oui equation to the following:

$$\text{Source} \times \text{Modifier} \times \text{Visual Process} = \text{Color}.$$

In a study of color it may be necessary to use mathematical or othe expressions for any part of a single section of this equation, but at no time can we ignore the existence and importance of all factors included in the equation.

[*] Presented May 20, 1948, at the SMPE Convention in Santa Monica.

It may be in order to repeat these statements in terms of common experience. Thus, in a natural outdoor situation radiant energy may come from the sun, may or may not be modified by scattering in the atmosphere before it is transmitted or reflected by the next modifier, which may be any object, and finally reaches the eye of an observer who makes an evaluation of color. In an indoor situation the energy may come from an electric lamp or other artificial source, may or may not be modified by filters or reflectors before it reaches the object for further modification, and then through the visual processes of an observer is evaluated as color. Ordinarily this is a single sequence of events but in photography this sequence occurs twice so that the scene in front of the camera may be translated into a picture for presentation to the observer. The picture is a specially prepared modifier which allows a representation of the original scene. This sequence is shown in the following equations:

$$\text{Source} \times \text{Modifier} \times \text{Photographic Process} = \text{Picture}$$
$$\text{Source} \times \text{Picture} \times \text{Visual Process} = \text{Color.}$$

The picture resulting from the first equation becomes the modifier in the second equation.

This definition of color includes white, gray, and black as colors, and the equation therefore accounts for results of black-and-white photography as well as of color photography. Since from a color standpoint a successful final result is a complex combination of many factors, color-reference points in the various stages of picture production are very welcome. Gray scales and other simple color charts are often used in testing the photographic process, and sometimes much more complete color charts are needed. The use of such test charts is a specialized procedure and so is their production.

All collections of color chips may be considered as modifiers of the radiant energy under which they are viewed or tested. Sometimes a single set of reference-color samples may serve a variety of purposes while at other times specially prepared material is necessary.

Color-reference material is needed by the photographer, designer, scene painter, lighting man, costumer, make-up artist and many others. This color-reference material usually takes the form of color charts in a particular medium such as fabrics, paint, filters, and cosmetics. The seeming complexity of the need for so many types of reference material has led many people on the search for a complete solution to the entire color problem in the form of a single color system.

There is one fundamental color system (International Commission on Illumination) in current use, but it does not provide for a single set of reference materials. Color materials may take a wide range of forms since factors such as gloss, transparency, texture, and their combinations indicate that no one collection of material samples will ever satisfy all requirements.

Because of the special requirements of a particular job at hand, it is very helpful to be able to appraise the color collections which are available so that the proper or best use of them may be made.

If, instead of studying available color collections, an attempt were made to gather a complete color range by continued assembly of existing samples, the result always would be the same, that is, there would be too many samples of similar colors, and many regions of color space not represented at all. At this point in the assembling of such a collection it usually becomes evident that a good series of color samples can be better prepared by special production rather than by continued assembly.

The production of a color collection is essentially the manipulation of the ingredients in each of the various material forms. The pigments and dyes in paints, papers, plastics, fabrics, and similar materials, are the ingredients whose variation produces the color variation and they are known as colorants. There are thousands of colorants but many are very similar in color. Color is only one property of colorants, and certain properties like stability, compatability, and transparency, often control their applications. Indeed, these other properties of the colorants often are the factors which must be considered first in producing a collection of color samples. Ease of manipulation, and the ability to control ingredients in the prepartion of any set of samples will determine whether the production of a collection can be completed in a reasonable time. This is the principal factor that explains the wide use of paint in making color collections. It is easy to apply paint to a surface, and a wide color range is possible with a small number of colorants. It is more difficult to produce specific colors in textiles, plastics, and ceramic materials.

It is natural to think of color collections in terms of their material ingredients but other important features must not be overlooked. In order for a collection of color samples to be called a color system, whether the samples constitute the system or merely represent it, it is usually expected that a certain orderliness of interrelation and presentation, as well as wide coverage of color range, be provided. Since the interrelation of the various samples in any single system is in part dependent on the geometrical pattern used in its construction, the geometry is often the first feature described. It may be thought of as the structure which holds the system together. However, the geometry of color systems should be kept separate from certain basic principles which control the derivation of the actual color samples in the collection. These derivation methods are of three general types and serve as the basis for a classification of color systems into three categories. One group includes collections of samples which illustrate the color gamut of colorants in prescribed mixtures, another group includes samples illustrating systems derived by color mixture (additive methods), and the third group includes samples that illustrate systems that deal exclusively with visual aspects of color space.

While at first glance this classification may seem obscure, it may be said that until these distinctions are understood, the entire subject of color will not be clear. These distinctions deserve emphasis because casual inspection fails to discose their importantly different features

One may, for example, start with color samples representing three end points, perhaps black, white, and a strong red. If half-and-half mixtures of a particular set of colorants are prepared the result will differ greatly from half-and-half mixtures of the same end colors mixed additively on disks, and both will differ greatly from color samples that are prepared to appear equally spaced between the same end points.

Unfortunately it is not practicable to reproduce in the JOURNAL the extensive series of color charts prepared to illustrate the discussion of this subject at the Santa Monica meeting. It can be stated, however, that all the points discussed are fully capable of illustration. The color differences illustrated at the meeting as produced by identical proportional treatment applied to the three categories of color systems are far too great for them to be considered similar. Wide differences exist in the three concepts and they should not be confused or considered alike in any way.

Most color problems call for consideration of these concepts in certain combinations and although it is often desirable to describe a sample illustrating one concept in terms of another, this does not dissolve the fundamental differences.

The evaluation of any color system or collection of color chips requires recognition of the color concepts it illustrates before attention is given to the factors of color range, stability, number, size, and form of the samples as well as the factors of body and surface properties of the samples.

The geometrical arrangement of the samples and the scales of unique variables control the interrelation of the samples in any system regardless of the concept it illustrates.

A knowledge of the three principal derivation methods of color-order systems enables one to make the best application of existing systems and also indicates how special systems may be constructed to satisfy special needs.

A number of the commercially available collections of color samples were described and illustrated at the meeting. In this report we shall content ourselves with listing and briefly describing several of the better-known systematic collections of color chips, listing each under the category into which its chief attributes place it.

Colorant-Mixture Systems

Although there are thousands of colorants, various reasons make it desirable to use a relatively small number when producing a color collection. The material usually thought of as the colorant is seldom used alone but almost always in connection with some other material whose color is to be altered. The use of a dye on textiles or film produces a color range or gamut depending upon the concentration or thickness of the colorant on the base whether clear or opaque. Thus the color range of a colorant may be thought of as produced by the variable ratio of two ingredients. The colorant attends to most of the selective modification while the other ingredients attends to the nonselective modification. This second ingredient is not ordinarily thought of as a colorant but rather as a diluent or extender when transparent or as a white when opaque. However, it always contributes significantly as a modifier. In paint, the colorant (selective modifier) is called a toner and is commonly shown in a variable-ratio gamut with white (nonselective modifier) to illustrate its principal coloration. possibilities.

Series or gamuts of this sort are widely used in all media and the general idea is usually the basis for color collections which illustrate the color range of the colorants in various combinations. While this is not the only useful method it is the one most commonly used.

Color chips of such systems may be defined in terms of percentages, by weight or volume, of the colorants used to make them.

Comprehensive systems of this sort require representation in three-dimensional form. In the construction of colorant-mixture systems the choice of component-mixture scales is usually dictated by visual considerations. This does not, however, alter the fact that they still represent the gamut of the particular colorant combinations.

Many charts and collections follow the general color gamut method and there will be many more. The collections range from extensive color coverage made under quantitatively controlled conditions to more limited coverage for requirements that are less exacting.

Each of the systems selected for description is currently available and represents a comprehensive collection of color chips based on the colorant-mixture method. Both show the extension gamut of colorants with white and provide formulas for duplicating the color of the samples in the material used in their production.

Plochere

In the Plochere color system,* as now available, a small number of selected chromatic colorants specially formulated as colors in oil are used to produce 26 series of mixed-base paints, each series having six steps from a near-neutral to the full chromatic step. Each series of six mixed-base paints is formulated to look like a constant-hue series and all 26 series of different hues are arranged in the usual sequence in radial order around a neutral point. This group of mixed paints, 26 × 6 = 156, represents the base of a cylindrical color solid which is developed by making a white-paint extension series containing eight steps from each of the 156 mixed or base paints. The white-paint extension scale is a progressive one which was empirically determined so that throughout all of the toner gamuts toward white, these series give excellent coverage of the color range of the colorants.

A total of 1248 samples result from this thorough development and they are presented as 3- by 5-inch cards in a file box. Each card has the formula on the back which shows how each color was made from the base paints. The base paints, ten in number, are essential in using the formula data given and they are offered for sale to those who wish to produce flat wall paint in quantities for average use.

The Plochere collection was produced to satisfy the demand for a relatively inexpensive collection of color cards showing actual formulations from paints which are currently available.

* Plochere Color System, G. Plochere, 1820 Hyperion Ave., Los Angeles 27 Calif., 1948.

The present collection is a simplification and revision of an earlier Plochere color guide which used a larger number of colorants.

Martin-Senour

The Nu-Hue system of the Martin-Senour Company* is based on a selection of six chromatic and two achromatic colorants. These basic paints are specifically designed for interior wall finishes. The system consists of one thousand samples presented in the form of a cone, the samples arranged in hexagonal closest packing, with planes parallel to the base, each plane representing a constant white-paint content. On the base, the six chromatic colorants are mixed in pairs of neighboring hues to provide hue coverage, and with black to provide color steps toward neutral. The remainder of the base is produced by combinations of two neighboring hue colorants and black and provides 271 samples on the base, 54 on the periphery. On this level there are nine rings around the black center.

Each chart above the base has a stated amount of white paint added to certain of the base colors, this amount increasing as the levels increase toward white. Each succeeding higher-level chart has one less ring decreasing from nine rings at the bottom to a single white point at the peak of the cone.

In this construction it is possible to show a continuous series from any paint on the base to the top, but the samples which occur in each series vary widely in number. It is a strict application of the prescribed mixtures and no deviations are made to include visual considerations other than variable ratios of the components in the scales.

The material is available on charts and in a 3- × 5-inch card-index file with formulas for each for obtaining a match with the limited series of base paints.

This collection was produced to provide a high-precision formulation technique for the production of paint in any quantity to match any one of the colors. This is usually done in establishments equipped with appropriate mixing devices.

COLOR-MIXTURE SYSTEMS

In this type of system the color range is determined by the end components in additive mixture. Quite often this is expressed in terms of the disk-mixture percentages of the components.

* Nu-Hue Color System, Martin-Senour Co., 2520 Quarry St., Chicago, Ill., 1946.

The general concept provides a structure for the development of a wide number of representations of color space. These can follow several patterns, as a collection of two-, three-, or four-component mixtures, the commonest one being a triangular array using two achromatic end points and one chromatic end point. By varying the chromatic end point, and repeating the procedure for a considerable number of such end points, wide color coverage may be obtained. These resulting solids are double conics.

Several internal geometries may be used, and the two that are described below illustrate the internal treatment of the most common multiple three-component cases. They have the same external shape but different internal co-ordinates.

The colors within solids determined by strict application of additive mixture of the components do not provide the color coverage that often is expected. Such deviations have led workers to alter the coverage although they still present the samples in a triangular array, or some derivation thereof. These altered color ranges can be included in a strictly rigorous additive treatment but would require a departure from the commonly used triangular arrangement.

Although scales used in additive methods are often adjusted to take account of visual considerations, this does not alter the fact that they still represent color mixture.

Ridgway

There are 1115 named samples in the Ridgway color charts[1] published in 1912. These charts are still in wide use, particularly in the biological and horticultural sciences where the color names used by Ridgway have become well known. It is understood that the publishers first planned a 5000-copy edition, but that not enough perfect color sheets were available to complete this when the first binding was made. A supply of enough matching sheets was made later to fill in the comparatively few colors (probably not more than 25 or so) which were necessary in order to bind more books. While there has been a second binding, there has been no really new edition since the 1912 publication, and the publisher's supply is now exhausted.

The Ridgway color solid is represented by a double cone with "pure colors" at the equator. The upper surface of the solid contains all "tints" which are produced by additive mixture with white. The lower surface of the solid contains all "shades" which are produced by additive mixtures with black. The internal sampling of the solid follows the same additive mixture paths to white and black from desaturated colors produced in turn by additive mixture of the "pure color" with a gray midway between the white and black. All of these broken colors lie on double-conic solids of successively smaller circumference.

These various series for a given saturation position are shown in succeeding sections of the book. The samples of each series are shown in vertical order and hue sequence in each section. There is a systematic abridgment as the samples become successively more desaturated.

From inspection there is considerable question as to whether the production of the samples followed a strict application of the premises of additive color mixture that are so clearly stated by Ridgway in the text. There is an indication that these paths tend to follow colorant gamuts rather than the disk mixture reported.

It is understood that botanical workers at the University of Toronto* have in process a publication that will provide a conversion from Ridgway to Munsell notation. This information will permit the careful analysis necessary to establish the actual compliance with color-mixture principles.

Ostwald

The ideal Ostwald system is a good example of a color-mixture system. Although it is not possible to obtain the ideal color specimens by Ostwald for his white and black, or for the semichromes specified for his full-color series, his ideal space representation has been computed.

The Ostwald color solid is represented by a double cone with full or saturated colors at the equator. The upper surface of the cone contains what Ostwald calls light clear colors, which are additive mixtures of full colors with white, and the lower surface the dark clear which are additive mixtures of full colors with black. The surface, except for the selection of the full colors and their angular placement, is the same as described for Ridgway.

Planes of constant Ostwald hue (actually constant dominant wavelength) radiate from a central neutral axis, black at the bottom, white at the top, each plane having triangular co-ordinates to represent the three components, full color, black, and white for any sample on the

* Professor D. H. Hamly and associates, Botany Department, University of Toronto, Toronto, Ontario, Canada.

plane. Visual considerations dictate the choice of logarithmic scales for the mixture ratios and partially control the placement of Ostwald-hue planes.

The hues are numbered beginning with 1 at yellow, either in a series of 100, or in the more usual abridgment of 24. The percentages of additive white-and-black content are indicated by letters, usually *a* to *p*, *pa* being the maximum color in the usual abridgment. The notation is written in the order of Ostwald-hue number, white content, black content, for example, a saturated red would be written as 7 *pa*.

Ostwald himself produced materials in many forms to illustrate his system and there are Ostwald color charts included with the authorized English translation of his work. In 1942, however, the Container Corporation of America produced in the charts of their Color Harmony Manual[2] what is perhaps the most carefully standardized set of Ostwald material that is available.

These samples were made so that on each Ostwald-hue chart it is intended that dominant wavelength be kept constant, and that ICI excitation purity be kept constant in the vertical, or "shadow," series, with opposite triangles complementary in dominant wavelength. The triangle arrangement is maintained, although the color range is altered in practice from the additive-color-mixture range to conform with the color range of available pigments.

There are 680 color samples in the Manual. Each is removable and is identified by notation. They were prepared by applying a pigmented film of appropriate color to a base of clear transparent cellulose acetate, thus providing a dull surface on one side, glossy on the other.

ICI specifications, based on spectrophotometric determinations, have been reported for this material.

COLORANT MIXTURE AND COLOR-MIXTURE PRINCIPLES

Certain methods of producing color collections contain within themselves the possibility of consideration according to different conceptual points of view.

A good example of this is the Maerz and Paul collection which is made by printing methods and is a combination of color and colorant mixture. The half-tone screens over white follow color mixture, but the half-tone screens using two colorants may be a combination, since in some places they provide an overprinting of two or more inks. As a rule, printing processes are combinations of color-mixture and colorant-mixture principles. It is possible, however, to consider them completely as color-mixture systems if the results of colorant mixture are taken as components in a color-mixture plan.

Maerz and Paul

The Maerz and Paul Dictionary of Color[3] contains several thousand colors arranged on pages designed for convenience in printing. Eight chromatic inks are used in paired combinations, with screens to provide a wide range between the starting points and white. Each is printed in a series of 8 charts in which succeeding pages are darkened by overprinting with a successively darker transparent gray ink. While there are a great many samples on the charts, certain regions are represented by very similar samples while larger steps occur between successively darker charts. The geometry of the color solid which these charts define is awkward in representation and too complicated to describe here.

This Dictionary is intended for use as an authority on color names. It is based on a wide survey of information regarding names.

COLOR-APPEARANCE SYSTEMS

Color collections based exclusively on visual evaluation fall into this category. Perhaps the most useful example calls for a uniform color space in which each color differs from its neighboring colors by some uniform amount.

There are many ways of sampling this space. One way illustrates the three unique attributes in terms of which the system is described. Another way could illustrate uniform sampling throughout the space. Collections representing these very different methods of sampling would have coexistent features.

The relation of the visual scales depends on the conditions of observation. Each collection of samples will be uniformly spaced only for the observer, illuminant, and background conditions used in its development.

The first method of sampling is the basis for the only collection of samples that has been produced to illustrate exclusively the color-appearance concept. No materials are yet available for the second type of collection described, although the general principles have been reported by the writer,[4, 5] and certain examples are now in preliminary production stages.

Munsell

The concept of uniform spacing upon which the Munsell system[6] is built is valid for all observers, illuminations, and conditions of observation. However, the collection of samples which illustrates this system must be differentiated from the concept itself, for it is built upon the single set of conditions which describe the average situation of normal observer, daylight illumination, and gray background. If these conditions are varied significantly, the feature of uniform spacing may no longer hold for the one set of materials now available.

The Munsell papers represent a sampling of color space in accordance with three unique attributes, hue, value, and chroma. Polar co-ordinates are used to represent the hue and chroma variables of color, and rectangular co-ordinates represent the relations of value and chroma.

The sampling of the solid may be considered as a series of constant-hue planes radiating from a central vertical axis, each hue plane showing its own range of value and chroma. Another equally useful and coexistent plan provides a series of constant-value planes, each showing its own range of hue and chroma. Value is uniformly spaced in parallel planes for the stated conditions. Hue planes are uniformly spaced angularly. Chroma is uniformly spaced cylindrically about the neutral axis of the solid.

The notation for any sample is written in hue, value, chroma sequence in terms of the number or letter attached to each scale, for example, a medium value, strong chroma red might carry the notation 5R 5/10. A decimal notation is provided for recording finer discriminations. About 1000 samples are available that illustrate regular positions on the Munsell scales. Many hundred more intermediate samples are available, usually produced to meet special requirements. The colors are available on charts or in disk or sheet form cut to various sizes.

ICI specifications, based on spectrophotometric determinations have been reported for most of the Munsell samples.

COLOR SYMPOSIA

In 1947 two color meetings, similar in purpose to the present symposium with the Society of Motion Picture Engineers, were arranged by co-operation of the Inter-Society Color Council, one for the annual meeting of the Technical Association of the Pulp and Paper Industry (February, 1947), and the other for the American

Ceramic Society, Design Division (April, 1947). Several of the papers contained in each series would be of interest in connection with the foregoing discussion. Each symposium contains at least one paper that discusses directly the subject of color-order systems. Bound copies of reprints of each symposium may be obtained by SMPE members without cost, as long as the supply lasts, by request to the Secretary, Inter-Society Color Council, Box 155, Benjamin Franklin Station, Washington 4, D. C.

Reference is also made to a discussion of color systems by the Colorimetry Committee[7] of the Optical Society of America and to three numbers of *The Journal of the Optical Society of America,* two of which[8,9] contain a series of papers on the Munsell system, and the third,[10] a series of papers on the Ostwald System.

REFERENCES

(1) R. Ridgway, "Color Standards and Nomenclature," A. Hoen and Company, Inc., Baltimore, Md., 1912.

(2) E. Jacobson, "Color Harmony Manual," Container Corporation of America, Chicago, Ill., 1942.

(3) A. Maerz and M. R. Paul, "Dictionary of Color," McGraw-Hill Publishing Company, New York, N. Y., 1930.

(4) Carl E. Foss, "Tetrahedral representation of the color solid," *J. Opt. Soc. Amer.,* vol. 37, p. 529 (A); 1947.

(5) Carl Foss, "Representations of color space and their applications," *Amer. Cer. Soc. Bull.,* vol. 27, pp. 55-56; 1947.

(6) "Munsell Book of Color"; standard library and pocket-size editions, Munsell Color Company, 10 E. Franklin St., Baltimore 2, Md. 20 hues, 1929; 40 hues, 1942.

(7) OSA Colorimetry Committee, "Colorimeters and color standards," *J. Opt. Soc. Amer.,* vol. 35, pp. 1-25; 1945.

(8) *J. Opt. Soc. Amer.,* vol. 30, pp. 574-645; 1940.

(9) *J. Opt. Soc. Amer.,* vol. 33, p. 355-422: 1943.

(10) *J. Opt. Soc. Amer.,* vol. 34, pp. 353-399; 1944.

System in Color Preferences[*]

By J. P. GUILFORD

University of Southern California, Los Angeles, California

Summary—Scientific studies of the effects of colors upon people have been very limited both in number and in scope. Preliminary investigations offer considerable promise that further scientific work along these lines will be fruitful. Some examples of results already obtained will be cited to show that preferences of people for colors and for color combinations can be predicted with satisfactory accuracy. Predictions are possible because there are definite relationships between some of the measurable aspects of colored objects and the degree of preference for colors as noted for the average person.

An architect recently asked, "What do you think *is* the limit to the use of color?" The reply to this was that there seems to be *no* limit. He was referring to the artistic uses of color and my reply was in the same vein. From the standpoint of the artistic uses of color, alone, recent developments in housing, home furnishing, landscaping, motion pictures, and clothing all demonstrate the growing freedom and increasing recognition of values in the employment of colors. Perhaps this is only one phase of the fact that our culture no longer frowns heavily upon sensory enjoyment or of the fact that our civilization is now past the pioneer stages, and our wealth and leisure foster attention to the arts as never before. However this may be, the new interest in color has called for the efforts of physicist and engineer, of artist and psychologist, and of manufacturer and producer, in attempts to gain more intimate understanding and improved control of colors and also to determine how the consumer is affected by them. The role of the psychologist in all this has been to assist in the classification of colors according to their appearance and more exclusively to learn how people react to color; how well they like single colors and color combinations and what kinds of emotional reactions they have in the presence of colors.

It is the subject of likes and dislikes for single colors that will be treated here. Some general conclusions have been drawn from experiences in the study of this problem. It can be shown that color preferences are not a matter of whim, caprice, or of fad; on the

* Presented May 20, 1948, at the SMPE Convention at Santa Monica.

contrary, they are consistent and orderly. Preferences are related to the inherent properties of colors and the preferences of groups of people can be predicted from those properties, not perfectly, but with a surprising degree of accuracy. By "color properties" are meant the three variables that psychologists have generally called "hue" (most obviously related to dominant wavelength of the stimulus); "chroma" (color strength, saturation, or richness, most obviously related to wave purity); and "tint" (the color's equivalence to some point or level on the black-white scale). This terminology agrees with that of the Munsell system except for the substitution of "tint" for the Munsell "value." The reason for this change here is that the author wishes to use the term "affective value" to indicate the position of an experience on the continuum that runs from pleasantness to unpleasantness. Affective value means degree of preference or degree of pleasure.

The early studies on color preferences suffered seriously from the lack of methods for color specification. In spite of this fact, certain consistent conclusions were often reached. The chief interest centered in preferences for different hues. The consensus seemed to be that the hues blue, green, and red were most preferred, in about that order, whereas the hues yellow, orange, and violet were among the least preferred. Where discrepancies occurred, there was no way of reconciling the results because one investigator's yellow sample, for example, might have differed from that of another investigator, with respect to either tint or chroma, or both.

The relations of preference to tint and to chroma as such received almost no attention, although when they did, the consensus seemed to be that ligher colors were preferred to darker ones, and the more saturated colors to the less saturated.

My own investigations of these problems began with two hypotheses: (1) that the affective value of a color is dependent upon *all* its inherent properties, its hue, tint, and its chroma, and (2) that the relationship of affective value to any one of these properties is a continuous function, in other words, degree of pleasure increases or decreases systematically as hue, tint, or chroma changes in a given direction. A serious test of these hypotheses requires intensive study of a liberal sampling of colors throughout the color solid. It also requires numerical specifications of the color sample used and a measurement of the degree of liking for each sample.

To meet the requirement of color specifications, it was decided to utilize the Munsell system as the basis. This decision was based on the assumption that it is the *appearance* of the color, not so much its stimulus composition, that determines the observer's feelings. This assumption may require special investigation. The Munsell system seems to offer the evaluations of color most consistent with the psychological point of view. To meet the requirement of measurement of affective value, the judgments of observers were utilized. There is insufficient time to go into the details of the technique for this here. Suffice it to say that the use of registration of physiological reactions was rejected on the ground that no such indicators of pleasure or displeasure have been demonstrated to be either sufficiently consistent or interrelated to justify their use for this purpose.

The experimental material consisted of swatches of colored paper two inches square.* Surface colors were used because it was felt that they represent the most common form of color experience. The 316 different colors used were sampled systematically from the color solid. First, ten alternate hues were selected in the Munsell Book of Color. Second, for each hue, samples were sought at alternate levels of tint or Munsell value. Third, samples were sought to match each chroma level represented. Munsell papers were not available, so it was necessary to collect papers from all other known sources. The resulting selections only approximated the specifications desired but since they could be evaluated in terms of the system, some discrepancies could be tolerated. Fig. 1 illustrates how the specimens were distributed for one of the chosen hues, namely, for red-purple.

Time does not permit going into details concerning all the experimental conditions and procedures.[1] Forty individuals of normal color vision, twenty men and twenty women, gave their judgments of preferences for each color sample on two different occasions. The colors were viewed each for a period of five seconds, on a uniform gray background rated at $N/5$ on the Munsell scale, under constant illumination. The Munsell specifications had been determined for each color under the same illumination. The resulting values for affective value of each color were in terms of an eleven-point scale extending from 0 to 10, inclusive, with the indifference point (at which experiences are rated as neither pleasant nor unpleasant) at 5.5. Values of

* The experimental results referred to in this article were obtained in the Psychological Laboratory at the University of Nebraska during the years 1935 to 1939. I am indebted to Ada P. Jorgensen and Patricia C. Smith for material assistance in the experiments.

6 and above therefore represented different degrees of pleasure and values of 5 and below represented different degrees of displeasure.

Having obtained the ratings of affective value for each color, it was important to know first whether those values were self-consistent. Although there were distinct differences of opinion among individuals as to the values of some colors, on the whole agreements far outweighed disagreements and the range of values for different colors far exceeded the range for a single color.[2] There were some differences of

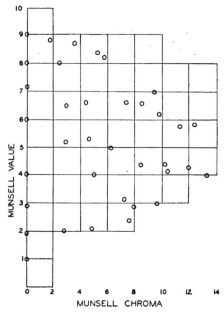

Fig. 1—Diagram of the red-purple plane of the Munsell color solid, showing by marked points the specifications of color samples of this hue used in the experiments.

opinion between the two sexes, on colors of certain hues, particularly, so that the data were treated separately for the two sexes throughout.

Next, let us try to obtain a picture of how the pleasure aroused by a surface color varies with each of the properties of that color: first, let us isolate each variable—hue, tint, and chroma—and note its unique effects and then study the combined or joint effects of all three variables. If we want to discover how preference varies with hue, and hue alone, we must hold constant the effects of tint and chroma while doing so.   Likewise, if we want to know the influence of tint, as such, we must hold constant the other variables, hue and chroma.   The influence of chroma can be ascertained only when hue and tint have been held constant.

Next, let us examine the relation of affective value to tint when chroma is zero and there is no hue specification.   Fig. 2 illustrates the average judgments of both men and women for samples of this kind.   The curves showing the supposed continuous regression of affective value upon tint have been drawn by inspection.   Assuming

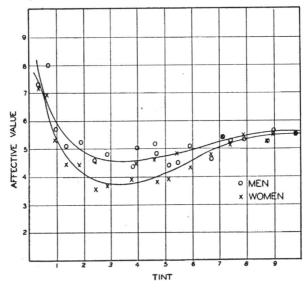

Fig. 2—Points and curves showing the relationship of affective value to tint (Munsell value) when chroma is zero.

that these curves have been properly drawn so as to represent best the observed values, which are represented by the points, one can see how much those observations differ from the supposed continuous trends.   Such deviations may be regarded as sampling or experimental errors.   In spite of their obvious sizes, the hypothesis of continuity seems fully supported by the results.   For achromatic samples, under these conditions of illumination, background, and size of sample, it appears that, on the average, white samples are neither pleasant nor unpleasant, that grays from Munsell value 1 through Munsell value 6 are mildly to moderately unpleasant, and the most extreme blacks tend to be distinctly pleasant. This kind of relationship, as shown in Fig. 2, carries over to many situations in which there is a hue and in which chroma is constant at some level other than zero, but it becomes modified at higher chroma levels, quite drastically, in fact, for colors of certain hues. More will be said on this point later.

Consider, next, the relation of affective value to hue when tint and chroma are constant. Earlier findings, that blue, green, and red are preferred to orange, yellow, and violet, are supported, but serious qualifications must be made. Not all blues are preferred to all yellows, nor all reds preferred to all violets. Fig. 3 shows smoothed

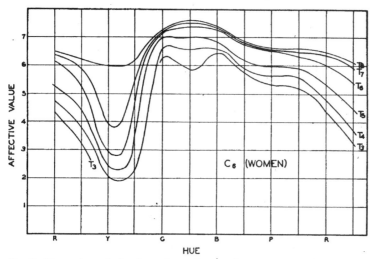

Fig. 3—Regressions of affective value on hue at chroma level 6 and at various constant tint levels when judges were women.

regressions of affective value upon hue when the chroma is held constant at 6 and as tint is held constant at various levels. The observed points are not shown, in order to avoid congestion in the illustration. Their discrepancies from the curves shown are comparable with those exhibited in Fig. 2. Smoothing had been done not only in the plane of this particular figure but also in planes of constant hue and constant tint, until we arrved at consistent results in all planes. Interpolated values thus became possible and a more complete picture can be given. In Fig. 3, not all the curves are complete, either because colors do not exist for some of the combinations of hue, tint, and chroma implied by the figure or because specimens and observations

fell short of complete sampling of those that do exist.   Similar figures
have been drawn representing other chroma levels.   Fig. 4 shows
what happens at the chroma level 2.   There is consistent evidence
that the judgments of women were more sensitive to variations in
hue than were those of men.   Women will not be surprised to hear of
this result.

Figs. 3 and 4 illustrate several principles.   Continuity is again
obvious.   The points of maximal preference for each combination of
constant tints and chromas come in about the same positions on the
hue scale—near red, green, and blue.   The minimal preferences come

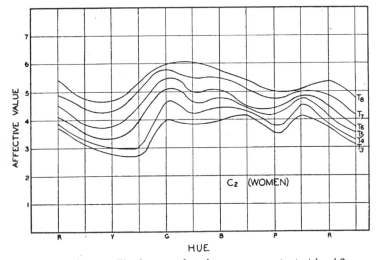

Fig. 4—Same as Fig. 3 except that chroma was constant at level 2.

near yellow, violet, and, in some instances, between green and blue.
The general level of any particular curve reflects clearly the influence
of tint and the range of affective values for any curve reflects relatively more the influence of chroma.   The influences of *these* variables,
as was said before, are best noted when hue is constant, however, and
we return to these variables later.   Here we are primarily interested
in the effects of hue.   In another publication,[3] the author suggested
that curves such as those in Figs. 3 and 4 can be regarded mathematically as periodic functions for which empirical equations could be
written.   A Fourier analysis of some data of this kind tended to show
that the chief components of one of these curves, when chroma is

approximately 8 and tint is approimately 6, are the first and third harmonics. There are interesting implications of this finding, but the use of such involved mathematical equations for predicting color preferences is not recommended; a much more practical method will be demonstrated.

Fig. 5 shows the regressions of affective value upon chroma when the hue is yellow and when tint is held constant at different levels.

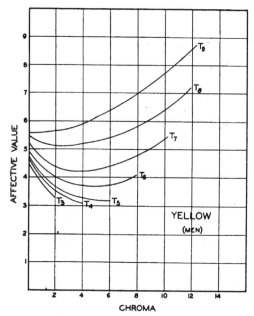

Fig. 5—Regressions of affective value on chroma for colors of yellow hue at different fixed tint levels when judges were men.

These curves are rather characteristic of those obtained for most hues. Earlier studies had brought out the general conclusion that affective value increases as chroma increases. We see in Fig. 5 that this is not true throughout the range of chroma. Near zero chroma, affective value usually *decreases* as chroma increases, reaching a minimum level when chroma is in the region of 2 to 4. It has been suggested that this may be caused by the fact that at such low chromas the nature of the hue is difficult for most observers to identify. This arouses annoyance and hence displeasure. However, this may be, for most hues at most tint levels the minimum preference is not at zero chroma but at some degree of saturation greater than zero. There was one noteworthy instance in which preferences decreased again for high chroma levels. This was for the hue purple-red, in judgments of women only. There was decreasing pleasure for colors above chroma 10 and above Munsell value 4. Whether this reflected a temporary fad or is a more fixed disposition of women is unknown.

It should be said that there may be certain individuals who have a

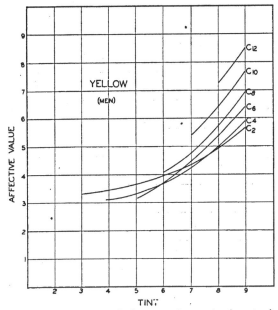

Fig. 6—Regressions of affective value on tint for a typical "warm" color at different fixed chroma levels when judges were men.

general preference for unsaturated colors and a dislike for "loud" or saturated colors. Some psychologists have maintained that there are two types of individuals with respect to reactions to saturated colors; one type likes them, the other does not. The results reported to you pertain to small colored surfaces and represent the average judgments of a number of supposedly normal observers. One should not generalize too far from these results. There is another finding in the psychological literature to the effect that unsaturated colors gain relatively more by increased size of stimulus whereas saturated colors gain less, and in fact may lose in affective value. This problem is very important and must be solved systematically and in thorough fashion before laboratory findings such as those reported here can be carried over to practical application.

The regression of affective value upon tint when hue is constant, and when chroma is constant and not zero, is illustrated by Figs. 6 and 7. It is necessary to show two illustrations for this because two groups of

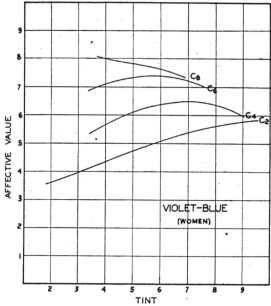

Fig. 7—Similar to Fig. 6 except that the results are for a "cool" color as judged by women.

colors yield different effects in this respect. The two groups may be roughly called the "warm" and the "cold" colors. The former tend to yield a regression with the lower preferences at moderate tint levels. This tendency is consistent with the relation of affective value to tint when chroma is zero (mentioned earlier), and holds generally for colors whose hues are yellow, red, or purple. (See Fig. 6.)·⁻ For the "cool" colors, on the other hand, the *higher* preferences may appear at moderate tint levels. At least, in most "cool" colors the samples of moderate tint are relatively more agreeable. This is true for the greens and blues, including violet-blue. A general principle which seems to embrace both types of regressions is that colors tend to be most preferred at tint levels at which they can be most saturated. The yellow-reds and yellows can be most saturated at the higher tint levels and at those levels one finds the most agreeable colors at each degree of chroma. The greens and blues can be most saturated at some moderate or low tint level, and, for any fixed degree of chroma, that is where the maximal preferences occur. An outstanding exception to this principle comes in the region of red and purple which can be most saturated, also at moderate or low tint levels and yet they yield results similar to those in the region of yellow.

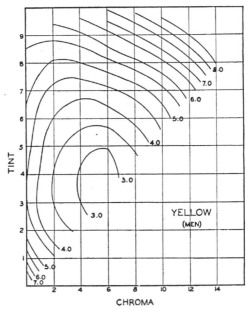

Fig. 8—An isohedon chart showing on the Munsell plane for yellow the lines of equal preference as judged by the average man.

From what has been briefly reported thus far, it can be seen that the relationships of preferences to hue, tint, and chroma are so continuous and systematic that, knowing the specifications of a color sample, we should be able to make a fairly accurate prediction of its affective value. With series of charts such as those already shown, one could start with the three Munsell designations for a color and look up its probable average affective value. One might even set up empirical equations in which the three values could be substituted and the equations solved to predict the expected affective value. This approach, as well as that using the charts of the types illustrated, however, would be too unwieldy for practical purposes. As a much more practical substitute, charts like those illustrated in Figs. 8 and 9 have been prepared; one for each of the ten alternate hues in the Munsell system. Twenty such charts could be prepared, but preferences for the hues not represented in the ten could be found by interpolation.

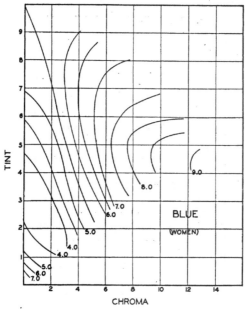

Fig. 9—Same as Fig. 8 except the plane is for Munsell blue and the judgments are for women.

Each chart, representing a constant hue, shows the lines of equal affective value in steps of one-half unit. By analogy to weather maps, the lines have been called "isohedons" (lines of equal pleasure) and the diagrams have been called "isohedon charts." Separate charts have been drawn for men and women, for, although similarities between reactions of the two sexes to colors generally outweigh differences, in the charts, for some hues in particular, the differences are so large that much accuracy in prediction is to be gained by recognizing those differences. The greatest sex differences occur in reactions to the hues of red-purple, purple, and yellow-red. Greatest similarity between the sexes occurs in the region from yellow to blue. For either sex taken alone, the predictions are most accurate when hues are red, yellow, green, or blue, and poorest when hues are purple, yellow-red, or green-yellow. The last statement suggests that familiarity with primaries may be a factor favoring stability of judgments. The over-all accuracy of prediction can be expressed in several ways. From one point of view, we can say that the predictions are nearly 80 per cent accurate for the average judgments of the women, and at least 85 per cent accurate for judgments of the men.[4] The men who read this will not be surprised at the somewhat lower predictablity of the reactions of women.

Having noted statements regarding the accuracy of predictions of average affective values from the known specifications of color samples, let us· be ready to recognize their limitations and to qualify the conclusions. First, it must be emphasized that we are talking about averages and not about single reactions in single individuals at particular moments of time. Fortunately we are usually called upon to make predictions of reactions of masses of individuals over periods of time, or of average people, and not of isolated reactions. The predictions hold under a set of conditions such as prevailed when these experimental results were being obtained. They are based upon the assumption that all other determiners of color preferences are held constant. They should hold for preferences of colored surfaces, of a given size, on a given background, under a given illumination, and for a population of individuals similar to those who rendered their judgments in the experiments. Before we are justified in predicting the preferences for groups of other kinds, with colors of different sizes, on different backgrounds, attached to different objects—clothing, houses, furnishings, and cars—and under different illuminations, much further experimental study will be needed in order to determine how these other factors influence preferences. It is quite possible that these other factors do not entirely overcome or even overshadow the effects of the color properties themselves as determiners of pleasure derived from colors, but this needs to be demonstrated experimentally.

Problems of the effects of color use and of color combinations are of even greater practical importance than the problems of preferences for single colors. These more practical problems cannot be fully solved, however, until those pertaining to single colors are also solved, and the most economical approach to these problems in the long run will be through the study of effects of single colors. It has already been shown experimentally that there is a strong relationship between the pleasure aroused by a color combination and by its components taken alone. Principles which hold for the preferences for single colors should also have an important bearing upon preferences for combinations of those colors. Although *uses* of color are very strongly determined by cultural and conventional forces, the range of these effects is undoubtedly limited by the influence of the properties of the colors themselves.

In conclusion, it may be said that a systematic study of preferences for colors, when the entire color solid is thoroughly sampled, shows that there are definite relationships between the degree of pleasure that a color arouses and the intrinsic properties of the color itself. As hue, tint (Munsell value), and chroma change continuously in a fixed direction, judgments of pleasure also change accordingly. The relationships are continuous but not simple. The relation of affective value (preference or pleasure) to any one property of color is modified by changes in any other property, but in a systematic manner. When other factors are held constant, including size, illumination, background, and type of observer, a prediction of average affective value can be quite accurately made from the knowledge of Munsell specifications of the color sample. By conversion from Munsell specifications to those of other systems, presumably just as accurate predictions of the same kind could be made. For fully useful predictions in practice, other determiners will need to be taken into account, determiners that were held constant in obtaining the results reported. The results already obtained suggest considerable promise for the fruitfulness of further systematic research on other determiners of liking for colors and also for their different uses and combinations. Such research should be amply rewarding for those who desire the solution of color problems as they affect the human observer.

REFERENCES

(1) More of the experimental details are described in an article, J. P. Guilford, "A study in psychodynamics," *Psychometrika*, vol. 4, pp. 1–21; 1939.

(2) Information on the reliability of the data is presented in an article, J. P. Guilford, "There is system in color preferences," *J. Opt. Soc. Amer.*, vol. 30, pp. 355–359; 1940.

(3) J. P. Guilford, "The affective value of color as a function of hue, tint, and chroma," *J. Exper. Psychol.*, vol. 17, pp. 342–370; 1934.

(4) For additional information on this point, see reference 2.

16-Mm Release Printing Using 35- and 32-Mm Film*

By FRANK LA GRANDE

PARAMOUNT PICTURES, NEW YORK, NEW YORK

C. R. DAILY AND BRUCE H. DENNEY

PARAMOUNT PICTURES, HOLLYWOOD, CALIFORNIA

Summary—This paper describes the method now used by Paramount in making 16-mm release prints from 35-mm original studio productions. The purpose of this method primarily is to produce 16-mm release prints comparable to the 35-mm sound and picture print quality and standards. A considerable advantage is gained by utilizing standard 35-mm facilities such as developing machines, rewinds, take-ups, and splicing equipment for most of the operations. Special 35-mm width films with 32-mm symmetrical 16-mm-type perforations are used for the sound and picture release negatives, with two tracks on each film so that two reels of 16-mm release are obtained with each developing and printing operation, thereby saving valuable time and equipment.

A specially designed optical-reduction printer is used to make the double-track picture release negative. A specially designed sound recorder is used to produce the highest possible quality of re-recorded sound negative. A specially designed contact release printer is used to print the 35-mm width, 32-mm perforated sound and picture double-track negatives to the 32-mm width fine-grain release positive stock. Twelve-hundred-foot rolls of the 32-mm print stock are used, corresponding to 3000 feet of the original 35-mm production. The 32-mm film is developed in a developing machine modified for this width and the finished print is then slit to make the two 16-mm release prints.

INTRODUCTION

IN CONSIDERING the problem of 16-mm release from 35-mm feature productions, it was concluded that the most economical and highest quality operation should include:

(a) Electrical re-recording of the 35-mm release sound track to permit making the necessary changes in frequency response and compression;

(b) The use of 35-mm width film with double 16-mm-type perforations for both the release sound and release picture negatives in order to utilize existing 35-mm film-developing and -handling facilities;

(c) The re-recording to *two* sound tracks on the release sound negative and the optical printing of *two* picture tracks or images on the release picture negative to save handling time in development. The sound and picture tracks run in opposite directions on the two sides of the film;

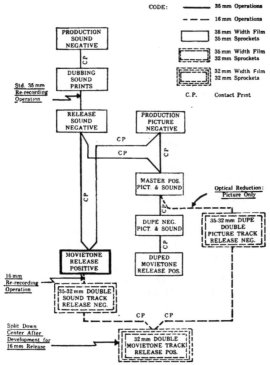

Fig. 1—New 16-mm film processing sequence (dashed lines), and conventional 35-mm regular and dupe processing (solid lines).

(d) The use of 32-mm release print stock to accommodate **two** complete 16-mm movietone prints. This film is split after development for assembly on 16-mm reels.

Film-Handling Sequence

Fig. 1 shows a block diagram of the various film steps involved in making (a) normal 35-mm release, (b) duped 35-mm release, and (c) the new operation for 16-mm release. Solid blocks show existing

35-mm operations while the three blocks with dashed outlines indicate the added operations required for 16-mm. The following special films are used:

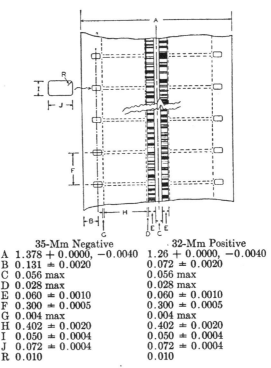

	35-Mm Negative	32-Mm Positive
A	1.378 + 0.0000, −0.0040	1.26 + 0.0000, −0.0040
B	0.131 ± 0.0020	0.072 ± 0.0020
C	0.056 max	0.056 max
D	0.028 max	0.028 max
E	0.060 ± 0.0010	0.060 ± 0.0010
F	0.300 ± 0.0005	0.300 ± 0.0005
G	0.004 max	0.004 max
H	0.402 ± 0.0020	0.402 ± 0.0020
I	0.050 ± 0.0004	0.050 ± 0.0004
J	0.072 ± 0.0004	0.072 ± 0.0004
R	0.010	0.010

The dimensions C, D, and E are those used by Paramount in recording and printing for 16-mm release: however, the track may be as wide as 0.080 (upper section of drawing), in which case C becomes 0.036 and D becomes 0.018. All dimensions are in inches.

Fig. 2

The "35- to 32-Mm Double-Sound-Track Release Negative" is made by electrical re-recording from the sound track of a regular 35-mm composite release print.

The "35- to 32-Mm Dupe Double-Picture-Track Release Negative" is made by optical reduction from the picture on the already existing 35-mm master fine-grain positive picture.

The "32-Mm Double-Composite-Track Release Print" is made by contact printing, first from the 35- to 32-mm sound negative and then

from the 35- to 32-mm picture negative.   This is practically the same
method, practice, and operation of making all present 35-mm standard release prints.

SOUND RE-RECORDING FOR 16-MM RELEASE

High-quality, 16-mm release recording requires different frequency
response, level, and compression characteristics than 35-mm release.
Therefore the use of optical reduction of the sound track was ruled
out.   Since electrical re-recording was indicated, a decision had to be

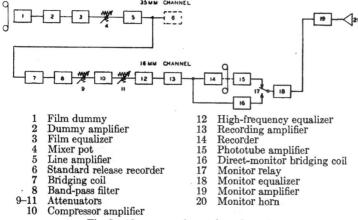

1 Film dummy	12 High-frequency equalizer
2 Dummy amplifier	13 Recording amplifier
3 Film equalizer	14 Recorder
4 Mixer pot	15 Phototube amplifier
5 Line amplifier	16 Direct-monitor bridging coil
6 Standard release recorder	17 Monitor relay
7 Bridging coil	18 Monitor equalizer
8 Band-pass filter	19 Monitor amplifier
9–11 Attenuators	20 Monitor horn
10 Compressor amplifier	

Fig. 3—16-mm sound-recording channel.

made between the use of 16-, 32-, or 35-mm width sound re-recording
negative.   As it was desirable to use standard 35-mm negative developing machines and accessory-handling equipment, 35-mm width
film was selected as shown in Fig. 2, with 32-mm symmetrical perforations to tie in with the printing of the final release track.   The extra
1.5 mm of film width G is located symmetrically on the outside edges
of the film.   This extra film is never removed, nor is this film ever
split, which permits the use of 35-mm width nitrate films for the 35-
to 32-mm release sound and picture negative tracks.

Re-Recording Channel

The sound pickup for 16-mm re-recording is made from a regular 35-mm release print, using a standard re-recording dummy, mixing
portion, and monitor horns of the normal 35-mm re-recording channel.

Fig. 3 shows the block schematic of the re-recording system used.
The over-all frequency characteristic of the entire 35-mm portion of

this channel, including a reproducer (1), reproducer amplifier (2), film equalizer (3), mixer (4), amplifier (5), the 35-mm film recorder (6), and the processed film is nearly flat as shown by curve A of Fig. 4. When re-recording for 16-mm, the film equalizer (3), in Fig. 3, is reduced in equalization to produce an over-all 35-mm channel characteristic shown by curve B in Fig. 4. The 16-mm portion of the re-recording channel is connected to the bridging bus of the 35-mm channel by the bridging coil (7) (Fig. 3). The regular monitor amplifier (19) and horn (20) of the 35-mm channel are connected to the direct and phototube-monitor outputs of the 16-mm channel as shown.

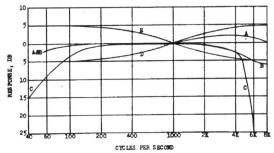

Fig. 4—Characteristics of 35-mm and 16-mm re-recording channel. (A) 35-mm over-all, including film, (B) 35-mm over-all, with reduced film equalization, (C) 16-mm high- and low-pass filters, (D) 16-mm film equalizer, (E) 16-mm monitor equalizer.

In accordance with the general suggestions of the Society of Motion Picture Engineers and the Academy of Motion Picture Arts and Sciences, the frequency characteristic of the 16-mm channel is restricted by high- and low-pass filters as shown by curve C in Fig. 4.

An RCA Type MI-10206C compressor-amplifier (10) (Fig. 3) provides the necessary compression and channel gain. This amplifier is equipped with a de-esser equalizer in the rectifier network to increase the compressor action for high-frequency speech components. The limiting characteristics of this amplifier at 400 and 4000 cycles are shown in Fig. 5. Since the film equalizer of the 35-mm channel is operating at reduced equalization to minimize high-frequency operation of the compressor, it is necessary to follow the compressor-amplifier by a 16-mm-type high-frequency equalizer (12) (Fig. 3) to obtain the desired over-all characteristic. This added equalizer characteristic is shown by curve D in Fig. 4.

The over-all characteristic from the input of the 35-mm film equalizer to the input of the 16-mm modulator is represented by Fig. 6.
It should be noted, however, that because of the variable-level and
frequency-compression characteristics of the compressor-amplifier
both ends of this characteristic are subject to change since the curve
shown was obtained for a single level input, without compression, i.e.,
the summation of curves B, C, and D in Fig. 4.

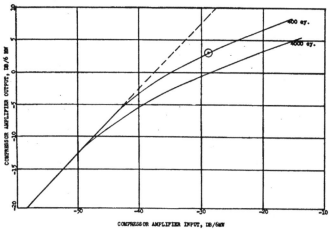

Fig. 5—Compressor-amplifier characteristics at 400 and 4000 cycles.

The 16-mm direct monitor circuit is obtained through the bridging
coil (16) in Fig. 3.   Since there is considerable high-frequency droop
in 16-mm reproducers, the monitor characteristic of this channel is
also reduced by the monitor equalizer (18) which has the characteristic shown by curve E of Fig. 4.

Recording Machine and Modulator

The chassis of a 35-mm Western Electric D-86715 recorder was
used as the basis of the special recording machine.   The sprocket
diameters were reduced in size in order to reduce the film speed from
90 feet per minute to 36 feet per minute.   The filtering was readjusted for excellent flutter characteristics at this speed.   Standard
35-mm width rollers were used but special sprockets had to be installed for pulling this special film which has 16-mm-type sprocket
holes as shown in Fig. 2.   The film magazine was specially designed
so that it could be reversed on the machine, permitting the film to be
run through the recorder first in one direction and then the other

without rewinding. The magazines were built to handle 2000 feet of film anticipating the future use of 1200- to 1600-foot rolls of negative stock.

The modulator is a modified Western Electric Type RA-1152 which produces variable-density intensity-type modulation. An equalizer used with this modulator gives the valve a flat amplitude-frequency-response characteristic over the desired range · A standard RA-1124 noise-reduction amplifier is used in the conventional manner, with 10 decibels noise reduction and 6 decibels margin. No reverse bias is employed.

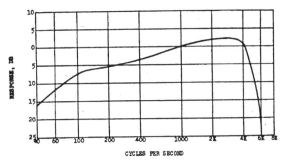

CYCLES PER SECOND

Fig. 6—16-mm-channel characteristic.

A modification of this modulator permits the recording of a sound track on *either* side of the film centerline without readjustment of the lamp, modulator, or film drive. In order to do this, the long-filament lamp normally employed with 200-mil push-pull recorders was used, together with *one side only* of an RA-1061, four-ribbon, push-pull-type valve, the other side being permanently masked off. Since the centerlines of the lamp, optical system, valve septum, and film coincide, it is only necessary to turn the valve end for end on the magnet to reverse the position of the sound track on the film. For the black-and-white film operation described in this paper, only one position of the valve is required. The ability to reverse the valve, however, may be needed in the future in connection with reversal or color-film operations.

The sound tracks are recorded near the center of the film as shown in Fig. 2, which permits the cylindrical lens of the modulator to be installed close to the film, between the recording sprocket teeth. The 60-mil width sound track is used for release. Masking exists in the light valve so no further masking is required in the printer.

The RA-1152 modulator is equipped with a high-quality deflector-type phototube monitor; therefore excellent monitor quality is obtained. This is an asset in this type of re-recording work. The excess illumination resulting from the 40 per cent reduction in film speed was reduced by using partially silvered deflector glass which attenuated the light to the film and at the same time the reflected light to the phototube was increased.

Re-Recording Synchronizing Procedure

The present practice is to re-record each reel twice on the same piece of negative film, once in one direction on the negative and the second

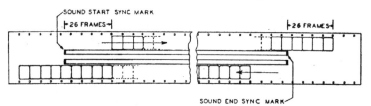

Fig. 7—32-mm release print.

time in the reverse direction. The start synchronizing mark on the 35-mm master film, from which the re-recording is to be made, must be at the standard Academy distance from the first splice and an end synchronizing mark is placed at an èqual distance from the end splice of each reel. The unexposed negative must carry both a start and end synchronizing mark corresponding to the master track. In re-recording the start mark is placed on the unexposed negative, the film is run through the recorder in one direction, an end synchronizing mark is placed on the film, the magazine is reversed without rewinding, the end mark is then used as the start synchronizing mark for the second re-recording as the same reel is made to run through the recorder in the reverse direction. (See Fig. 7.)

This procedure of double re-recording of the same reel is used to facilitate the release printing and to avoid film losses which otherwise would exist because of differences in reel length. The same emulsion-type film is used for the re-recording negative as is used for regular 35-mm release negative, the only difference being in the type of perforation employed. Therefore, *the same release sound negative developing machines and solutions can be used for both 16-mm and 35-mm requirements.*

OPTICAL-REDUCTION PICTURE PRINTER

In order to reduce the picture optically from the 35-mm fine-grain master positive to the special 35- to 32-mm dupe picture negative, a specially designed daylight optical-reduction printer was manufactured by the Bell and Howell Company (Fig. 8). The components are 1, the main machine pedestal; 2, an accurately machined base plate; 3, complete lamphouse assembly with small-size exhaust fan; 4, a light-shielding tube; 5, the 35-mm intermittent mechanism to accommodate the master positive; 5A, the 35-mm master positive

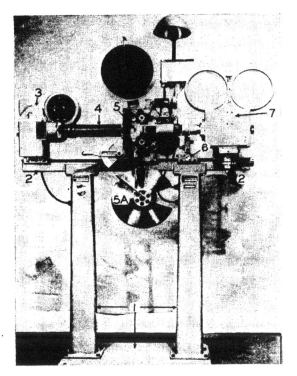

Fig. 8—Step optical reduction picture printer.

take-up; 6, critical objective lens adjusting device; and 7, a standard 35-mm camera head modified to accommodate 35-mm width negative with double 16-mm perforation pitch.

In two operations, the two picture tracks of reel 1A, for example, are laid down on this film in the positions shown by the dotted lines in Fig. 2, running in opposite directions like the sound track. The offsets of the picture tracks longitudinally on the film are shown by Fig.

7, which illustrates one reel of release print before splitting. While the sound tracks are exactly opposite each other, with the "start" synchronizing mark of one recording corresponding in position with the "end" synchronizing mark of the other recording, the picture tracks are offset 26 frames on each side so that the film will play in synchronization on a 16-mm reproducer.

Fig. 9—Modified printer head for printing double 32-mm picture or sound.

The 35- to 32-mm dupe picture negative is the same type of emulsion as used for the regular 35-mm dupes but has of course the special 32-mm perforations. The regular 35-mm dupe-negative developing machine and developer are used which represent an additional economy in operation since no new equipment of this type need be installed.

REDUCTION NEGATIVE PICTURE AND TRACK ASSEMBLY

Reels 1A, 1B, and 2A of both picture and sound negatives are assembled eliminating identifications between reels 1A and 1B and between reels 1B and 2A thereby permitting the printing of three regular release reels in proper continuity for mounting on standard 1200-foot 16-mm reels after slitting.

CONTACT RELEASE PRINTER

The sound and picture 35- to 32-mm negatives are contact-printed to the 32-mm width standard fine-grain release positive using a specially designed Bell and Howell continuous printer (Fig. 9) with a modi-

Fig. 10

fied Model D printer head for printing picture or sound from a 35-mm width double-16 perforated negative to a double-16 perforated 32-mm width standard fine-grain release positive, which is slit after development. The sprocket holes are of the same type as on the 35- to 32-mm negatives but this print stock is only 32 mm wide. Therefore, this printer uses guide rollers for film positioning which keep the centerlines of the two films together within 1 mil which is sufficiently accurate for release purposes. Both sound tracks on the sound negative are printed at the same time as are both picture tracks on the picture negative when it is run through.

32-Mm Developing Machine

This double-track composite print is developed in standard release positive developer in a 35-mm-type developing machine which has been slightly modified to handle the narrower 32-mm width film. This is the only modified developing machine used in the entire operation. Please note, however, that the modifications are such as to permit 32- or 35-mm width film development without interruption of the continuous operation.

Slitting and Assembly

After development, the double-track composite print is run through a specially designed slitter which splits the film down the center as shown in Fig. 10. The weave of this slitter is held within 1 mil. The film is then ready for final assembly. Since 1200-foot rolls of release positive are used, they represent three 1000-foot rolls of the original 35-mm production. Synchronizing marks are removed between rolls and the film spliced together in slightly less than 1200-foot shipping reels.

Acknowledgment

The complete planning and method of handling the 16-mm negative and release project was suggested and worked out by Frank La Grande; and from his specifications, the optical-reduction printer, the release printer, and slitter were engineered and manufactured by the Bell and Howell Company.

The special recording machine and facilities for re-recording and obtaining the sound negative were worked out by Bruce Denney and C. R. Daily under the supervision of Loren L. Ryder, all of the Hollywood Paramount Studio Sound Department. •

Addenda

The procedure described in this paper is now being used by Paramount in the United States and is also being installed in the new Paramount Laboratory in London, England. Mr. La Grande has submitted this procedure to the British Kinematograph Society for general acceptance.

With improved techniques of optical reduction it is also possible to obtain the sound negative by the optical-reduction method in place of sound rerecording. This presupposes that the frequency characteristic and volume range on the 35-mm sound record are correct for 16-mm reproduction. If the original recording is a density recording, printer-light changes can be made as a part of the duping process in which case some reduction in volume range may be obtained.

Three Proposed American Standards

THREE PROPOSED American Standards for cutting and perforating 32-mm film appear on the following pages. They have been developed by a subcommittee on film dimensions, of the SMPE Standards Committee.

Film of this type has been used since 1934 although there never has been a formal standard. During the intervening years a number of changes have been made in the dimensions. Debrie, who was the originator of this film, was aware that the slitting of the 32-mm film into two 16-mm widths might be inaccurate. This inaccuracy would make one half wider than the other half, and might cause trouble since the wide half might stick in the projector gate. Therefore, he made the original French film narrower than twice the width of 16-mm film. The first French film was about 1.252 inches in width. Manufacturers in this country made film of this width for some time but later widened it to 1.257 inches, an increase of 0.005 inch.

It appears that there have been four or five slightly different styles of perforating in use at various times. The values currently adopted for the width of the film and for the transverse pitch of the perforations are believed to be acceptable to all manufacturers. The differences between the present standards and the earlier dimensions are so slight, it is doubtful that the users can perceive them. The dimensions of the perforation, the longitudinal pitch, and the like, are the same as that of the current 16-mm film and the dimensioning of the drawing is in keeping with those standards (Z22.5-1947 and Z22.12-1947).

It will be noticed that the new standards include one for 35-mm film with 32-mm perforations. The reason for the existence of this film is that it can be processed on 35-mm sprocketless developing machines with consequent saving in equipment. This film is commonly used for sound recording and reduction negatives. The negative thus made is printed in the usual fashion. In general, this 32- on 35-mm film is not used for release purposes. However, the fact that people other than manufacturers can perforate 35-mm film in this way has led to some concern. If 35-mm nitrate film were to be perforated with 32-mm perforations, it might later be slit to 16-mm size and be used in projection equipment. Therefore, the standard includes a proviso, "This film should not be made on nitrate base because if this material were slit to 16-mm it might be used on a projector with consequent danger of fire."

No proviso of this sort has ben indicated in other standards for the reason that it is an unwritten law in film-manufacturing companies that no nitrate base should ever be slit to 8-, 16-, or 32-mm widths. The manufacturers do, however, slit both nitrate and acetate film to 35-mm dimensions. Other film users sometimes buy unperforated film and perforate it as they see fit. It was thought, therefore, that special attention should be called to the danger that might result if nitrate film were perforated to any dimensions that might make it usable on 16-mm projectors.

These proposed standards are being published for a ninety-day period for your comment and criticism. If no adverse comment is received before the end of this period, these proposals will be submitted to the Standards Committee for final approval.

Letter Symbols for Physics

Another new standard that will be of interest to many motion picture engineers is ASA Z10.6-1948 "Letter Symbols for Physics" recently announced by the American Standards Association.

The standard suggests that authors who are preparing manuscripts give careful attention to the use of symbols which should always be clearly defined to avoid errors in interpretation. "Letter symbols are to be distinguished from abbreviations, mathematical signs and operators, graphical symbols, and chemical symbols:

  (a) Abbreviations are shortened forms of names and expressions employed in texts and tabulations and should not be used as symbols in equations.

  (b) Mathematical Signs and Operators are characters used with letter symbols to denote mathematical operations and relations.

  (c) Graphical Symbols are conventionalized diagrams and letters used on plans and drawings.

  (d) Chemical Symbols are letters and other characters designating chemical elements and groups."

Copies are now available from the American Standards Association, 70 East 45 Street, New York 17, N. Y., at the price of $1.00.

| Proposed American Standard
Cutting and Perforating Dimensions for
32-mm Sound Motion Picture
Negative and Positive Raw Stock | **Z22.71**
December 1948 |

Page 1 of 2 Pages

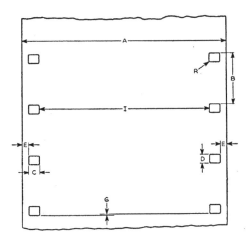

Dimensions	Inches	Millimeters
A	1.257 ± 0.001	31.93 ± 0.025
B*	0.300 ± 0.0005	7.620 ± 0.013
C	0.0720 ± 0.0004	1.83 ± 0.01
D	0.0500 ± 0.0004	1.27 ± 0.01
E	0.036 ± 0.002	0.91 ± 0.05
G	Not > 0.001	Not > 0.025
I	1.041 ± 0.002	26.44 ± 0.05
L**	30.00 ± 0.03	762.00 ± 0.76
R	0.010 ± 0.001	0.25 ± 0.03

These dimensions and tolerances apply to the material immediately after cutting and perforating.

* In any group of four consecutive perforations, the maximum difference of pitch shall not exceed 0.001 inch and should be as much smaller as possible.

** This dimension represents the length of any 100 consecutive perforation intervals.

Proposed American Standard
Cutting and Perforating Dimensions for

32-mm Sound Motion Picture

Negative and Positive Raw Stock

Appendix

The dimensions given in this standard represent the practice of film manufacturers in that the dimensions and tolerances are for film immediately after perforation. The punches and dies themselves are made to tolerances considerably smaller than those given, but owing to the fact that film is a plastic material, the dimensions of the slit and perforated film never agree exactly with the dimensions of the punches and dies. Shrinkage of the film, due to change in moisture content or loss of residual solvents, invariably results in a change in these dimensions during the life of the film. This change is generally uniform throughout the roll.

The uniformity of perforation is one of the most important of the variables affecting steadiness of projection.

Variations in pitch from roll to roll are of little significance compared to variations from one sprocket hole to the next. Actually, it is the maximum variation from one sprocket hole to the next within any small group that is important. This is one of the reasons for the method of specifying uniformity in dimension B.

Thirty-two-millimeter release print stock is slit, after printing and developing, to 16-mm. width. Since a possible error is involved in this slitting, the width of 32-mm. film is made 0.001" narrower than twice the width of standard 16-mm. film. This narrowing gives a tolerance of 0.001" in this secondary slitting operation. If the error of slitting exceeds this tolerance, one of the 16-mm. halves may exceed the width allowed for 16-mm. film and cause interference in the gate of a projector. In addition to errors of centering, there are errors caused by recurring variations in width. These errors will cause weave on the screen even though the maximum width of the film may not be great enough to cause interference in the projector gate.

Proposed American Standard Cutting and Perforating Dimensions for **32-mm Silent Motion Picture** Negative and Positive Raw Stock	Z22.72 December 1948

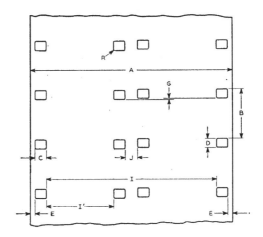

Dimensions	Inches	Millimeters
A	1.257 ± 0.001	31.93 ± 0.025
B*	0.300 ± 0.0005	7.620 ± 0.013
C	0.0720 ± 0.0004	1.83 ± 0.01
D	0.0500 ± 0.0004	1.27 ± 0.01
E	0.036 ± 0.002	0.91 ± 0.05
G	Not > 0.001	Not > 0.025
I	1.041 ± 0.002	26.44 ± 0.05
I'	0.413 ± 0.001	10.490 ± 0.025
J	0.071 ± 0.001	1.803 ± 0.025
L**	30.00 ± 0.03	762.00 ± 0.76
R	0.010 ± 0.001	0.25 ± 0.03

These dimensions and tolerances apply to the material immediately after cutting and perforating.

* In any group of four consecutive perforations, the maximum difference of pitch shall not exceed 0.001 inch and should be as much smaller as possible.

** This dimension represents the length of any 100 consecutive perforation intervals.

Proposed American Standard
Cutting and Perforating Dimensions for
32-mm Silent Motion Picture
Negative and Positive Raw Stock

Appendix

The dimensions given in this standard represent the practice of film manufacturers in that the dimensions and tolerances are for film immediately after perforation. The punches and dies themselves are made to tolerances considerably smaller than those given, but owing to the fact that film is a plastic material, the dimensions of the slit and perforated film never agree exactly with the dimensions of the punches and dies. Shrinkage of the film, due to change in moisture content or loss of residual solvents, invariably results in a change in these dimensions during the life of the film. This change is generally uniform throughout the roll.

The uniformity of perforation is one of the most important of the variables affecting steadiness of projection.

Variations in pitch from roll to roll are of little significance compared to variations from one sprocket hole to the next. Actually, it is the maximum variation from one sprocket hole to the next within any small group that is important. This is one of the reasons for the method of specifying uniformity in dimension B.

Thirty-two-millimeter release print stock is slit, after printing and developing, to 16-mm. width. Since a possible error is involved in this slitting, the width of 32-mm. film is made 0.001" narrower than twice the width of standard 16-mm. film. This narrowing gives a tolerance of 0.001" in this secondary slitting operation. If the error of slitting exceeds this tolerance, one of the 16-mm. halves may exceed the width allowed for 16-mm. film and cause interference in the gate of a projector. In addition to errors of centering, there are errors caused by recurring variations in width. These errors will cause weave on the screen even though the maximum width of the film may not be great enough to cause interference in the projector gate.

Proposed American Standard Cutting and Perforating Dimensions for **32-mm on 35-mm Motion Picture** Negative Raw Stock	**Z22.73** **December 1948**

Page 1 of 2 Pages

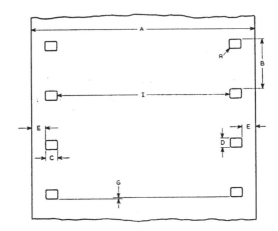

Dimensions	Inches		Millimeters	
A	1.377	± 0.001	34.98	± 0.025
B*	0.300	± 0.0005	7.620	± 0.013
C	0.0720	± 0.0004	1.83	± 0.01
D	0.0500	± 0.0004	1.27	± 0.01
E	0.096	± 0.002	2.44	± 0.05
G	Not	> 0.001	Not	> 0.025
I	1.041	± 0.002	26.44	± 0.05
L**	30.00	± 0.03	762.00	± 0.76
R	0.010	± 0.001	0.25	± 0.03

These dimensions and tolerances apply to the material immediately after cutting and perforating.

* In any group of four consecutive perforations, the maximum difference of pitch shall not exceed 0.001 inch and should be as much smaller as possible.

** This dimension represents the length of any 100 consecutive perforation intervals.

Proposed American Standard
Cutting and Perforating Dimensions for
32-mm on 35-mm Motion Picture
Negative Raw Stock

Appendix

The dimensions given in this standard represent the practice of film manufacturers in that the dimensions and tolerances are for film immediately after perforation. The punches and dies themselves are made to tolerances considerably smaller than those given, but owing to the fact that film is a plastic material, the dimensions of the slit and perforated film never agree exactly with the dimensions of the punches and dies. Shrinkage of the film, due to change in moisture content or loss of residual solvents, invariably results in a change in these dimensions during the life of the film. This change is generally uniform throughout the roll.

The uniformity of perforation is one of the most important of the variables affecting steadiness of projection.

Variations in pitch from roll to roll are of little significance compared to variations from one sprocket hole to the next. Actually, it is the maximum variation from one sprocket hole to the next within any small group that is important. This is one of the reasons for the method of specifying uniformity in dimension B.

This kind of 32 mm. film is made on 35 mm. stock so that it may be processed on 35 mm. sprocketless negative developing machines.

This film should not be made on nitrate base, because if this material were slit to 16 mm. it might be used on a projector with consequent danger of fire.

Joseph H. McNabb

Joseph H. McNabb, president and chairman of the Board of the Bell and Howell Company, died on January 5, 1949, in Chicago. Mr. McNabb was born in Canada, and received his technical education at the Collegiate Institute, St. Thomas, Ontario. He joined Bell and Howell in 1916, and in 1922, he was made chairman of the Board.

Mr. McNabb is well remembered for his activities in the professional and amateur motion picture fields, as an organizer and executive, and for his contributions to the war effort, particularly in the field of optics. He was a Fellow of the Society of Motion Picture Engineers.

65th Semiannual Convention

Hotel Statler, New York, N. Y., April 4-8, 1949

—PAPERS PROGRAM

The 65th Semiannual Convention of the Society will be held at the Hotel Statler (formerly Hotel Pennsylvania) in New York City from April 4 to 8, 1949, inclusive. Authors should submit complete manuscripts of their papers to Miss Helen Stote, SMPE JOURNAL Editor, Room 912, 342 Madison Ave., New York 17, N. Y., by March 1. All abstracts should be sent to Mr. Seeley before February 15. The Chairman and Vice-Chairmen of the Papers Committee are as follows:

N. L. Simmons, *Chairman*
6706 Santa Monica Blvd.
Hollywood 38, California

Joseph E. Aiken, *Vice-Chairman*
116 N. Galveston
Arlington, Va.

Lorin Grignon, *Vice-Chairman*
20th Century-Fox Films Corp.
Beverly Hills, California

R. T. Van Niman, *Vice-Chairman*
4331 West Lake Street
Chicago 24, Illinois

H. S. Walker, *Vice-Chairman*
1620 Notre Dame Street, W.
Montreal, Que., Canada

E. S. Seeley, *Vice-Chairman*
Altec Service Corp.
161 Sixth Avenue
New York 13, N. Y.

—SPECIAL FEATURES

The Monday afternoon session will be a forum on the many aspects of the use of films in television, including statements by several qualified experts, followed by a question-and-answer period. There will be a demonstration Monday night.

Tuesday will be devoted to technical and economic aspects of television.

On Wednesday, April 6, there will be held a Symposium on High-Speed Photography. Authors and delegates from several foreign countries will attend this second SMPE Symposium on this subject.

Thursday and Friday will be devoted to papers dealing with Motion Picture Studio and Theater Equipment, Recording, Films, and Processing.

—LUNCHEON—COCKTAIL HOUR—BANQUET

The usual get-together Luncheon will be held in the Georgian Room at 12:30 P.M., Monday, April 4.

The Cocktail Hour and Banquet will be in the Georgian Room on Wednesday evening, April 6.

PRELIMINARY PROGRAM

Monday, April 4, 1949

9:30 A.M. REGISTRATION
Conference Room 9,
18th floor
ADVANCE SALE OF LUNCH-
EON AND BANQUET
TICKETS

12:30 P.M. GET-TOGETHER LUNCH-
EON
Georgian Room

3:00 P.M. OPENING SESSION
Salle Moderne

8:00 P.M. TECHNICAL SESSION
Salle Moderne

Tuesday, April 5, 1949

9:30 A.M. REGISTRATION
Conference Room 9,
18th floor
ADVANCE SALE OF BAN-
QUET TICKETS

10:00 A.M. TECHNICAL SESSION
Salle Moderne

2:00 P.M. TECHNICAL SESSION
Salle Moderne

OPEN EVENING

Wednesday, April 6, 1949

9:30 A.M. REGISTRATION
Conference Room 9,
18th floor
ADVANCE SALE OF BAN-
QUET TICKETS

10:00 A.M. TECHNICAL SESSION
Salle Moderne

2:00 P.M. TECHNICAL SESSION
Salle Moderne

7:15 P.M. COCKTAIL HOUR
Georgian Room Foyer

8:30 P.M. BANQUET (DRESS OP-
TIONAL)
Georgian Room

Thursday, April 7, 1949

OPEN MORNING

2:00 P.M. TECHNICAL SESSION
Salle Moderne

8:00 P.M. TECHNICAL SESSION
Salle Moderne

Friday, April 8, 1949

10:00 A.M. TECHNICAL SESSION
Salle Moderne

2:00 P.M. TECHNICAL SESSION
Salle Moderne

5:00 P.M. ADJOURNMENT

●

—LADIES' PROGRAM

Mrs. Earl I. Sponable will serve as Hostess, and Mrs. W. H. Rivers will assist her. Ladies' reception and registration headquarters will be located in Room 129.

—HOTEL RESERVATIONS

Room-reservation cards have been mailed to the membership. These should be filled in and returned promptly to the hotel.

Motion Picture Test Films

THE TEST-FILM PROGRAMS of the Society of Motion Picture Engineers and the Motion Picture Research Council have been combined in recent years in an attempt to avoid unnecessary duplication of sources and catalogs that had been confusing to those not familiar with both organizations. The result has been improved service to test-film users in the form of better integrated programs for developing new films and a single catalog. In 1947 the first catalog, listing the twenty-nine films then available, was published as part of the August issue of the JOURNAL and was later reprinted by most motion picture trade magazines. It is now out of print but a revised version will be off the press shortly with copies available to all film users.

All of the test films, both 35-mm and 16-mm, are now supplied on safety stock. They are furnished as a service to the industry and because neither organization has provision for extending credit, all test films are sold on a "cash-with-order" basis at the prices shown here. Orders received without payment enclosed will be shipped C.O.D.

Prices shown do, however, include shipping charges to all points within the United States. Where required by purchasers in other countries, shipping charges will be billed to the customer. If necessary, pro forma invoices will be supplied promptly.

When placing an order with either the Society of Motion Picture Engineers, Inc., 342 Madison Avenue, New York 17, N. Y., or the Motion Picture Research Council, Inc., 1421 North Western Avenue, Hollywood 27, California, be sure to specify clearly—

1. Name of the film
2. Code Number
3. Price
4. Enclose payment with order.

TEST FILMS

Test Film	Code No.	Length (in Feet)	Price
35-Mm Visual Test Film	VTF-1	450	$25.00
Focus-and-Alignment Section	VTF-FAS	100	8.00
Travel-Ghost Target Section	VTF-TGS	100	8.00
Jump-and-Weave Target Section	VTF-JWS	100	8.00
35-Mm Theater Sound Test Film	ASTR-3	500	17.50
35-Mm Multifrequency Test Film			
Type A—Laboratory Type	APFA-1	450	25.00
Type B—Service Type	ASFA-1	300	17.50
35-Mm Transmission Test Film	TA-1	250	17.50
35-Mm Buzz-Track Test Film	ABZT-1	50 min*	0.04/ft

Motion Picture Test Films

35-Mm Scanning-Beam Illumination Test Film		·	
Type A—17-Position Track	A17P-1	230	12.50
Type B—Snake Track	AST8-1	8	0.50
35-Mm Sound-Focusing Test Film			
Type A—9000-Cycle Track	A9KC-1	50 min	0.04/ft
Type B—7000-Cycle Track (Area)	A7KC-1	50 min	0.04/ft
Type C—7000-Cycle Track (Density)	D7KC-1	50 min	0.04/ft
35-Mm 3000-Cycle Flutter Test Film	A3KC-1	50 min	0.05/ft
35-Mm 1000-Cycle Balancing Test Film			
For Two Machines	ABL2-1	14	0.50
For Three Machines	ABL3-1	21	0.75
1000-Cycle Test Film	ABLN-1	· 50 min	0.04/ft
35-Mm Multifrequency Warble Test Film	APWA-1	450	25.00
16-Mm Sound-Projector Test Film	Z52.2	200	12.50
16-Mm Multifrequency Test Film	Z22.44	150	41.25
16-Mm Buzz-Track Test Film	Z22.57	100	27.50
16-Mm Scanning-Beam Illumination Test Film			
Laboratory Type	Z52.7-L	100	27.50
Service Type	Z52.7-S	100	27.50
16-Mm Sound-Focusing Test Film			
Laboratory Type	Z22.42–7000	100	27.50
Service Type	Z22.42–5000	100	27.50
16-Mm 3000-Cycle Flutter Test Film	Z22.43	100	27.50
16-Mm 400-Cycle Signal-Level Test Film	Z22.45	100	27.50

* Minimum.

Book Review

An Introduction to Color, by Ralph M. Evans

Published (1948) by John Wiley and Sons, Inc., 440 Fourth Ave., New York 16, N. Y.   324 pages + X pages + 3-page general bibliography + 12-page index. 269 illustrations + 15 color plates.   7¹/₂ × 9³/₄ inches.   Price, $6.00.

Color may be defined as a hue of the rainbow or spectrum, or as a tint produced by a mixture of such hues, or as a paint or a pigment, or in various other ways for ordinary usage.   To the scientist, however, color has a more specific meaning depending upon its use and application.

To the physicist color is one of the characteristics of the radiant energy known as light.   He is concerned with the properties of light in terms of wavelengths or frequencies, and with its intensity, both absolute and comparative.   He has devised many instruments, such as monochromers, spectrophotometers, radiometers, photoelectric cells, and the like, and employs many different light sources for the emission of light, all for the purpose of describing and defining light and color in terms of his accepted constants such as lumens, foot-candles, ergs, joules, and so forth.

The psychophysicist, on the other hand, takes the "color" of the physicist and relates it to the sensitivity of the human eye.   In so doing he evaluates the physicist's "color" in terms of an average observer usually by the use of mechanical means.   He is interested in the entire system composed of the light source, the radiant energy, and the human eye.

The psychologist studies the action of the mind in perceiving and interpreting light and color, and explains, for example, why the same physical light is not always seen as the same color.

Some time ago the Committee on Colorimetry of the Optical Society of America, recognizing the complexities and interrelationship of the many variables involved in the field of color, tabulated them in chart form under the head of physics, psychophysics, and psychology.   For the first time, however, to this reviewer's knowledge, a book written by an expert in the field has been published in order that the reader interested in color may understand, correlate, and interrelate all of these many variables.

It would be difficult to find a man better suited for the task than Mr. Evans. An authority in the field of color, he has for many years been conducting research on visual effects in photography, and has delivered many lectures, and written numerous papers on the subject of color.   His first book on the subject is a milestone.

In "An Introduction to Color" Mr. Evans covers the physics of color in the first six chapters with headings as follows: Chapter I, "Color and Light"; Chapter II, "The Physical Nature of Light"; Chapter III, "Light Sources"; Chapter IV, "Illumination"; Chapter V, "Colored Objects"; Chapter VI, "The Physics of Everyday Color."   Without delving too deeply into the subject, the reader is given a brief but adequate grasp of the basic physics of light and color.

Book Review

The psychophysics of light and color are discussed in Chapter VII, entitled, "Color Vision," including a brief review of the elementary physiology of the eye and the functions of the various brain centers. Mr. Evans discusses light and color as evaluated by the eye as a fixed sensitivity receptor.

Having treated the physical and psychophysical aspects of light and color in earlier chapters, Mr. Evans then directs his attention to the psychological aspects of vision. In Chapters VIII through XI, entitled, Chapter VIII, "The Visual Variables of Color"; Chapter IX, "Perception and Illusion"; Chapter X, "Brightness and Perception"; Chapter XI, "Color Perception." Mr. Evans reveals the part which the mind plays as the final interpreter of light and color phenomena. He makes clear why the same physical light can appear different to the eye under different conditions and in different environments. Illusions and other unexpected phenomena which exist in the form and shape of objects can also exist in light and color, and some of the bases of their interrelationship are shown in these chapters.

In the subsequent eleven chapters of the book Mr. Evans warms up to problems which are more purely those of color such as the measurement of color, the specifications of color, a thorough review of the various color systems, their assumptions, tools and methods of use and application, color differences and color names, additive and subtractive mixtures of colors, and transparent and nontransparent color mixtures (paints). The author uses the same approach in discussing these problems as he did in considering the relationship of the physics, the psychophysics, and the psychology of light and color.

In the last three chapters of the book the role of color in photography, art, design, and abstraction is studied and interpreted in the light of the previous chapters.

The book is extensively illustrated, and noteworthy are the clear explanations and titles of the many graphs and pictures. "An Introduction to Color" treats difficult subjects with clarity and simplicity. Definitions and usages of words unique to their field of science are adequately explained.

The author's emphasis through explanation and re-explanation, even if sometimes repetitious, will be helpful and pleasing to the many teachers and students who will unquestionably use this book for study and reference purposes.

The absence of much of the mathematical background of the many concepts discussed will be of benefit to the relatively untrained reader interested in color, and will not prove a serious deficiency for the trained scientist.

The book is highly recommended by this reviewer for the serious worker in any field in which color is an important factor, and is a "must" for the artist, technician, and scientist in color photography and related arts.

HERBERT T. KALMUS
Technicolor Motion Picture Corporation
Hollywood, Calif.

1949 Nominations

THE 1939 NOMINATING COMMITTEE, as appointed by the President of tl Society, was confirmed by the Board of Governors at its January meetin

D. E. HYNDMAN, *Chairman*
Room 626—342 Madison Avenue
New York 17, N. Y.

HERBERT BARNETT
General Precision Equipment Corp.
63 Bedford Road
Pleasantville, N. Y.

F. T. BOWDITCH
Research Laboratories
National Carbon Company
Box 6087
Cleveland 1, Ohio

F. E. CAHILL, JR.
Warner Bros. Pictures, Inc.
321 W. 44 Street
New York 20, N. Y.

R. E. FARNHAM
General Electric Company
Nela Park
Cleveland 12, Ohio

G. R. GIROUX
Technicolor Motion Picture Corp.
6311 Romaine Street
Los Angeles 28, Calif.

A. N. GOLDSMITH
597 Fifth Avenue
New York 17, N. Y.

T. T. GOLDSMITH
Allen B. Du Mont Laboratories
2 Main Avenue
Passaic, N. J.

K. F. MORGAN
Western Electric Company
6601 Romaine Street
Hollywood 38, Calif.

All voting members of the Society who wish to submit recommendations fo candidates to be considered by the Committee as possible nominees, are requeste to correspond directly with the Chairman or any of the members of the Nominat ing Committee. Active, Fellow, or Honorary Members are authorized to mak these suggestions which must be in the hands of the Committee by May 1, 1949

There will be eight vacancies on the Board of Governors as of January 1, 1950 which must be filled. Those members whose terms of office expire are:

Financial Vice-President...............................D. B. JOY
Engineering Vice-President............................J. A. MAURER
Treasurer...R. B. AUSTRIAN
Governor....A. W. COOK   Governor....P. J. LARSEN
Governor....JAMES FRANK, JR.   Governor....G. E. SAWYEI
Governor....L. T. GOLDSMITH

The recommendations of the Nominating Committee will be submitted to th Board of Governors for approval at the July meeting. The ballots will then b prepared and mailed to the voting members of the Society forty days prior to th Annual Meeting of the Society. This is the business session held during the Fal Convention, which this year will be in Hollywood, California, October 10–14.

D. E. HYNDMAN, *Chairman*
Nominating Committee

Optical Society Meeting

THE OPTICAL SOCIETY OF AMERICA will hold its Winter Meeting in New York City, on March 10, 11, and 12, 1949, with headquarters at the Hotel Statler.

A special feature of the meeting will be a symposium on luminescence, to be held on the first day of the meeting under the guidance of G. R. Fonda of the General Electric Company. This symposium will be high-lighted by "A Survey of Present Methods Used to Determine the Optical Properties of Phosphors," by Wayne B. Nottingham, Massachusetts Institute of Technology, and "Review of the Interpretations of Luminescence Phenomena," by Fred E. Williams, Research Laboratory, General Electric Company.

Anyone desiring to attend this meeting should notify Dr. G. R. Fonda, General Electric Company, Schenectady, N. Y., at least three weeks in advance of the meeting.

The sessions for contributed papers on Friday and Saturday will be preceded by the following invited papers: "The Ruling of Large Diffraction Gratings," by G. R. Harrison, Massachusetts Institute of Technology; "Measurements of Size and Shape of Large Molecules by Light Scattering," by P. Debye, Cornell University; and "A Color Translating Ultraviolet Microscope," by E. H. Land, Polaroid Corporation.

This meeting will be open to nonmembers of the Society, and all interested persons are cordially invited to attend.

Current Literature

THE EDITORS present for convenient reference a list of articles dealing with subjects cognate to motion picture engineering published in a number of selected journals. Photostatic or microfilm copies of articles in magazines that are available may be obtained from The Library of Congress, Washington, D. C., or from the New York Public Library, New York, N. Y., at prevailing rates.

American Cinematographer
29, 12, December, 1948
Latensification (p. 409) H. W. MOYSE
Mobile Camera Lab. (p. 410) W. M. CLINE
Karl Freund Introduces the Contract Lineup System (p. 413) R. LAWTON

Audio Engineering
33, 1, January, 1949
Psycho-Acoustic Aspects of Higher Quality Reproduction (p. 9) C. J. LEBEL

International Photographer
20, 12, December, 1948
New "Spectra" Meter (p. 7) L. MOEN

International Projectionist
23, 12, December, 1948
The New British SUPA Projector (p. 21) H. HILL

Radio and Television News
40, 6, December, 1948
The Recording and Reproduction of Sound. Pt. 22 (p. 48) O. READ

～ New Products ～

Further information concerning the material described below can be obtained by writing direct to the manufacturers. As in the case of technical papers, publication of these news items does not constitute endorsement of the manufacturer's statements nor of his products.

Brenkert Film Projector

A new 35-mm Brenkert film projector, especially designed for the medium size theater which must operate on a conservative budget, was introduced recently by the **RCA Theater Equipment Section**, 36 W. 49 St., New York 20, New York.

Engineering and performance features of the new model include a design for the rear shutter blade which supplies good ventilation to the projection

240

aperture for cooling purposes, and an operating compartment that is oilfree and roomy, providing maximum space for threading the projector.

An important feature of the BX-60 is the automatic lubrication system. All rotating shafts running through the main case casting are equipped with oil baffles, so that shaft bearings are continuously lubricated throughout their length, but no oil can leak into the operating compartment.

A large door on the operating side of the projector exposes the entire film compartment and two glass-covered openings permit the operator to observe the film loops above and below the film trap while the mechanism is in operation. Quick access to the shutter blades and the rear of the film trap is gained by removal of a panel on the operating side which is held in place by two thumbscrews. A filter glass is provided in this panel for viewing the light on the aperture.

The intermittent mechanism in the BX-60 is identical to that in the larger Brenkert BX-80 projector. Unit construction is used to facilitate easy, quick, and accurate servicing. All units are doweled to the main frame for correct alignment of parts, thereby maintaining the accuracy built into the mechanism.

Journal of the
Society of Motion Picture Engineers

VOLUME 52 · MARCH 1949 NUMBER 3

Subscription to nonmembers, $10.00 per annum; to members, $6.25 per annum, included in
their annual membership dues; single copies, $1.25. Order from the Society's General Office.
A discount of ten per cent is allowed to accredited agencies on orders for subscriptions and
single copies. Published monthly at Easton, Pa., by the Society of Motion Picture Engineers
Inc. Publication Office, 20th & Northampton Sts., Easton, Pa. General and Editorial Office,
342 Madison Ave., New York 17, N. Y. Entered as second-class matter January 15, 1930,
at the Post Office at Easton, Pa., under the Act of March 3, 1879.

Society of
Motion Picture Engineers ·

342 MADISON AVENUE—NEW YORK 17, N. Y.—TEL. MU 2-2185
BOYCE NEMEC . . . EXECUTIVE SECRETARY

OFFICERS

1949–1950

PRESIDENT	PAST-PRESIDENT	SECRETARY
Earl I. Sponable	Loren L. Ryder	Robert M. Corbin
460 W. 54 St.	5451 Marathon St.	343 State St.
New York 19, N. Y.	Hollywood 38, Calif.	Rochester 4, N. Y.

EXECUTIVE VICE-PRESIDENT	EDITORIAL VICE-PRESIDENT	CONVENTION VICE-PRESIDENT
Peter Mole	Clyde R. Keith	William C. Kunzmann
941 N. Sycamore Ave.	120 Broadway	Box 6087
Hollywood 38, Calif	New York 5, N Y	Cleveland 1, Ohio

1948–1949   1949

ENGINEERING VICE-PRESIDENT	TREASURER	FINANCIAL VICE-PRESIDENT
John A. Maurer	Ralph B. Austrian	David B. Joy
37-01—31 St.	1270 Avenue of The	30 E. 42 St.
Long Island City 1, N. Y.	Americas New York 20, N. Y.	New York 17, N. Y.

Governors

1948–1949	1949	1949–1950
Alan W. Cook	James Frank, Jr.	Herbert Barnett
25 Thorpe St.	426 Luckie St., N. W.	Manville Lane
Binghamton, N. Y.	Atlanta, Ga.	Pleasantville, N. Y.
Lloyd T. Goldsmith	William H. Rivers	Fred T. Bowditch
Warner Brothers	342 Madison Ave.	Box 6087
Burbank, Calif.	New York 17, N. Y.	Cleveland 1, Ohio
Paul J. Larsen	Sidney P. Solow	Kenneth F. Morgan
508 S. Tulane St	959 Seward St.	6601 Romaine St.
Albuquerque, N. M.	Hollywood 38, Calif.	Los Angeles 38, Calif.
Gordon E. Sawyer	R. T. Van Niman	Norwood L. Simmons
857 N. Martel Ave.	4431 W. Lake St.	6706 Santa Monica Blvd.
Hollywood 46, Calif.	Chicago 24, Ill.	Hollywood 38, Calif.

Theater Television

INTRODUCTION

D URING THE YEARS 1944, 1945, and 1946, the Theater Television
Committee of the SMPE worked on many of the various engineering problems related to placing television in the motion picture
theater. Considerable time and effort were expended in collecting information on such subjects as existing theater architecture, available
projection equipment, picture quality to be expected, and many other
problems which would be of interest to the motion picture industry if
it intended to use this new medium. Also during these years, Paul
J. Larsen, together with other representatives of the Society, appeared before the Federal Communications Commission and succeeded in obtaining for the motion picture industry frequency allocations for theater-television use on an experimental basis only.

The Motion Picture Association was then approached with repeated requests that it co-operate in the television work of the Society if it had any reason to believe that theater television would be
practical. Neither producers nor distributors, ·however, were interested in theater television at that time nor were they particularly
concerned about television as a competitive entertainment medium.
The exhibitors on the other hand, showed some concern but did not
wish to take any active measures either on their own or with the
Society. The general attitude seemed to be that it might be possible
to buy into the television industry at some future date and thereby
save the high cost of research and development.

In November, 1946, a point had been reached in the technical work
where it was believed a definite statement of interest by the motion
picture industry was required if the work were to continue. In addition, public hearings before the FCC were scheduled for early 1947
at which it was proposed to reallocate to other services the frequencies
formerly provided for experimental theater use.

In spite of the lack of interest shown, the Society again undertook
having a brief prepared and Mr. Larsen, then chairman of the Theater
Television Committee, appeared before the FCC on February 4, 1947.
(See FCC Docket No. 6651.) Immediately preceding the hearings a
telegram was received from Eric Johnston, president of the Motion
Picture Association, endorsing the SMPE's stand. A similar telegram
was sent to the FCC three weeks following the hearings by Donald

Nelson, then president of the Society of Independent Motion Picture Producers.

The decision of the FCC was handed down early in 1948 and while it did not provide specific frequency allocations for theater use, it did make available certain frequencies which could still be used by the motion picture industry for experimental purposes.

Consequently, even though the motion picture industry's position is weaker from the standpoint of obtaining a permanent part of the radio-frequency spectrum, the Theater Television Committee decided to continue its engineering work.

The members agreed to draw up as comprehensive a report as possible outlining the present state of the art in so far as it regards the motion picture theater and again to seek the co-operation of the industry as a whole. Such a report is contained in the following pages.

In this report an attempt has been made to present the information in as nontechnical language as possible so that those not closely associated with the television industry will be aware of the tremendous strides made in this new medium during the last few years.

Government regulations involved, types of theater equipment available, and distribution facilities now in existence are described in a manner which it is hoped will be of value to all branches of the motion picture industry.

GOVERNMENTAL ASPECTS

The following is a general statement as to the regulations and procedures which in part govern the establishment of a theater-television system of urban, intercity, or national scope. It is intended only as a general and partial guide and must be supplemented by detailed information prior to any definite action on the part of those planning theater-television-program syndication either locally or on a nation-wide scale. Detailed information on the rules and regulations of the Federal Communications Commission can be secured by addressing the Secretary of the Commission in Washington.

No Federal license or governmental permission is required for the establishment of a theater-television *receiving* station, either for the reception of programs by wire or coaxial cable or by radio. However, municipal regulations may control the placement of masts or other structures on roofs, the guying of reinforcements of such structures, and the safety of any electrical wiring of permanent nature installed in the theater. If high towers are erected for reception, and if these are so located that they may become an obstacle or hazard to aerial navigation, it is possible that the Civil Aeronautics Authority must grant approval prior to the erection of such facilities.

If one or more theaters in a given city are to receive a television program from central studios, means must be provided for carrying the program from the central studios to the theaters in question. The program may originate from live-talent presentations in the studio (or at remote pickup points such as sports arenas, legitimate theaters, or the like), or they may originate from film records previously made. They may be carried to the theaters by means of specially equalized telephone lines or by coaxial cables, either of which presumably will be furnished by the local telephone company or other public-utility common carrier in the communications service. Alternatively, the programs may be sent by narrow radio beams from a central transmitting station to the individual theaters where they are received on highly directional antennas or aerials.

If radio beams are to be used, it becomes necessary to secure the approval of the Federal Communications Commission and to receive a construction permit and, thereafter, a station license to permit the operation of the transmitter which sends the studio program to the various theaters.

The transmissions in question are *not* broadcast (that is, addressed to the general public), and a broadcasting license would not be required from the FCC. The transmissions are rather of the type known as multiple-addressee messages which are private communications addressed by a single sender to a group of recipients, each of whom receives the same message. Such messages, unlike broadcasts, are private in nature, and are not legally available to the public.

Any central transmitter erected to send television programs to a group of theaters will require a suitable tower to support the transmitting equipment. If this tower is a potential aeronautical hazard, authorization will be required from the Civil Aeronautics Authority for its erection. Municipal codes relative to the establishment of towers in residential districts must also be considered.

The FCC has taken the present stand that the distribution of material of a private nature, such as is here contemplated, falls within the scope of the common carriers and that it should, therefore, be handled by a telephone or telegraph company. It is not known whether the FCC will permanently adhere to this policy, particularly in the case of urban television studios and transmitters for syndication of programs to theaters. In any case, the channels (frequencies) assigned to such television transmissions necessarily would be different from those used for television *broadcasting*. The channels would probably be at considerably higher frequencies, would be wider (to accommodate possibly higher-detail pictures, or color pictures, or both), and might differ from the broadcasting channels in other respects as well.

The SMPE has previously requested allocations of channels from the FCC for commercial theater television, but that request has not as yet been granted. It is not known whether the Commission ultimately would grant such channels but it is believed that their grant would require the following steps.

An individual theater owner planning to establish a television service to its theaters would first apply for an *experimental license* from the FCC to permit him to transmit his programs, purely experimentally and noncommercially, for a specific period. He would be obligated to describe his plans clearly, and to report from time to time to the Commission on his technical progress.

If his experiments were successful, he might then ask that his experimental license be converted into a *commercial license* permitting normal and continued operation during the period of the license (which might be set at three years, or some similar period). The Commission then doubtless would hold hearings to determine the need for and desirability of the service in question. If it found that the service was useful and necessary and that channels were available, it would then grant the corresponding commercial license. City-allocation hearings would also be held.

It should be added that each theater chain in the same city would require its own transmitting facilities or wire network (unless an interchange of programs or the common use of a single transmitter were acceptable to all involved). That is, for completely independent service each theater group would require its own transmitter, its own receivers, and its own channel allocations from the FCC. An individual theater owner or theater circuit might arrange for transmission of its programs by an existing television station or licensee in a manner similar to current commercial-television broadcasts.

In addition, for remote pickups outside the main studios, it becomes necessary (if radio is to be used to carry the program from the remote point to the central studios for retransmission to the theaters) to secure an FCC license for the radio-beam transmitter which will carry such programs. Channels are presently available for that purpose and might, for good reason, be secured from the FCC.

When nation-wide syndication is involved, it becomes necessary to interconnect the central studios of each network in the cities in which it serves theaters by means of coaxial cables (of the telephone company) or by radio relay (either supplied by the telephone company or other common carrier, or else established by special permission of the FCC as the result of a change in its present policy). Such radio-relay systems consist of a number of repeater stations about thirty miles apart, each of which receives the program from the preceding station and automatically carries it forward to the next station. . These relay or "booster" stations may be unattended and subject only to occasional inspection and the replacement of expendable material.

In brief, those considering the use of radio in theater television are particularly directed to the following basic points which may be novel to those not familiar with the field of radio communication.

1. Unlike the motion picture field, television by radio is subject to numerous governmental regulations and controls. Consequently, those entering theater television, and using radio transmission, must be thoroughly familiar with governmental rules and procedures and be governed thereby for their own protection.

2. In the second place, television by radio requires so-called wide-channel assignments by the FCC. Such channels are scarce and much sought. Accordingly, nonuse of such channels amost inevitably leads to their pre-emption by others.

3. Accordingly, if theater television is to secure such radio channels, it must promptly request their assignment. However, a mere request is not generally sufficient to persuade the FCC to grant channels. Usually financial responsibility, definiteness of construction and operating plans, nature of ownership and affiliation, willingness to report all technical (and perhaps program) progress, and other obligations must be made sufficiently clear and definite to the Commission to justify the assignment of channels. There seems little likelihood that a vague expression of general interest or intent will lead to channel assignments.

4. In any case, even an otherwise satisfactory application for channels must be denied if no available and noninterfering channels are any longer existent, because of prior assignments. The conclusions to be drawn are evident.

PROJECTION SYSTEMS

Two basic systems of large-screen theater television are currently being evaluated in this country. One is the instantaneous or direct-projection system by which high-brilliance cathode-ray-tube images are projected by means of an efficient reflective optical system; the other is the storage or intermediate-film system using standard motion picture projection technique, after television images have been photographed or transcribed on motion picture film and suitably processed.

Although neither type is commercially available in production quantities, the rapid progress of the art warrants description of equipment in experimental installations.

DIRECT-PROJECTION SYSTEM

The direct-projection system consists of three major optical elements: 1. the projection cathode-ray tube which is the source of the light image; 2. the optical system which projects the image into the screen; and 3. the screen from which the final image is viewed. In addition to the optical elements of the system, which are housed in a projection barrel, the electronic auxiliaries include a control console containing the associated television equipment, and a power-supply rack. A typical experimental arrangement of this equipment as installed in a theater is shown in Fig. 1.

Projection Cathode-Ray Tube—The cathode-ray tube used in the direct-projection system is similar to the direct-viewing tube used in the conventional television receiver, except that projection tubes have a much greater light output resulting from higher voltage operation for which they are designed.

Since available television for commercial operation is not adapted, at the present state of the art, to the use of supplementary light sources as are motion pictures, the brightness of the image available for projection depends upon the efficiency of the cathode-ray tube and the operating potentials. An average picture on a projection tube will draw a beam current of approximately 1 milliampere at a potential of 80 kilovolts. This is a power of only 80 watts. With a screen efficiency of 5 candle power per watt this represents a light output of 400 candle power.

Optical System—Typical optical systems employed in large-screen television have been discussed in various issues of the SMPE JOURNAL, and will therefore be only briefly reviewed here. The familiar refractive projection optics used in motion picture film projectors deliver approximately 6 per cent of the light from the arc-light source to the screen. On the other hand, the reflective optics developed for television deliver 30 per cent of the light output from the cathode-ray tube to the screen.

Reflective optics have been designed for large-screen projection of pictures up to 18 × 24 feet. One system for a 7'/₂ × 10-foot screen uses a 21-inch mirror, a 14-inch "lens" (correction plate), and a 7-inch, 50-kilovolt, cathode-ray tube. The largest system built so far con-

Fig. 1—Typical direct-projection system.

sisted of a 42-inch mirror, a 26-inch "lens," and 12- and 15-inch projection tubes operating at 80 kilovolts. The throw was fixed at 40 feet and by changing the cathode-ray tube either a 15- × 20- or an 18- × 24-foot picture was shown. The magnification is fixed by the mirror radius. High present production cost of large-mirror systems seemingly indicate the advisability of concentrating on smaller optics and increasing the voltage capabilities of smaller cathode-ray tubes (7-inch) in order to make a compromise system which might be successful commercially.

Viewing Screen—The viewing screen forms the third and final optical element of direct-projection television. Standard motion picture screens have a diffuse surface which distributes the light more or less uniformly in all directions. Since the distribution is nondirectional a great deal of light is lost to the ceiling and floor. Directivity, if it could be obtained in the vertical plane, would concentrate the light where it would be most useful and effect an important increase in efficiency. Beaded screens have been made to control the direction of the reflected light from the screen, but the directivity pattern, while showing a gain of 2, restricts the horizontal reflective pattern and tends to reflect a great deal of illumination back into the optical system where it reduces the contrast of the projected image. Developments in directional screens now underway promise gains as high as 3. A lenticular screen of this type was successfully used in the Fox Theater in Philadelphia where a 15- × 20-foot picture was shown featuring the 1948 Louis-Walcott fight. This screen is embossed on an aluminized surface. with small convex-lens elements to control the directivity pattern. The observed results were excellent and a gain of two and one-half times was measured.

Such is not the case with a normal translucent screen (rear projection) since the light comes from a relatively small source and is a diverging cone of light at the screen. (The usual translucent screen receives direct rays which are normal to the center of the screen but diverge nearer the edges resulting in a bright spot in the center of the screen.) A field lens can be used on the rear of the translucent screen to direct the rays in a parallel pattern, and hence give more uniform illumination over the entire screen, or by a modification the pattern may be made to suit almost any application. Such a field lens may be applied only in small screens as in the home type of projection receiver where a molded-plastic screen can be used. A compromise screen of high-density translucent material can be made, but the gain will be low and the directivity pattern becomes very sharp.

Equipment Elements—The current design trend for direct-projection systems is to break the equipment into several discrete units: 1. The optical housing containing the mirror, lens, cathode-ray tube and its associated deflection coil, and a cooling system for the cathode-ray tube; 2. The control console containing the critical television elements such as the video amplifier and deflection circuits as well as the operating control panel; 3. The auxiliary power equipment consisting of a power-supply rack and a high-voltage power unit.

Equipment Location—Various locations have been suggested and tried for this type of projection-television equipment. The present throw limitation makes the normal booth installation impracticable. Longer throw systems up to 65 feet, can be made but again the cost and size factor rule them out. Rear projection might seem ideal for short-throw systems, but the screen directivity is too sharp to make this practicable. If it were economical to waste a great deal of light on a very dense screen there might be some compromise possible in this direction. Another important consideration in selecting the location is the projection angle because the limited depth of focus of the short optical system demands operation with the screen normal to the projection axis.

The installation requirements are peculiar to the optical system employed and ideally would locate the optical housing on the front of the balcony. Alternatively in a nonbalcony house, the optical housing may be located either on a special ceiling suspension or in the orchestra. The control console should not be more than 15 feet from the optical housing because of circuit requirements, which usually dictates its placement at the balcony rail. The balance of the equipment can be remotely placed at any convenient point, but cost will probably indicate a location less than 100 feet from the optical housing.

Performance—Picture quality from large-screen television projectors is now limited by the quality of the transmitted signals. The capabilities of the projection system are equal to the best studio television equipment and any deterioration of the signal between the camera and the projector causes an inferior picture on the screen. Experience has shown that with a picture of suitable quality it is possible to produce results acceptable to critical audiences. Present transmission of television pictures on standard channels is limited in bandwidth so that the projected pictures actually have about 300 lines resolution. If the pictures were transmitted by microwave relay, the entire capability of the projection system of approximately 350 to 450 lines could be utilized.

A television system specifically designed for theater use will no doubt be a private system using ultrahigh-frequency channels and all of the equipment and techniques of operation will be improved to utilize the present standards to the fullest extent.

STORAGE-PROJECTION METHODS

Two basic image storage television projection systems are being investigated. The first uses motion picture film as the intermediate storage medium while the second employs electronic means.

Film-Storage Method—The film-storage method of large-screen television projection is the only storage system available even on an ex-

Fig. 2—Rapid processing unit.

perimental basis in this country. The system described here was developed by Paramount Pictures and has been used on several occasions in the Paramount Theater in New York City. While developed by Paramount, the fundamentals are similar in many respects to equipments designed and built by others and may give the motion picture industry an insight into the problem involved in setting up such a system.

The film-storage system consists of four basic elements: 1. television receiving equipment; 2. recording camera; 3. rapid film-processing equipment; 4. a conventional 35-mm motion picture projector. Illustrations of such equipment are shown in Figs. 2 and 3. In practice, Paramount has used mobile cameras together with microwave radio-relay equipment to bring the program material to the theater. The mobile cameras with associated control equipment and microwave-relay unit are of the conventional type used by television broadcasters for remote pickup and cost approximately $55,000.

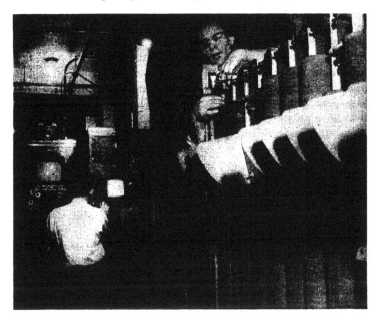

Fig. 3—Complete equipment in operation.

Receiver—All receiving equipment* is housed in one unit. This includes all video and audio equipment together with high- and low-voltage supplies. Two screens are provided. One employs a 15-inch cathode-ray tube for monitoring; the other is a 10-inch cathode-ray tube having an aluminum-backed, flat-face screen. This 10-inch cathode-ray tube is of the blue, short-persistence type and provides the received image which is photographed. This screen has the polarity reversed and the received image is a negative. Audio portions of the program are monitored by a loudspeaker included in this unit.

* NOTE: The total cost of receiver, camera, and processing unit is approximately $35,000 plus installation.

Camera—A special recording camera* is employed having no mechanical shutter but having its pulldown mechanism synchronized at the standard film rate of 24 frames per second with an electronic shutter incorporated in the circuits of the 10-inch cathode-ray tube. Twenty frames following exposure of the picture, the film passes through the sound modulator. A film magazine mounted directly above the recording camera holds sufficient unexposed film for two hours' continuous recording.

Processing—Exposed film from the recording camera passes through a chute directly to a high-speed processing unit.* A maximum of 66 seconds is required to develop, fix, wash, and dry the exposed film. Facilities are provided either to wind the processed film on reels or feed it directly to the projectors. The processing unit requires a hot- and cold-water supply of approximately 20 gallons per minute. The hot-water supply must have a minimum temperature of 140 degrees Fahrenheit. Cold-water supply at conventional tap temperature is adequate. Automatic mixing is provided within the unit to attain a resultant temperature of approximately 125 degrees Fahrenheit. A slop-sink should be provided for disposal of spent photographic chemicals.

Power—The total power required to operate the three units (receiver, camera, and processing) is 100 amperes, 3-phase, 208 volts, alternating current.

Space Requirements—The space required to house the receiving, recording, and processing units is 200 square feet. To facilitate operation and maintenance, a room 10 × 20 feet is recommended with the equipment set up in a straight line allowing at least a 2-foot aisle on all sides.

Electronic-Storage Methods—Equipment in this category is not currently available for use in American theaters and it does not appear that such equipment will be available in the immediate future.

Two basic systems are described, however, in this report. The first uses the dark trace or Skiatron types of screens which are known in the American market as P-10 phosphors. Manufacturers in this country do not plan in the near future to market a tube which has the proper characteristics for television, and some of them express the opinion that this screen is not feasible for such use.

This fact is, of course, well known to the industry from the results of published research by many independent investigators as well as the

---

* NOTE: The total cost of these three units (receiver, camera, and processing unit) is approximately $35,000 plus installation.

engineers from some of the companies contacted in this survey. Generally speaking, the Skiatron tube, at the present development in the art, produces an image which does not permit sufficient contrast and low persistence to compete successfully with phosphorescent screens or with photographic emulsions. It is also difficult to produce a screen which produces true blacks and whites. Similarly, its decay time is a complex phenomenon, and although it can be controlled to some extent in manufacture, satisfactory performance in this regard has not been obtained to date. It is entirely possible, however, that future developments may reverse present thinking in this regard. Any such trend will be noted and presented in a future report of this committee.

The second storage system was developed in Switzerland and is known as the AFIF Method of Large-Screen Television Projection. This system was developed by Dr. F. Fischer at the Swiss Federal Institute of Technology. Since it was known that this system was not currently available for sale, no contact was established with this Institute. Because the operation of this system is not well known in this country, a brief description is given in the Appendix.

Although not commercially available, a laboratory model of a theater projector using this system was demonstrated in Zürich, Switzerland, during the week of September 5, 1948. Reports from those viewing the demonstration were that screen brightness was equivalent to present motion picture practice and picture definitions were adequate for theater use. The demonstration was conducted, however, using 729 lines rather than 525 now currently standard for broadcast purposes in this country.

DISTRIBUTION FACILITIES

Coaxial Cables and Radio Relays—The Bell System coaxial cable and radio-relay facilities installed and in operation are as follows:

ROUTE	CHANNELS (EXCLUDING STAND-BY)	INTERMEDIATE TERMINALS
EASTERN SEABOARD NETWORK		
New York–Washington	4	Philadelphia
New York–Newark	1	
New York–Boston	2	
Washington–Richmond	1	
MIDWEST NETWORK		
Chicago–Cleveland	2	Toledo
Chicago–St. Louis	2	Danville (pickup only)
Chicago–Milwaukee	1	
Cleveland–Buffalo	1	
Toledo–Detroit	2	

It is planned to have two channels in operation between Philadelphia and Cleveland by January 1, 1949, with an intermediate terminal at Pittsburgh, which will join the eastern and midwest networks.

Tentative additional facilities planned for completion in 1949 are as follows:

ROUTE	CHANNELS (EXCLUDING STAND-BY)	INTERMEDIATE TERMINALS'
New York–Washington	1	As required
Los Angeles–San Francisco	2	
Milwaukee–Madison	1	

Also planned for 1949 are additional connecting facilities to approximately six cities on existing routes.

In 1950 the following routes are planned:

ROUTE	CHANNELS	INTERMEDIATE TERMINALS
New York–Boston	3	As required
New York–Chicago	3	As required
New York–New Haven	1	
Toledo–Detroit	3	
Toledo–Cincinnati	3	Dayton
Dayton–Columbus	3	
Dayton–Louisville	2	Indianapolis
Philadelphia–Wilmington	1	
Buffalo–Rochester	1	
Boston–Providence	1	

The additional facilities planned for 1949 and 1950 between New York and Washington and between New York and Boston will be of the same type now installed between these points. In the case of the New York–Chicago route, however, the three additional channels will be provided by radio relay.

At the present time, there are no definite plans for a transcontinental network. Development of such a network or other routes will depend upon the needs of the television industry.

Charts showing the existing and planned facilities of the Bell system are shown in Figs. 4 and 5.

Rates—The following are the proposed rates for intercity video channel services which are now under review by the Federal Communications Commission.

A. Monthly Service—Where Allocation of Usage Is Not Required

INTEREXCHANGE CHANNEL

Service Seven Days per Week

Per airline mile, per month:
    Eight consecutive hours or fraction
      thereof per day   • $ 35.00 per month
    Each additional consecutive hour or
      fraction thereof per day $ 2.00 per month

Additional Hours, per Occasion of Use

When the additional hours precede or
    succeed and are consecutive with the
    daily service period:
    Per airline mile, per hour or fraction
      thereof $ 0.25 per hour
When the additional hours are not con-
    secutive with the daily service period:
    Per airline mile, per hour or fraction
      thereof $ 0.50 per hour

STATION CONNECTIONS

Service Seven Days per Week

Each station connection, per month:
    Eight consecutive hours or fraction
      thereof per day $500.00 per month
    Each additional consecutive hour or
      fraction thereof per day $ 35.00 per month

Additional Hours, per occasion of Use

When the additional hours precede or
    succeed and are consecutive with the
    daily service period:
    Each connection, per hour or fraction
      thereof $ 5.00 per hour
When the additional hours are not con-
    secutive with the daily service period:
    Each connection, per hour or fraction
      thereof $ 10.00 per hour

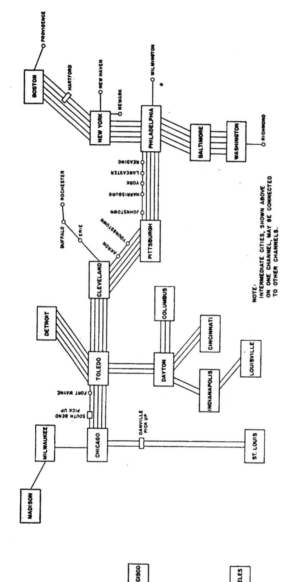

Fig. 4.—Tentative Bell System program for provision of intercity television channels 1948–1950.

B. Monthly Service—Where Allocation of Usage Is Required

When allocation of usage of available facilities is necessary to meet the requirements of two or more customers for monthly service, such service will be furnished over the facilities involved only at the following rates. Subject to thirty days' notice to customers receiving monthly service under **A** above, such service will be terminated when allocation of usage is required. Subject to thirty days' notice to customers receiving monthly service under the following rates, such service will be terminated when allocation of usage is no longer required. Monthly service usage of available facilities will be equitably allocated by the Telephone Company to meet in so far as practicable the reasonable requirements of all monthly service customers.

INTEREXCHANGE CHANNEL

Service Seven Days per Week

Per airline mile, per month:
First four hours or fraction thereof per day (composed of consecutive or nonconsecutive periods in multiples of 15 minutes) $ 25.00 per month
Each additional hour or fraction thereof per day (composed of consecutive or nonconsecutive periods in multiples of 15 minutes) $ 4.00 per month

Additional Hours, per Occasion of Use

When the additional hours are consecutive with the daily service period:
Per airline mile, per hour or fraction thereof $ 0.25 per hour
When the additional hours are not consecutive with the daily service period:
Per airline mile, per hour or fraction thereof $ 0.50 per hour

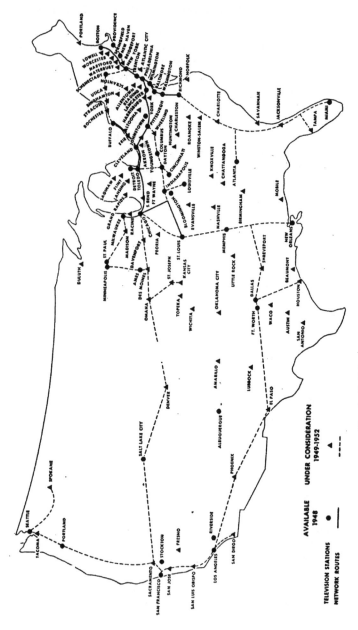

Fig. 5—Bell System television-network routes.

B. Monthly Service—Where Allocation of Usage is Required (continued)

STATION CONNECTIONS

Service Seven Days per Week

Each station connection, per month:
First four hours or fraction thereof per day (composed of consecutive or non-consecutive periods in multiples of 15 minutes) $350.00 per month

Each additional hour or fraction thereof (composed of consecutive or nonconsecutive periods in multiples of 15 minutes) $ 60.00 per month

Additional Hours, per Occasion of Use

When the additional hours are consecutive with the daily service period:
Each connection, per hour or fraction thereof $ 5.00 per hour

When the additional hours are not consecutive with the daily service period:
Each connection, per hour or fraction thereof $ 10.00 per hour

Maximum Allocation Charges for a Route or Interconnected Routes

The charge for each customer's allocated service will be the total charges computed at these allocated usage rates less the proportion that the total allocation charges for all such customers exceeds, if any, the charges which would obtain if one customer had used the entire service under the regular monthly service rates, **A** above, including the charges applicable to switches (**D** below).

C. Occasional Service—Minimum One Hour

The maximum charge for occasional service will be that for monthly service under **A** above.

C. Occasional Service—Minimum One Hour (continued)

INTEREXCHANGE CHANNEL

Per airline mile:

First hour or fraction thereof	$ 1.00 per hour
Each additional 15 minutes or fraction thereof consecutive with the initial period	$ 0.25 per hour

STATION CONNECTIONS

Each station connection, per month (plus $10.00 per hour of use or fraction thereof)	$200.00 per month

D. Switches

Each switch of a section of a network (except where allocation of service is required)	$ 1.00

SUMMARY

From the foregoing report, it is clearly evident that theater-television equipment has been developed which is capable of providing pictures of continuing entertainment value. While not equal in quality to present 35-mm film, evidence has been presented which indicates such quality will be approached in the future. Methods of distribution of program material by coaxial cables or radio channels also have reached a stage of development where satsifactory television pictures can be transmitted over necessary distances.

Further development of equipment as well as provision by the Federal Communications Commission of suitable radio channels is now mainly dependent upon the interest shown by the motion picture industry. Active participation by theater owners and related organizations is essential if the opportunity to use this new medium is not to be lost.

The FCC, however, does not grant channel allocations on a vague request that they may be needed at some future date. Concrete evidence must be presented that the group requesting such allocations is prepared financially and technically to provide a service in the public interest. Only by such action can it be hoped that the request will receive favorable consideration.

The radio-frequency spectrum is very rapidly becoming overcrowded. If the motion picture industry ever hopes to use television in the theater, action must be taken now. A year from now may be too late. Producers, distributors, and exhibitors alike must unite and approach the FCC with a well-formulated plan that they seriously intend immediate experimental operation.

APPENDIX

AFIF Television Projection System—The television projection system described below was developed by the Gesellschaft der Förderung der Forschung auf dem Gebeite der Technischen Physik an der Eidgenös-

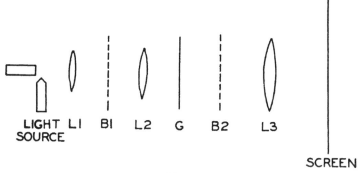

LIGHT L1   B1   L2   G   B2   L3
SOURCE

SCREEN

Fig. 6

sischen Technischen Hochschule of Zürich, Switzerland. For brevity, the system will be referred to simply as the "Swiss System."

The method by which this system forms a large-scale television picture can be explained most easily by first considering the optical system shown in Fig. 6.

Light from the source at the left (an ordinary projection arc lamp) passes through the condensing-lens system *L1*, the baffle plate *B1*, a second lens *L2*, a transparent plate *G* (which will be seen to act as a light valve), a second baffle plate *B2*, a projection-lens system *L3*, and finally reaches the screen at the right.

For simplicity, the optical system can be considered to consist of the following 3 component systems:

1. The light source and *L1* provide uniform illumination of the glass plate *G*.

2. The lens *L2* forms an image of *B1* in the plane of *B2* (assuming for the moment that *G* is a glass plate with parallel surfaces) with a magnification of unity. The baffle plates *B1* and *B2* are formed as shown in Fig. 7.

They consist of a series of narrow opaque bars separated by gaps equal to the width of the bars. The two baffles are so aligned that the images of the gaps in *B1* fall on the bars of *B2*. The result is that no light passes through the system as so far described.

3. The projection lens *L3* forms an enlarged image of the plate *G* on the projection screen at the right of Fig. 6. Of course, since no light was transmitted past *B2*, the screen is uniformly dark except for a certain unavoidable amount of scattered light.

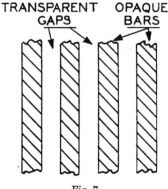

TRANSPARENT GAPS   OPAQUE BARS

Fig. 7

It has been assumed that the plate *G* consists of a sheet of glass with parallel surfaces. Actually this is not the case. The construction and manipulation of this plate are the vital elements of the entire system. To understand the function of the plate consider the effect of a small irregularity on its surface (the nature of this surface will be described later). Let the irregularity be restricted to an area *A*, which has the dimensions of a television-picture element. Further let the irregularity have the form of a slight bulge in the surface, the height of the bulge being denoted by *H*.

A light ray which passes through *A* will be deflected from its path due to the curvature of the surface resulting from the bulge. Some of the rays will be deflected enough to clear the bars in the barrier *B2* and enter the projection system. By increasing the curvature of the surface (increasing the value of *H*), the fraction of the total light passing through *A* which is deflected into the projection system is increased.

It should be noted here that rays which are deflected in a direction parallel to the bars of the barrier are still stopped by the barrier, while rays deflected at right angles to the bars have an excellent chance of getting through. For this reason, a cylindrical distortion of *A* with the axis of the cylinder parallel to the bars is more efficient than a spherical distortion since the sphere produces deflection in all directions and the cylinder causes deflection only in the preferred direction.

Since the projection lens forms an image of the plate on the screen, all light passing through A and through the gaps in $B2$ is focused into an image area A' on the screen. Further, the brightness of A' is roughly proportional to H, the amplitude of the distortion of A.

At this point it is clear that a television picture can be formed on the screen if the surface of G can be divided into a sufficient number of small elements, each of which is distorted in cylindrical fashion with an amplitude of distortion proportional to the brightness of the corresponding element of the original scene. This effect is achieved as follows:

The plate consists of a solid sheet of transparent conducting material coated with a liquid layer. An electrostatic charge is placed on the surface of the liquid by mounting the entire plate in a vacuum tube and scanning it in normal television fashion with a beam of electrons from an electron gun which is provided with suitable deflection and intensity-control devices. Electrostatic forces between the conducting-base plate and the charges thus placed on the surface of the liquid cause a deformation of the surface. By somewhat complicated manipulation of the electron beam this deformation can be made to take on the required form outlined above.

Now consider the cycle of events at one element of the surface. The scanning beam passes over the element and establishes the proper charge distribution. The liquid surface flows into a corresponding distortion pattern and light passes through to the screen. As long as this distortion remains, light reaches the screen so that the tube is effectively a storage device. However, the time duration of the storage must be limited to something less than a scan period so that a new value of distortion can be established when the scanning beam returns. To this end the liquid is made slightly conducting so that the charge variation slowly smooths itself out and the distortion disappears. Then the beam returns and the cycle repeats. In practice, the conductivity is such that the distortion amplitude drops to about 10 per cent of its peak value in the period of one scan. This gives an effective storage period of perhaps one half the scan period.

The liquid used to coat the plate must also meet other requirements. Since the plate is in a vacuum, a very low vapor pressure is necessary. Also the electrostatic forces on the liquid tend to set up a flow of material from the center of the scanned area to the edges. Thus a rather high viscosity is needed to resist these forces. Such a liquid is obtained by mixing apiezon oil, Canada balsam, and a conducting material.

In addition, the liquid layer is continually smoothed and cooled by using a disk for the base plate and rotating the disk slowly and continuously. Thus the scanned area is removed from the scanning position, passed through smoothing and cooling devices, and returned to the scanning position. This motion is very slow so that no blurring of the image is produced.

This system is to some extent similar to the Scophony system. For one thing, both use "schlieren" optics, that is, optical systems which transmit light only when some disturbance is set up in a transmitting medium. In both cases the transmitting medium is liquid, but the nature of the disturbance is quite different. The Scophony system uses a compression wave which travels through a liquid cell. Thus the storage period is limited to the travel time of the compression wave through the cell, and can never be longer than the period of one television line. Also, rotating mirrors are an essential part of the Scophony system.

The Swiss method creates a surface distortion such that each element of the surface is responsible for just one element of the image. Theoretically, this would permit storage during the period of a television frame, and in practice storage periods of about one half this value are realized.

The result is a much higher optical efficiency than can be realized with the Scophony system. In fact, the screen illumination that can be produced is stated to be about 30 per cent of that obtained with a motion picture projector using the same light source and projection lens.

ACKNOWLEDGMENT

This work has been prepared as an interim report of the Theater Television Committee of the Society of Motion Picture Engineers. It is a statement of the present state of the art written in nontechnical language. Information on government regulations, types of equipment available, and approximate costs has been included. It is directed to those of the motion picture industry who wish to take advantage of this rapidly expanding entertainment medium. The membership of the committee directly responsible for this work is as follows:

D. E. HYNDMAN, *Chairman*
Eastman Kodak Company

G. L. BEERS
RCA Victor Division

F. E. CAHILL, Jr.
Warner Brothers Pictures

A. W. COOK
Ansco Division

JAMES FRANK, JR.
United Photo Supply Corporation

R. L. GARMAN
General Precision Laboratories

E. P. GENOCK
Paramount Pictures

A. N. GOLDSMITH
Consulting Engineer

T. T. GOLDSMITH, JR.
Allen B. Du Mont Laboratories

C. F. HORSTMAN
RKO Theaters

L. B. ISAAC
Loew's, Inc.

A. G. JENSEN
Bell Telephone Laboratories

P. J. LARSEN
Los Alamos Scientific Laboratory

N. LEVINSON
Warner Brothers Pictures

H. RUBIN
Paramount Pictures

OTTO SANDVIK
Eastman Kodak Company

E. SCHMIDT
E. I. du Pont de Nemours and Co.

A. G. SMITH
National Theater Supply Company

E. I. SPONABLE
Twentieth Century-Fox Film Corporation

J. E. VOLKMANN
RCA Victor Division

THEATER TELEVISION

BIBLIOGRAPHY

(1) William H. Forster, "6000 mc television relay system,' *Electronics*, vol. 22, pp. 80–85; January, 1949.

(2) Fred G. Albin, "Sensitometric aspect of television monitor tube photography," *J. Soc. Mot. Pict. Eng.*, vol. 51, pp. 595–613; December, 1948.

(3) Richard Hodgson, "Paramount Pictures' system of theater television," Presented SMPE 64th convention, Monday, October 25, 1948, *J. Soc. Mot. Pict. Eng.*, to be published.

(4) Roy Wilcox and Hubert J. Schafly, "Theater installation of instantaneous large-screen television," Presented SMPE 64th convention, Monday, October 25, 1948, *J. Soc. Mot. Pict. Eng.*, to be published.

(5) Ernest H. Schreiber, "Video distribution facilities for television transmission," *J. Soc. Mot. Pict. Eng.*, vol. 51, pp. 574–589; December, 1948.

(6) Otto H. Schade, "Electro-optical characteristics of television systems," *RCA Rev.*, vol. 9, pp. 5–37, March, 1948; pp. 246–286, June, 1948; pp. 491–530, September, 1948; pp. 653–686, December, 1948.

(7) J. L. Boon, W. Feldman, and J. Stoiber, "Television recording camera," *J. Soc., Mot. Pict. Eng.*, vol. 51, pp. 117–126; August, 1948.

(8) Thomas T. Goldsmith, Jr., and Harry C. Milholland, "Television transcription by motion picture film," *J. Soc. Mot. Pict. Eng.*, vol. 51, pp. 107–116; August, 1948.

(9) A. G. D. West, "Development of theater television in England," *J. Soc Mot. Pict. Eng.*, vol. 51, pp. 127–168; August, 1948.

(10) H. L. Logan, "Brightness and illumination requirements," *J. Soc. Mot. Pict. Eng.*, vol. 51, pp. 1–12; July, 1948.

(11) Ralph V. Little, Jr., "Developments in large-screen television," *J. Soc. Mot. Pict. Eng.*, vol. 51, pp. 37–51; July, 1948.

(12) I. G. Maloff, "Optical problems in large-screen television," *J. Soc. Mot. Pict. Eng.*, vol. 51, pp. 30–36; July, 1948.

(13) Robert M. Fraser, "Motion picture photography of television images," *RCA Rev.*, vol. 9, pp. 202–217; June, 1948.

(14) Boyce Nemec, "Review of SMPE work on screen brightness," *J. Soc. Mot. Pict. Eng.*, vol. 50, pp. 254–259; March, 1948.

(15) "Report of Screen Brightness Committee," *J. Soc. Mot. Pict. Eng.*, vol. 50, pp. 260–273; March, 1948.

(16) Alfred N. Goldsmith, "Theater television—A general analysis," *J. Soc. Mot. Pict. Eng.*, vol. 50, pp. 95–121; February, 1948.

(17) F. M. Deerhake, "2000 mc television program chain," *Electronics*, vol. 21, pp. 94–97; February, 1948.

(18) Harry R. Lubcke, "Effect of time element in television program operations," *J. Soc. Mot. Pict. Eng.*, vol. 48, pp. 543–547; June, 1947.

(19) L. G. Abraham and H. I. Romnes, "Television network facilities," *Elec. Eng.*, vol. 66, p. 47; May, 1947.

(20) "Statement of SMPE on revised frequency allocations," *J. Soc. Mot. Pict. Eng.*, vol. 48, pp. 183–202; March, 1947.

(21) W. Boothroyd, "UHF television relay system," *Electronics*, vol. 20, p. 86; March, 1947.

(22) Lester B. Isaac, "Television and the motion picture theater," *J. Soc. Mot. Pict. Eng.*, vol. 47, pp. 482–486; March, 1947.

(23) W. J. Poch and J. P. Taylor, "Microwave equipment for television relay services," *Broadcast News*, vol. 44, p. 20; October, 1946.

(24) Allen B. Du Mont, "The relation of television to motion pictures," *J. Soc. Mot. Pict. Eng.*, vol. 47, pp. 238–247; September, 1946.

(25) Paul J. Larsen, "Report of Committee on Television Projection Practice," *J. Soc. Mot. Pict. Eng.*, vol. 47, pp. 118–119; September, 1946.

(26) C. F. White and M. R. Boyer, "A new film for photographing the television monitor tube," *J. Soc. Mot. Pict. Eng.*, vol. 47, pp. 152–164; August, 1946.

(27) Ralph B. Austrian, "Film—The backbone of television programming," *J. Soc. Mot. Pict. Eng.*, vol. 45, pp. 401–413; December, 1945.

(28) A. H. Rosenthal, "Problems of theater television projection equipment," *J. Soc. Mot. Pict. Eng.*, vol. 45, pp. 218–240; September, 1945.

(29) "Frequency allocations for theater television," *J. Soc. Mot. Pict. Eng.*, vol. 45, pp. 16–19; July, 1945.

(30) D. W. Epstein and I. G. Maloff, "Projection television," *J. Soc. Mot. Pict. Eng.*, vol. 44, pp. 443–455; June, 1945.

(31) Harold S. Osborne, "Coaxial cables and television transmission," *J. Soc. Mot. Pict. Eng.*, vol. 44, pp. 403–418; June, 1945.

(32) Ralph B. Austrian, "Some economic aspects of theater television," *J. Soc. Mot. Pict. Eng.*, vol. 44, pp. 377–385; May, 1945.

(33) J. Huber, "Theoretical inquiries on distortions undergone by electrical television signals in vacuum-tube connections," *Schweizer Archiv.*, vol. 11, p. 77, March, 1945; p. 115, April, 1945.

(34) "Report of the British Television Committee 1943," London, His Majesty's Stationery Office, 1945.

(35) "Statement of the SMPE in opposition to the brief of the CBS as it relates to theater television," *J. Soc. Mot. Pict. Eng.*, vol. 14, pp. 263–274; April, 1945.

(36) "Statement of the SMPE on allocation of frequencies in the radio frequency spectrum from 10 kilocycles to 30 million kilocycles for theater television service," *J. Soc. Mot. Pict. Eng.*, vol. 44, pp. 105–117; February, 1945.

(37) Paul J. Larsen, "Statement presented before the Federal Communications Commission relating to television broadcasting," *J. Soc. Mot. Pict. Eng.*, vol. 44, pp. 123–127; February, 1945.

(38) Alfred N. Goldsmith, "Future of theater television," *Television*, February, 1945.

(39) Ralph R. Beal, "Electronic research opens new frontiers," *Proc. I.R.E.*, vol. 33, pp. 5–9; January, 1945.

(40) T. M. C. Lance, "Some aspects of large screen television," *J. Telev. Soc.*, vol. 4, p. 82; December, 1944.

(41) L. de Forest, "Early beginnings of large screen television," Program, First Annual Conference, Television Broadcasters Association, p. 37; December 11–12, 1944.

(42) Paul J. Larsen, "Theatrical television," Program, First Annual Conference, Television Broadcasters Association, p. 39; December 11-12, 1944.

(43) Paul J. Larsen, "Panel meeting on theater television," Proceedings of First Annual Conference, Television Broadcasters Association, p. 108; December 11-12, 1944.

(44) E. K. Carver, R. H. Talbot, and H. A. Loomis, "Film distortions and their effect upon projection quality," J. Soc. Mot. Pict. Eng., vol. 41, pp. 88–93; July, 1943.

(45) E. E. Masterson and E. W. Kellogg, "A study of flicker in 16-millimeter picture projection," J. Soc. Mot. Pict. Eng., vol. 39, pp. 232–244; October, 1942.

(46) P. C. Goldmark, J. N. Dyer, E. R. Piore, and J. M. Hollywood, "Color television," J. Soc. Mot. Pict. Eng., vol. 38, pp. 311–352; April, 1942.

(47) M. R. Null, W. W. Lozier, and D. B. Joy, "Color of light on the projection screen," J. Soc. Mot. Pict. Eng., vol. 38, pp. 219–228; March, 1942.

(48) F. Fischer and H. Thiemann, "Theoretical considerations on a new process for television large-screen projection," Schweizer Archiv., vol. 7, p. 1, January, 1941; p. 33, February, 1941; p. 305, November, 1941; p. 337, December, 1941.

(49) "Television report—Order, rules, and regulations," J. Soc. Mot. Pict. Eng., vol. 37, pp. 87–97; July, 1941.

(50) I. G. Maloff and W. A. Tolson, "A résumé of the technical aspects of RCA theater television," RCA Rev., vol. 6, p. 5; July, 1941.

(51) "Baird high definition color television," J. Telev. Soc., vol. 3, p. 171; 1941.

(52) E. A. Williford, "Twenty-four years of service in the cause of better projection," J. Soc. Mot. Pict. Eng., vol. 36, pp. 294–301; March, 1941.

(53) C. Frederick Wolcott, "Problems in television image resolution," J. Soc. Mot. Pict. Eng., vol. 36, pp. 65–81; January, 1941.

(54) "Report of Television Committee," J. Soc. Mot. Pict. Eng., vol. 35, pp. 569–583; December, 1940.

(55) F. H. Richardson, "Advancement in projection practice," J. Soc. Mot. Pict. Eng., vol. 35, pp. 466–483; November, 1940.

(56) M. W. Baldwin, "The subjective sharpness of simulated television images," Proc. I.R.E., vol. 28, pp. 458–468; October, 1940.

(57) P. C. Goldmark and J. N. Dyer, "Quality in television pictures," J. Soc. Mot. Pict. Eng., vol. 35, pp. 234–253; September, 1940.

(58) "A new projection system for large-screen television receivers," Electronics and Telev. and Short-Wave World, vol. 13, p. 372; August, 1940.

(59) W. C. Kalb, "Progress in projection lighting," J. Soc. Mot. Pict. Eng., vol. 35, pp. 17–31; July, 1940.

(60) A. H. Rosenthal, "A system of large-screen television reception based on certain electron phenomena in crystals," Proc. I.R.E., vol. 28, pp. 203–213; May, 1940.

(61) F. Fischer, "On the paths to television large-screen projection," Schweizer Archiv., vol. 6, p. 89; April, 1940.

(62) Alfred N. Goldsmith, "Future development in the field of the projectionist," J. Soc. Mot. Pict. Eng., vol. 34, pp. 131–142; February, 1940.

(63) T. Mulert, "Viewpoints in the construction of television projection tube receivers," Fernseh A. G. Hausmitteilungen, vol. 1, p. 217; December, 1939.

(64) G. Schubert, W. Dillenburger, and H. Zschau, "The intermediate film process," *Fernseh A. G. Hausmitteilungen*, vol. 1, Part I, p. 65; April, 1939. Part II, p. 162; August, 1939. Part III, p. 201; December, 1939.

(65) "The latest Scophony big-screen projector," *Electronics and Telev. and Short-Wave World*, vol. 12, p. 654; November, 1939.

(66) W. C. Harcus, "Screen color," *J. Soc. Mot. Pict. Eng.*, vol. 33, pp. 444–448; October, 1939.

(67) Lorin D. Grignon, "Flicker in motion pictures," *J. Soc. Mot. Pict. Eng.*, vol. 33, pp. 235–247; September, 1939.

(68) D. M. Robinson, "The supersonic light control," *Proc. I.R.E.*, vol. 27, p. 483; August, 1939.

(69) J. Sieger, "The design and development of television receivers using the Scophony television optical scanning system," *Proc. I.R.E.*, vol. 27, pp. 487–492; August, 1939.

(70) G. Wikkenhauser, "Synchronization of Scophony television receivers," *Proc. I.R.E.*, vol. 27, pp. 492–496; August, 1939.

(71) H. W. Lee, "Some factors involved in the optical design of a modern television receiver using moving scanners," *Proc. I.R.E.*, vol. 27, pp. 496–501; August, 1939.

(72) R. Moeller and G. Schubert, "Further developments in receiver and picture pickup equipment in the year 1939," *Fernseh A. G. Hausmitteilungen*, vol. 1, p. 153; August, 1939.

(73) C. H. Bell, "Facts and problems for the cinema," *Telev. and Short-Wave World*, vol. 12, p. 467; August, 1939.

(74) R. Moeller and G. Schubert, "Ten years of television technique," *Fernseh A. G. Hausmitteilungen*, vol. 1, p. 111; July, 1939.

(75) E. Schwartz, "The development of television Braun tubes by Fernseh A. G.," *Fernseh A. G. Hausmitteilungen*, vol. 1, p. 123; July, 1939.

(76) Allen B. Du Mont, "Design problems in television systems and receivers," *J. Soc. Mot. Pict. Eng.*, vol. 33, pp. 66–74; July, 1939.

(77) "Report of Television Committee," *J. Soc. Mot. Pict. Eng.*, vol. 33, pp. 75–79; July, 1939.

(78) "The EMI cinema projector," *Telev. and Short-Wave World*, vol. 12, p. 389; July, 1939.

(79) "Report of Projection Practice Committee," *J. Soc. Mot. Pict. Eng.*, vol. 33, pp. 101–105; July, 1939.

(80) O. Reeb, "A consideration of the screen-brightness problem," *J. Soc. Mot. Pict. Eng.*, vol. 32, pp. 485–494; May, 1939.

(81) R. Moeller, "The lens array screen," *Fernseh A. G. Hausmitteilungen*, vol. 1, p. 73; April, 1939.

(82) "Baird cinema equipment," *Telev. and Short-Wave World*, vol. 12, p. 199; April, 1939.

(83) "Kolorama television" *Telev. and Short-Wave World*, vol. 12, p. 137; March, 1939.

(84) "Making history (circumstances of the televising of the Boon-Danaher prize fight)," *Telev. and Short-Wave World*, vol. 12, p. 131; March, 1939.

(85) P. W. S. Valentine, "The Philips large screen television receiver," *J. Telev. Soc.*, vol. 3, p. 7; January, 1939.

(86) "Report of the Projection Practice Committee," J. Soc. Mot. Pict. Eng., vol. 31, pp. 480–510; November, 1938.

(87) E. Schwartz, H. Struebig, and H. W. Paehr, "Beam generation in television cathode-ray tubes for projection purposes," Fernseh A. G. Hausmitteilungen, vol. 1, p. 5; August, 1938.

(88) W. Dillenburger, "Radio large projection receiver with 1.5 square meter picture surface," Fernseh A. G. Hausmitteilungen, vol. 1, p. 29; August, 1938.

(89) "Report of the Projection Practice Committee," J. Soc. Mot. Pict. Eng., vol. 30, pp. 636–650; June, 1938.

(90) F. H. Richardson, "A discussion of screen-image dimensions," J. Soc. Mot. Pict. Eng., vol. 30, pp. 334–338; March, 1938.

(91) "Report of the Projection Practice Committee," J. Soc. Mot. Pict. Eng., vol. 29, pp. 614–621; December, 1937.

(92) A. Goetz and W. O. Gould, "The objective quantitative determination of the graininess of photographic emulsions," J. Soc. Mot. Pict. Eng., vol. 29, pp. 510–538; November, 1937.

(93) V. K. Zworykin and W. H. Painter, "Development of the projection kinescope," Proc. I.R.E., vol. 25, pp. 937–954; August, 1937.

(94) R. R. Law, "High-current electron gun for projection kinescope," Proc. I.R.E., vol. 25, p. 954–977; August, 1937.

(95) "Report of the Projection Screen Brightness Committee," J. Soc. Mot. Pict. Eng., vol. 27, pp. 127–139; August, 1936.

(96) "Report of the Projection Screen Brightness Committee," J. Soc. Mot Pict. Eng., vol. 26, pp. 489–504; May, 1936.

(97) R. P. Teele, "Photometry and brightness measurements," J. Soc. Mot. Pict. Eng., vol. 26, pp. 554–569; May, 1936.

(98) S. K. Wolf, "An analysis of theater and screen illumination data," J. Soc. Mot. Pict. Eng., vol. 26, pp. 532–542; May, 1936.

(99) M. Luckiesh and F. K. Moss, "The motion picture screen as a lighting problem," J. Soc. Mot. Pict. Eng., vol. 26, pp. 578–591; May, 1936.

(100) W. F. Little and A. T. Williams, "Résumé of methods of determining screen brightness and reflectance," J. Soc. Mot. Pict. Eng., vol. 26, pp. 570–577; May, 1936.

(101) C. M. Tuttle, "Density measurements of release prints," J. Soc. Mot. Pict. Eng., vol. 26, pp. 548–553; May, 1936.

(102) A. A. Cook, "A review of projector and screen characteristics, and their effects upon screen brightness," J. Soc. Mot. Pict. Eng., vol. 26, pp. 522–531; May, 1936.

(103) B. O'Brien and C. M. Tuttle, "An experimental investigation of projection screen brightness," J. Soc. Mot. Pict. Eng., vol. 26, pp. 505–521; May, 1936.

(104) E. M. Lowry, "Screen brightness and the visual functions," J. Soc. Mot. Pict. Eng., vol. 26, pp. 490–504; May, 1936.

Research Council
Small Camera Crane*

By ANDRÉ CROT

MOTION PICTURE RESEARCH COUNCIL, HOLLYWOOD, CALIFORNIA·

Summary—The Research Council small camera crane was designed from requirements and specifications set down by the Council's Photographic Committee. While similaɪ in principle to other cranes, it embodies an entirely new design and built-in safety features. The crane dolly is electrically driven. The boom arm is manually operated and can be panned through 360 degrees. The crane is large enough to have a lens height of from 2 to 10 feet from floor level and small enough so that, fully equipped, it will pass through a doorway 6 feet high and 36 inches wide.

THIS CRANE, known as the Research Council small camera crane, is similar in principle to other cranes used in the motion picture industry, but it embodies an entirely new design and built-in safety features. It can be built as a standard production piece of equipment at a reasonable price.

The specifications of the Camera Crane Committee, composed of members from each studio, called for a self-propelled, remote-controlled crane, large enough to have a lens height of from 2 to 10 feet from the floor level, but small enough so that fully equipped it would pass through a doorway 36 inches wide and 6 feet high. The crane was designed to accommodate the Technicolor equipment; the weight to be of minimum weight possible in order to permit its use on the present stage floors, preferably without track; and deflection and distortion to be kept to a minimum.

From the specifications, the present Research Council crane was designed and built with the use of aluminum alloy wherever possible, in order to keep weight to a minimum and castings were used to keep deflections to a minimum. After load-testing it thoroughly, we observed that deflection under weight was negligible.

Although the cost of patterns was high, we have shown this method to be the best and most economical way of building a production boom. The weight of the crane itself is about 1200 pounds.

The drive unit is actuated by a 2-horsepower, 110-volt, direct-

* Presented May 18, 1948, at the SMPE Convention in Santa Monica.

current, series-wound motor, especially designed for this crane by the General Electric Company. This motor is supported on rubber mounts and is coupled to the differential drive by a more-flex coupling. The differential is a 10:1 cone worm-drive unit, with all driving surfaces lapped in assembly. This design was considered necessary to insure smooth and silent operation. The differential housing is a full-floating axle type and the full weight of the crane is carried by the housing itself, leaving the driving axles torque-loaded only.

The motor can be remotely controlled if desired. The control itself is a complete and separate unit consisting of resistors, reversing switch, and solenoid brake. With this control, various degrees of

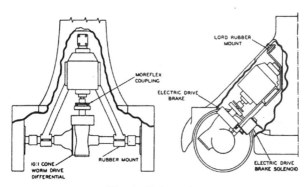

Fig. 1—Drive unit.

acceleration or deceleration can be obtained at will. A Cannon plug is used to connect the control to the crane.

The brake is a friction, air-cooled, disk-type, actuated by a 110-volt, direct-current solenoid.

The brake solenoid is actuated by a carbon pile to provide a sort of "feel control" which gives the operator a proportional amount of braking power to the pressure applied to the *carbon pile*.

The steering unit is of an unusual design which permits the crane to be completely revolved within a 6-foot radius. It also allows the crane to be placed squarely against a wall with the least maneuvering. This device is provided with a lock-preventing arm, allowing a very sharp turn with the least amount of effort.

The following units are mounted on the base casting: (a) the drive unit, (b) the steering unit, (c) the center post, (d) the hydraulic pump and take-up tank, and (e) four jacks which are interconnected to

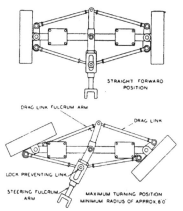

STRAIGHT FORWARD POSITION

DRAG LINK FULCRUM ARM

DRAG LINK

LOCK PREVENTING LINK

STEERING FULCRUM ARM   MAXIMUM TURNING POSITION
MINIMUM RADIUS OF APPROX. 6'0"

Fig. 2—Steering unit.

permit the whole crane to be lifted for service or to protect the wheels during storage periods.

The center post is a 7-inch telescoping tube mounted on ball bearings. It permits the boom to be panned 360 degrees around the crane base, tilted up 55 degrees, and down 45 degrees from the horizontal position. The boom arm and parallel bar are mounted on it. A hydraulic cylinder of 16-inch extension is mounted in the center of the telescoping tube. A flow restrictor is located on the cylinder base, providing added safety to equipment and personnel, as it limits the downstroke to a predetermined speed in case of accidental breaking of any hydraulic line or mishandling of equipment.

The panning brake is hand-operated by moving a lever in either direction from the centerline and can be adjusted to any degree of friction required by the operator, but cannot be locked rigidly. This is to prevent damage to the crane in case of accidental shock to the boom. A special positioning device operates automatically to keep a uniform degree of friction; another automatic locking pin prevents using the panning brake when the hydraulic pump is in use or the use of the hydraulic pump when the panning brake is not in the neutral position.

The tilting brake is located in the center post and the boom casting itself is relieved to form the brake drums. The brake is hand-operated. Handles are provided on each side of the boom and can be set to any degree of friction. Adjusting screws are provided to limit the maximum braking force.

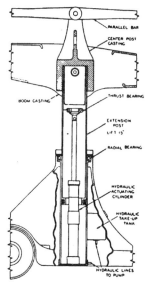

PARALLEL BAR

CENTER POST CASTING

BOOM CASTING

THRUST BEARING

EXTENSION POST
LIFT 15"

RADIAL BEARING

HYDRAULIC ACTUATING CYLINDER

HYDRAULIC TAKE-UP TANK

HYDRAULIC LINES TO PUMP

Fig. 3—Center post.

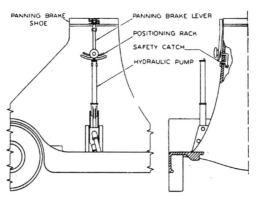

Fig. 4—Panning brake.

The camera platform is mounted on the forward or long end of the boom. It is kept in a constant horizontal position by the parallel bar. The platform is machined to receive the camera table, the camera-table-leveling device, and a series of common and twist plugs. These are to accommodate cables for the camera motor, the remote-focusing device, and lights needed by the camera crew.

The counterweight inner and outer boxes are mounted on the rear or short end of the boom. The outer box is mounted directly on the boom and parallel bar, and the inner base is closely mounted in the interior of the outer box. The inner box is retractable and can be extended out about 16 inches which gives a greater degree of safety for heavy loads such as the Technicolor equipment. This provides extra room for lead counterweights if needed.

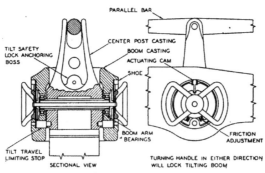

Fig. 5—Tilt brake,

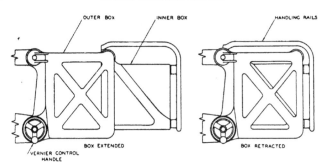

Fig. 6—Counterweight boxes.

The Vernier counterweight is a hand-operated device which can be operated from each side of the boom and corrects for an unbalance up to 35 pounds. It is located in the shell of the boom.

The safety-tilt device is entirely enclosed in the shell of the forward or long end of the boom. It is composed of a 2-inch hydraulic cylinder anchored on one end to the center-post casting and on the other end to the boom itself. Any vertical change of position of the boom shortens or lengthens the distance between these two points, causing a displacement of fluid from one end of the cylinder to the other. This flow is controlled in one direction only; i.e., *on the upward movement*, thus allowing the cameraman or his assistant to leave their respective seats without endangering the balance of the boom.

The control of this device is accomplished by the use of a hydraulic solenoid valve which, in turn, is energized whenever the cameraman or his assistant or both leave their seats. All action returns to normal

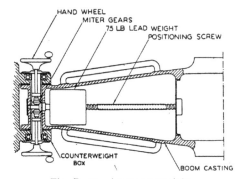

Fig. 7—Vernier counterweight.

when the seats are reoccupied or an equal amount of counterweight is removed from the opposite end of the boom. The pressure-relief valve, which is part of this safety-tilt mechanism, goes into action whenever too much weight is removed from the camera table. As the boom is allowed to rise slowly, this prevents a sudden drop of the boom arm or an unsafe unbalance of the crane.

The camera-turret table is an integral part of the crane and is permanently mounted on the camera platform. It consists of a camera-leveling table allowing 7 degrees of correction in all positions. On this table the following are mounted:

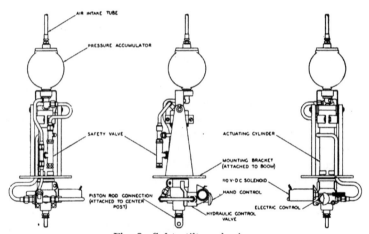

Fig. 8—Safety-tilt mechanism.

(a) An operator's seat, capable of 360-degree rotation, with adjustable height, length, and back rest.

(b) An assistant's seat with two possible locations permitting the focusing of the lens from either side of the camera.

(c) A compensating counterweight to offset the natural centrifugal effect during a fast-panning shot.

(d) A two-speed panning mechanism which will carry the turret fully equipped to any position and at any chosen acceleration.

(e) Two-gear head adapters for use with all existing gear heads.

(f) Means are provided so that the camera-turret table and the camera can be rotated as a unit or separately, permitting an overshot and correcting the camera position without the necessity of moving the cameraman and his assistant, therefore giving a much smoother correction.

(g) A positive camera turret lock at easy reach of the cameraman's hand. It can be set to any degree of friction desired.

When the Research Council camera crane was designed, it was foreseen that the crane could and would be used off the stage; that is, on street sets or on location. As a single piece of equipment could not meet this and the other requirements, it was found necessary to design a subchassis which is now called a "trailer." When the crane is mounted on the trailer, it is elevated $4^{1}/_{2}$ inches from the floor level.

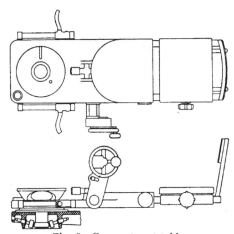

Fig. 9—Camera-turret table.

The trailer is motorized by direct friction drive from the crane unit, thus eliminating the necessity of a second driving unit. It is equipped with wheel brakes, either hand- or foot-operated, a drive seat for the crane operator on each side of the trailer, and a platform for the boom operator.

No equipment other than the crane and trailer is necessary to combine the two for use. The assembling can be done by two men in less than 5 minutes. A towing attachment is provided with each trailer for transportation to location.

Experiment in Stereophonic Sound*

By LORIN D. GRIGNON

Twentieth Century-Fox Films, Beverly Hills, California

Summary—This paper reports an extension of the theory and methods of stereophonic recording and reproduction, as particularly applicable to motion pictures. Microphone technique becomes very different from that previously used because of the manner of staging, the use of varied angles of view by the camera, and a fixed theater picture size. Typical microphone technique is illustrated and re-recording with added sound effects is described. Resulting conclusions and observations establish a good foundation for further work in this field.

Introduction

THE THEORETICAL BASIS of stereophonic recording and reproduction is rather generally known. However, for convenience, it will be restated as follows: If an infinite number of ideal microphones could be placed within a three-dimensional region bounding a source of sound energy, each microphone being connected to a distortionless transmission system terminated in an ideal reproducer at some other location in surroundings identical to those at the source location and in space relation to each other as their corresponding microphones, then an observer at the distant location would experience the same sensation as an observer at the source point. The first compromise to the ideal situation considers microphones and reproducers in two-dimensional space as an acoustically transparent curtain between the source and the observers. The second compromise employs an infinite number of microphones in a single straight line. It has been found that three complete systems give a good subjective approximation when three microphones are equally spaced along some straight line in relation to the source. Complete descriptions of experiments in this direction are given by a Bell Telephone Laboratories monograph.[1]

The work reported here is an extension of the theory and methods for the use of stereophonic sound in motion pictures. The opportunity to investigate this possibility came about by a desire on the part of Twentieth Century-Fox management to evaluate possible technical

---

* Presented May 18, 1948, at the SMPE Convention in Santa Monica.

improvements in motion pictures. Western Electric Company cooperated through Electrical Research Products Division by supplying film recording and reproducing equipment and other technical assistance.

Methods were devised for recording dialog and music for use in motion pictures, without basically changing accepted fundamental forms which include the use of long, medium, and close shots and intercutting techniques. This is not to say that present cutting philosophy for stereophonic motion pictures is entirely suitable, as there is evidence that indicates some new approach is needed. Re-recording, with added sound effects, prescore, and playback methods were all used. The end result of the experiment to be described was the production of two single-act plays, several full-orchestra numbers (one with picture), and a vocal rendition with accompanying orchestra.

It was concluded that stereophonic methods, with suitable modifications, can be applied to motion picture technique and result in a sound presentation considerably superior to methods now in use.

MICROPHONE TECHNIQUE DEVELOPMENT

The problem of pickup, particularly for dialog, was first approached by setting up a three-channel monitoring system, using the amplifier and horn apparatus as installed in the test theater, and providing microphones and mixer equipment in an adjacent stage. On this stage a small living room set was constructed and pickup tests were begun using stock players from the studio roster.

It was natural first to try the accepted method of having three microphones equally spaced and placed in some straight line in relation to the actors. This method failed immediately—the reasons being as follows:

1. The action was taking place in a restricted space.

2. Seminondirectional microphones were essentially useless because of the proximity to the sources and to acoustic reflections from parts of the set which produced false apparent origins.

3. Actors generally play to other actors and do not face an audience as do public speakers. Methods were needed for giving sound placement to actors who are speaking at right angles to the camera axis and within a few feet of one another.

4. Since various camera lenses are used to give emphasis or localize action, magnifications or distance distortion exists and a similar effect was necessary in the sound pickup.

Using adjustable directivity unidirectional microphones and separate microphone booms for all further work, several basic microphone setups resulted.

In all of the following illustrations, actors will be depicted by a "V" within a circle, the apex of the letter indicating the speaking direction; microphones by a circle with the protruding arrowhead indicating the direction of maximum pickup and the extended tail the direction of minimum reception, the inscribed letter stating the connected channel as left *L*, center *C*, or right *R;* movement of actors or microphones by dashed lines with arrowheads giving direction.

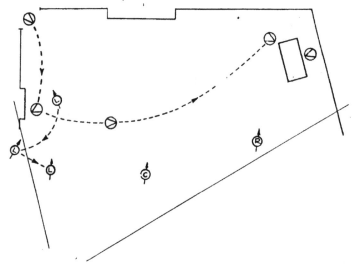

Fig. 1—Typical microphone arrangement for large-set long shot with broad actor movement.

For purposes of definition, consider a source which emits sound continuously and which is in constant movement, then the reproduced sound must also move continuously in the same way without obvious dwells or jumps from position to position and this characteristic will be termed smooth sound placement transition.

The usual equidistant, in-line microphone technique can be used in long shots of large sets with wide separation of actors and broad movements. Even then, to have smooth dialog transition, some microphone movement may be required, as illustrated in Fig. 1.

This method is also generally used for recording of effects. One experiment using such a pickup consisted in recording airplane take-offs, wherein the microphones were placed along the runway. The microphones were spaced 150 feet apart!

When actors are not disposed closer than six to seven feet and are speaking directly to each other, the setup of Fig. 2 is used. Should either turn or move away and speak lines, then microphones must be moved accordingly. For example, should the right-hand actor turn 180 degrees and speak lines, then the R microphone must be moved sufficiently to the right to give good pickup and the C microphone readjusted to a somewhat central position, probably favoring the right-hand actor. Also, dependent upon set conditions, it is some-

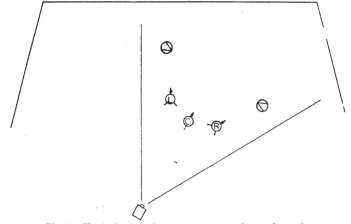

Fig. 2—Typical microphone arrangement for medium shot.

times necessary to adjust the null directions of the L and R micro-phones on the opposing person. This change in pattern does not eliminate pickup into any microphone so adjusted, because the null is imperfect and sufficient energy arrives from other directions, thus satisfactorily meeting necessary pickup requirements.

In many instances microphones were grouped in a cluster tighter than shown in Fig. 2. This condition carried to its limit, occurs where a close-up of two actors is used who are physically separated by only three to four feet and farther, they will appear on the screen to be eight to ten feet distant from one another, resulting in the crossed-over configuration of Fig. 3. Note that the L microphone is placed on camera right but is actually picking up the left-hand actor, provided of course that it is adjusted to minimize pickup from the right-hand performer. The R microphone is, of course, also reversed as indicated. This, obviously, is one of the most difficult types of pickup, as the microphone positions must be carefully chosen, the directivity nulls correctly used and, in particularly bad cases, the relative channel gain adjusted. It is sometimes wisest to abandon sound placement under these conditions and use the condition of Fig. 4, which is an effective means to maintain stereophonic quality without the feature of sound-origin placement. By proper choice of dimensions, unwanted sound placements can be eliminated. Average dimensions might be five feet on each side of the triangle and three and one-half feet on the base.

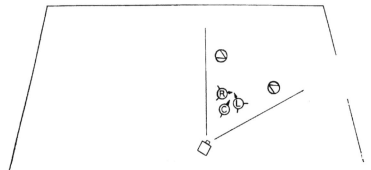

Fig. 3—Typical microphone arrangement for close shot of two actors requiring "sound magnification" or distance distortion. Note crossed-over placement.

Manifestly, it is no longer permissible to revert to the demands of early sound motion pictures that actors be fixed at specific positions for the delivery of dialog. Therefore, motion of microphones is required. Combinations of all of the microphone configurations shown in the preceding figures have been used in many of the scenes recorded. The principal problem is one of smooth transition and proper apparent sources.

Two interesting and useful effects, not possible without stereophonic methods, have been used. The first creates the illusion of an actor talking and moving within the set but never being seen while the camera showed a small portion of the set and another player. This offstage illusion can be created by sound unassisted by the visible actor who, without stereophonic sound, must describe the unseen action by following with his eyes. The second is the ability to make sounds from either side apparently very much offstage.

RECORDING AND REPRODUCING EQUIPMENT

The equipment used consisted of three essentially identical recording channels with a common film recording machine, placing three 200-mil push-pull variable-density tracks on a single 35-mm film. The three modulators were arranged in an arc, the two outside optical paths brought parallel with two sides of a front surface prism, and the center path passed through a hole in the same prism. Separate objective lenses were used for each modulator with a single cylindrical

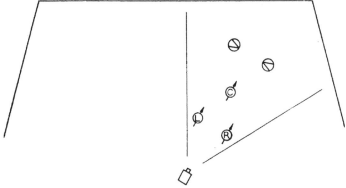

Fig. 4—Microphone arrangement to maintain stereophonic quality but without sound placement other than center screen.

lens near the film for all channels. Pre- and postequalization were utilized.

Monitoring was provided by earphones where each side channel was connected to single reproducers on respective ears and the center channel was split to both ear receivers in such a way as to supply 3 decibels less power to each ear than the corresponding side channel, the total power from the center channel being the same as the side-channel outputs. Further, the side-to-side cross-feed was not permitted higher than −25 decibels from the direct source. This means, for monitor, was used on all work and found satisfactory.

The film reproducers used a single exciter lamp, a fixed aperture plate with large radius of curvature over which the film passed to maintain contact for focus, a single objective lens, three engraved slits, and associated photocells arranged in an arc to provide equal optical path length. Filtering was similar in principle to the newer mechanisms now being supplied and essentially equal in effectiveness.

The sound reproducers were two-way systems of good characteristics. One was located at screen center and the other two placed with their axes $^2/_3$ screen width off-center. These dimensions are not inviolate. The best arrangement is determined by existing conditions and desired effect.

RE-RECORDING

In connection with one two-reel playlet which was used, it was necessary to add horse-hoof sounds, footsteps, cup crashes, and shots.

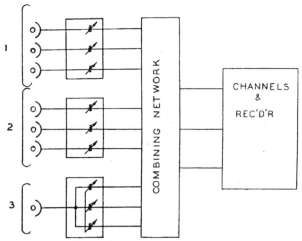

Fig. 5—Block schematic of re-recording equipment.

All of these effects, except the cup crashes, were re-recorded from stock library material. The equipment arrangements are given by Fig. 5. Note that mixers 1 and 2 are conventional 3-channel stereophonic units. Mixer 3 is a special control designed to transfer the single input to any of the three channels or to any two adjacent channels, meanwhile maintaining constant total power. With this control it is possible to move a single source smoothly back and forth to create any desired illusion.

By the use of the special control, offstage horses were made to sound as though they approached from a distance to the left and came to a point just off stage, also gunshots and footsteps were added and properly placed.

The use of re-recording retained the advantages of level smoothing and permitted a small amount of placement correction. Dependence cannot be placed on re-recording for changes in placement of original material because placement, except in certain special cases, is not primarily due to intensity differences. This point is developed more fully later.

MUSIC AND VOCALS

Large orchestras (90 to 100 pieces) were recorded within a regular scoring stage. It was desired to obtain good separation of instruments and due to the compactness of the arrangement, unidirectional microphones were again used, placed in a relatively close group.

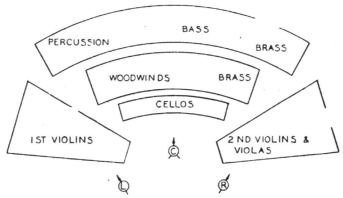

Fig. 6—Typical orchestra and microphone arrangement for a large group in a particular scoring stage.

This also helped to minimize troubles due to room acoustics causing false origins. A typical setup is shown in Fig. 6.

The recordings obtained from one such session were used for playback and the orchestra photographed in a large set. Various cuts and angles were used and it was necessary to exercise caution to select angles and musical passages which were compatible. In some instances minor sacrifices to correct sound placement were made to provide adequate camera freedom.

One vocal number was recorded experimentally at the time a regular production prescore and vocal-recording session was in progress. The vocalist was performing in a small vocal booth with the orchestra in the adjoining scoring stage. A separate microphone was provided for the vocalist and a stereophonic pickup of the orchestra arranged.

The vocal and center-channel music microphone outputs were mixed at the time of the performance, thus obtaining monaural vocal always on the center channel and a stereophonic record of·the accompanying orchestra. This track was later used for playback and the actress photographed, but since the vocal existed only on the center track it was necessary to frame the action so that the performer was nearly always center screen.

. It was noted that orchestra levels greater than normal could be used without destroying the effectiveness of the vocal selection. This might be explained as follows: The vocal is always reproduced as a single direct source and is audibly compared to a stereophonically reproduced accompaniment, thereby increasing the perceptible aural differences and subjectively providing greater separation. No work of this type was done with both sources recorded by stereophonic methods and until this is done and compared to the method reported here, no final conclusion can be reached.

GENERAL OBSERVATIONS

Since this experiment was an integrated project involving all present motion picture production methods, demanding close correlation between each contributing group, it was possible to evaluate the effect that stereophonic-recording application might have on motion picture production and presentation in general and various phases in particular. Those effects, and other observations based on the work herewith reported and which presently seem of the greatest importance, will now be discussed.

From the microphone-pickup work come three cardinal points; sound placement matching corresponding picture, smooth placement transition of the sound from a moving source, and a third point not previously mentioned, that to avoid major changes in quality some sound must be picked up in all microphones at all times.

The requirement of correct sound placement is obvious. It has been found that sound-intensity differences do not play the major role in determining placement except under unfavorable acoustic conditions. Those situations in which high-intensity directive reflections occur and are then picked up by a microphone other than the one closest to the source create the exceptions. Under such circumstances there exists only a small intensity difference between the nearest and other microphones and an otherwise minor change in intensity adjustment can introduce a change in placement. With suitable acoustic conditions, intensity differences due to equipment maladjustment of 6 to 10 decibels do not destroy localization but loudness is of course affected. These observations would indicate that the greater contribution to sound placement is caused by phase differences which are a complex function of acoustics, frequency, and ratio of microphone spacing to frequency.

Smooth sound transition is necessary, otherwise sudden placement jumps occur which are very disturbing to any observer after a short acquaintance with stereophonic reproduction.

The third point concerning quality is related to the inherent improvement in stereophonic over monaural methods. It has been demonstrated that a two-channel stereophonic system does not provide the quality improvement afforded by a three-channel arrangement, as might be expected since the former approaches closer to the inferior monaural condition. The quality difference between two- and three-channel systems is such as to establish the foregoing statement concerning pickup in all microphones.

In connection with recording on a production basis two specific items of equipment were greatly needed. The sound mixer should have a picture monitor displaying the scene the camera is photographing. This apparatus will shorten rehearsal time and guarantee sound and picture match. Such devices are now available by television technique and are rapidly approaching practicality for motion picture use. Second, a better mechanical device than presently used microphone booms must be devised. Some combination of mechanisms which could permit microphone movements with fewer personnel is highly desirable, not only for the sake of reducing manpower costs but to minimize errors caused by lack of co-ordination.

Film editing must be considered when using stereophonic sound. Directors, photographers, and editors are ever watchful of camera angles and actor movement to facilitate smooth cutting, and stereophonic recording demands the same consideration. Much of this problem is automatically solved when the visual action is properly done, but consider the effect of an offstage voice from the left when the offstage person is shown, in the very next cut, on screen right. When cutting pictures editors always strive for action which flows smoothly and logically unless spectacular effects are desired and these are then introduced deliberately. Exactly the same situation exists with stereophonic recording. With proper understanding "jumpy" effects can be eliminated or purposely used when apt. At present, picture editing is hampered very little by sound. It may. very well be that some types of cuts or cutting practice cannot be used successfully in connection with stereophonic methods, and picture may have to concede somewhat more importance to sound.

How will production costs be affected by stereophonic recording? Any answer to this question at this stage of the art requires some speculation since so many factors contribute to such costs. Expense can be greatly influenced by the degree of perfection in the result desired. Considering some of the more tangible factors, it does not appear that production costs need be increased excessively. At no time were more than three complete rehearsals required to satisfy sound-pickup requirements. Usually one or two sufficed; one of these being necessary to observe the action through the camera finder to establish relative placements. Rehearsal time could have been cut in half had the remote finder mentioned above been available. Undoubtedly, with greater experience, the demands of original stereophonic sound would not be much different than at present.

Dubbing costs would of course increase, since greater time would be involved to match both lip movements and action. However, the use of dubbing, while absolutely necessary in some cases, is a dodge and is best minimized.

Demands on set construction are no different for stereophonic than for monaural recordings. An acoustically good set monaurally is still a good set stereophonically and good acoustics, in general, are now sought. Actually, a saving may result in using stereophonic methods because of the poor records frequently obtained in portions of otherwise acceptable sets which, if sufficiently inferior, are dubbed or re-recording time is used up in attempted correction.

Equipment costs, which are a small percentage of production charges, would be raised two to two and one-half times.

Stereophonic reproduction in the theater naturally will require additional equipment. In the event that stereophonic methods are applied to motion pictures, some technical and economical method must be devised to supply both stereophonically and monaurally recorded film during conversion. It is even likely that such practice would continue for some time in order to supply the very small low-income theaters and, in some cases, for reduction for 16-mm release.

CONCLUSIONS

1. Greatly improved sound quality can be obtained by the use of

stereophonic methods. It is easily demonstrable that recordings made in sets which give unnaturally "boomy" or otherwise poor results monaurally result in records which more nearly reproduce the true conditions in that set when recorded stereophonically. This is still true when disregarding subjective sound placement.

2. Sound placement is affected only to a small degree by individual system gain differences indicating that phase and not intensity differences play the major role in determining placement.

3. The three important points of stereophonic pickup are: (1) sound placement matching visible or desired implied action; (2) smooth sound-placement transition and (3) some pickup in all microphones at all times.

4. Many more illusions can be created by sound alone, opening new dramatic, effective avenues for motion picture story presentation.

5. Just as the directions of visual action must be properly done to permit cutting, so must stereophonic sound directions be considered. Of a similar nature, since it pertains to camera angles and editing, prescoring for playback purposes should be planned to match the intended action and anticipated cutting. There is evidence that present editing practices would need modification.

6. With sufficient experience and certain desirable auxiliary equipment, production cost need not be greatly increased. Two of such auxiliaries are a picture monitor (remote viewfinder) for the sound mixer and more suitable microphone-handling equipment. The degree of perfection desired would be the largest cost factor.

7. Re-recording, technically, is no more difficult than at present but having introduced one additional degree of freedom, more manipulation will be required. Many stock library monaural tracks may be used, provided equipment is available for controlling placement of the desired sound.

8. Increased effectiveness of stereophonic sound is obtained if used with a picture of greater aspect ratio than presently used. Given a picture in which the ratio of width to height is approximately 1.75 instead of the existing 1.33 a somewhat closer approach to the horizontal angle of human vision is obtained and the relatively greater width assists sound placement by simplifying the original pickup and giving better coverage in the theater.

As with any other subject of similar complexity, no one experiment answers all the questions. Much work remains to be done. Reproduction in various kinds of auditoriums has only been superficially explored. Some of the questions will only be fully answered by actual production experience.

Contemplation of the results obtained from the described project and with a realization of remaining problems, it is concluded that stereophonic recording can be used for motion pictures and will provide a superior sound presentation which is one step closer to technical perfection and realism on the screen. Unfortunately, stereophonic sound cannot be introduced overnight but it can be made available to the industry if wanted.

ACKNOWLEDGMENT

The welcome assistance of E. I. Sponable of Twentieth Century-Fox Films and K. F. Morgan of Western Electric, as well as several others, is acknowledged.

REFERENCE

(1) Bell Telephone System, Monograph B-784, "Auditory Perspective," Symposium of six papers, 1934.

DISCUSSION

QUESTION: Have you ever tried to balance two microphones in a derby hat and use this as a pickup for sound? We have found in this way, we can get excellent reproduction, even better than with three microphones placed at such long distances. I think the reason for this is that, as you see in your picture, you will have only time differences between the microphone.

MR. L. D. GRIGNON: No, we have not tried that particular combination. We assumed that a good starting point wonld be based on the extensive previous work referred to in a Bell System Monograph. It was our job to try to adapt stereophonic as it was then known to the motion picture problem.

QUESTION: During the war we were able to do much work on stereophonic sound in Holland, and I think we found a better principle to start from than you did here in America.

DR. J. G. FRAYNE: If you make an error in the original, how far can it be corrected in re-recording process? How can you switch people from right to left if you do not get the original track straight?

MR. GRIGNON: Possible corrections depend upon the degree of the error. If it is a minor error, you can push the intensity difference enough to make some correction, but if it is a major error, nothing can be done.

CHAIRMAN C. R. DAILY: With regard to release track in stereophonic, does one require the same quality of reproduction from each individual track that we now feel is desirable from single theater tracks?

MR. GRIGNON: Probably not, although let us put it this way: If a major change of this kind were to be made, certainly we should take advantage of the opportunity to try to increase the fidelity of recording and reproduction. These three channels were somewhat better in quality arrangement than are commonly used. With three channels, let us say, of the ordinary type now used, the subjective quality in reproduction would still be much better than we now have. On that basis, we could take less and still come out equally well.

Single-Element
Unidirectional Microphone*

By HARRY F. OLSON AND JOHN PRESTON

RCA LABORATORIES, PRINCETON, NEW JERSEY

Summary—A single-element unidirectional microphone has been developed for use in sound motion picture recording with the following characteristics: single-ribbon type; the back of the ribbon is coupled to a damped folded pipe and an acoustical impedance in the form of an aperture; improved cardioid directional pattern; greater output; reduced weight; and reduced wind-noise response.

INTRODUCTION

FROM THE INCEPTION of sound reproduction it was apparent to those associated with the problems of sound pickup that some form of directivity was desirable in the sound-collecting system to improve the ratio of direct to reflected sounds and thereby improve the reverberation characteristics and otherwise discriminate against undesirable sounds. Horns and reflectors were used for the early directional sound-collecting systems. As the fidelity of reproducing systems was improved, it became apparent that considerable distortion in the form of frequency discrimination in both the direct and reflected sounds was introduced because the directional characteristics of the horn and reflector systems varied with frequency. About two decades ago the velocity directional microphone[1] was developed which exhibited uniform directional characteristics over the entire audible spectrum. This microphone established the usefulness and superiority of a sound-collecting system with uniform directional characteristics.

The conventional velocity microphone consists of a single mass-controlled ribbon with both sides freely accessible to the medium. The many desirable performance characteristics exhibited by this microphone may be attributed to the obvious simplicity of the vibrating system. The constants of the system may be chosen so that response and directional characteristics will be uniform over the entire audible-

* Presented May 17, 1948, at the SMPE Convention in Santa Monica.

frequency range. The nonlinear distortion for the intensity range of
the human ear is a small fraction of a per cent. The light-mass-con-
trolled system insures good transient response.

The polar-directional characteristic of the velocity microphone is
bidirectional. For certain sound-pickup problems, unidirectional
characteristics are more desirable. Accordingly, shortly after the

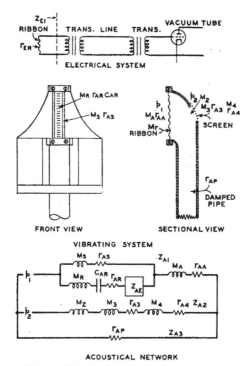

Fig. 1—Electrical system, vibrating system,
and acoustical network of the single-element
unidirectional microphone.

development of the velocity microphone, a unidirectional microphone[2,3] with uniform directional and frequency response over a wide
frequency range was developed. The conventional unidirectional
microphone consists of the combination of a pressure element and a
velocity element. The original unidirectional microphone employed
ribbon elements for both the velocity and pressure elements. Employing ribbon elements makes it possible to maintain uniform phase

relations between the velocity and pressure elements without resorting to correcting networks. The acoustic fidelity of the unidirectional microphone is essentially the same as that of the velocity microphone.

In view of the importance of directional microphones and the high fidelity of ribbon transducers, work has been continued on these systems with the object of increasing the scope and simplifying the vibrating system. In particular, a microphone has been developed for sound motion picture recording with the following characteristics:

Fig. 2—Assembled single-element unidirectional microphone (production model).

(1) Single-ribbon type.
(2) Improved cardioid directional pattern.
    (a) In the high-frequency range due to a smaller vibrating system and suitable magnetic structure.
    (b) In the low-frequency range due to the use of an additional acoustical element.
(3) Greater output.
(4) Reduced weight.

It is the purpose of this paper to describe the single-element unidirectional microphone with the characteristics listed above.

Acoustical System

The electrical system, the acoustical system, and the acoustical network of the single-element unidirectional microphone are shown in Fig. 1. Referring to the acoustical network, it will be seen that this is of the bridge type.

The phase relations between the two actuating pressures varies with the direction of the incident sound. For example, for sound incident upon the back of the microphone, the time required for the sound to pass through the hole in the pipe to the back of the ribbon is the same as the time required for the sound to pass around the magnet structure to the front of the ribbon. Under these conditions, the same pressure with the same phase is exerted on the front and back of the ribbon, and as a result the ribbon does not move. For sound incident upon the front of the microphone, the sound which appears at the back suffers a delay caused by the path around the magnet structure and a delay

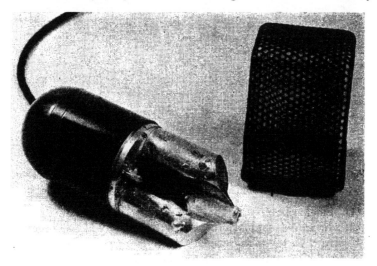

Fig. 3—Front view with the screen removed.

through the hole in the pipe. Therefore, there is a relatively large phase angle between the sound pressures on the two sides of the ribbon which means that the ribbon will move because of the difference in pressure due to the difference in phase. For the particular phase relations that exist between these pressures for different angles of incidence, it is possible to choose the constants of the system so that a cardioid characteristic will be obtained.

It was found that the deviation from a cardioid pattern at the lower frequencies was caused by the unbalance produced by the acoustical resistance in the branch z_{A1}. By introducing a corresponding acoustical resistance in the branch z_{A2}, the balance can be restored. This state of affairs can be deduced from the acoustical network and the following theoretical considerations.

In the acoustical circuit the elements are as follows:

In branch z_{A1}

M_S = inertance of the slit between the ribbon and pole
r_{AS} = acoustical resistance of the slit between the ribbon and pole piece
M_R = inertance of the ribbon
r_{AR} = acoustical resistance of the ribbon
C_{AR} = acoustical capacitance of the ribbon
M_A = inertance of the air load upon the ribbon
r_{AA} = acoustical resistance of the air load upon the ribbon
z_{AE} = acoustical impedance due to the electrical circuit.

It is given by

$$z_{AE} = \frac{(\beta l)^2}{z_{E1} + r_{ER}}$$ (1)

where

β = flux density in the air gap
l = length of the ribbon
z_{E1} = electrical external load on the ribbon (see electrical circuit, Fig. 1)
r_{E1} = electrical resistance of the ribbon.

z_{A1} is the acoustical impedance of the branch composed of the above elements.

In branch z_{A2}

M_2 = inertance of the aperture
M_3 = inertance of the screen covering the hole
r_{A3} = acoustical resistance of the screen covering the hole
M_4 = inertance of the air load upon the screen and aperture
r_{A4} = acoustical resistance of the air load upon the screen and aperture.

z_{A2} is the acoustical impedance of the branch composed of the above elements.

In branch z_{A3}

$z_{A3} = r_{AP}$ = acoustical resistance of the acoustically damped pipe.

The sound pressure acting on the open side of the ribbon may be written

$$p_1 = p_{01}\ \epsilon^{j(\omega t + \phi_1)}$$ (2)

where

p_{01} = amplitude of the pressure
ω = $2\pi f$
f = frequency
t = time
ϕ_1 = phase angle with respect to a reference point.

The sound pressure acting on the aperture in the labyrinth connector may be written

$$p_2 = p_{02}\, \epsilon^{j(\omega t + \phi_2)} \tag{3}$$

where

p_{02} = amplitude of the pressure
ϕ_2 = phase angle with respect to a reference point.

The reference point for the phase may be changed so that

$$p_1 = p_{01}\, \epsilon^{j(\omega t)} \tag{4}$$

and

$$p_2 = p_{02}\, \epsilon^{j(\omega t + \phi_3)}. \tag{5}$$

The phase angle ϕ_3 is a function of the angle of the incident sound as follows:

$$\phi_3 = \phi \cos \theta \tag{6}$$

where

θ = angle between the normal to the surface of the ribbon and the direction of the incident sound, and
ϕ = phase angle determined by the dimensions and geometry of the ribbon and the structure surrounding the ribbon.

The volume current of the ribbon due to the pressure p_1 is

$$\dot{X}_1 = \frac{p_1(z_{A2} + z_{A3})}{z_{A1}z_{A2} + z_{A1}z_{A3} + z_{A2}z_{A3}}. \tag{7}$$

The volume current of the ribbon due to the pressure p_2 is

$$\dot{X}_2 = \frac{p_2(z_{A3})}{z_{A1}z_{A2} + z_{A1}z_{A3} + z_{A2}z_{A3}}. \tag{8}$$

The resultant volume current X_R of the ribbon is the difference between (7) and (8).

$$\dot{X}_R = \dot{X}_1 - \dot{X}_2. \tag{9}$$

The internal voltage generated by the motion of the ribbon is given by

$$e = \frac{\beta l \dot{X}_R}{A} \tag{10}$$

where

β = flux density in the air gap
l = length of the ribbon
A = area of the ribbon
$\dot{X}_R$ = volume current of the ribbon given by (9).

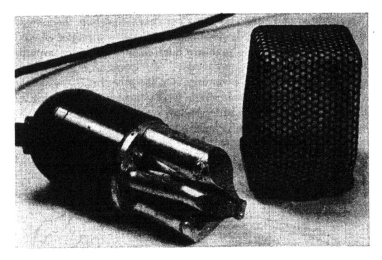

Fig. 4—Rear view with the screen removed.

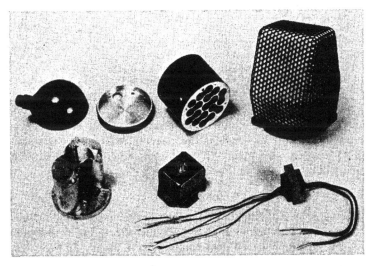

Fig. 5—Exploded view showing bottom cover, cover plate, folded and damped pipe, screen, magnetic and vibrating structure, transformer case, and transformer.

A consideration of the above system shows that it is theoretically possible to obtain a true cardioid characteristic by a proper choice of constants. However, in a practical microphone certain incompatible factors appear to make this impossible. For example, high sensitivity requires a large magnetic structure. On the other hand, the simplest way to obtain a true cardioid characteristic is to use a vibrating system and surrounding structure which is small compared to the wavelength. The sensitivity of a microphone with such relatively small dimensions would be too low. However, it is possible to reduce the phase effects in a relatively large structure by a suitable design. In

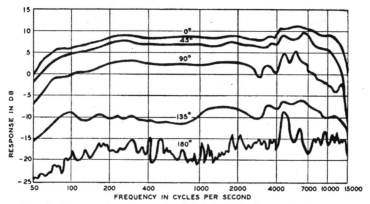

Fig. 6—Response-frequency characteristics of the single-element unidirectional microphone for the incident angles of 0, 45, 90, 135, and 180 degrees.

order to determine the optimum compromise between sensitivity and directivity, a study was made of the magnetic and acoustic characteristics of twenty structures. The selected structure for this microphone will be described in the next section.

CONSTRUCTION

Photographs of the final model of the single-element unidirectional microphone are shown in Figs. 2 to 5, inclusive. Fig. 2 shows the completed microphone with the wind screen in place. In Figs. 3 and 4 the wind screen has been removed. Fig. 3 shows a front view, depicting the magnetic structure and ribbon. Fig. 4 is a rear view, depicting the magnetic structure and labyrinth connector. The

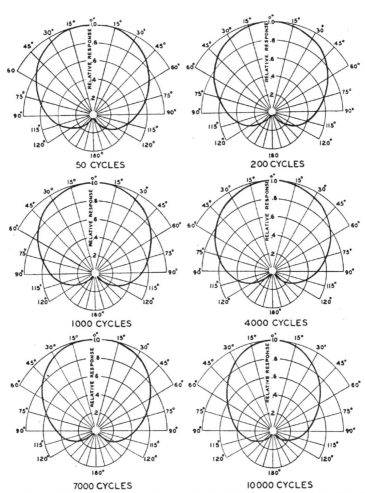

Fig. 7—Polar-directional patterns of the single-element unidirectional microphone for the frequencies of 50, 200, 1000, 7000, and 10,000 cycles per second.

acoustical resistance, in the form of a screen over the hole in the labyrinth connector, can also be seen in Fig. 4.   Fig. 5 is an exploded view, showing all the component parts of the microphone.

Performance Characteristics

The measured response-frequency characteristics and directional characteristics of the single-element unidirectional microphone are shown in Figs. 6 and 7.   It will be seen that the directional pattern is practically independent of the frequency of the incident sound over the frequency range from 50 to 8000 cycles.   Furthermore, a high order of cancellation for sounds arriving from the rear is obtained from 50 to 14,000 cycles.   The high order of cancellation at the low frequencies is due to the addition of the acoustical-resistance element. The higher order of cancellation at the high frequencies is caused by the design of magnetic structure which exhibits uniform sound diffraction over the frequency range up to 8000 cycles.   The uniform directivity has not been obtained at the expense of sensitivity because the open-circuit output for an impedance of 250 ohms is 260 millivolts per dyne per square centimeter at 1000 cycles.   The light weight of two pounds is quite low for this order of sensitivity and frequency range.

References

(1) H. F. Olsen, "Mass controlled electrodynamic microphone: The ribbon microphone," J. Acous. Soc. Amer., vol. 3, pp. 56–68; July, 1931.

(2) H. F. Olson, "A unidirectional ribbon microphone," J. Acous. Soc. Amer., vol. 3, pp. 315–316; January, 1932.

(3) H. F. Olson and J. Weinberger, U. S. Patent, Reissue 19115, 1934.

16-Mm Film Phonograph for Professional Use*

By CARL E. HITTLE

RCA VICTOR DIVISION, HOLLYWOOD, CALIFORNIA

Summary—Superior performance of new 16-mm film phonograph designed for the motion picture industry permits its use as a standard to determine the sound quality of 16 mm films and to check performance of 16-mm projectors.

Additional features include a self-contained preamplifier, rewind at accelerated speed, dependable operation, compactness, accessibility for servicing, and attractive styling.

SINCE THE INCEPTION of sound on film, 35-mm film has been the accepted standard of the motion picture industry. For several reasons, 16-mm film has not been used where the best in picture and sound were required, but 16-mm film production is on the increase and there is a tendency in this to judge the quality of 16-mm sound in direct comparison with that obtained with 35-mm film. With the increase in the number of 16-mm productions has come an increase in the use of 16-mm equipment exclusively in these productions. Much of the original recording is being done on 16-mm film. Therefore, not only is the quality of the performance of the 16-mm film recorder important to the achievement of high-quality sound print, but also the quality of performance of the 16-mm film phonographs used in the re-recording of the productions and in determining the quality of the original recording. A paper by Collins[1] described the construction and performance of a new 16-mm film recorder. The sound quality obtainable with it compares favorably with that obtainable using 35-mm recorders of modern design. The PB-176, 16-mm film phonograph (Fig. 1) has been designed as a companion unit using all the applicable features of the PR-32, 16-mm recorder described in the aforementioned paper.

The principal features which the two units have in common are:

(a) A base assembly (Fig. 2) containing the receptacles for electrical connections, lamp rheostat, and cast-in recesses that form the handles for carrying.

* Presented May 20, 1948, at the SMPE Convention in Santa Monica.

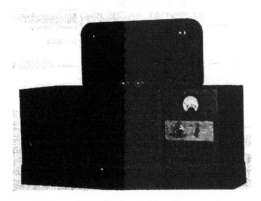

Fig. 1—PB-176 16-mm film phonograph.

(b)   Covers for the optical, motor, and gear compartments.

(c)   A head assembly (Fig. 3) containing all of the gearing and film-handling equipment.

(d)   A driving motor of synchronous or interlock type for either 60- or 50-cycle power.

(e)   A control-panel assembly containing lamp-rheostat adjust-ment, lamp ammeter, and switches for motor and lamp.

(f)   Enclosed-type beltless take-up and rewind assembly.

Fig. 2—Rear view.

The film phonograph differs from the recorder mainly in that it contains a reproducing optical system plus a phototube preamplifier in place of the recording optical system (Fig. 4).

Principal mechanical features of the PB-176 film phonograph which deserve special notice include:

(a) Film motion having total flutter less than 0.1 per cent.

(b) Lateral film weave of less than 0.001 inch.

(c) Machine noise of 53 decibels below 100 per cent modulated 1000-cycle film.

Fig. 3—Rear view, covers removed.

(d) Film compartment of enclosed type having a wide-opening door for easy access to sprockets, rollers, and that portion of the optical system installed within this compartment. A wide slot on the upper side of the compartment simplifies threading to and from the take-up and rewind reels mounted on reel spindles above the compartment. Guide rollers are provided to direct the film into the compartment.

(e) Take-up of enclosed positive-drive type capable of handling 800-foot reels. (For those special applications where 1600- or 2000-foot reels must be used, a take-up drive assembly will be available for reels of that size.)

(f) Rewind which is enclosed with take-up mechanism and which automatically utilizes a portion of take-up gearing driven in the reverse direction. To rewind, the film is threaded directly from the right-hand reel to the left-hand reel and the motor is then run in reverse rotation by throwing the reversing switch mounted on the control panel to the right of the film compartment. In the synchronous-motor-driven units the reversing switch reverses two motor leads thus reversing motor phasing and causing the motor rotation to be be reversed. In the Selsyn-motor-driven units, even though the

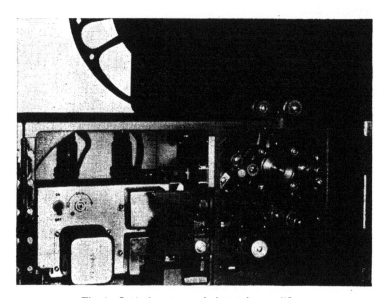

Fig. 4—Optical system and phototube amplifier.

motor circuit is more complex than that of the synchronous motor, operation of a single switch on the control panel also will cause the motor rotation to be reversed. This switch is a four-pole double-throw type with a positive neutral or off position. In the up or normal running position, the motor is connected to the Selsyn distributor. In the down or reverse position of the switch, the stator of the motor is connected to a separate three-phase power supply whose phasing is the reverse of that for Selsyn operation and the rotor leads are short-circuited together through a relay mounted in the base of the film phonograph. This permits the motor to operate as an induction-type motor at nearly one and one-half times normal speed. The rewind time for a film phonograph equipped with a 60-cycle Selsyn motor is $2^1/_2$ minutes for 800 feet of film.

The new improved reproducing optical system is of advanced design in keeping with the remainder of the film phonograph. This unit is mounted to the side of the film-compartment portion of the main case. Mounting is in such a manner as to provide easy lateral adjustment of the scanning-light beam by means of an adjusting screw. This assembly includes:

(a) The slit and objective-lens assembly which is of comparable quality to those used in theater-type soundheads. Azimuth adjustment is provided by two opposing setscrews which rotate only the slit-condenser and slit-plate assembly. Focusing of the objective is accomplished by turning the lens mounted in a keyed barrel and an instantaneous change-over focus adjustment is provided, by rotating a calibrated ring equipped with a finger pin, for films having their emulsion either toward or away from the scanning-light source.

(b) An exciter lamp which is of the standard prefocus type with curved filament designed to operate at 10 volts, 5 amperes. Convenient vertical adjustment of the lamp filament is provided. A quick release lever on the lamp socket permits easy removal of a defective lamp.

(c) The phototube, RCA Type 927 Radiotron, is mounted on the optical bracket and extends from it into the film compartment above the sound drum. The phototube is readily accessible from the film compartment and may be removed after removing its slip-on-type light shield. The scanning-light beam, after passing through the film, is efficiently directed to the phototube by a single-unit prism and lens assembly mounted adjacent to the sound drum.

A compact phototube amplifier is provided as an integral part of the film phonograph. It is mounted in the optical compartment behind the optical system and is accessible for servicing or removal from the film phonograph by removing the optical compartment cover which is held in place by two thumbscrews. In order to obtain maximum efficiency, the phototube mentioned above is directly connected, by a short low-capacitance cable, through a high-impedance circuit to the grid of the first preamplifier stage. Electrical connections to the phototube amplifier are made at conveniently located externally mounted terminal boards. The preamplifier contains a power switch

and gain control.   Electrical characteritics of the phototube amplifier
are as follows:

(a) Output impedance:	150/250 ohms (balanced or unbalanced)
(b) Filament:	Adjustable from 6.3 to 14 volts direct current
(c) Plate:	250 volts direct current
(d) Output level:	-2 dbm* for normal operation and $+8$ dbm maximum
(e) Noise level:	53 decibels below 100 per cent modulation
(f) Frequency response:	Flat characteristic within ±1 decibel from 30 to 6000 cycles using $1/2$-mil optical slit and flat film
(g) Polarizing voltage:	80 volts direct current
(h) Tube complement:	Two RCA 1620 Radiotrons
(i) Distortion:	Less than $1/2$ per cent from 50 to 7500 cycles at $+4$ dbm output
(j) Gain control:	7-decibel range

* Decibels with respect to 0.001 watt.

The many self-contained features such as a conveniently located
exciter lamp and motor controls consisting of an ammeter, rheostat,
and switches, plus phototube amplifier; the selectivity offered of
take-up and rewind sizes, the compactness of the design, the pleasing
styling, and the convenience and dependability of operation make the
PB-176, 16-mm film phonograph a distinctive unit physically.   Its
equally distinctive performance characteristics further enhance its
desirability by those engaged in 16-mm work for re-recording activity
or for carefully checking recorded material.

ACKNOWLEDGMENT

Recognition is given to J. L. Pettus, L. T. Sachtleben, K. Singer,
and E. P. Ancona for their contributions to the design and testing of
this equipment.

REFERENCE

(1) M. E. Collins, "Lightweight recorders for 35- and 16-mm film," J. Soc.
Mot. Pict. Eng., vol. 49, pp. 415–425; November, 1947.

DISCUSSION

MR. HARRY ERICKSON: Would you please repeat the distortion effect figure?
MR. CARL E. HITTLE: The figure is less than one-half per cent, from 50 to 7500
cycles per second at plus four dbm output.

Frequency-Modulated Audio-Frequency Oscillator for Calibrating Flutter-Measuring Equipment*

By P. V. SMITH and EDWARD STANKO

THEATER SERVICE DIVISION, RCA SERVICE COMPÀNY, CAMDEN, NEW JERSEY

Summary—This paper describes an electronically frequency-modulated oscillator of the resistance-capacitance type which has been developed to facilitate rapid and accurate calibration and testing of flutter-measuring equipment; specifically, of the portable flutter indicator used in theater service work. It supersedes mechanically driven capacitor devices for producing the frequency-modulated 3000-cycle signal used for flutter-indicator calibration. It possesses the advantages of better frequency stability, purer wave form, and absence of spurious output impulses.

FOR MANY YEARS one of the major problems inherent in all types of sound recording and reproducing methods has been that of attaining constant speed of the record medium, whether film, flexible tape, wire, or disk, at the recording or scanning point. The extent of this problem has been shown by the many articles which have been published in the JOURNAL of the Society of Motion Picture Engineers, the *Proceedings* of The Institute of Radio Engineers, the *Journal* of the Acoustical Society of America, and others, concerning the work done in the effort to minimize speed variations.

In many present-day recording and reproducing equipments such undesirable speed variations are reduced below the level of audible disturbance. The attainment of this desirable level in operation requires care, both in making the original installation and in the maintenance of the equipment. To ensure satisfactory maintenance, various instruments and test methods have been develped to assist the field engineer. Among the first of these was the portable flutter indicator, developed by the Radio Corporation of America for measuring flutter in theater sound equipment and made available to its theater service field men in 1938.

* Presented May 20, 1948, at the SMPE Convention in Santa Monica.

Previously, the calibration of these flutter-indicator instruments was accomplished by the use of a frequency-modulated oscillator whose 3000-cycle fundamental frequency could be varied (by means of a rotating motor-driven capacitor) plus or minus a maximum of 2 per cent in selectable increments of 0.1 per cent. The design of this motor-driven capacitor presented some interesting problems. Previous experience with mechanically operated, frequency-modulation devices indicated that connections to the rotating plates would be troublesome, and therefore, a symmetrical double rotor and stator, with the circuit connections made to the two stators, would be necessary. While the symmetrical double rotor aided in obtaining mechanical balance of the rotating parts, the shape of the plates required to obtain symmetrical sine-wave modulation when paralleled with the remaining circuit capacitance was rather unique. Further, as the required amount of frequency modulation was, of necessity, a variable ranging from 0 to 2 per cent, the addition of the required series and/or shunt capacitors to the rotating capacitor circuit changed the symmetry of the modulation. Mechanical vibration of the moving parts, together with hysteresis of the iron core of the oscillator coil, introduced unwanted and random modulation. Any looseness or wear in the bearings of the rotating capacitor also contributed to instability of the output. As the rate of frequency change was controlled solely by the shape of the rotating capacitor plates, various modulation patterns could be obtained only by designing and making a special capacitor for each pattern desired.

After considering all of the deficiencies of the motor-driven capacitor oscillator, it was felt that an electronic type of oscillator which would be free from the inherent limitations of the mechanically variable type should be designed. Similar oscillators have been developed in recent years for various applications, notably the RCA facsimile transmitter. The circuit shown in Fig. 1 is the outcome of considerable research work and study applied to the basic circuit. Scrutiny of this circuit will reveal that a resistance-capacitance phase-shift oscillator is the heart of the device.

To vary the frequency of this oscillator at the desired rate, one of the resistances in the network is composed partly of a physical resistor and partly of the plate resistances of a push-pull pair of triodes. Control voltage applied to the grid of the lower triode section results in practically instantaneous change of plate resistances and, therefore, of the frequency of the oscillator. The amplitude of the oscillation remains substantially unchanged through considerable variation in frequency, and therefore, the output of the oscillator is free from unwanted amplitude modulation.

The 2000-ohm control rheostat connected in parallel with the 200-ohm cathode resistor of V-3 (which together furnish grid bias for the control tube), is provided to set the oscillator frequency initially to 3000 cycles per second. Plate voltage to the circuit is stabilized by the voltage regulator tube V-5. The frequency-modulated output of the oscillator is fed into an audio amplifier to obtain the re-

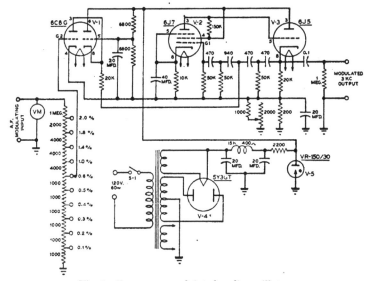

Fig. 1—Frequency-modulated audio oscillator.

quired amount of driving power for proper indication of the flutter bridge being calibrated.

The input to the grid of the control triode may be of any frequency from zero to several hundred cycles per second and of any wave shape from extremely peaked through sine-wave to rectangular. For sine-wave frequency modulation, an audio-frequency oscillator of the familiar beat-frequency or resistance-capacitance type may be used. For other types of modulation, such as square wave, appropriate generators are available.

To obtain the precise amount of frequency modulation desired within the range of 0 to 2 per cent, a thermocouple-type voltmeter and an attenuator are built into the instrument. The voltmeter indicates the root-mean-square voltage applied to the voltage-divider attenuator. The attenuator taps correspond to the percentage points for which the flutter-bridge scale is marked.

The use of a thermocouple voltmeter insures that regardless of the wave shape of the modulating voltage used, the percentage modulation of the oscillator will be correct root-mean-square values as stipulated elsewhere.[1]

A further advantage of. the use of a thermal voltmeter is the fact that at modulation frequencies approaching zero the thermal inertia of the couple prevents unsteady indication of the voltmeter pointer. As this instrument is a laboratory device, the extra sturdiness of other types of voltmeters is considered unnecessary.

Since the thermocouple voltmeter indicates root-mean-square values directly, standardization of the instrument is easily made with direct current. This is done by applying direct and reversed direct current to the grid of the control tube, and measuring the voltage required to give the required frequency deviation as indicated on a Conn stroboscope. The required attenuator resistors are calculated and the input voltmeter scale marked at one scale point corresponding to the required attenuator input voltage. The accuracy of the attenuator taps is then checked by comparison with the Conn stroboscope, and by any necessary readjustments made to the attenuator resistors.

This instrument is free from amplitude-modulation effects, gives excellent frequency stability, good output wave form, long-time constancy of calibration, and provides rapid checks of flutter-bridge scale markings. It may be modulated readily at any low audio frequency, and with any wave shape desired by using an external modulation generator. Its use has enabled extensive investigation of flutter-indicator characteristics, which will in turn permit more accurate field tests of sound-reproducing equipment.

REFERENCE

(1) "Proposed standard specifications for flutter or wave as related to sound records," J. Soc. Mot. Pict. Eng., vol. 49, pp. 147–160; August, 1947, section 4.1.

Silent Playback and Public-Address System*

By BRUCE H. DENNEY AND ROBERT J. CARR

PARAMOUNT PICTURES, HOLLYWOOD, CALIFORNIA

Summary—The silent playback and public-address system provides a unique time-saving aid in photographing and recording dialog sequences having musical backgrounds. By means of a music- or speech-modulated 100-kilocycle transmitter, a supersonic-frequency magnetic field is generated throughout the scene area. The magnetic field includes secondary currents in concealed loops worn by the actors. These signals are demodulated in miniature batteryless units and energize small hearing-aid earphones. The earphones may be worn without concealment on distant and medium scenes. On close-ups the earphones are concealed and the sound is coupled to the ear through small plastic tubing camouflaged by make-up. Several types of loops, demodulators, and earphone units meet the varying needs of the scenes to be photographed and recorded. The stage microphone and sound-on-film recording channel is free from 100-kilocycle interference.

IT HAS BEEN generally recognized that the cost and complexity of photographing and recording many motion picture scenes would be reduced if the actors could be cued, or directed, or could be enabled to hear a scene's music, by means of earphone units, small enough to be disguised by make-up, energized by miniature radio receivers concealed in the clothing. The recording-system microphone would hear only the desired sounds while the actors would be guided through complex scenes by the director, this, in a manner reminiscent of the silent motion picture method of direction. In the case of dialog lines read through a musical background, certain actors could speak their lines to the recording-system microphone while other actors might dance through the scene to music heard by way of disguised earphones and receivers. Or an actor wearing the miniature equipment could sing a song to the silent accompaniment of a temporary piano recording which later would be replaced by a recording of a full orchestra. In each of the previous cases, the proper balance between the music and speech would be restored in the re-recording operations.

* Presented May 18, 1948, at the SMPE Convention in Santa Monica.

An assignment to design and construct this type of equipment was given the authors. Considering only the studio-production requirements, it was decided that while making the receivers as small as possible, also to make them batteryless, expendable in case of trouble, and to build a transmitter powerful enough to cause the receiver to operate with good volume. Receivers were constructed using germanium-crystal detectors and the miniature hearing-aid-type capacitors and

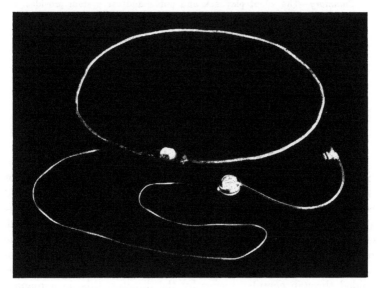

Fig. 1—Loop receiver, earphone, and coupling unit.

connectors. The receivers energize hearing-aid-type phones coupled to the auditory canal with either commercial-type adaptors or especially designed adaptors coupled by plastic tubing.

However, sufficient transmitted energy to insure an adequate output of the crystal-detector receivers, would cause a large radio-frequency signal to be radiated. To avoid this, a novel scheme is used. Using a speech-modulated, 100-kilocycle-per-second transmitter, a current is sent through a *closed* single turn of wire surrounding the set. A high-frequency magnetic field is thereby generated throughout the set area. This field induces secondary currents in multiturn loops worn by the actors. These loops of from 6 to 10 inches in diameter are worn in the hairdress or under the clothing of the actors. In effect, the single turn of wire about the set is the primary of an air-core transformer while each loop is a separate secondary. The induced secondary currents, demodulated in the germanium-crystal-detector receiver, energize the earphone units.

Fig. 1 shows the completed receiver with its protective covering. The miniature Western Electric Type 400-C germanium rectifier is used for detection. The tuning capacitor, rectifier, load-resistor, and blocking capacitor are built into a small terminal cord to which the loop is fastened and connected. The receiver circuit is conventional except that no by-pass capacitor is used across the earphones. The advantage in quality with a by-pass capacitor was small and, in the interests of size, it was removed. The loop has an inductance of 1.25 millihenries. A 0.002-microfarad capacitor in parallel with the loop tunes the circuit to the transmitter frequency. After each unit is tested, the loop is covered with split rubber tubing that has been recemented, and the equipment cord is covered with an air-drying rubber compound. The hearing-aid phone unit is of the crystal type and is connected to the loop receiver by flexible leads equipped with miniature connectors.

Connection to the auditory canal of the ear is made in several ways depending upon the conditions of use. If the actor is in a close-up camera angle a small plastic elbow fitting is adapted to the canal by means of various sized rubber inserts. The inserts are available commercially. The elbow terminates a thin walled plastic tube that is coupled to the phone unit. The tubing, properly coupled, has very little voice-frequency attenuation, and lengths up to 30 inches may be used.

Upon occasion, especially fitted earmolds have been made for a particular actor or actress by a hearing-aid-equipment laboratory. The earmolds are designed to insure the best possible degree of invisibility. They are comfortable and easily fitted to the ear of the user. Plastic tubing couples the earmold to the phone unit. When the actors who are to receive directional or music cues are at some distance from the camera, they may be equipped with commercial adaptors. The adaptors are available in a range of sizes for either ear. These are of the same type worn for test purposes by hard-of-hearing persons while being fitted with hearing-aid equipment.

A woman's formal attire does not help with the disguise of the equipment. To solve this problem a smaller loop receiver was designed for use in the hairdress (Fig. 2). This loop receiver is almost as sensitive as the larger receivers. A hairdresser assists in concealing the elements of the system. Splicing equipment is available to enable the fitter of the ear-coupling equipment to shorten or lengthen the plastic tubing to the most desirable length.

The 100-kilocycle-per-second transmitter, Fig. 3, does not differ in many ways from conventional radio-frequency units. The inductances, tuning capacitors, and by-pass capacitors are larger than those used in broadcast-frequency equipment. The plate circuit of the 100-kilocycle-per-second power amplifier is tuned by a variable induct-

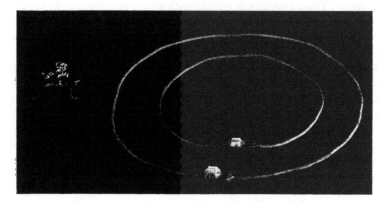

Fig. 2—Standard loop receiver and small hairdress loop receiver.

ance and a group of fixed capacitors. A tuning capacitor for this frequency, and for the voltage used, would be of an awkward size; therefore, the capacitor values are selected by a tap switch. Fine tuning is accomplished by a variometer-type rotor in the electrical center of the inductance coil.

The 100-kilocycle-per-second oscillator is crystal-controlled and uses Type 6F6 tubes in push-pull. The oscillator drives a buffer stage that uses Type 807 tubes push-pull pentode-connected. Frequency multiplication is not used, all amplification of the high frequency being at the frequency of 100 kilocycles per second. The power stage is a Class C stage using Type 812 tubes. A step-down close-coupled secondary winding, connected to the loop surrounding the set area, is wound over the power-amplifier plate-inductance coil.

The dimensions and resistance of the loop change the effective inductance of the tuning circuit, making it necessary to retune for different sized loops. This is the transmitter's only critical adjustment.

The speech amplifier is conventional in most respects. Dual inputs, one for the microphone and one bridging the playback circuit, are provided. Provision is made, by means of a relay operated from the "push-to-talk" push button on the microphone, not only to make operative the microphone circuit, but also to reduce the playback volume when the microphone is used. This prevents the masking of the director's voice by the playback music. The amount of this change is adjustable. A switch changes the microphone amplifier stage to a 400-cycle-per-second resistance-capacitance oscillator. This tone is used to check the strength and distribution of the set-

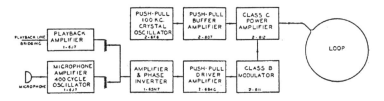

Fig. 3—100-kilocycle-per-second transmitter block schematic.

area magnetic field and the sensitivity of the receivers. The maximum signal level heard in the phones has been more than sufficient to provide the wearer of the equipment a loud, clear signal under all conditions. The field is uniform enough that no volume control is necessary on the receivers; the audio level being adjusted in the speech amplifier to a level suitable to the actors.

The modulator is a conventional Class B amplifier using Type 811 tubes, coupled to a modulation transformer which plate-modulates the Class C high-frequency power stage. The modulator plate-current meter is used as a speech-volume indicator.

Although the sound-recording microphone is used in the field, no interference from the "silent" unit has been heard in the sound-recording system, except in instances of cable-shield trouble, in which cases the device acts as a tool to locate such difficulties.

The transmitter is powered from 115 volts alternating current. The power-switching circuit includes a time-delay relay to prevent application of the plate voltages until filaments are heated; an overload relay to prevent damage to the high-power tubes in the case of overmodulation or mistuning; a series ballast resistor to reduce power to the high-power tubes during the tuning of the transmitter; and door-interlocks to turn off power when high-voltage-area protective covers are removed. The steps of power-switching are indicated by pilot lamps. The exciter, power amplifier, and modulator stages are equipped with plate-current meters, and a thermocouple ammeter reads the loop current. The transmitter unit (Fig. 4), assembled from the experimental equipment, has been built into a portable dolly. Storage facilities for the loop receivers, phones, and

Fig. 4—Transmitter and auxiliary equipment dolly.

phone attachments, and for cables and spare parts, have been built into the dolly.

From a production standpoint, this silent playback and public-address system has proved itself to be a time-saving tool. The recording of certain musical sequences has been remarkably simplified. In one dancing sequence, twenty loop receivers were in operation at one time; forty are available. It has also found service in the recording of tap-dance sound effects where the usual earphones and their connecting cords have limited the dancers' movements. In its first four months of operation, the equipment was used on six different pictures with gratifying compliments from the production office. A second transmitter has been built to meet increased demands for this type of service.

DISCUSSION

QUESTION: What is the power output of the transmitter?

MR. BRUCE H. DENNEY: It is a 200-watt transmitter; I believe the actual power output is probably in the neighborhood of 50 watts, due to the inefficiency of the final stage coupling. It was a bit of a problem. We tried many different methods to gain the greatest efficiency.

MR. GEORGE LEWIN: Are the loops around the set in a vertical plane?

MR. DENNEY: Horizontal plane, either on parallels ten feet above the floor or just laid in a carpet around the stage or directly outside of the stage wall. It is really not critical, but it is essentially parallel to the floor.

QUESTION: Is there not apt to be a tremendous change in level?

MR. DENNEY: We thought that ought to be true, but surprisingly it is not noticeable to the people wearing the receivers. I might mention that the receivers are worn parallel to the floor, usually in the form of a necklace.

MR. W. S. STEWART: Why did you choose 100 kilocycles instead of a higher frequency?

MR. DENNEY: Our problem was to concentrate a magnetic field in the interest of a minimum amount of radiation. We felt that the longer the wavelength, the less radiation there would be in the different sides of the loop. Actually there is such a small part of one wavelength in the loop that no standing waves exist.

QUESTION: For lower power and higher efficiency, would not a somewhat higher frequency be better?

MR. DENNEY: That may be true, but we fixed upon the frequency in our minds. We went to work and happily enough it worked very well. On the second unit, we hope to make measurements indicating the exact degree of field intensity, and so on.

New Automatic
Sound Slidefilm System*

By W. A. PALMER

W. A. PALMER AND COMPANY, SAN FRANCISCO, CALIFORNIA

Summary—A new completely automatic sound slidefilm system has been devised which operates by virtue of low-frequency tones recorded on the disk along with the program material. The system permits an uninterrupted flow of sound without audible gongs or bells. It is positive in action and immune to record wear, scratches, or blunt needles. A tone generator is available to record the disks required for the system.

THERE HAVE BEEN a number of systems proposed from time to time to synchronize automatically the movement of a film strip with a disk recording, thereby eliminating the usual signal bell or gong as used on the present sound slidefilm systems. Most of the proposed devices have made use of so-called "inaudible" frequencies recorded along with the regular program material. A typical device, tried some years ago, made use of a 40-cycle tone recorded at the proper intervals on the disk to energize a resonant circuit and operate a relay, causing the magnetic picture-shifting mechanism to trip. This was found to be impractical because it operated purely on a marginal basis and if the automatic unit were made sensitive enough to be sure to operate from the control tone, it became extremely vulnerable to false tripping by external forces such as accidental shocks, flicking of the needle, motor rumble, and other parasitic low-frequency disturbances.

Other devices made use of high-frequency tones, usually in the 6000-cycle region, but this did not prove practical because, since the disks are recorded at constant velocity, a 6000-cycle tone must be recorded at a very high relative level in order to become at all usable. The tone then became almost impossible to eliminate from the audio system unless the audio channel was muted for the duration of the tone. This defeated the purpose of the automatic device and introduced an interruption of silence in place of the bell or gong, and re-sulted in no great improvement over the manual sound slidefilm.

* Presented May 18, 1948, at the SMPE Convention in Santa Monica.

THE 30/50 AUTOMATIC SLIDEFILM SYSTEM

In the development of the 30/50 automatic sound slidefilm system, J. T. Mullin has incorporated a new approach which has the advantages of a tone-on-the-record system and eliminates the erratic behavior of the older single tone. In this system, a 50-cycle tone, called the "lock-out" tone, is recorded throughout the disk, along with the program material, at such a level that it does not cause intermodulation. This 50-cycle tone is used to immobilize or "lock out" any tendency for the unit to function except when a picture change is desired, then the 50-cycle "lockout" tone is interrupted and a 30-cycle or "operate" tone is introduced. This passes readily through the automatic circuit, operates a relay, and causes the magnetic picture-changing mechanism to function. By having one tone (50 cycles) as an inhibiting factor and a second (30 cycles) as a positive or "operate" factor, the device becomes completely reliable.

A projector with the device actually is simpler to operate than the nonautomatic type, since the threading of the projector is the same, that is, the strip is placed in the projector and focused on the first frame. Then with the turntable running, the needle is slid into the first groove of the record. The projector then carries on and shifts all the frames automatically at the proper cues. The device will not be tripped falsely by flicking the needle nor is it necessary that the needle be placed in the very first groove.

Fig. 1 shows a block diagram of the unit incorporated in a sound slide projector. The signal from the pickup, which can be an inexpensive crystal cartridge, is fed to a normal volume control from which the audio frequencies go to an ordinary amplifier and are delivered to a loudspeaker. It is desirable to provide a low-frequency falloff in the audio-frequency amplifier to minimize any possible pickup of the low-frequency tones, although usually the small baffle and speaker used will not deliver appreciably anything under 150 cycles. Another lead from the pickup is connected directly to the automatic analyzing circuit which consists first of a low-pass filter which eliminates all frequencies above 100 cycles and passes those in the 30- to 50-cycle region on to an ordinary amplifying stage. The signals then go to another amplifier stage which is driven to saturation and becomes a limiting means. Therefore the output from this last amplifier will be the same, regardless of which of the two frequencies is handled, for the amplification is of such an order that the lowest output from the crystal pickup under the most unfavorable temperature is sufficient to saturate the amplifier fully. When this limiter stage is saturated with the 50-cycle lockout tone, it becomes very insensitive to any other components of a lower magnitude and any 30-cycle components from rumble or accidental shocks are compressed to the extent that they cannot cause false operation.

The signal from the limiting amplifier is then carried to a band-pass filter which has a high Q in the 30-cycle region and will pass very little of the 50-cycle tone. From the band-pass filter the signal passes to a rectifier combined with a time-delay circuit. Thus the direct voltage developed after the tuned circuit is dependent only on

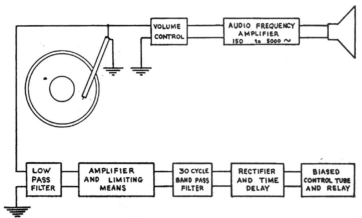

Fig. 1—Block diagram of 30/50 automatic analyzing circuit.

the frequency of the applied tone as long as the amplitude is sufficient to saturate the tube. The direct voltage developed as the result of a 50-cycle tone is very low whereas the voltage developed from a 30-cycle tone is some 30 volts, at least five times as great. This differential in voltage, which is independent of the variations in record characteristics, pickup temperature variations, and needle conditions, is used to operate a relay which in turn causes the operation of the magnetic drive on the film-strip projector.

There have been a number of circuits worked out to accomplish this series of functions. In Fig. 2 is shown one satisfactory circuit. The signal is fed into the resistance-capacitance network consisting of two resistors and two capacitors which results in enough of a rolloff beyond approximately 70 cycles, to prevent interference from the audio frequencies. The first tube acts as an amplifier and passes the signal on to the pentode which receives enough signal to be overdriven at all times by either the 30- or the 50-cycle frequencies. The pentode then feeds into a transformer which has its secondary tuned by a capacitor X to peak at 30 cycles and this acts as the band-pass filter. It will pass the 30-cycle components with a great deal more efficiency than the 50-cycle ones although the output from the pentode is the same regardless of which frequency is being picked up. The capacitor

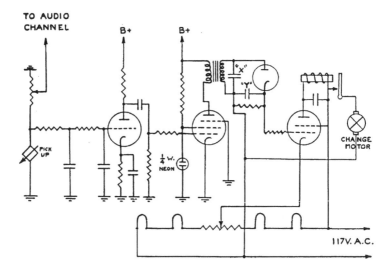

Fig. 2—Schematic diagram of 30/50 circuit.

Y serves as a time-delay means to prevent false operation by an occasional transient such as a bump against the machine or the flicking of the needle.

The final tube is biased to cutoff when the direct voltage, developed as a result of the 50-cycle lockout tone, is in force and therefore current is prevented from flowing in the relay except when the 30-cycle tone occurs. At that time, sufficient plate current is developed to close the relay contacts positively.

A quarter-watt neon tube is placed between the screen grid of the pentode and ground and serves to maintain the screen at a constant potential of 65 volts. In this way the circuit becomes practically immune to line-voltage fluctuations, the operation being substantially the same from 85 to 140 volts line potential.

The system requires no change in the technique of the film-strip production or printing nor is the pressing of the disks changed in any way. The one point requiring the maintenance of certain new techniques is in the recording of the disks.

THE TONE GENERATOR

In order to facilitate the making of records for the 30/50 automatic system, a tone generator has been designed to supply the two tones and mix them with the program material. This is a photoelectric oscillator, making use of a revolving photographic disk, driven by a synchronous motor, and having the two tones recorded in concentric rings upon it. An electronic switching circuit operates a relay to shift the current from one small exciter lamp which generates the 50-cycle tone, to another similar exciter lamp which generates the 30-cycle tone. Normally the generator delivers a 50-cycle tone until the remote push button is pressed. This shifts the frequency to 30 cycles for one and one-half seconds. It is then restored to 50 cycles automatically. Another push button is located on the panel which will cause a 30-cycle tone to be generated as long as the push button is pressed.

In use, the program material is connected to the input of the tone generator and the output from the generator is patched to the amplifier feeding the disk-cutter recording head. The tone generator discards all frequencies below 150 cycles in the audio channel and mixes others with the 50- and 30-cycle tones. There are two gain controls, one for each of the two tones, since each must be adjusted to the proper value, which will depend upon the amplifier and cutter characteristics. To simplify the making of this adjustment of relative levels between the 30- and 50-cycle tones, a test disk is furnished which has the two tones recorded at the optimum level. The two gain controls are adjusted to deliver tones at approximately 15 decibels under full modulation as shown by the volume indicator on the disk cutter. Then test cuts of both frequencies are made and compared with the respective levels on the test disk. Any indicated level changes are made and further test cuts are compared until they are within 1 decibel of the standards on the test record. The recording can then be made without further problems, the cues being placed at the proper intervals in the continuity by depressing the remote push button on the end of a portable cord supplied with the instrument. It is important that the 50-cycle tone be recorded on the disk from the very start as well as continued through to the tailoff and closed groove at the finish of the record so that it will serve its function as a lockout to prevent false operation.

CONCLUSION

By the use of two low-frequency tones, one to serve as an operation-preventing factor and the other to insure operation positively, an automatic sound slidefilm system has been devised which is completely reliable in operation. The elimination of the interrupting gong makes possible a vastly improved medium approaching a talking motion picture in its ability to sustain audience interest and transfer information efficiently.

FORTY YEARS AGO
Something for Nothing

The managers of the moving picture shows in the big theaters have become so thoroughly imbued with the idea that they should get everything they want in the way of slides for nothing, that they are the most parsimonious lot ever known when it is necessary to buy something. Most of them make a cheap show of themselves when they throw an announcement on the screen. Instead of buying a beautifully painted photographic slide, they use plain glass coated over with opaque, with the message scratched through, which to a person who desires to see a perfect show causes a thrill of disgust. The managers think because some music publisher has furnished them a few sets of song slides free they should get announcement slides free also. The meanest looking announcement slides, poorly written and almost illegible, are used at the Grand Opera House.

—*The Moving Picture World, June 13, 1908*

Magnetic Device for Cuing Film*

By JAMES A. LARSEN

ACADEMY FILMS, HOLLYWOOD, CALIFORNIA

Summary—The magnetic cuing device was designed to eliminate the necessity of cutting notches in the edges of motion picture originals used in film printers.

The method of magnetic cuing consists in painting a small dot of magnetic material on the edge of the film between two perforation holes. The paint may be placed on either the emulsion or base side of the film and still be detected by the low impedance (50-ohm) magnetic detector.

The equipment as designed contains two independent cuing channels, one for controlling printer-light changes and the other for controlling a fade-in or fade-out device which is built in on some printers. A mounting for two extremely small pickup heads has been designed to allow direct replacement of existing notch actuator switches in film printers.

THE MAGNETIC CUING device was designed primarily for use on 16-mm color printers, however, it is believed that it will have use in the printing of 35-mm film and possibly also in re-recording of sound tracks.

Present-day requirements of 16-mm printing call for printer-light-intensity changes to correct for camera-exposure discrepancies and for installation of fade-ins and fade-outs and dissolves in the printer rather than in the camera. In Kodachrome-film printing, where it is desirable, if not essential, to print each duplicate from the Kodachrome original, two separate sets of cuing marks must be used; one to control the device for light changes and the other to control a fade-in and fade-out device. When the problem arose of how to cue the fade-in—fade-out device, the obvious solution was to use the same method of cuing fades as has been used during light changes, namely, a notch cut in the edge of the film. However, it has been obvious for some time that the method of cutting notches even on one edge of the film for controlling light changes had several disadvantages. These disadvantages would be more than doubled by the necessity of cutting a second set of notches on the other edge of the film to control fade-ins and fade-outs. Some of the most apparent disadvantages of the method of notching films are these:

* Presented May 18, 1948, at the SMPE Convention in Santa Monica.

1. Notches once cut in the film are permanent. They cannot be removed. If it becomes necessary to make changes in the position of the notches, the notched section must be filled in laboriously.

2. Cutting notches definitely weakens the film and, we believe, shortens the working life of the film, thus reducing the maximum number of duplicates which can be made from a single original. In the commercial- and educational-film field, the need may arise for making several hundred copies from a single Kodachrome original.

Fig. 1—Close-up of two heads mounted on a Bell and Howell Model J 16-mm continuous printer showing relation of pickup heads to film and film-guide rollers.

3. In some printers, the cutting of a notch in the edge of the film causes a definite sideways jump of the picture in the frames adjacent to the notch. Cutting of notches in both edges of the film conceivable could cause sideways jumps in both directions.

The need for some systems of cuing film, other than cutting notches, has long been realized. Two other systems have been developed to accomplish the same result. One uses a separate cue track, running at one quarter the speed of the picture original.

All the necessary "cuing" notches are cut in the cue track, in synchronism with the picture. This method requires a special synchronizer and major modification of the printer to handle the cue track. Also, the operation of making up the cue track is slow and laborious and is subject to considerably more errors than other systems.

Another method, which has been described in detail,[1] uses a small white dot painted on the film. Then the film is scanned with a light beam and a photocell.

The most recent system, a magnetic cuing device, is believed to be simpler than either of these previous systems. In this device, a small dot of magnetic material is painted on the edge of the film. Then

Fig. 2—Amplifier chassis in which all amplifying circuits and the control or pilot relay are located.

the edge of the film is scanned with a magnetic detector or pickup head. The resultant electrical pulse is utilized to control electronically either light changes or fades or both.

The physical equipment for magnetic cuing takes the form of separate pickup heads operating at 50 ohms output impedance. The two pickup heads are shown in Fig. 1, in their relation to the film and to the film-guide rollers. One head is used to control light changes and the other to control fades. In the photograph, the upper pickup head controls light changes and the lower pickup head controls the fade-in—fade-out device. The pulse-signal output of each pickup head is fed to a conventional preamplifier. Signal levels and input-transformer shielding conform to present-day high-standard microphone preamplifier practice. Normally, a 10-foot length of twisted-pair shielded cable is utilized for connecting the pickup head to the amplifier. The amplifier chassis, in which all amplifying circuits and the control or "pilot" relay are located, is shown in Fig. 2. However, the amplifier could be located at distances up to 100 feet from the pickup head if care is taken in routing of the cable similar to that used in routing microphone cable. The input transformer is multiple-shielded with permalloy and copper shields to preserve a good signal-to-noise level and to allow mounting of two separate pre-amplifier channels on the same chassis with the equipment power supply.

The signal is a low-frequency pulse in the neighborhood of from 50 to 100 cycles, depending upon printer speed. Since it was found that mechanical shock excitation contained considerable high-frequency components, the amplifier high-frequency response was limited to increase further the operational signal-to-noise ratio and to provide a safety factor against mechanical excitation of the pickup head.

Following the conventional two-stage, high-gain, signal amplifier, a biased clipper tube is provided to reject tube hiss, random noise, mechanical shock noises, and to provide a general safety factor and sensitivity adjustment. The clipper tube operates in reverse fashion to those normally associated with speech circuits. This clipper removes and discards the lower portion of the signal and all noise. It allows only the signal peak to pass and to actuate the "one-shot" multivibrator circuit which operates the control relay. The multivibrator consists of two triodes in a single envelope. The first triode normally conducts continuously and acts as a signal amplifier for the small clipped signal. When the multivibrator operation cycle begins, the first triode is cut off. The second triode normally is cut off and fires only when the multivibrator operates.

The control relay is in the cathode circuit of the second triode and is actuated for an interval of time determined by the constants of the multivibrator circuit and not by the amplitude or length of the signal pulse. This latter feature removes the random effect of varying signal strength and pulse length brought about by different size magnetic-paint dots and by placing the pickup heads at varying distances from the film. The over-all pickup head and circuit sensitivity is such that the pickup head at no time touches the film. A spacing of about $1/16$ inch is normally maintained between the pickup head and the film. The pickup head is contoured to fit next to a roller over which the film passes. The normal film path is not disturbed. In order that chance mechanical motion of the pickup head with respect to the near-by mass of the roller does not result in a spurious signal, the roller must be made of a nonmetallic material. A phenolic linen roller with a metal bearing insert is used to replace the usual metal roller.

On printers having alternating-current-operated solenoids near the film path, where the pickup heads must be installed, it is necessary to install Mu-metal or Permalloy slip-on shields over the outer body of the solenoids. This is required to overcome interaction of the solenoid field with the pickup head.

The magnetic paint consists of a mixture of hydrogen-reduced powdered iron and clear fingernail polish with acetone added. Hydrogen-reduced iron is readily available at any prescription pharmacy and is pure iron in its most finely divided commercial state. Clear fingernail polish was chosen because it dries very rapidly, is easily obtainable, and comes in a small bottle complete with applicator brush. When the iron, polish, and acetone are properly proportioned, the magnetic paint dries very rapidly, in from 15 to 20 seconds. This paint adheres to either side of the film, emulsion, or base. The paint may be removed from the film by scraping with either a razor blade or a retoucher's knife. If the space between the cuing dots and the printer aperture can be standardized, it will greatly simplify the making of prints when originals are transferred from one laboratory to another. This new cuing device offers the possibility of setting up standards of spacing which have never existed in the notching of films. We suggest that the pickup heads be mounted 2 frames apart and that the light change head be set 16 frames from the printing aperture with the fade-in—fade-out head at 14 frames from the aperture. Also, the light-change dots should be placed on the edge of the film opposite to where the sound track prints. This would place the fade dots on the same side as the sound track.

It is hoped that the introduction of this new magnetic cuing device will help to standardize the 16-mm printing work in laboratories all over the country. If a standard number of frames between the magnetic dot and the printing aperture can be agreed upon, so that a film printed in one laboratory can be printed in any other laboratory without changing the position of the magnetic dots, a very great advance toward standardization and simplification will have occurred.

REFERENCE

(1) Irwin A. Moon, "A photoelectric film cuing system," *J. Soc. Mot. Pict. Eng.*, vol. 49, pp. 364–372; October, 1947.

DISCUSSION

Mr. Lloyd Thompson: Mr. Larsen, what I have is not exactly in the nature of a question, but I should like to add some of my own comments to this. We are very much in favor of this type of cuing device. As a matter of fact, about a year ago, we built up a model like this, or very similar, and tried it out and convinced ourselves that it worked very well. There were a few things we did a little differently than you did. We tried, for instance, the lacquer with the iron powder, which is about the same thing as the fingernail polish. We had some objections to it because we found it was possible to peel it off. It would not always peel off, but you could not depend that it would always stick, either. We feel that that is one of the things that should be standardized before it is used, in some sort of ink or paint, that will stay on the film and there will be no question about it.

Then, we concentrated on a little simpler amplifier than you did. We ended up with a one-tube amplifier and we had what we called a reluctance-type circuit. We think it would be possible to use some metal, other than steel, so that you could buy cellophane that has a coating of iron. It makes a little difference in your pickup head, perhaps, and your circuits; or you could use a magnetic device as you have suggested, where you actually record a signal on some of this dope that they use on magnetic recording of film and have that trip the circuit. Many of those things are a little expensive and we prefer to have something that will work with a one-tube amplifier if we can, since it would be much cheaper to build and simpler to maintain so you can afford to put them on all the printers you have.

Aside from all of that, we think this question of where the dot should be placed on the film should be very definitely standardized before anyone uses it commercially because anyone with 16-mm laboratory business today knows what a mess this notching is. We get films that have been to other laboratories that have notches on one side or both sides, and not one set of notches but maybe two or three sets of notches. A long time ago we decided it would never be possible to standardize these notches because.too many people have built their own printers and you can never get them all to agree to place them at one place on the film. I think that is just out. With this device it seems to us you can place them anywhere on the film. That is, anybody can take any printer and put this device on it, so if you have a standard anyone can use it on any existing printer.

When we got this experimental model made, instead of putting it in operation we thought it should be standardized so we wrote to the SMPE Standards Committee, Maurer, Hancock, and Kodak and told them about it and said, "We should like for you to get together and set up a standard before we go any further with this," and that they agreed to do. As time went on nothing much happened to it, but not too long ago John Maurer wrote to me and said, "If you want to get something started on standardization you better suggest a standard and let them start from there." We did not want to do that. We wanted the Standards Committee to originate the standards, but they said somebody had to originate it first. So we did a little investigation and wrote to several people that we knew that had different types of printers and we came up with a suggested standard of having 32 frames instead of 16. In that way we figured that it could be used on the present Bell and Howell printer, the Depue printers, the printers we use ourselves, and we also wrote to George Colburn and he said it would work on his printers. Now, that is only a suggestion, but we should like to have the Standards Committee take it and set up a standard as to where this mark should be placed, what sort of ink or paint should be used so that it is standard; and we have always felt that the size of the dot should be standardized. With your device, where you can change the amplifier so that it will work at different printer speeds, maybe you have the answer to that so that it will only be a dot. We felt maybe it ought to be an elongated thing instead of a dot.

MR. LARSEN: When I said "dot" I was speaking rather loosely. It is actually a dash.

MR. THOMPSON: I know what you mean, because there are different printer speeds and there has to be some way of taking care of that because a slow printer will actuate it easier than one running fast.

MR. LARSEN: That is one of the advantages of the more-complicated circuit we came up with, because it is entirely independent of the thickness of the dash.

MR. HUMPHREY: Just how long do these dots stay on? Do you find that they have a tendency to wear off?

MR. LARSEN: No. That is one of the advantages of the concoction we have used, the mixture of fingernail polish and acetone. They are very, very permanent. What actually happens is this: A film is soluble in acetone and when they get together they act as they do with film cement. The iron paint actually cements itself to the base side if you paint it on the base side, which we do, although it is not necessary to do so.

MR. HUMPHREY: I am interested in the positioning of the cuing marks. Evidently your cuing marks are in the same position as the Bell and Howell notch cuer.

MR. LARSEN: Yes. As a matter of fact, this device is built so you can remove the Bell and Howell notcher and slide this in place. That is the reason we suggest using 16 frames.

MR. HUMPHREY: If you get a film that is already notched you will not have much room.

MR. LARSEN: All we do in that case is to put our paint on the next frame after the one that happens to have the notches cut out of it where we want to put them, which does not make any difference in the actual result. With this device we can take a film that has numerous sets of notches cut in it and ignore all of them and put a set of dots on where we have to put it with our particular setup, which happens to be 16 frames, and proceed from there, ignoring the previous set of notches. However, we think it is better if the notches are not there in the first place.

Improved 35-Mm
Synchronous Counter*.

By ROBERT A. SATER and JAMES W. KAYLOR

CINECOLOR CORPORATION, BURBANK, CALIFORNIA

Summary—The types of synchronous counters, or heads, in general use present several problems in the cutting of the multiplicity of negatives used in color processes. The necessity of constantly. threading and unthreading negatives to match the work print invites scratches and other damage to the negatives.

An improved type of synchronous counter has been developed which accommodates either three or six negatives (six being used in dissolves and fades) together with a work print. A special positioning lock allows the work print to be advanced or retarded in relation to the negative, and relocked in frame. To facilitate threading, the keeper rollers are designed to lift into a vertical position, thus making it possible simply to set the film onto the sprocket without having to slide it under or over rollers or sprockets. A novel arrangement of a lucite stripping-shoe allows illumination of the film from beneath in order to locate frame lines.

AN IMPROVED 35-mm synchronous counter has been designed and put into operation. The design was such as to eliminate some of the more objectionable features commonly found in available synchronous machines, outstanding among which may be listed:

1. An arrangement of keeper rollers which makes it necessary to slide film under the rollers and over the sprockets when threading or unthreading, a situation which is conducive to scratching.

2. The necessity of unthreading and rethreading the work print in order to advance or retard it in relation to the negative being cut.

3. The lack of proper illumination for viewing the frame lines and the subject matter on the film directly over the sprocket.

Fig. 1 shows the synchronous counter with the keeper rollers closed. Fig. 2 is a close-up of the forward end of the machine, with the various parts numbered for easy reference. In operation, the work print is threaded on the front sprocket (1), after releasing the keeper rollers (2). This is done by pressing the spring release (3) forward. When the film has been placed on the sprocket, the keeper rollers are closed simply by pressing them down into position against

* Presented May 18, 1948, at the SMPE Convention in Santa Monica.

the film. The rear sprockets (1A) are threaded in the same way. It can be seen that with the keeper rollers lifted into a vertical position when open, film scratches due to threading are practically eliminated.

When matching negatives to a cut work print, frame to frame, the necessity of removing the work print from the machine at every cut is eliminated. First, the work print is rolled down to the desired edge number or splice where the negative cut is to be made. Then button (8) is pushed, which applies a brake (7) against the inside of sprocket (1). This holds the work print in the desired position. Next, handle (5) is pulled out, sliding a key out of spline (6), which leaves the remainder of the sprockets freewheeling for advancing negatives to the place where the cut is to be made. Then handle (5) is pushed in,

Fig. 1—Four-gang synchronous film counter for cutting multiple-color negatives.

the brake is released, the cut is made, and both work print and negatives are rolled to the next scene. The counter, a five-figure foot and frame counter commonly used on motion picture equipment is geared to the shaft carrying the three rear sprockets (1A). The counter may thus be used for counting the footage from the beginning of a scene to the point where a negative cut is to be made, and for scene-to-scene footage in making continuity cards.

To obtain a field of illumination directly over the sprockets for viewing frame lines and checking subject material, ports were provided in the base of the machine so that light from underneath could be projected up to the sprockets. To carry the light around the sprocket hub and under the film, a strip of lucite (9) was formed into a

U shape and mounted at each sprocket as shown. The area of the lucite strip directly over the sprocket was ground, in order to diffuse the light. The lucite member is thus made to serve both as an illuminating device, and as a stripper shoe.

Fig. 3 shows an extension of the design shown in Figs. 1 and 2. It is a longer version of the same machine, embodying three additional sprockets at the rear. The mechanical make-up of this machine is

Fig. 2—Close-up of four-gang synchronous film counter showing locking hand wheels, brake lock, keeper rollers, and lucite stripper shoe.

identical with that of the machine shown in Figs. 1 and 2 with the exception that there are six gang sprockets on the rear shaft gear to the counter. This facilitates the cutting of dissolves, montages, and superimposed titles for foreign release. The counter is used primarily for taking foot and frame counts and making cue sheets.

Fig. 4 shows a special machine, designed and built for cutting the skip-frame negatives used in color-cartoon photography. It is constructed so as to accommodate both two-color and three-color successive frame negatives. The work print is threaded on the front

Fig. 3—Seven-gang synchronous film counter for cutting dissolves, montages, and superimposed titles.

sprocket (1B), and the successive-frame negative on the rear sprocket (1). Sprocket (1) is rotated at either two or three times the speed of sprocket (1B) by means of a small gear transmission (3). The shift from one ratio to the other is accomplished by means of knob (4). Counter (7) is connected to sprocket (1B) so that print or screen footage is counted.

A significant interest in the Cinecolor synchronous counters has been shown by other studios and laboratories, and several of these machines have been built for use in other laboratories.

Fig. 4—Skip-frame synchronous film counter for cutting either 2-to-1 or 3-to-1 cartoon negatives.

Proposed Standards for 16-Mm and 8-Mm Picture Apertures

IN CONNECTION with the broad program of reviewing all standards at the close of the war, the Standards Committee of the Society of Motion Picture Engineers was asked to restudy the six American Standards for picture apertures in 16-mm and 8-mm cameras and projectors. Since it appeared that a thorough revision was in order, a subcommittee, with John A. Maurer as chairman, was organized for this assignment. By the end of 1947, the pattern and most of the details of the new proposals were well established. Because of other changes at that time, it became desirable to disband the special subcommittee and to transfer to the Society's standing Committee on 16-mm and 8-mm Motion Pictures the task of ironing out the remaining few, but important, controversial points. Agreement was reached in October, 1948, and the new proposals were passed along to the Standards Committee with a recommendation for favorable action. The new proposals, four in number, are shown on the following pages. This is in keeping with the policy of the Standards Committee of publishing in the JOURNAL all proposals involving new material or major revisions before taking action on the question of submitting them to the American Standards Association. Your comments are invited.

Specifically, the four proposed standards are entitled:

Z22.7   Location and Size of Picture Aperture of 16-mm Motion Picture Cameras.

Z22.8   Location and Size of Picture Aperture of 16-mm Motion Picture Projectors.

Z22.19   Location and Size of Picture Aperture of 8-mm Motion Picture Cameras.

Z22.20   Location and Size of Picture Aperture of 8-mm Motion Picture Projectors.

When these are finally approved, they will replace the same Z22 numbers that were issued in 1941. Since the two 16-mm proposals cover sound, as well as silent, equipment, it is intended that they will also supplant 1941 American Standards Z22.13 and Z22.14 which related to 16-mm sound cameras and projectors.

In the drafting of these proposed standards, an effort was made to dimension the drawings so that they will be of the most direct use to the engineer. The introduction of the "K" dimensions, showing the distance from the centerline of the aperture to the registering edge of the film perforation, is a case in point.

During the evolution of these proposals, there was a good deal of debate on the question of specifying which edge of 16-mm film should be guided. The Committee on 16-mm and 8-mm motion pictures finally concluded that this question properly should be left to the designer, but the proposal does call attention to the factors involved. A similar statement is made relative to the problem of assigning a definite value to dimension "G."

In the past, standards dealing with this type of subject matter generally have been limited to dimensioned drawings. The Standards Committee feels it is desirable to include explanatory text or notes to make the intention and application of the standard more certain. This practice is followed in these four proposals to a degree that makes any further discussion of technical points appear to be unnecessary.

———————●———————

SEVERAL STANDARDS developed by subcommittees of ASA Sectional Committee Z10 listed below are available. Authors are encouraged to follow these standard symbols and abbreviations.

Title of Standard		Price
Abbreviations for Scientific and Engineering Terms	Z10.1-1941	$0.45
Letter Symbols for Hydraulics	Z10.2-1942	0.45
Letter Symbols for Mechanics of Solid Bodies	Z10.3-1948	0.30
Letter Symbols for Heat and Thermodynamics	Z10.4-1943	0.65
Letter Symbols for Physics	Z10.6-1948	1.00
Letter Symbols for Chemical Engineering	Z10.12-1946	0.50

Proposed American Standard Location and Size of Picture Aperture of **16-Millimeter Motion Picture Cameras**	**Z22.7-** **February 1949**

Page 1 of 3 pages

This standard applies to both silent and sound 16-mm. motion picture cameras. It covers the size and shape of the picture aperture and the relative positions of the aperture, the optical axis, the edge guide, and the film registration device. The notes are a part of this standard.

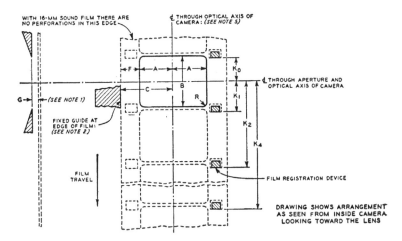

Dimension	Inches	Millimeters	Note
A (measured perpendicular to edge of film)	0.201 minimum	5.11 minimum	1
B (measured parallel to edge of film)	$0.292 \begin{array}{l}+0.006\\-0.002\end{array}$	$7.42 \begin{array}{l}+0.18\\-0.05\end{array}$	1
C	0.314 ± 0.002	7.98 ± 0.05	2
F	0.110 minimum	2.79 minimum	3
K_0	0.125 ± 0.002	3.18 ± 0.05	4
K_1	0.175 ± 0.002	4.44 ± 0.05	4
K_2	0.474 ± 0.002	12.04 ± 0.05	4
K_3	0.773 ± 0.002	19.63 ± 0.05	4
K_4	1.072 ± 0.001	27.23 ± 0.03	4
R	0.020 maximum	0.51 maximum	1

Proposed American Standard Location and Size of Picture Aperture of 16-Millimeter Motion Picture Cameras	Z22.7- February 1949

The angle between the vertical edges of the aperture and the edges of normally positioned film shall be 0 degrees, ± ½ degree.

The angle between the horizontal edges of the aperture and the edges of normally positioned film shall be 90 degrees, ± ½ degree.

Note 1: Dimensions A, B, and R apply to the size of the image at the plane of the emulsion; the actual picture aperture has to be slightly smaller. The exact amount of this difference depends on the lens used and on the separation (dimension G) of the emulsion and the physical aperture. G should be no larger than is necessary to preclude scratching of the film. The greatest difference between the image size and aperture size occurs with short focal-length, large diameter lenses.

Dimensions A and B are consistent with the size of the images on a 16-mm. reduction print made from a 35-mm. negative with the standard 2.15 reduction ratio.

It is desirable to hold the vertical height of the actual aperture to a value that will insure a real (unexposed) frameline. This results in less distraction when the frameline is projected on the screen than is the case when adjacent frames overlap.

Note 2: The edge guide is shown on the sound-track edge. This location for it has the advantage that the rails bearing on the face of the film along this edge and also between the sound track and picture area can be of adequate width. Disadvantages of this location for the edge guide are that, because film shrinkage and tolerances affect the lateral position of the perforations, the pulldown tooth must be comparatively narrow and will not always be centered in the perforation.

The guide can be on the other edge, adjacent to the perforated edge of sound film. With the guide at this edge, the width of the pulldown tooth does not have to be decreased to allow for shrinkage. However, because of variations introduced by shrinkage of film, this location for the edge guide has the important disadvantage that it makes extremely difficult the provision of rails of adequate width to support the sound-track edge without encroaching on, and consequently scratching, the picture or sound-track area. (See Section 3, Proposals for 16-mm. and 8-mm. Sprocket Standards, Vol. 48, No. 6, June 1947, Journal of the Society of Motion Picture Engineers).

Proposed American Standard Location and Size of Picture Aperture of **16-Millimeter Motion Picture Cameras**	**Z22.7- February 1949**

The film may be pressed against the fixed edge guide by a spring, by the tendency of the film to tilt in the gate, or by other means. In the second case, there is a fixed guide for each edge of the film. The important point is to have the film centered laterally on the optical axis.

Dimension C is made slightly less than half the width of unshrunk film so that the film will be laterally centered if it has a slight shrinkage at the time it is run in the camera. This is the normal condition. As indicated by the above discussion, C may be measured in either direction from the vertical centerline.

Note 3: Dimension F must be maintained only when a photographic sound record is to be made on the film that passes through the camera; otherwise F may be disregarded.

Note 4: The K dimensions are measured along the path of the film from the horizontal centerline of the aperture to the stopping position of the registration device. Both the dimensions and tolerances were computed to keep the frameline within 0.002 to 0.005 inch of the centered position for films having shrinkages of 0.0 to 0.5 per cent at the time they are exposed in the camera. For any given camera, use the value of K corresponding to the location of the registration device.

If the film does not stop exactly where the film registration device leaves it, because of coasting or some other cause, a slight adjustment of the value of K will be necessary. This will be indicated if film that has a shrinkage of 0.2 to 0.3 per cent when it is run in the camera does not show a properly centered frameline. From such a test, the amount and direction of the adjustment can be determined.

Note 5: "Optical axis of camera" is defined as the mechanical axis or centerline of the sleeve or other device for holding the picture-taking lens. Except for manufacturing tolerances, it coincides with the optical axis of the lens.

Proposed American Standard Location and Size of Picture Aperture of **16-Millimeter Motion Picture Projectors**	**Z22.8-** **February 1949**

This standard applies to both silent and sound 16-mm. motion picture projectors. It covers the size and shape of the picture aperture and the relative positions of the aperture, the optical axis, the edge guide, and the film registration device. The notes are a part of this standard.

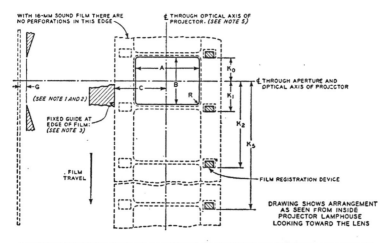

Dimension	Inches	Millimeters	Note
A (measured perpendicular to edge of film).	0.380 ± 0.002	9.65 ± 0.05	1
B (measured parallel to edge of film)	0.284 ± 0.002	7.21 ± 0.05	1
C	0.314 ± 0.002	7.98 ± 0.05	3
K_0	0.124 ± 0.005	3.15 ± 0.13	4
K_1	0.174 ± 0.005	4.42 ± 0.13	4
K_2	0.473 ± 0.005	12.01 ± 0.13	4
K_3	0.771 ± 0.005	19.58 ± 0.13	4
K_4	1.070 ± 0.005	27.18 ± 0.13	4
K_5	1.368 ± 0.005	34.75 ± 0.13	4
R	0.020 maximum	0.51 maximum	1

Proposed American Standard Location and Size of Picture Aperture of 16-Millimeter Motion Picture Projectors	Z22.8- February 1949

The angle between the vertical edges of the aperture and the edges of normally positioned film shall be 0 degrees, ± ½ degree.

The angle between the horizontal edges of the aperture and the edges of normally positioned film shall be 90 degrees, ± ½ degree.

Note 1: Dimensions A, B, and R apply to the portion of the image on the film that is to be projected; the actual opening in the aperture plate has to be slightly smaller. The exact amount of this difference depends on the lens used and on the separation (dimension G) of the emulsion and the physical aperture. To minimize the difference in size and make the image of the aperture as sharp as practicable on the screen, G should be no larger than is necessary to preclude scratching of the film. When the reduction in size from the image to the actual aperture is being computed, it is suggested a 2-inch f/1.6 lens be assumed unless there is reason for doing otherwise.

Note 2: The limiting aperture is shown as being between the film and the light source so that it will give the maximum protection from heat. If other factors are more important, it may be on the other side of the film.

Note 3: The edge guide is shown on the sound-track edge. This location for it has the advantage that the rails bearing on the face of the film along this edge and also between the sound track and picture area can be of adequate width. Disadvantages of this location for the edge guide are that, because film shrinkage and tolerances affect the lateral position of the perforations, the pulldown tooth must be comparatively narrow and will not always be centered in the perforation. Also, in some prints the sound-track edge is slit after processing, in which case there is likely to be some lateral weave between this edge and the pictures.

The guide can be on the other edge, adjacent to the perforated edge of sound film. With the guide at this edge, the width of the pulldown tooth does not have to be decreased to allow for shrinkage. Also, slitting the sound-track edge after processing will not introduce lateral unsteadiness. However, because of variations introduced by shrinkage of film, this location for the edge guide has the important disadvantage that it makes extremely difficult the provision of rails of adequate width to support the

Proposed American Standard Location and Size of Picture Aperture of 16-Millimeter Motion Picture Projectors	Z22.8- February 1949

sound-track edge without encroaching on, and consequently scratching, the picture or sound-track area. (See Section 3, Proposals for 16-mm. and 8-mm. Sproket Standards, Vol. 48, No. 6, June 1947, Journal of the Society of Motion Picture Engineers).

The film may be pressed against the fixed edge guide by a spring, by the tendency of the film to tilt in the gate, or by other means. In the second case, there is a fixed guide for each edge of the film. The important point is to have the film centered laterally on the optical axis.

Dimension C is made slightly less than half the width of unshrunk film so that the film will be laterally centered if it has a slight shrinkage at the time it is run in the projector. This is the normal condition. As indicated by the above discussion, C may be measured in either direction from the vertical centerline.

Note 4: The K dimensions are measured along the path of the film from the horizontal centerline of the aperture to the stopping position of the registration device. It is customary to provide a framing movement of 0.025 inch above and below this nominal position. For any given projector, use the value of K corresponding to the location of the registration device.

If the film does not stop exactly where the film registration device leaves it, because of coasting or some other cause, a slight adjustment of the value of K will be necessary.

Note 5: "Optical axis of projector" is defined as the mechanical axis or centerline of the sleeve for holding the projection lens. Except for manufacturing tolerances it coincides with the lens axis.

Proposed American Standard Location and Size of Picture Aperture of 8-Millimeter Motion Picture Cameras	Z22.19- February 1949

This standard applies to 8-mm. motion picture cameras. It covers the size and shape of the picture aperture and the relative positions of the aperture, the optical axis, the edge guide, and the film registration device. The notes are a part of this standard.

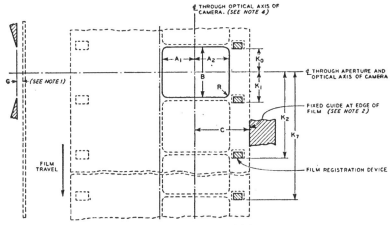

DRAWING SHOWS ARRANGEMENT AS SEEN FROM
INSIDE CAMERA LOOKING TOWARD THE LENS

Dimension	Inches	Millimeters	Note
A_1 (measured perpendicular to edge of film)	0.094 min., 0.104 max.	2.39 min., 2.64 max.	1
A_2	0.094 min.	2.39 min.	1
B (measured parallel to edge of film)	$0.138 \begin{array}{l}+0.008\\-0.001\end{array}$	$3.51 \begin{array}{l}+0.20\\-0.03\end{array}$	1
C	0.205 ± 0.002	5.21 ± 0.05	2
K_0	0.050 ± 0.002	1.27 ± 0.05	3
K_1	0.100 ± 0.002	2.54 ± 0.05	3
K_2	0.249 ± 0.002	6.32 ± 0.05	3
K_3	0.399 ± 0.002	10.13 ± 0.05	3
K_4	0.549 ± 0.002	13.94 ± 0.05	3
K_5	0.698 ± 0.002	17.73 ± 0.05	3
K_6	0.848 ± 0.002	21.54 ± 0.05	3
K_7	0.998 ± 0.002	25.35 ± 0.05	3
R	0.010 maximum	0.25 maximum	1

Proposed American Standard Location and Size of Picture Aperture of **8-Millimeter Motion Picture Cameras**	**Z22.19-** **February 1949**

The angle between the vertical edges of the aperture and the edges of normally positioned film shall be 0 degrees, ± ½ degree.

The angles between the horizontal edges of the aperture and the edges of normally positioned film shall be 90 degrees, ± ½ degree.

Note 1: Dimensions A, B, and R apply to the size of the image at the plane of the emulsion; the actual picture aperture has to be slightly smaller. The exact amount of this difference depends on the lens used and on the separation (dimension G) of the emulsion and the physical aperture. G should be no larger than is necessary to preclude scratching of the film. The greatest difference between the image size and aperture size occurs with short focal-length, large diameter lenses.

It is desirable to hold the vertical height of the actual aperture to a value that will insure a real (unexposed) frameline. This results in less distraction when the frameline is projected on the screen than is the case when adjacent frames overlap.

Note 2: The film may be pressed against the fixed edge guide by a spring, by the tendency of the film to tilt in the gate, or by other means. In the second case (generally used in pre-loaded magazines), there is a fixed guide for each edge of the film. The important point is to have the film located in the correct lateral position with respect to the optical axis.

The value of dimension C has been chosen on the assumption that the film will have a slight shrinkage when it is run through the camera. This is the normal condition.

Note 3: The K dimensions are measured along the path of the film from the horizontal centerline of the aperture to the effective stopping position of the registration device. Both the dimensions and tolerances were computed to keep the frameline within 0.002 to 0.005 inch of the centered position for films having shrinkages between 0.0 and 0.5 per cent at the time they are exposed in the camera. For any given camera, use the value of K corresponding to the location of the registering device.

If the film does not stop exactly where the film registration device leaves it, because of coasting or some other cause, a slight adjustment of the value of K will be necessary. This will be indicated if film that has a shrinkage of 0.2 to 0.3 per cent when it is run in the camera does not show a properly centered frameline. From such a test, the amount and direction of the adjustment can be determined.

Note 4: "Optical axis of camera" is defined as the mechanical axis or centerline of the sleeve or other device for holding the picture-taking lens. Except for manufacturing tolerances, it coincides with the optical axis of the lens.

Proposed American Standard Location and Size of Picture Aperture of **8-Millimeter Motion Picture Projectors**	**Z22.20-** **February 1949**

This standard applies to 8-mm. motion picture projectors. It covers the size and shape of the picture aperture and the relative positions of the aperture, the optical axis, the edge guide, and the film registration device. The notes are a part of this standard.

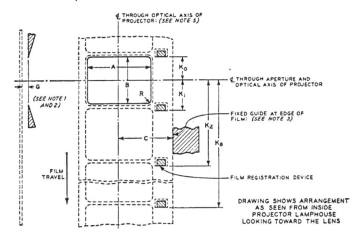

Dimension	Inches	Millimeters	Note
A (measured perpendicular to edge of film)	0.172 ± 0.001	4.37 ± 0.03	1
B (measured parallel to edge of film)	0.129 ± 0.001	3.28 ± 0.03	1
C	0.205 ± 0.002	5.21 ± 0.05	3
K_0	0.050 ± 0.005	1.27 ± 0.13	4
K_1	0.100 ± 0.005	2.54 ± 0.13	4
K_2	0.249 ± 0.005	6.32 ± 0.13	4
K_3	0.398 ± 0.005	10.11 ± 0.13	4
K_4	0.547 ± 0.005	13.89 ± 0.13	4
K_5	0.696 ± 0.005	17.68 ± 0.13	4
K_6	0.846 ± 0.005	21.49 ± 0.13	4
K_7	0.995 ± 0.005	25.27 ± 0.13	4
K_8	1.144 ± 0.005	29.06 ± 0.13	4
R	0.010 maximum	0.25 maximum	1

Proposed American Standard Location and Size of Picture Aperture of **8-Millimeter Motion Picture Projectors**	**Z22.20-** **February 1949**

The angle between the vertical edges of the aperture and the edges of normally positioned film shall be 0 degrees, $\pm$ ½ degree.

The angle between the horizontal edges of the aperture and the edges of normally positioned film shall be 90 degrees, $\pm$ ½ degree.

Note 1: Dimensions A, B, and R apply to the portion of the image on the film that is to be projected; the actual opening in the aperture plate has to be slightly smaller. The exact amount of this difference depends on the lens used and on the separation (dimension G) of the emulsion and the physical aperture. To minimize the difference in size and make the image of the aperture as sharp as practicable on the screen, G should be no larger than is necessary to preclude scratching of the film. When the reduction in size from the image to the actual aperture is being computed, it is suggested a 1-inch f/1.6 lens be assumed unless there is reason for doing otherwise.

Note 2: The limiting aperture is shown as being between the film and the light source so that it will give the maximum protection from heat. If other factors are more important, it may be on the other side of the film.

Note 3: In 8-mm. projectors the edge guide should bear on the edge of the film adjacent to the perforations. The other edge of the film usually is slit after processing and so is more likely to weave laterally with respect to the pictures.

The value of dimension C has been chosen so that film having a slight shrinkage when it is projected will be properly centered. This is the normal condition.

Note 4: The K dimensions are measured along the path of the film from the horizontal centerline of the aperture to the stopping position of the registration device. It is customary to provide a framing movement of approximately 0.025 inch above and below this nominal position. For any given projector, use the value of K corresponding to the location of the registration device.

If the film does not stop exactly where the film registration device leaves it, because of coasting or some other cause, a slight adjustment of the value of K will be necessary.

Note 5: "Optical axis of projector" is defined as the mechanical axis or centerline of the sleeve for holding the projection lens. Except for manufacturing tolerances, it coincides with the lens axis.

65th Semiannual Convention
SOCIETY OF MOTION PICTURE ENGINEERS

Hotel Statler ● April 4–8, 1949 ● New York, N. Y.

● ● ●

OFFICERS OF THE SOCIETY

EARL I. SPONABLE	*President*
LOREN L. RYDER	*Past-President*
PETER MOLE	*Executive Vice-President*
JOHN A. MAURER	*Engineering Vice-President*
CLYDE R. KEITH	*Editorial Vice-President*
DAVID B. JOY	*Financial Vice-President*
WILLIAM C. KUNZMANN	*Convention Vice-President*
ROBERT M. CORBIN	*Secretary*
RALPH B. AUSTRIAN	*Treasurer*

General Office, New York

BOYCE NEMEC	*Executive Secretary*
HELEN M. STOTE	*Journal Editor*
WILLIAM H. DEACY, JR.	*Staff Engineer*
SIGMUND M. MUSKAT	*Office Manager*

Chairmen of Committees for the Convention Program

Convention Vice-President	WILLIAM C. KUNZMANN
Atlantic Coast Section and Local Arrangements	WILLIAM H. RIVERS
Papers Committee Chairman	NORWOOD L. SIMMONS, JR.
Vice-Chairmen	EDWARD S. SEELEY AND R. T. VAN NIMAN
Publicity Committee	HAROLD DESFOR
Assisted by	LEONARD BIDWELL, GEORGE DANIEL, AND HAROLD WENGLER
Registration and Information	WILLIAM C. KUNZMANN
Assisted by	ERWIN R. GEIB, L. H. WALTERS, PAUL D. RIES, AND WILLIAM F. JORDAN
Luncheon and Banquet	OSCAR F. NEU
Vice-Chairman	LESTER B. ISAAC
Public-Address Equipment	R. E. WARN
Hotel and Transportation	WILLIAM F. JORDAN
Ladies' Reception Committee	MRS. EARL I. SPONABLE, Hostess
Cohostess	MRS. WILLIAM H. RIVERS
Membership and Subscription	LEE JONES
Projection Program—35-Mm	HENRY F. HEIDEGGER
* Assisted by officers and members New York Local 306, I.A.T.S.E.	
Projection Program—16-Mm	FRANK B. ROGERS, JR.

349

GENERAL INFORMATION

Hotel Reservations and Rates

The Hotel Statler (formerly Hotel Pennsylvania) will be the headquarters of the 65th Semiannual Convention. Room reservation cards were mailed to the membership early in February. If you plan to attend the Convention, check your desired accommodations and return the card immediately to the hotel so your reservation can be booked and confirmed by the hotel management.

Booked reservations are subject to arrival date change or cancellation prior to March 20.

Rail, Pullman, and Plane Travel

Eastern travel conditions still remain acute; therefore if attending the Convention in New York, book your travel accomodations at least a month prior to April 4, with your local travel agent.

Convention Registration and Papers Program

Members and others within the motion picture industry who are contemplating presenting papers during the 65th Semiannual Convention dates are urged to submit the title of the paper to be presented, name of author, and a complete manuscript to the Journal Editor, Society of Motion Picture Engineers, 342 Madison Avenue, New York 17, N. Y. A second copy, without illustrations, should be sent to Mr. E. S. Seeley, Altec Service Corporation, 161 Sixth Avenue, New York 13, N. Y.

The Convention Registration Headquarters will be located in Conference Room 9, on the 18th floor of the hotel; all business and technical sessions will be held in the Salle Moderne, on the same floor.

The Publicity Committee and Press Headquarters will also be located on this floor.

You can register at the Registration Headquarters on the morning of April 4, also procure your desired luncheon tickets prior to 10:30 A.M. on this date, to be assured table seating.

Your registration is requested since this revenue from collected registration fees is used to defray the Convention expenses.

Convention Get-Together Luncheon

The usual convention get-together luncheon will be held in the Georgian Room of the hotel on Monday, April 4, at 12:30 P.M. Eminent speakers will address the luncheon gathering.

Most Important—Procure your desired luncheon tickets at the Registration Headquarters prior to 10:30 A.M. on April 4, otherwise no table seating is guaranteed.

Checks or money orders for registration fees and luncheon or banquet tickets should be made payable to W. C. Kunzmann, Convention Vice-President, and *not to the Society.*

65th Semiannual (Informal) Banquet and Cocktail Hour

The Convention's social cocktail hour for holders of banquet tickets will be held in the hotel's Georgian Room foyer, on Wednesday evening, April 6, between 7:15 and 8:15 P.M.

The 65th Semiannual Convention informal banquet (dress optional) will be held in the Georgian Room on Wednesday evening, April 6, promptly at 8:30 P.M. There will be dancing and entertainment. Tables for the banquet can be reserved at the registration headquarters.

Ladies' Reception and Registration Headquarters

The Ladies' Reception and Registration Headquarters will be located in Room 129 in the hotel and open daily during the Convention. Mrs. Earl I. Sponable will serve the Convention as hostess and Mrs. William H. Rivers, cohostess, to the ladies attending the Convention.

The ladies' entertainment program will be announced in later released convention bulletins.

Motion Pictures and Recreation

The Convention-issued identification cards to registered members and guests will be honored at the following de luxe motion picture theaters in New York during the Convention: Capitol, Paramount, Radio City Music Hall, Roxy, and Warner Strand.

Literature and information pertaining to places of interest to visit in New York and vicinity can be obtained at the hotel's information bureau or the Convention registration headquarters.

PROGRAM SCHEDULE

Monday, April 4, 1949

9:30 A.M.   Registration, Conference Room 9, 18th floor
            Advance sale Luncheon and Banquet tickets
            Procure your luncheon tickets prior to 10:30 A.M., this date
12:30 P.M.   Get-Together Luncheon, Georgian Room
3:00 P.M.   Films in Television (Forum), Salle Moderne
8:00 P.M.   Theater Television Demonstration, Georgian Room

Tuesday, April 5, 1949

9:30 A.M.   Registration, Conference Room 9, 18th floor
            Advance sale of Banquet tickets
10:00 A.M.   Television Session, Salle Moderne
2:00 P.M.   Television Session, Salle Moderne

OPEN EVENING

Wednesday, April 6, 1949

9:30 A.M.   Registration, Conference Room 9, 18th floor
            Advance sale of Banquet tickets
10:00 A.M. High-Speed Photography Session, Salle Moderne
2:00 P.M.   High-Speed Photography Session and Equipment Demonstration,
            Salle Moderne
7:15 P.M.   Cocktail Hour, Georgian Room foyer
8:30 P.M.   Informal Banquet, Georgian Room (dress optional).   Dancing and
            Entertainment

Thursday, April 7, 1949

OPEN MORNING

2:00 P.M.   Films and Film Laboratories, Discussion of 16-Mm Reproducer
            Characteristics; Sound Recording, Salle Moderne
8:00 P.M.   Popular Lecture and Demonstration, Salle Moderne

Friday, April 8, 1949

9:30 A.M.   Registration, Conference Room 9, 18th floor
10:00 A.M. Theaters and Projection, Salle Moderne
2:00 P.M.   Photographic Equipment and Studio Techniques, Salle Moderne
5:00 P.M.   Adjournment

JOURNAL SUBSCRIPTION RATES

Most printing costs in recent years have advanced markedly and
Society publications are not immune because increased labor charges
and the cost of paper to our printer have been passed on and are
charged against the JOURNAL. The Society works on a remarkably
tight publication budget and it has been necessary to increase the
JOURNAL subscription rate to nonmember subscribers from $10.00
to $12.50 per year. All subscriptions renewed prior to March 15,
1949, were billed at the old rate but subscriptions received after
March 15 are being charged for at the $12.50 rate. In 1949 we
expect to print 1622 JOURNAL pages as against 1376 pages for 1948,
1264 pages for 1947, and 1200 pages for 1946.

A variety of cost-cutting methods are being investigated with the
hope that further advances in printing costs may be offset and make
it unnecessary for additional increases in our subscription rates.

CHARLES G. WEBER

CHARLES G. WEBER, assistant chief of the Paper Section, National Bureau of Standards, died on January 18, 1949.

He was graduated from the New York State College of Forestry of Syracuse University in 1916, and was a veteran of World War I.

A member of the Bureau's staff for 20 years, Mr. Weber directed researches in papermaking materials and processes, printing, packaging materials, low-cost housing materials, the standardization of testing methods for paper and related products, and the use of motion picture films for records.

During World War II, he assisted the Army Map Service in the development of wet-strength map paper for field use; he was successful in the development of improved methods in multicolor offset printing that greatly reduces losses caused by misregister of the prints. He supervised research on low-cost housing materials in co-operation with the Federal Housing Administration, and at the time of his death, he was directing research on the melamine-resin bonding of offset papers.

Mr. Weber served on standardizing committees of the Technical Association of the Pulp and Paper Industry, the American Society for Testing Materials, and the Society of Motion Picture Engineers. He was a member of the Technical Committee of the Lithographic Technical Foundation, and an Active Member of the SMPE.

Section Meeting

Central

R. T. Van Niman presided over the December 9, 1948, meeting of the Central Section, which was held at the Lincolnwood Plant of the Bell and Howell Company. Nearly 200 members, guests, and members of the new optical group were present. Charles E. Phillimore welcomed the group and invited those interested to a tour of the plant following the meeting.

The first paper on "Design Considerations for Television Studio Motion Picture Projectors," by James V. Starbuck and Elmer Enke, projection engineers, WGN-TV, and H. J. Daly and Dudley A. Howell, projectionists, television station WBKB, was read by Mr. Howell. He outlined the arrangement and problems of the television projection rooms of WBKB and WGN-TV. He indicated that in the main, the 35-mm equipment was satisfactory but with the syncrolite lamp source it was impossible to frame the picture during projection. The 16-mm projection equipment has many faults to make it a practical instrument in the projection booth. These faults were outlined as follows: (a) equipment too light, (b) needs real focusing adjustment, (c) sprockets too small, (d) should have brake for instantaneous stopping, (e) replacement of lamp should be easier, a matter of seconds instead of minutes, (f) douser should be developed, (g) improved mounting for preamplifier is needed, and (h) claw intermittent should be eliminated.

"Continuous Stereoscopic Aerial Strip Camera Photography," was presented by Colonel George W. Goddard, chief of the Photographic Laboratory, United States Air Force, Wright Field, Dayton, Ohio. With a series of excellent slides, Colonel Goddard traced the history of Aerial Reconnaissance Photography from 1880 to the present time. This included the planes used up to the very latest designs of jet planes flying at 420 miles per hour photographing at an altitude of 4000 feet; 240-inch lenses are used. In the model now being developed the camera will use 40-inch wide film in 400-foot lengths.

The last part of Colonel Goddard's talk was a demonstration of stereoscopic color photography taken from the air with a slit camera. The film travel in this camera is automatically governed by the ground speed of the plane, insuring high resolution of the twin images. Some very startling and interesting scenes of the cities of Germany were viewed with the use of polaroid glasses. These films were projected from the rear on a 12- by 16-foot special screen.

---●---

Armed Forces Communications Association

Brigadier General David Sarnoff, president of the Armed Forces Communications Association and also chairman of the Board of the Radio Corporation of America, announced that the third annual meeting of the Armed Forces Communications Association will be held in Washington, March 28 and 29, 1949.

Business meetings will be held the first day and will be climaxed by the annual banquet at which it is expected nearly 1000 members will attend. This year's meeting will feature the Navy's communications and photographic activities. Navy leaders and other distinguished government figures will be the principal speakers at the banquet. The second day and perhaps part of a third will be devoted entirely to exhibits and demonstrations planned and directed by the Navy at its stations and aboard ships in the Washington area.

The Association, made up of civilians and military members, is dedicated to the purpose of insuring that the Navy, Army, and Air Force will have the best in communications, radar, and photography.

PROGRAM

Monday, March 28

9:00 A.M.

REGISTRATION—Shoreham Hotel.

10:00 A.M.

ANNUAL BUSINESS MEETING—Shoreham Hotel.

12:00-1:30 P.M.

LUNCHEON—Shoreham Hotel.

1:30-4:00 P.M.

GENERAL BUSINESS MEETING, all members. Addresses by the Chief of Naval Communications, the Chief Signal Officer, the Director of Air Communications, the President of AFCA, and his successor. Presentation of AFCA certificates of merit. Orientation by Navy of exhibits and demonstrations.

6:00 P.M.

COCKTAILS—Shoreham Hotel.

7:30 P.M.

BANQUET—Shoreham Hotel. Addresses by Admiral Louis E. Denfeld, Chief of Naval Operations, and by AFCA's president, David Sarnoff.

Tuesday, March 29

ARRANGEMENTS for the second day's meeting will all be made by the Navy Department, Captain Robert J. Foley, USN, in charge.

9:30 A.M.

TRANSPORTATION from Shoreham Hotel to Navy Exhibits.

10:00 A.M.

EXHIBITS of Naval communications equipment.

1:00 P.M.

LUNCHEON—Navy Station.

2:30-5:00 P.M.

DEMONSTRATION of Navy communication, photographic, and combat equipment.

355

1949 Nominations

THE 1939 NOMINATING COMMITTEE, as appointed by the President of the Society, was confirmed by the Board of Governors at its January meeting.

D. E. HYNDMAN, *Chairman*
Room 626—342 Madison Avenue
New York 17, N. Y.

HERBERT BARNETT
General Precision Equipment Corp.
63 Bedford Road
Pleasantville, N. Y.

F. T. BOWDITCH
Research Laboratories
National Carbon Company
Box 6087
Cleveland 1, Ohio

F. E. CAHILL, JR.
Warner Bros. Pictures, Inc.
321 W. 44 Street
New York 20, N. Y.

R. E. FARNHAM
General Electric Company
Nela Park
Cleveland 12, Ohio

G. R. GIROUX
Technicolor Motion Picture Corp.
6311 Romaine Street
Los Angeles 28, Calif.

A. N. GOLDSMITH
597 Fifth Avenue
New York 17, N. Y.

T. T. GOLDSMITH
Allen B. Du Mont Laboratories
2 Main Avenue
Passaic, N. J.

K. F. MORGAN
Western Electric Company
6601 Romaine Street
Hollywood 38, Calif.

All voting members of the Society who wish to submit recommendations for candidates to be considered by the Committee as possible nominees, are requested to correspond directly with the Chairman or any of the members of the Nominating Committee. Active, Fellow, or Honorary Members are authorized to make these suggestions which must be in the hands of the Committee by May 1, 1949.

There will be eight vacancies on the Board of Governors as of January 1, 1950, which must be filled. Those members whose terms of office expire are:

Financial Vice-President..............................D. B. JOY
Engineering Vice-President...........................J. A. MAURER
Treasurer..R. B. AUSTRIAN
Governor....A. W. COOK   Governor.....P. J. LARSEN
Governor....JAMES FRANK, JR.   Governor....G. E. SAWYER
                    Governor....L. T. GOLDSMITH

The recommendations of the Nominating Committee will be submitted to the Board of Governors for approval at the July meeting. The ballots will then be prepared and mailed to the voting members of the Society forty days prior to the Annual Meeting of the Society. This is the business session held during the Fall Convention, which this year will be in Hollywood, California, October 10–14.

D. E. HYNDMAN, *Chairman,*
Nominating Committee

Book Reviews

Sound and Documentary Film, by K. Cameron (Foreword by Cavalcanti)

Published (1947) by Sir I. Pitman and Sons, Ltd., Pitman House, 39–41 Parker Street, Kingsway, London, W.C. 2, England. Also distributed by Pitman Publishing Corporation, 2 W. 45 St., New York, N. Y. 157 pages + XV pages + 3-page index. 77 illustrations and diagrams. $5^1/_2 \times 8^1/_2$ inches. Price, 15 shillings.

This little book of 157 pages represents an analysis of some of the problems that face the producer and the sound engineer when making a documentary film. While the primary emphasis is on British films, the discussion can be applied freely to films made in the United States. According to the author, "in the perfect sound film the actual sound should be so perfectly wedded with the picture that the illusion of reality is complete." Documentary films, as we know them today, are about twenty years old. Two general classes of documentary films are described: (1) the straightforward description of an incident with a simple commentary, music, and sound effects, and (2) an imaginative, human exposition of how ordinary people live and work. The latter type is more difficult to make well, and is represented by such films as "Target for Tonight" and "Listen to Britain." The book describes the planning of a documentary or "realist" film, the problems facing the sound crew, the use of music and sound ·effects, postsynchronizing and dubbing, re-recording, and finally showing the film. The last 50 pages are devoted to brief technical abstracts of some of the processes involved in recording sound, and to a glossary of technical terms.

GLENN E. MATTHEWS
Eastman Kodak Company
Rochester, N. Y.

Discharge Lamps, by H. K. Bourne

Published (1948) by Chapman and Hall, Ltd., 37 Essex St., W.C. 2, London, England. 417 pages + 7 pages + XV pages. 186 figures. $5^3/_4 \times 8^3/_4$ inches. Price, $12.00. Book available from American Photographic Publishing Co., 353 Newbury St., Boston 15, Mass.

The reader who is primarily interested in the characteristics of light sources will find this book a convenient reference. As the title indicates, the author is principally concerned with discharge sources, but he also describes in considerable detail the characteristics of tungsten-filament lamps, carbon arcs, and photoflash lamps. One may assume that data on these types are included in order better to establish the effectiveness of discharge sources for many photographic and certain projection applications.

While discharge lamps may be constructed employing any of several gases, it is appropriate that lamps employing mercury vapor should be the principal theme of this book. The relatively high efficiency of such sources, particularly in terms

Book Reviews

of their effect on photographic materials, has established their pre-eminent position among discharge lamps. Their increasing usefulness is further indicated by the wide range of design possible with various combinations of operating pressure and electrical loading, as well as by modification of spectral quality through the use of fluorescent material on the surrounding envelope of the low-pressure arc, or the introduction of additional metallic vapors in those of higher pressure.

Only the last chapter of the book is devoted to applications of the sources, a treatment which might have been considerably expanded with great advantage to such groups as are represented by the readers of this JOURNAL.

The author is to be complimented for including an excellent index as well as an Appendix of valuable supplementary information.

F. E. CARLSON
General Electric Company
NELA PARK, OHIO

Standards Recommendation

On December 16, 1948, the Standards Committee recommended reapproval of the American Standard for cutting and perforating 35-mm negative raw stock, Z22.34-1944.

Reapproval or revision of this standard has been under consideration since October, 1946, when review of all the existing Z22 standards was undertaken by the ASA Sectional Committee, Z22. The long delay in determining what action should be taken was caused by the difference which exists between the 35-mm negative and positive perforations. This difference has caused considerable trouble in connection with printing 35-mm color release prints where extremely accurate registration is necessary. Consequently, the whole matter of 35-mm perforating was reinvestigated with the thought that perhaps a universal positive-negative perforation could be agreed upon. Such a perforation based upon the original proposal by Dubray and Howell is now in the process of development and a proposed standard will be published in the JOURNAL in the near future.

The use of the negative perforation, however, has become so firmly established in the industry that elimination as a standard at this time does not seem possible. Therefore, the foregoing recommendation that it be reaffirmed as an American Standard has been made. Technically, the draft which is now being submitted to Sectional Committee Z22 is identical with the 1944 edition. The method of dimensioning, however, has been modified so as to be in accord with present industrial practice.

Journal Exchange

Mr. G. D. Murphy wishes to dispose of a complete set of JOURNALS of the SMPE starting with the October, 1930, issue. Persons interested in purchasing these copies should write to him at R.F.D. 2, Rockville, Md.

~ New Products ~

Filmgraph

The Miles Reproducer Company, 812 Broadway, New York 3, New York, recently announced their latest Filmgraph Model RTD. This machine makes it possible to produce synchronized personal talking pictures

directly on developed black-and-white or color 16-mm family motion pictures for permanent playback, without processing or darkroom when used in conjunction with the standard 16-mm silent projector.

Filmgraph employs a jewel stylus to indent permanently a fine sound groove 0.002 inch wide and to reproduce it instantly through the same apparatus. The sound track is made at either edge, between the sprocket holes and the frames of the picture, on the glossy side of the film. Several sound tracks may be indented side by side, if desired, and will not show on the screen, since they are hidden by the frame of the aperture in the projector.

To make a recording, the Filmgraph is placed in front of the projector in such a manner that it will not interfere with the picture on the screen. The film is threaded through the projector in the usual way except that the film is brought up to the reel of the recorder instead of the feeding reel (upper magazine) of the projector. Sufficient leader film is used to bring the sight and sound in synchronism.

The new model is portable, measuring $4^1/_2 \times 8 \times 9$ inches and weighs less than 10 pounds. A microphone and an external loudspeaker mounted on a baffle are supplied as part of the system. The loudspeaker is equipped with a 15-foot cable to permit placing it behind or near the screen.

For synchronizing talking pictures in conjunction with 8-mm picture films Model MRC is recommended. This model has a recording capacity of approximately one hour, using 100 sound tracks on each face of the safety film.

～ New Products ～

Further information concerning the material described below can be obtained by writing direct to the manufacturers. As in the case of technical papers, publication of these news items does not constitute endorsement of the manufacturer's statements nor of his products.

Single-Case Filmosound

Bell and Howell Company, 7100 McCormick Road, Chicago 45, Illinois, announces the development of a new single-case, sound motion picture projector. The single-case filmosound, weighing only 43³/₄ pounds, is designed with a 6-inch speaker mounted on a removable door in the side of the projector case for carrying convenience. The door may be swung out at right angles to the case and the speaker operated from this position, or it may be removed from the case and placed near the screen.

Speaker and projector are connected by a 40-foot cable, and up to 60 feet of

additional cable may be added, if necessary. A 10-watt amplifier is provided allowing the use of a larger speaker as an accessory, if desired.

———————●———————

EMPLOYMENT SERVICE

POSITION WANTED

HIGH-SPEED
photography

THIS ISSUE IN TWO PARTS

Part I—March, 1949, Journal • Part II—High-Speed Photography

MA

1 9

CONTENTS

FOREWORD

●The Committee on High-Speed Photography of the Society of Motion Picture Engineers was organized in January, 1948, to further existing knowledge in high-speed photography and to disseminate that lore. The Committee also undertook to sponsor the development of equipment to make high-speed photography more usable through portability and better performance.

In order that the photographic engineering profession be brought up to date on the advancement of the art, a symposium was held in Washington, D. C., on October 29, 1948. At this symposium various high-speed cameras were described as well as techniques used by governmental and industrial agencies.

The Committee has been striving to have the issuance of papers on high-speed photography concentrated so people who are interested in the subject will not have to peruse many publications in order to find out what are the latest developments in the science of high-speed photography. This supplement to the Society's JOURNAL has all of the papers that were presented in Washington with the exception of one. The paper on the design of rotating-prism-type cameras is being revised in the light of comments made by John Kudar which appeared in an earlier issue of the JOURNAL.

It is felt that the papers contained herein form the foundation for high-speed photographic work which is being continued at the first International Symposium on High-Speed Photography to be held at the Hotel Statler in New York on April 6, 1949. The subjects covered not only include high-speed motion picture photography but high-speed still photography as well. There is material enough for a third symposium if the reaction is as favorable to the second as it was to the original one held in Washington.

The Chairman of the Committee wishes to express his appreciation to all members of the Committee and to those who have assisted the Committee in providing the active program as they have. If there is sufficient interest in the subject, plans can then be made with the Headquarters of the Society of Motion Picture Engineers to take care of the special needs of those interested in high-speed photography and photographic engineering.

JOHN H. WADDELL
Chairman
High-Speed Photography Committee

What Is High-Speed Photography?

By MAYNARD L. SANDELL

EASTMAN KODAK COMPANY, ROCHESTER, NEW YORK

PREFACE

IN SPITE OF THE FACT that a great deal of work has been done in the field of high-speed photography, much of it has been isolated, and the research has not been centralized for the convenience of everyone. Your Society has realized this shortcoming, and early in 1948 proposed that a permanent Committee on High-Speed Photography should be made a part of its organization. It was decided that this activity should embrace both still and motion picture photography because of the close relationship between the two in this instance.

In order to understand what is meant by high-speed photography, it is necessary to define it. In the field of still photography, with the knowledge and materials at hand, it is estimated that no conventional mechanical shutter is likely to be designed with a speed in excess of $1/1000$ of a second. Any still picture made at a rate exceeding this will be considered a high-speed photograph. In the field of motion pictures, the mechanical limitations imposed on an intermittent mechanism will probably prevent its use at speeds in excess of 250 frames a second. For this reason, your Committee has decided that motion pictures at a rate in excess of this will be considered high-speed motion pictures.

High-speed still photography is not new. Nearly a hundred years ago, Henry Fox Talbot, who pioneered in so many phases of photography, successfully photographed a section of the *London Times* while it was being rapidly whirled on a disk. Talbot's light source was the spark generated by the discharge of a Leyden jar through an air gap. The spark's duration, probably less than a microsecond, effectively stopped the motion of the disk during exposure, permitting the finest print of the newspaper to be read when the plate was developed.

From this humble beginning, the technique of high-speed still photography steadily improved with the increasing demand for its use in the science of ballistics. High-speed silhouettes, or shadowgraphs, became commonplace in the literature, and photographs, by the schlieren method, of sound waves and thermal disturbances in gases appeared. These pictures were made in a matter of millionths

5

of a second. However, as an indication of how far the science has progressed, it might be well at this point to cite an interesting anomaly in this phase of photography. By using a Kerr cell as an electro-optical shutter, photographs have been made at a speed as high as 4×10^{-9} second. To illustrate how incredibly short this period of time is, let us consider it in terms of space. If the distance from Washington, D. C., to Hollywood, California, which is something less than 3000 miles, were represented as one second, then 4×10^{-9} second would be only about seven tenths of an inch along the road from the Nation's Capital to the film capital.

With the development of gaseous discharge tubes for electronic flash, high-speed still photography by reflected light in both black and white and color was made possible. What was once a laboratory curiosity became the everyday tool, not only of science and industry, but of news, commercial, and portrait photographers.

High-speed motion pictures progressed more slowly because of the lag in developing cameras and projectors. In the early days, it was not convenient to produce a series of pictures in rapid sequence. The attempts of Muybridge required that a separate camera be used for each picture in the sequence. Furthermore, there was no satisfactory solution to the problem of viewing such pictures if they were produced. The Zoetrope, or wheel of life as it was called, was a temporary expedient. It consisted of a shallow, slotted drum inside of which was mounted the strip of pictures to be viewed, the number of slots corresponding to the number of pictures. As this drum was rotated, a fleeting view of each picture was obtained in sequence as the slots passed by, and if the motion was rapid enough to overcome the eye's persistence of vision, an illusion of motion resulted.

It was not long after the motion picture camera and projector were invented that attempts were made to alter the natural speed of subjects on the screen by changing the camera speed, the projector speed, or both. Action was generally accelerated to achieve ludicrous performances in comedies, while for better analysis of motion, as in sports events, the action was slowed down. Since projection speeds soon became more or less constant, these changes of tempo were obtained by varying the taking speed.

As the value of slowing down action for analytical purposes became more evident, attempts were made to increase the taking rate, until speeds in excess of 150 frames a second were attained. This resulted in slowing up action as much as ten times when the film was projected at 16 frames a second. Although this was adequate for analysis of most human or animal action, it became apparent that much higher taking speeds would be necessary for studying fast mechanical movements. To accomplish this, some means had to be found for supplanting the conventional intermittent mechanism of the camera because of its mechanical limitations. When this was achieved, high-speed motion picture photography was born.

Motion picture cameras capable of speeds greater than 250 frames a second generally employ a rotating prism or mirror, a series of rotating lenses, or a stroboscopic light source instead of the usual shutter and intermittent mechanisms. In the latter case, the camera is equipped with a commutator to synchronize the flashing light with the film movement, and the camera is operated in a dark or semi-darkened room. Such cameras cannot be used for photographing self-luminous subjects. Cameras employing the rotating-prism principle usually bracket the range from 250 to 10,000 frames a second. Speeds in excess of 10,000 frames a second fall in the classification of ultrahigh-speed photography and are usually obtained in what are known as strip cameras.

Of special interest among ultrahigh-speed motion picture cameras of the strip type is that developed by Brian O'Brien of the University of Rochester. Though the results obtained with this camera are somewhat lacking from the standpoint of resolution, the camera is capable of making fifteen million pictures a second, but for a period of only $^1/_{200}$ of a second. If this camera could be operated for as long as a full second, the resulting sequence of pictures when projected continuously at the normal speed of 16 frames a second would last for nearly 11 days.

Despite the handicaps under which the makers of high-speed still and motion pictures worked, the art has grown steadily since its inception. Perhaps the greatest growth has taken place during the war and the few years following it, until today this branch of photography has become a frequently used and invaluable tool of science and industry.

Specialists in every field of high-speed photography, including X-ray, infrared, ultraviolet, and color, have been invited to participate in the activities of this new Committee of the Society. Representatives of industry and the universities as well as members of the Armed Forces are contributing from their experience and knowledge. The papers presented on the following pages will cover various methods employed in still and motion picture high-speed photography.

Electrical-Flash Photography

By HAROLD E. EDGERTON

MASSACHUSETTS INSTITUTE OF TECHNOLOGY, CAMBRIDGE, MASSACHUSETTS

Summary—Within the past ten years the use of electrically produced flashes of light for photography has become widespread. This type of light has some very important technical advantages which eventually will win it a place in everyday photography in addition to those applications where high speed is required. Modern electronic lighting equipment is particularly suited for color photography since the color of the light closely approximates daylight, the color remains constant, and the quantity of light per flash can be closely controlled. Furthermore, the electronic system can produce large quantities of light for large-scale color photography in studios with results that cannot be duplicated with any other known type of lighting equipment.

The object of this paper is to outline the circuits and components that are now in common use. The principles of light production are given as well as methods of calculating exposures, especially for color films. A method of measuring integrated incident light from flash sources and a meter for that purpose are also described.

LIGHT PRODUCTION

WHEN ELECTRICAL ENERGY from a capacitor is discharged into an open spark a visual flash of light of very short duration is produced. Modern electrical flash systems (speedlights, stroboscopic lights, and others) use the same principle of the spark, except the efficiency of light production is greatly increased. The improvement resides in the use of a noble gas, such as argon, krypton, or xenon, which is constrained in a glass or quartz tube from which air has been removed.

Xenon is the preferred gas, even if the most expensive, because it has a high efficiency of conversion of the stored energy from the capacitors into visual and photographic light. Likewise, of the gases that can be used, xenon gas produces a spectral distribution of energy close to daylight and therefore the xenon flashtube is used as a source for photography with daylight color film. The spectra of Fig. 1 show a comparison between daylight and the xenon-filled flashtube. Note that the unfiltered xenon spectrum has an excess of radiation in the near-ultraviolet and blue-violet portion of the spectrum. A filter is used to reduce this radiation by the right amount when such sources

8

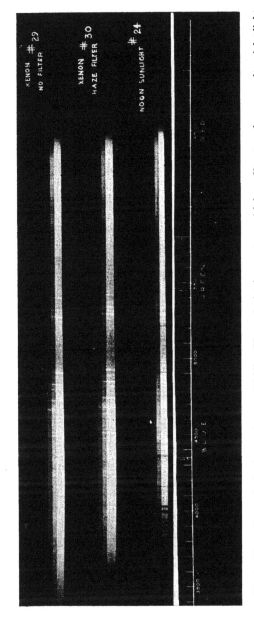

Fig. 1.—Top—Xenon flashtube spectrum. Middle—Xenon flashtube spectrum with haze filter to reduce near-ultraviolet light at the left end of the scale. Bottom—Noon sunlight. Note absorption lines caused by the sun's and earth's atmospheres. The cutoff at the right-hand side of all three spectrograms is caused by the lack of sensitivity of the photographic film in the infrared.

are used to expose daylight-type color emulsions. Different types of color films respond differently to the short flashes of light; therefore, a different filter is required for each type of film.

The xenon flashtube has several very important properties that are of great interest to those who photograph in color. First of all, the light output can be controlled accurately since the electrical energy can be determined accurately. Second, the color quality is not materially changed with slight voltage changes. Third, the color quality of the light is not changed with the life of the tube.

Those who have used tungsten sources for color photography will appreciate the above xenon-tube advantages since the tungsten-lamp output and color temperature are a function of voltage and age of the lamp. The xenon flashtube, unlike the tungsten lamp, is not subject to spectral changes during the life of the tube. In other words, the color distribution is the same for a new lamp and for one that has experienced thousands of flashes. This color constancy is a characteristic of the electronic system of producing light where a noble gas is excited electrically. The color of light from a tungsten lamp changes during life since the filament changes its dimensions due to evaporation. Furthermore, the evaporated filament material that is deposited on the bulb acts as a filter.

One of the most important characteristics of a flashtube is its efficiency, that is, its ability to receive electrical energy from a capacitor and convert much of that energy into a radiant form of which a part is light. Fig. 2 gives the efficiency of a typical small flashtube as a function of voltage and capacitance. The output of the flashtube under given conditions can be obtained from this diagram by reading the efficiency and calculating, as follows:

$$Q = \frac{CE^2}{2}\, n$$

where

$$
\begin{aligned}
Q &= \text{light output of flashtube in lumen-seconds}\\
C &= \text{capacitance in farads}\\
E &= \text{initial voltage of capacitor in volts}\\
CE^2/2 &= \text{energy storage in watt-seconds}\\
n &= \text{efficiency of flashtube (function of } E \text{ and } C \text{ as given in Fig. 2),}\\
&\quad \text{lumens per watt.}
\end{aligned}
$$

The highest efficiency is attained in the region of the "damage limit." Since glass cannot be held accurately to dimensions, this damage-limit region will vary for individual tubes of the same type.

Therefore, it is advisable not to operate too close to the border. The manufacturers of flashtubes rate the tubes according to the maximum energy at a specified voltage.

For any given shape of flashtube, the efficiency with xenon-gas filling can be varied by changing the pressure. Below a few centimeters of pressure, efficiency is almost a direct linear function of pressure. The spectral characteristics of the light output for low pressure (about 1 centimeter for the tube whose characteristics are shown in Fig. 2)

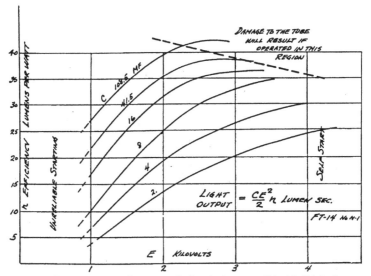

Fig. 2—Efficiency curves for xenon-flash spiral source $^{13}/_{16} \times 1^{5}/_{8}$ inches as used in General Electric flashtubes FT-210, FT-214, and FT-220.

are characterized by less continuum and stronger line spectra. For pressures above 5 centimeters the efficiency-versus-pressure relationship flattens out to give a constant efficiency. It is in this region that most flashtubes are designed.

Operating Limits

The operating limits for the tube whose characteristics are shown in Fig. 2 are boxed in by several factors. The left side is limited by the "unreliable-starting" region. On the other side is the "self-start" limit where the voltage is enough to start the discharge without the ionizing assistance of a high trigger potential on the outside of the tube. The "unreliable-starting" and "self-starting" limits are influenced by pressure. Both become higher with increased pressure gas. Slight impurities of gas likewise increase these two limits. The top limit is the "damage limit" where crazing of the tube is caused by the arc.

Relay-operated flashtubes are designed so that the "self-starting" limit is below the voltage to which the capacitor is charged. This is accomplished for any particular design by reducing the pressure until the desired limit value of voltage is attained.

Life of a flashtube is a function of the energy used per flash and the total number of flashes. At present, the life of the existing flashtubes is in excess of 10,000 flashes and thus seldom a problem. In fact, the flashtubes might be soldered or permanently wired into the reflector mountings since lamp replacement is a rare event.

A flashtube reaches the end of its useful life for several reasons such as:

(1)   It becomes a "hard-starter" due to the release of contaminating gases.

(2)   It becomes a "self-starter" due to the absorption of gas.

(3)   It becomes inefficient because of electrode sputtering and absorption of light by the darkened walls.

(4)   It is broken physically or has a cracked seal.

Of these four factors, the last is often the most important especially where the flash equipment is used on location.

For single-flash work the designer can load up the tube so that it works in the high-efficiency region. However, if the tube must be flashed at a rapid rate then *average heating* may be the limiting feature and lower energy inputs per flash must be used. For example, the tube whose characteristics are shown as Fig. 2 will operate at about 15 watts for continuous operation. Thus, at one flash per second the allowable input is 15 watt-seconds, that is, 7.5 microfarads at 2000 volts. At this loading, Fig. 2 shows that the efficiency is about half of the maximum that can be attained with the tube. At higher flashing rates the efficiency will be still lower.

Tubes to be flashed repetitively should be cooled by a blast of air so that the average power can be increased. Also, the tubes for this purpose should be made of quartz since quartz is capable of withstanding higher temperatures than glass.

ELECTRICAL CONDITIONS IN A FLASHTUBE CIRCUIT

Figs. 3 and 4 show the circuit elements used in two of the common types of electrical-flash systems. There must be a capacitor-charging system which is capable of supplying enough current to charge the capacitor in the time available between flashes. This time usually is 5 or 10 seconds for single-flash equipment. For the very large flash units it can be 30 seconds or more if the power input is limited, for example, to 30 amperes peak from a 110-volt lighting circuit. For multiflash applications the time may need to be $1/100$ of a second or less. The flashing rate of a given tube is limited both by the charging rate and by deionization of the tube as well as by tube temperature. Special circuits[1-3] are required when very high-frequency operation is desired.

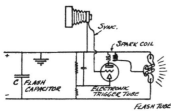

Fig. 3—Elementary circuit of the "electronic," "instantaneous," or "trigger" type of electrical-flash speedlight.

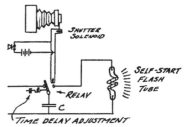

Fig. 4—Elementary circuit of a "relay"-operated flashtube. The lamp-operating relay is connected in parallel with the shutter solenoid. Adjustment of the relay time delay is made until light can be seen through the lens.

The power to operate a flashtube at a given frequency of F cycles per second is approximately the following:

$$\text{power} = \frac{(CE^2)F}{2} \text{ watts.}$$

For the majority of photographic problems, a flash unit is not required to operate continuously. Therefore, the electrical designer can overload his circuit elements, such as transformers, but he needs to design for adequate current-delivering properties. Power supplies for electrical-flash equipment are evaluated according to (1) final voltage or open-circuit voltage, (2) short-circuit current to charge the capacitor at the instant after the lamp flashes, and (3) time to charge a given capacitor to 90 per cent of the final charge.

Actual power to the power system may be double the value given by the above equation due to losses in the power supply. The electrical efficiency is seldom of importance in design except in the case of portable equipment where the number of available flashes from a given battery are concerned, or where the power circuit is overloaded.

The *average* current I to charge a capacitor C, in T seconds to a voltage E, is approximately

$$I = \frac{EC}{T} \text{ amperes.}$$

At the beginning of the charge cycle the peak current will need to be several times this value.

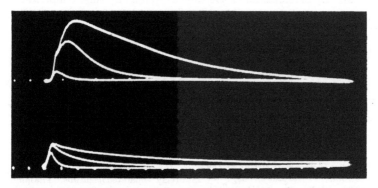

Fig. 5—Oscillographic traces of the light output (top) and current (bottom) for three different discharge capacitors, 120, 38, and 10 microfarads. The peak light for the 120-microfarad trace is at the top and has a value of about 27×10^6 lumens. The peak current is about 375 amperes. The above tests were made on the spiral as used in the FT-210, FT-220, and FT-214 (General Electric tubes) at 2000 volts. Timing dots are spaced 61.2 microseconds apart.

The discharge current, Fig. 5, resembles that of a resistance-capacitance circuit in form, especially for the large values of capacitance. The peak current for the 120-microfarad example is about 375 amperes. Assuming the capacitor initial voltage of 2000 holds, the apparent initial resistance of the flashtube is 2000/375 = 5.3 ohms. The time constant for this example is $120 \times 10^{-6} \times 5.3 =$ 635 microseconds, which checks roughly with the time required for the current to decrease to 37 per cent of the peak. The light at this time is 20 per cent of the peak.

The main differences in commercial single-flash units are brought out by these questions:

(1) How are the units supplied with power? Studio units are

connected to ordinary 60-cycle power outlets but portable equipment must be battery-operated.

(2) How much light output is produced by each unit? Eventually standards will be set up so that the lumen-second or beam-candle-power-second output of different tubes or tubes and reflectors can be measured. Likewise the beam-candle-power-second distribution of light from the reflectors as a function of angle will be evaluated since this is always an important factor.

(3) How is the flash lamp caused to flash? How is synchronization of the flash with a camera shutter accomplished? Two types are in common use (a) the "trigger" type usually started instantaneously by an electronic trigger tube, and (b) the "relay" type whereby the tube is connected directly across the capacitor by the relay. Both methods will be discussed later in this article since this synchronizing question should be foremost in the mind of anyone who desires to connect a camera and flash unit together.

FLASH DURATION

The exposure time for photographs taken with electrical flash lighting is determined entirely by the flash duration if the continuous light that is present does not also contribute to the exposure. The shutter time plays no part in determining the exposure time with flashtubes since the flash is always much shorter than the shutter-open interval. In fact, the function of the shutter, when using flashtubes, is to prevent extraneous continuous light from exposing the picture.

Exposure time with electrical flash can be defined as the time that the light is above $1/3$ of the peak light. When studies of duration, as in Fig. 5, are made with a specific flashtube, a curve such as Fig. 6 results. In general the duration is shorter with high voltage and with a small capacitor.

Flash duration or exposure time is seldom of importance except where rapidly moving objects are to be photographed or where close-up photographs of moderately fast motions are made. Any specific case can be checked for acceptable flash duration when the velocity of the object is known and when the minimum acceptable blur is expressed. For example, a .22-caliber bullet with a velocity of 1100 feet per second requires an exposure of about 3×10^{-6} second (3 microseconds) if the maximum allowable blur motion is 0.04 of an inch. With a flash duration of 100 microseconds ($1/10,000$ second) the bullet will move more than an inch.

Special equipment is available for bullet photography with a 2-microsecond exposure time. A short flash is obtained by the use of a small flash capacitor ($\frac{1}{3}$ microfarad) charged to high voltage (7000 volts). The flashtube also is designed with a short discharge path with a fairly large cross section so that the tube resistance is low but still sufficient to prevent any oscillations of the discharge current. A small amount of hydrogen mixed with argon gas has been found to give some quenching of the afterglow in the gas after the current has ceased to flow.

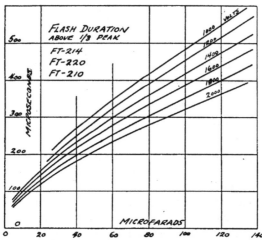

Fig. 6—Exposure time or flash duration as measured while light is above $\frac{1}{3}$ of peak light as a function of capacitance and voltage for the flashtube as used in the General Electric tubes FT-210, FT-220, and FT-214. Data from General Electric Company.

The very large energy discharges (3000 to 30,000 watt-seconds) into the quartz flashtubes as used for sources for large-scale color photography, would produce an annoying amount of noise if flashed without a small series inductance (about 5 millihenries). Such a series inductance does not materially increase the flash duration since its main function is to reduce the rapid rise of current at the start of the discharge. The series inductance also aids in deionization since it forces the current to flow longer than usual and thus brings the residual capacitor voltage to a lower value.

CAMERA SYNCHRONIZATION

The first problem after an electric-flash unit is obtained is how to connect it to a camera. Fortunately, many of the manufacturers of cameras and shutters are aware that the electric-flash system is due for growth and have been designing electrical contacts in their devices for flashing the light. These contacts need not be capable of carrying very much current when used with electronically triggered flashtubes.

Unfortunately the synchronizing requirements are different for the various types of chemical and electrical flashbulbs that may be used with a shutter. Three synchronizing requirements meet most of the problems that arise.

(1) *Instantaneous* contacts that close an electrical circuit at the exact instant that the shutter blades reach a full-open position as required for triggered speedlights of the electronic types. Such contacts are called "X" types.

(2) *A 5-millisecond shutter delay.* Here the lamp circuit is required to close a few milliseconds before the shutter reaches its open position so that the peak light output of the SM flashbulb at 5 milliseconds may occur while the shutter is open.

(3) *A 20-millisecond shutter delay.* Here the lamp circuit closes 15 to 17 milliseconds before the shutter reaches its open position. This delay is required for the average chemical flashbulb to reach its peak light output after the primer circuit is closed.

The delay mechanisms are more difficult to manufacture and to adjust and keep in adjustment. For many years, in fact since the advent of the chemical flashbulb, various external magnet and relay devices for firing the chemical bulb and obtaining a delay have been in widespread use. These devices as well as shutters including delay mechanisms will continue to be used to synchronize chemical flashbulbs. It is interesting to note, however, that a camera with an instantaneous synchronizer "X" set of contacts can be used with any 5- or 20-millisecond chemical bulb by slowing down the shutter speed to $1/25$ second. Eventually all shutters and cameras should be supplied with instantaneous types of "X" synchronizing contacts since such devices are easily arranged in shutters.

However, electric-flash speedlight equipment of the electronically triggered ionization types cannot be flashed with a shutter delay since the light flash will be extinguished by the time the shutter opens.

The "relay" type of electrical-flash equipment does have a time delay that is usually adjustable by the user. The delay is a function of battery voltage, temperature, position, circuit resistance, and so forth. When the "relay" system is used, *two* delay mechanisms must function reliably in order for the light to occur when the shutter is open. With long shutter times, this can be done. However, difficulty may arise at fast shutter speeds if either the shutter delay or the firing relay exhibits any time variability. Unlike the chemical flashbulb the electrical flash is short and a blank will result if the timing is off.

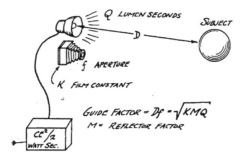

Fig. 7—Method of calculating the guide factor
for front-lighted photography.

The current required to operate the relay in relay-fired flash units may be too high for the internal contacts in some shutters. Either the relay will not operate or the contacts will be damaged by the excessive current.

GUIDE FACTOR

The photographer who obtains an electrical-flash unit is mainly interested in the *guide factor*, which is the product of the lamp-to-subject distance and the camera aperture (see Fig. 7), for a suitable photographic result. Guide factors for flash units are experimentally obtained by trial and error. An equation is given below which shows how the various factors influence the guide factor for the simple light-on-the-camera type of lighting.

$$Df = \sqrt{KMQ} = \text{guide factor}$$

where

D = the lamp-to-subject distance in feet
f = the camera aperture
Q = the lamp output in lumen-seconds
M = the reflector factor, a number that gives the ratio of light at the center of the beam with the reflector to the light without the reflector. Commercial reflectors often have an M factor of from 4 to 12.
K = an experimentally determined constant that corresponds to conditions suitable for satisfactory photography with any particular type of emulsion. See Table I for approximate values of K.

TABLE I

PRELIMINARY VALUES FOR K FOR A FEW EMULSIONS ARE GIVEN BELOW TO
SERVE AS A ROUGH GUIDE UNTIL MORE ACCURATE DATA ARE AVAILABLE

Film Type	Suggested Filter	Incident Exposure at Subject for $f/3.5$ Foot-Candle-Seconds	K
Kodachrome 35-mm daylight	CC15	80	0.012
Kodachrome professional cut-film type	CC15	100	0.01
Ektachrome	CC33	100	0.01
Ansco Color tungsten type	Conversion 12	32	0.032
Super Panchro Press— Type B	None	1.5 to 0.5	0.3* to 1.0*

* Value used depends upon processing and type of negative desired.

Actually it is not as simple as it appears from above since photographs are very seldom taken with front lighting of the subject. Therefore, the actual guide factor for any given equipment and condition should be found by direct experiment after preliminary estimates have been made using the theory that has just been presented.

WEIGHT OF FLASH EQUIPMENT

Whenever one carries portable flash electric equipment on an assignment the question is always asked, "Why not make it lighter?"

For portable work every effort should be made by the user to get the maximum out of his film-processing and lens equipment so that he can get his assignment with a minimum of light.

With the minimum light output specified, it is then up to the equipment designer to get a flashtube of the greatest efficiency so that his watt-second storage capacity will be as small as possible. Watt-second storage capacity means weight and for this reason the watt-second value should be as small as possible.

A very considerable saving of weight could be obtained by over-volting the electrical capacitors that serve as storage devices for the required watt-seconds. For example, doubling the voltage would result in a saving of weight by a factor of 4. However, the capacitor may *break down*, and probably will, since such flash capacitors are already rated as high as possible. It then comes to a decision of what chances one can take on capacitor failure. Many of the designers of flash equipment have erred on the side of overstress and capacitor failures have been the result.

For professional flash equipment in a studio, weight is of no consequence compared to the inconvenience caused by capacitor failure. It behooves the user of large studio equipment to insist upon conservatively designed capacitors even at the cost of weight and price in order to enjoy failure-free performance.

The weight of capacitors of a particular voltage rating is almost a direct function of the energy storage. From the equation for the guide factor, the weight is then proportional to the square of the guide factor, considering the case where the lamp efficiency and the reflector factor are constant. The above rule enables a flash-circuit designer to estimate the weight of a proposed flash unit in terms of an existing unit.

For example, a 20-pound flash unit has been found to have a guide factor of 20 when used with color film. Suppose a guide factor of 200 is desired, that is, a tenfold increase. The weight of the new flash unit will be ten squared or 100 times that of the first. A 2000-pound unit is indicated.

LIGHT MEASUREMENT FOR EXPOSURE DETERMINATIONS

The light meter for integrating light, which has been described more completely elsewhere,[4,5] uses a vacuum photoelectric tube to produce a current that is proportional to instantaneous incident light. This current is integrated against time in an electrical capacitor and an electronic voltmeter measures the integrated value as voltage. The voltage is proportional therefore to the integral of the incident light against time, which is exposure.

An incident-light meter that measures foot-candle-seconds (lumen-seconds per square foot) has two main uses. The first is to measure the lighting level at the subject for exposure determination, especially with color emulsions.   For any given material and camera aperture the time-light product at the subject needs to be of a specific value. The lighting is then arranged until the lighting level is correct.   A further refinement of this system is to use the meter to measure the light received on the ground glass from a white target card at the subject.   The advantage of this system is that all factors such as lens

Fig. 8—Exposure meter which reads incident light in foot-candle-seconds (lumen-seconds per square foot) from a xenon electronic-flash source.   The meter also accepts a photocell probe for use on the ground glass of a camera. (See Fig. 9.)

absorption and bellows extension are included.   With this arrangement there is only one correct reading regardless of aperture for any specific type of film.

The second use of the meter is to measure the output of flashtubes and of the tubes in reflector systems.   The horizontal candle-power-second output of a lamp is obtained by multiplying the incident lumen-seconds per square foot by the lamp-meter distance squared. Beam candle-power-seconds from a tube in a reflector is calculated in the same manner where the distance is greater than about ten reflector diameters.

The beam candle-power-seconds (b.cp.s.) is approximately equal to $$\text{b.cp.s.} = \frac{Q}{10} M$$

for the average flashtube. The spherical distribution from the lamp is not uniform in space, resulting in a factor of about 10 instead of 4π for the ratio of the total lumen output to horizontal candle-power output.

USES FOR FLASH UNITS

The various photographic problems that seek solution by means of electric-flash lighting can be roughly classified as follows. Those that require:

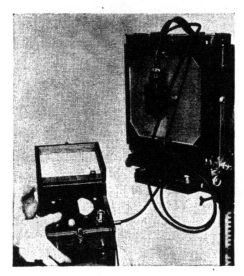

Fig. 9—Probe measurements of exposure on the ground glass is illustrated above. The photocell in the probe measures the light imaged from a white card at the subject.

(1) *Large quantities* of light of consistent color quantity and quality for professional studio color photography. Units of up to 20,000 watt-second capacity are now in use.

(2) *Short flashes* of light for rapidly moving objects such as bullets. Equipment with 2-microsecond flashes is in active use. A shorter flash with adequate light output appears to be difficult to obtain due to afterglow in the gas.

(3)   A series of pictures at a rapid rate for studying objects that move quickly. A companion paper by K. J. Germeshausen* describes the hydrogen-thyratron modulator high-speed stroboscope light which seems to have great promise as a pulsed light source.

(4)   *Monochromatic* light for use in interferometers. A short flash is also desired.

(5)   *Small intense* source for silhouette and microscope photography.

In conclusion, it should be pointed out that the electric-flash system is a most powerful method for both everyday photography and difficult scientific research problems. Many new applications and uses will undoubtedly result in the next few years.

* See pages 24–35.

REFERENCES

(1) F. E. Carlson and D. A. Pritchard, "The characteristics and application of flashtubes," *Illum. Eng.*, vol. 42; February, 1947.

(2) H. E. Edgerton, K. J. Germeshausen, and H. E. Grier, "Multiflash photography," *Photo Technique*, vol. 1, p. 14; October, 1939.

(3) H. M. Lester, "Electronic flashtube illumination for specialized motion picture photography," *J. Soc. Mot. Pict. Eng.*, vol. 50, pp. 208–233; March, 1948.

(4) H. E. Edgerton, "Photographic use of electrical discharge flashtubes," *J. Opt. Soc. Amer.*, vol. 36, pp. 390–399; July, 1946.

(5) H. E. Edgerton, "Light meter for electric flash lights," *Electronics*, p. 78; June, 1948.

(6) John S. Carrol, "Principles and design factors of electronic photo-flash units," *Elec. Manufacturing*, vol. 39, p. 47; April, 1947.

New High-Speed Stroboscope for High-Speed Motion Pictures

By KENNETH J. GERMESHAUSEN

MASSACHUSETTS INSTITUTE OF TECHNOLOGY, CAMBRIDGE, MASSACHUSETTS

Summary—This paper describes a high-speed stroboscopic light for use with high-speed motion picture cameras. It provides flashing rates up to 7000 per second and an effective flash duration of 1.5 microseconds. Experimental results are described showing the improvement in picture quality obtained when the light is used in conjunction with a Fastax camera.

INTRODUCTION

HIGH-SPEED MOTION PICTURE PHOTOGRAPHY is finding a constantly increasing use in the solution of industrial and scientific problems. The need for a tool to analyze motions too rapid for the eye to follow is readily apparent, especially one that does not require the attachment of measuring devices to the object under study. A number of high-speed motion picture cameras have been developed, all of which employ a continuously moving film and in general two methods are employed to stop the motion of the image with respect to the film. The first of these methods is primarily an optical one involving the use of rotating lenses, prisms, or mirrors; the second is the use of stroboscopic light to illuminate the subject, the flashes of light being of such short duration that the film does not move appreciably during the exposure.

Typical cameras of the first type are the Eastman high-speed camera and the Western Electric Fastax. Both of these cameras employ a rotating prism to compensate for the continuously moving film. An example of the second type is the General Radio high-speed motion picture assembly.

Each of these two types of high-speed cameras has certain advantages not possessed by the other. The optical types are generally smaller and simpler to operate. They may be used to photograph self-luminous subjects and under some conditions pictures may be taken outdoors in bright sunlight without additional sources of illumination. The stroboscopic type has the outstanding advantage of a very short exposure time, on the order of 2 to 10 microseconds, which

24

prevents blur when photographing rapidly moving objects. Furthermore, stroboscopic illumination is relatively *cool*, a distinct advantage when photographing subjects liable to damage by heat.

Stroboscopic illumination is limited to small areas, not more than 4 to 8 square feet for a single lamp, and cannot be used in the presence of strong general illumination. However, the majority of subjects requiring short exposure times are small and can be illuminated with a single lamp. Optical-type cameras are limited in definition, partly because of motion of the subject and partly because of distortions introduced by the moving optical system.

An ideal combination would be an optical-type camera that could be synchronized readily with stroboscopic light. For self-luminous subjects or large subjects particularly in sunlight, the camera could be used as is. For subjects requiring short exposure time, or where the intense heat associated with continuous light is objectionable, the same camera could be used with stroboscopic light, giving improved definition in the picture.

In the past, stroboscopic lighting has been criticized as bulky and complicated, often requiring the services of an expert to operate it, and, in addition, limited in maximum operating frequency. Modern designs of equipment overcome these objections to a large extent.

Fundamental Circuit

The new high-speed stroboscope represents a further development of equipment designed during the last war for the California Institute of Technology. Their problem required the illumination of a large water tank for studies of underwater projectiles. Stroboscopic lighting was indicated because of the short exposure required but none of the equipment then in existence would meet their requirements of light output, flashing rate, and duration of exposure.

The development of hydrogen thyratrons and their application to radar modulators[1] offered interesting possibilities as a solution to the above problem if a lamp of suitable characteristcis could be designed to match the modulator. Basically the circuit of the radar modulator as applied to the stroboscope is as shown in Fig. 1.

A capacitor C is charged from a direct-current supply through an inductance L_1, a rectifier T_1, and an inductance L_2. In this type of circuit the voltage on C will rise to nearly twice the supply voltage. Once the capacitor is charged to this voltage it will remain charged, since the rectifier T_1 prevents it from discharging back into the supply and thus returning to supply voltage. If the impedance of L_2 is low compared to the impedance of L_1, very little voltage will be developed across L_2 during the charging cycle.

When it is desired to flash the lamp the hydrogen thyratron T_2 is triggered. This effectively connects the charged capacitor across the lamp and the inductance L_2. If the voltage on C is sufficient to cause breakdown of the lamp, the capacitor will then discharge through the lamp causing a flash of light. In order that most of the energy in the capacitor be discharged into the lamp, the impedance of the lamp must be low compared to the impedance of L_2. During the discharge of the capacitor and while the thyratron T_2 is conducting, the relatively high impedance of L_1 prevents any appreciable current flowing from the supply through T_2. Immediately after the capacitor C becomes discharged the thyratron T_2 extinguishes and thus permits C to become charged again in the manner previously described.

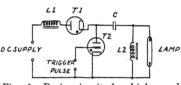

Fig. 1—Basic circuit for high-speed stroboscope.

The ability of the hydrogen thyratron, as compared to other thyratrons, to carry heavy short-duration discharge current and to extinguish very quickly, make the circuit of Fig. 1 feasible.

One of the major difficulties with the older type of high-speed stroboscope circuits[2] was the failure of the discharge circuit to become nonconducting or to deionize rapidly enough. This failure to deionize prevented recharging of the capacitor C at very short intervals and thus limited the maximum flashing rate of the stroboscope. The ability of the hydrogen thyratron to deionize rapidly permits much higher flashing rates, up to as high as 10,000 per second in properly designed circuits.

LAMP DESIGN

To secure maximum power output from the hydrogen thyratron the load, or the lamp, should have 30 to 50 ohms impedance. Furthermore, for a specific circuit, proper operation will be secured only if this impedance is held to rather close limits.

From the work of earlier experiments[3] it appeared that the impedance of flashlamps was generally much lower than the required 30 to 50 ohms. It also appeared that lamp impedance was not a constant but depended upon the length of the discharge path, the diameter of the tube, gas pressure, and the value of the discharge capacitance.

An investigation of these various parameters was made, extending them outside the range explored by previous experimenters. As a result of these experiments a lamp design evolved having an impedance of approximately 30 ohms with the values of discharge capacitance employed in the new stroboscope. This lamp has an arc length of approximately 4 inches. By means of a new type of construction, this 4-inch arc length is coiled into a cylinder $^3/_8$ by $^3/_8$ of an inch, giving a relatively small source size that may be used with high optical efficiency. A photograph of one of the lamps is shown in Fig. 2.

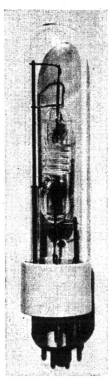

A second problem in lamp design for high flashing rates is one of efficiency. Flashlamp efficiency is a function of energy per flash and in general, to obtain maximum efficiency, the energy per flash should be high. If the loading per flash is high, however, then at high flashing rates the average power into the lamp becomes excessive, since average power is the product of energy per flash and flashing rate.

It was found by experiment that it was impossible to employ optimum loading per flash without destruction of the lamp by overheating. To maintain the loading per flash at the highest possible level two steps were taken. One, the lamp is made of quartz which will withstand much higher operating temperatures than glass, and two, the operating time of the lamp is limited to the minimum needed to expose a roll of film.

Fig. 2—High - speed stroboscope lamp.

For a 100-foot roll of 16-mm film at 4000 pictures per second, the total duration of the film is one second. With the loading per flash used, operating at the above speed for more than 2.0 seconds will result in overheating of the lamp.

THE HIGH-SPEED STROBOSCOPE

Description

The general circuit of the stroboscope is shown in Fig. 3. Components to the left of the dotted line comprise the power supply and components to the right of the dotted line comprise the modulator portion of the stroboscope. As constructed, these sets of components are on separate rack and panel chassis but may be mounted together in a single cabinet. The complete unit (Figs. 4 and 5) mounted in a

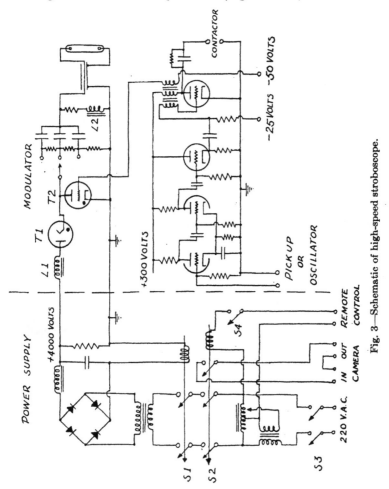

Fig. 3—Schematic of high-speed stroboscope.

cabinet measures 22 inches wide, 15 inches deep, and 28 inches tall and weighs 180 pounds.

The power supply is a conventional bridge-circuit rectifier with certain added control features. It supplies up to 500 milliamperes direct current at 4000 volts to the modulator panel. To maintain minimum size and weight all components are specially designed for this particular application, keeping in mind that the load on the supply is intermittent.

Input power is normally 220 volts, single-phase, 60 cycles, at 15 amperes maximum. By means of a built-in Variac the voltage on the primary of the power transformer can be held at 220 volts over a range of line voltages of 195 to 250 volts. Considerable thought was given to the choice of input voltage; 220 volts was decided upon be-

Fig. 4—Front view of high-speed stroboscope with accessory cables and the lamp.

cause the power drain was too great for most 115-volt lines. Relay S_2 in the primary of the power transformer can be remotely controlled by means of a switch connected to the remote-control terminals. On this same relay are a set of contacts which may be used to control camera motors simultaneously with the stroboscopic light.

In series with the remote-control circuit is a timer switch S_4 which may be adjusted to shut off the power after an interval of from 1 to 15 seconds. An overload relay S_1, also in the primary of the power transformer, shuts off the power instantaneously if excessive direct current is drawn from the power supply.

The lamp circuit is essentially the same as shown in Fig. 1 with the addition of adjustable discharge capacitances. Three values are provided, 0.01, 0.02, and 0.05 microfarad. Because of the voltage doubling obtained, as previously explained, the capacitors $C_1, C_2,$ and C_3 are charged to 8000 volts. The choice of discharge capacitance depends upon the flashing rate. In Table I are listed the energy storage per flash for each capacitor, maximum flashing rate, and average power delivered to the lamp at that rate.

Fig. 5—Rear view of high-speed stroboscope showing method of construction. The upper chassis is the modulator portion of the unit and the lower chassis the power supply.

A coaxial cable connects the lamp to the modulator. As much as 75 feet of cable may be used without materially affecting the light output or duration of the flash.

Voltage pulses to trigger the thyratron are obtained from an amplifier as shown in Fig. 3. The amplifier may be driven in several ways. A contactor or switch such as is used on the General Radio camera can be connected to the contactor terminals. The light will flash when contact is made. A sine-wave voltage of approximately 10 volts applied to the oscillator terminals will control the flashing rate over a range of 50 to 7000 flashes per second. If a steeply rising voltage pulse is applied to the input of the amplifier less voltage is required to trigger the thyratron. For sharp pulses approximately 0.5 volt at the input of the amplifier is adequate for satisfactory operation.

TABLE I

Capacitor Value, Microfarad	Watt Seconds per Flash	Flashing Rate per Second	Average Watts
0.01	0.32	7000	2250
0.02	0.64	3500	2250
0.05	1.60	1400	2250

To synchronize the flashing of the lamp with the rotation of the prism in cameras such as the Eastman and Fastax, a magnetic or reluctance pickup has been developed. This device consists of a small, permanently magnetized iron armature mounted in a coil of wire. The armature is mounted so the teeth of the film sprocket pass near one end. Each time a tooth goes by the armature a voltage pulse is developed in the coil. The pulse

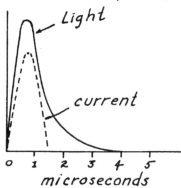

Fig. 6—Light and current versus time for stroboscopic lamp.

developed is of sufficient amplitude when applied to the pickup terminals of the amplifier to control the firing of the thyratron.

Curves of light output and lamp current versus time are shown in Fig. 6. These curves are for a value of discharge capacitance of 0.02 microfarad. As can be seen from the curves the duration of the light for the above capacitance is 1.5 microseconds to $1/3$ peak light and the total duration is 4 microseconds. Characteristically the light persists an appreciable time after the current has stopped. This persistence or afterglow is held to a minimum by the addition of hydrogen to the gas filling.

The total light output and the duration will depend upon the choice of discharge capacitance. In general the duration is approximately proportional to the capacitance; however, due to the change in lamp efficiency with loading, relatively more light is obtained with the larger values of capacitance.

Precision of firing of the lamp with respect to the triggering means is very good. From the time a signal is applied to the grid of the thyratron there is a delay of approximately 0.5 microsecond to the flashing of the lamp. The variation in flashing of the lamp with respect to the triggering signal is less than 0.1 microsecond. This same precision applies to signals fed through the amplifier but the over-all delay in firing is somewhat greater.

Light output in terms of effective exposure on the film depends on a number of factors such as the type of film, reflectors, and discharge capacitor. With a 5-inch diameter, diffuse aluminum reflector, adequate exposure is obtained at lens apertures of $f/4$ on areas 1.5 feet by 1.5 feet, using 16-mm super XX film and a discharge capacitance of 0.02 microfarad. Larger reflectors, or specularly finished reflectors, will increase the light on the subject. For small subjects, where the light can be focused on a small area, it is possible to work at apertures of $f/16$.

Light output is independent of flashing rate and depends only on the value of the discharge capacitance.

PHOTOGRAPHIC RESULTS

The high-speed stroboscope was used in conjunction with a Western Electric 16-mm Fastax to determine what improvement could be obtained by the use of stroboscopic light in contrast to incandescent illumination. Synchronization of the light with the rotating prism was accomplished by means of the magnetic pickup previously described.

It was found that the position of the pickup had to be set very carefully to insure that the prism was vertical when the light flashed. If this was not done a serious loss in definition resulted. Placement of the pickup was checked by observing the reflection of the image of the flashlamp. The flashlamp was placed on the axis of the lens-prism assembly and the position of the pickup adjusted until the image of the light was reflected back along this same axis.

Figs. 7A and B show the improvement in definition obtained by eliminating blurring due to motion of the subject. Both are pictures of a 10-inch disk rotating at 7200 revolutions per minute, the camera speed in each case was 3200 frames per second, and the lens aperture $f/4$. Fig. 7A was taken with incandescent light and Fig. 7B with stroboscopic light. From these two pictures it can be seen that for rapidly moving subjects stroboscopic light offers a considerable improvement in definition.

It should be pointed out that the sharp images obtained with stroboscopic light are most advantageous where it is desired to make a frame-by-frame inspection of the film. For a purely qualitative analysis of films by projection, blurring due to motion may not be objectionable. If the pictures are taken at a sufficiently high rate to give a good slow-motion effect, at least 100 frames per cycle of the event, then blurring due to motion of the subject is not too noticeable during projection and may even aid the appearance of the picture.

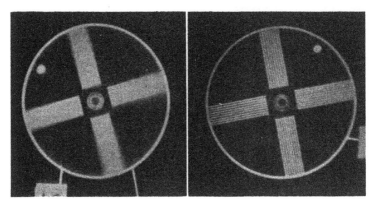

A—With incandescent light.　　　　B—With stroboscopic light.

Fig. 7—Enlargements from 16-mm high-speed motion pictures of a 10-inch disk rotating at 7200 revolutions per minute. Taken at 3200 pictures per second.

Figs. 8A and B are photographs of National Bureau of Standards 25X resolution charts. Fig. 8A was taken with incandescent light and Fig. 8B with stroboscopic light, the lens aperture in both cases being $f/4$. These pictures are enlargements of a portion of the 16-mm frame in order to show the resolution obtained; the numbers on the pictures correspond to resolution in lines per millimeter.

As can be seen from these two pictures, resolution is somewhat improved by the use of stroboscopic light. In Fig. 8B, 40 lines per millimeter can be resolved as against 28 lines per millimeter in Fig. 8A. This improvement in definition results in a better over-all appearance of pictures used for qualitative analysis and, again, is of considerable advantage where frame-by-frame inspection is to be made.

CONCLUSION

In conclusion it may be stated that the use of stroboscopic light with rotating prism-type high-speed cameras such as the Fastax will result in better definition. The improvement is of particular advantage where frame-by-frame inspection of the film is to be made for purposes of analysis by measurement. Stroboscopic light will stop the motion of rapidly moving subjects and permit accurate measurements to be made of form, velocity, and acceleration. A further advantage is gained in those cases where the subject is liable to damage by heat because of the relative coolness of the stroboscopic light.

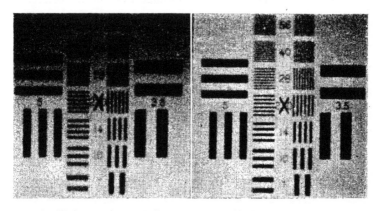

A—With incandescent light.   B—With stroboscopic light.

Fig. 8—Enlargements for 16-mm high-speed motion pictures of a resolution chart.

For many problems the results obtained with incandescent light are adequate and the improvement obtained by the use of stroboscopic light may not justify the additional complexity and cost of the lighting equipment.

REFERENCES

(1) The Hydrogen Thyratron, Sect. 8.11 Pulse Generators, M.I.T. Radiation Laboratory Series, 1st ed., McGraw-Hill Book Company, New York, N. Y., 1948.

(2) H. E. Edgerton, K. J. Germeshausen, and H. E. Grier, "High-speed photographic methods of measurement," J. Appl. Phys., vol. 8, pp. 2–9; January, 1937.

(3) P. M. Murphy and H. E. Edgerton, "Electrical characteristics of stroboscopic flash lamps," J. Appl. Phys., vol. 12, pp. 848–855; December, 1941.

Lamps for High-Speed Photography

By R. E. FARNHAM

General Electric Company, Cleveland, Ohio

Summary—This paper first discusses in general the requirements of light sources for high-speed motion picture photography, and in detail, five types of light sources that might be used for this work. Included with the discussion of each of these sources, specific lamp types are recommended and the detailed characteristics are given. A new type of reflector bulb lamp, developed specifically for high-speed motion picture photography, is explained.

SPECIFICATIONS

A S A RESULT of considerable discussion by the subcommittee on illuminants for high-speed photography, the following requirements for an illuminant have been established.

1. It should cover an area to be photographed of 4×4 inches.

2. The minimum distance from the source to the area is 18 inches.

3. The variation in the illumination from the center of the area to the corners should not exceed 2 to 1.

4. It should produce a minimum value of illumination at the center of the target of 50,000 foot-candles, although 100,000 foot-candles may be required.

5. A color temperature of the source of approximately 3500 degrees Kelvin is desirable.

6. Lamps would be operated at full power for approximately 10-second periods.

Other desirable features include the following:

1. To permit the light source to be kept close to lens-subject line it should be as compact as possible.

2. The required illumination levels should be obtained with preferably two lamps.

3. In the case of incandescent lamps, the use of tungsten cleaning powder is suggested to minimize the effect of bulb blackening.

Upon converting the 18-inch minimum lamp-to-subject distance and the 4- $\times$ 4-inch area to light-source-distribution requirements, we find approximately an 18-degree beam spread. This is based on the beam candle power dropping to 50 per cent of maximum, 9 degrees each side of the center of the distribution pattern.

35

APPRAISAL OF LIGHT SOURCES

The remainder of this paper discusses the advantages and disadvantages of available illuminants. Among those considered are:

Incandescent lamps (with separate reflector and reflector-bulb types).

Electrical-discharge lamps: mercury, fluorescent, and flashtubes.

Combustion sources.

The Incandescent Lamp

Without doubt this type of illuminant has been most widely used thus far and when its advantages and disadvantages are compared with those of other sources, it stands pre-eminent.

Incandescent lamps can be manufactured in sizes from a few watts to 10,000 or more watts. They can be operated from storage batteries or standard lighting circuits of almost any voltage, without auxiliary control or regulating equipment. They operate equally well on alternating or direct current. They light and extinguish quickly, and can be designed for a wide range of lives consistent with the requisite color temperature or efficiency. The light source itself can be made quite concentrated, thus producing high brightness and permitting accurate control of the light by means of reflectors. The spectral-energy distribution of the incandescent source is continuous and makes possible satisfactory color pictures with materials adapted to the tungsten radiation. The variation of the light occurring with alternating-current operation is less than 5 per cent for lamps above 100 watts on 60-cycle circuits.

Among its disadvantages are the high temperatures that accompany high levels of illumination. While heat-absorbing filters can be used, they are usually bulky, and where high wattages are employed, circulating, cooling water with its complications, generally is necessary. It is more often the practice to operate without filters and ignore the heat. A stream of air on the area being photographed will help materially.

Compared to other illuminants the incandescent lamp is not the most efficient in terms of photographic effectiveness versus input watts. On the other hand, the light of tungsten-filament lamps can be efficiently directed on to the area being photographed, thereby often more than compensating for the lesser actinicity of the light.

The wattages generally required for high-speed motion picture photography (at least 1000 to 1500) are such as to necessitate adequate power facilities, either in the form of ample-size wire or sufficient storage batteries plus heavy conductors, to insure that the lamps operate at their correct voltage. The photographic effectiveness of the light from tungsten-filament lamps is quite sensitive to variations in the applied voltage.

Incandescent lamps applicable to high-speed photography may be listed in two groups, those provided with reflectors integral with the bulb and those requiring external reflectors. (See Table I.)

TABLE I

INCANDESCENT LAMPS FOR HIGH-SPEED MOTION PICTURE PHOTOGRAPHY

Item	Watts	Bulb	Hours Life	Degrees Spread to 50% Max.	Max. Beam Candle Power	Ordering Code
				External Reflector Types		
1	750	T-12	25	..	..	750T12P
2	1000	T-12	10			1MT12P
3	1200	T-12	10			1200T12/49
4	1500	T-12	25	..		1500T12/P
				Integral Reflector Types		
5	150	PAR-38	1000	17	12,500	150PAR/SP
6	300	R-40	1000	18	12,500	300R/SP
7	500	R-40	6	84	6,500	RFL2
8	500	R-40	6	16	50,000	RSP2
9	750	R-40	6	18	75,000	750R40

Items 1 to 4 are provided with medium prefocus bases and projection-type biplane filament construction.

Items 5 to 9 are provided with medium screw bases.

All lamps available in 115, 120, and 125 volts rating.

Several of the lamps listed in Table I do not meet all the requirements outlined at the beginning of this paper. They are lamps that have been used in high-speed motion picture photography, plus one new type. Items 1, 2, 3 are motion picture projection lamps having highly concentrated light sources and operating at high efficiency; they must be used in conjunction with a reflector. Item 4 is a lamp developed a number of years ago for high-speed photography and is intended to be used with a deep-bowl polished-aluminum reflector of an elliptical contour.

Items 5 and 6 are standard projector and reflector spotlamps used in many applications where a lamp of this type is desirable. They are mentioned simply because some photographers have taken lamps of 115 volts rating and operated them on 220 volts for short periods of time, thus obtaining a severalfold increase in light output. Item 7 is the reflector floodlamp of the photoflood type and is included in the table because it is a more convenient lamp where the area to be photographed is relatively large. Obviously, quite a number of lamps would be required to obtain the requisite intensity of illumination. Item 8 is the reflector photospot lamp and is one of the most used lamps for high-speed motion picture photography.

Item 9 refers to a new lamp developed especially for high-speed motion picture photography, keeping in mind the specifications discussed at the start of this paper. Two of these lamps will produce 65,000 foot-candles on an area 4 × 4 inches at a distance of 18 inches. Naturally, experience with this lamp is limited but it appears to be capable of producing quite good results. It is intended for intermittent burning, and for best performance the operating periods should be no longer than 15 or 20 seconds at full power.

ELECTRIC-DISCHARGE TYPES

Mercury

Mercury lamps, with the exception of the H6, have such fundamental disadvantages that they can seldom be used. Nevertheless, it is well to include a brief discussion of this source.

The advantages of mercury lamps, while few in number, are as follows: The color of the light is good photographically where black-and-white materials are involved but it is not suited for color photography. Likewise, on the basis of equal photographic effectiveness, there is considerably less heat.

The light output follows the cyclic variation of an alternating current quite closely, with the result that there would be a periodic series of underexposed pictures interspersed with overexposed pictures; consequently, it is necessary in using mercury lamps for this application to operate them from direct-current sources. They require auxiliary ballast equipment whether operated from direct- or alternating-current circuits, which becomes an important factor in the event that the lighting equipment is transported to the job. The majority of them require from four to five minutes to come up to full brightness and an even longer period in the event that the light has been turned off. The source brightness of mercury lamps ranges from approximately 175 candles per square inch to about 6000 candles per square inch for the commercially available types. This might be compared with the source brightness of about 20,000 to 24,000 candles per square inch for concentrated-filament incandescent lamps mentioned in items 1 to 4 of Table I. Thus, it is much more difficult to obtain high illumination levels with these more extended sources.

The one exception mentioned above, namely the H6 lamp, has a source brightness of approximately 200,000 candles per square inch, more than 10 times that of the incandescent source. Furthermore, the lamp has a light source of about $1/_{16}$ of an inch in diameter and 1 inch long, thus making it possible with suitable reflecting equipment to obtain enormously high illumination levels. One of the types of the H6 lamp, the A-H6, operates in a water jacket which removes a large part of the infrared radiation and thus a comparatively cool light can be obtained, an important factor in certain types of high-speed photography. The lamp comes up to full brilliancy within 1 or 2 seconds, which removes an objection characteristic of the other types. However, its light follows the cyclic variation of the alternating current and it must, therefore, be operated on a direct-current supply, capable of producing about 1250 to 1300 volts. While its light output includes some more red than the other types of mercury lamps, it still is not entirely suited for color photography. It requires considerable auxiliary equipment in addition to the transformer in order to control the water flow.

Fluorescent Lamps

This source is not at all practical for high-speed photography because of its relatively low brightness; in fact, it is not possible to obtain more than about 600 foot-candles on a surface even quite close to the lamps, irrespective of the number of lamps employed.

Flashtubes

In order to make this summary of light sources complete, flashtubes have been included in the electric-discharge types of lamps; however, their characteristics are quite completely covered by Edgerton,[1] and no further mention need be made of them here.

Photoflash Lamps

It may seem a little odd to mention the photoflash lamp which is essentially a flash source in a paper dealing with continuously burning illuminants. However, one investigator[2] has taken advantage of the long flash duration of the FP-Type of flashlamp and combined the light output of a number of lamps, flashing in succession, to produce a reasonably constant source of very high value.

The General Electric Type 31 flashlamp has a flash duration of approximately 55 milliseconds between $1/2$ peak values of the flash. It is possible, by a suitable motor-driven switch to initiate the flash of a second lamp in such a manner that the decay of the first lamp combines with the increase in light output of a second lamp to produce a continuous source ranging in value from $1^1/_4$ to $1^1/_2$ million lumens. This process is continued with a third and a fourth lamp, and so on; thus 17 lamps will produce illumination lasting for one second.

Equipment developed by Lester[2] mounts these 17 lamps on the periphery of a wheel which is driven by the same motor that operates the flashing contacts and which makes one revolution per second. Each lamp in succession passes through a slot at the back of a large deep-bowl reflector while going through its flashing cycle.

This type of illuminant possesses several unique advantages:

An average of 1,400,000 lumens is emitted during this prolonged flash which, if obtained from lamps of the photoflood type, would require 45 kilowatts of lamps.

A couple of dry cells are all that are required to ignite the flashes, plus additional power to drive the small motor rotating the wheel on which the lamps are mounted.

Thus, the use of photoflash lamps permits making high-speed pictures when the usual or sufficient power supply is not available.

The heat received on the area being photographed is negligible, because of the short time that the lamps are on.

Two disadvantages might be mentioned.

1. The lamps move some little distance through the reflector during the flash; thus, the direction of the light beam from the reflector changes.

2. The time required to relamp after each shot.

REFERENCES

(1) Harold E. Edgerton, "Electrical flash photography," J. Soc. Mot. Pict. Eng., this issue, Pt. II, pp. 8–24.

(2) Henry M. Lester, "Continuous flash lighting—An improved high-intensity light source for high-speed motion picture photography," J. Soc. Mot. Pict. Eng., vol. 45, pp. 358–369; November, 1945.

DISCUSSION

QUESTION: When will the lamp be commercially available?

MR. R. E. FARNHAM: It will be available very soon.

QUESTION: What is the code number?

MR. FARNHAM: 750R40.

MR. W. T. WHELAN: Do you plan to bring this lamp out with a diffuser face like your 500-watt photoflood? Are you going to manufacture it with a diffuser face?

MR. FARNHAM: The face of the lamp is clear. This has been done to produce extreme concentration. The filament is of such a character that it will give a reasonable uniformity over the limited spot it is designed to cover.

MR. WHELAN: We have an application where a very large area must be covered with fairly uniform diffused light and we use the 500-watt photoflood behind a secondary diffusing screen.

I suggest that it might be useful to some users of the light sources to manufacture that particular type of lamp in the diffuser type. We use the one that is right above it on your chart.

CHAIRMAN L. R. MARTIN: As Mr. Farnham explained, I think that this specific lamp was designed to specific requirements at the request of the Society's Committee on High-Speed Photography. I feel that we all recognize that everything fits into the pattern, and I would not say Mr. Farnham would want not to recommend it for the type of service of which you speak.

MR. FARNHAM: You are quite right. This lamp was designed for a specific purpose and you will find other lamps shown in that table that will meet the requirements for illumination on a larger area.

Motion Picture Equipment for Very High-Speed Photography

By BRIAN O'BRIEN AND GORDON G. MILNE

INSTITUTE OF OPTICS, UNIVERSITY OF ROCHESTER, ROCHESTER, N. Y.

Summary—With the increased use of high-speed photography to study various commonplace phenomena which take place at extremely high speeds, it has been found necessary to develop motion picture camera equipment capable of exposing film at increasingly greater rates. A brief description is given of various types of cameras capable of speeds up to a million frames per second.

MOTION PICTURE CAMERAS such as the Eastman high-speed or the Bell Laboratories Fastax, which use a rotating prism as optical compensator, possess so many advantages for studies at speeds up to about 12,000 frames per second that they are the almost universal choice for this range. However, a significant number of events of technical importance produce motion too fast·to record properly at 12,000 frames per second. For example, the simple spreading of a crack in a plate of glass occurs at a speed of about one mile per second. If a 10-inch square of window glass be struck in the center, the cracks will travel to the edge of the plate in $^1/_{12,000}$ second or in the interval between successive frames. Many commonplace electrical phenomena take place at even higher speeds. A piece of fine fuse wire subjected to a very heavy overload can expand at a speed of 10 miles per second! For the study of these and other very fast events, motion picture cameras operating at speeds of 50,000 to several million frames per second are required. This is the range covered in this paper.

Methods of optical compensation other than the rotating prism may be used to permit continuous motion of the film, as in the rotating lens turret of the Merlin-Gerin camera and a camera very recently described by Partch and Wyckoff,[1] but these also suffer from the limitation in the speed at which the film may be moved. Consider a standard 8-mm motion picture film, with 80 frames per foot. At 12,000 frames per second such film must be moved through the camera at 150 feet per second, a speed approaching the present limit for perforated film on sprockets. If greater speed is required some other method of handling the film must be provided.

42

A significant gain in linear film speed may be obtained by attaching a single turn or a spiral of film to the outside of a rotating drum. For a given linear speed the centrifugal force varies as the inverse power of the radius of the drum, but even with large drums the problem of holding the film in place against the centrifugal force is quite serious. A marked mechanical improvement results from placing the film on the inside of a rotating drum, the centrifugal force holding the film in place. The optical arrangement with this system is not in general as convenient, but centrifugal force upon the film is no longer a problem, and the speed limitation is imposed only by the strength of material forming the drum. In this case no gain in linear speed results from increasing the diameter of the drum. The mass of the

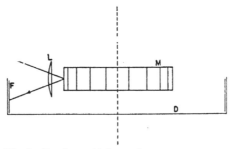

Fig. 1—Scophony high-speed camera.
D, rotating drum; F, film; M, multiple mirror; L, camera lens.

drum, considered as a spinning ring, increases with increasing diameter so as just to offset the decrease in centrifugal force. Cameras handling film by this method provide the highest linear film speeds attainable to date.[2] Very recently a camera handling film in this manner has been reported by Scophony Limited.[3] This camera is illustrated diagrammatically in Fig. 1. A rotating multiple-face mirror M rigidly connected to the drum D carrying the film provides optical compensation for the motion of the film and yields an actual motion picture record. The Scophony camera operates at a film speed of more than 600 feet per second resulting in standard 35-mm motion picture frames at a speed of 10,000 per second. If 8-mm instead of 35-mm frames were used, with an appropriate alteration in the mirror, it would seem possible to obtain 50,000 frames per second by this principle. In effect this has been accomplished in the Suhara[1] camera illustrated diagrammatically in Fig. 2.   Here the optical compensation is provided by a multiple-face mirror M, gear-driven from the main shaft which, again, uses a shallow open drum D carrying within it a single turn of film.   Speeds in excess of 100,000 frames per second have been reported from this camera, but the device is very large and cumbersome and has found little application.

Quite recently Baird[4] has described a moving-film camera with optical compensation provided by a multiple-faced mirror in which the

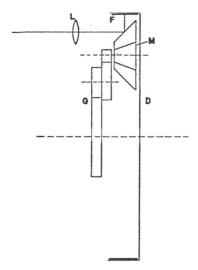

Fig. 2—Suhara camera.
D, rotating drum; F, film; M, multiple mirror; G, gear-drive to multiple mirror; L, camera lens.

limitation of linear film speed is not so serious.   By increasing the width of the film and providing a number of identical objectives side by side, Baird has in effect connected a number of moving-film mirror-compensation cameras together, side by side.   Suppose that there are five objectives and thus five rows of pictures on the moving film.   The multiple mirrors which provide the optical compensation, although mounted together rigidly on a single shaft, are stepped apart in angular position by one fifth the angular separation between successive faces of the multiple mirror for any particular row.   Thus objective 1 and multiple mirror 1 produce frames number 1, 6, 11, 16, etc., of the final motion picture, while objective 2 and multiple mirror 2 produce frames number 2, 7, 12, 17, etc. In this way five times as many frames per second are obtainable for a given film speed and size of image.

For nonluminous events which may be illuminated by a spark or repetitive flashing lamp, satisfactory high-speed motion pictures may be secured without any compensation for movement of the film. Edgerton and his associates have produced excellent photographs by this method, but again the number of frames per second is limited by the speed at which it is possible to move the film. The repetitive flashing lamp has another use which has not been so generally recognized. If linked with any of the optical compensation cameras so as to flash in synchronism with the camera mechanism, a very short exposure time is provided for any nonluminous portions of an event. It is common experience that a frame speed of only a few thousand per second is quite adequate to analyze many fast motions, yet the exposure time of individual frames may have to be reduced to $1/_{50,000}$ second or less to eliminate blur resulting from motion of the object during a single exposure. By the addition of such a lamp system cameras operating in the range of the Eastman high-speed or the Fastax can be applied successfully to events which, at first inspection, would seem to be beyond the speed range of these cameras.

An obvious way to overcome the difficulty of fast film movement is to allow the motion picture film to remain stationary and to sweep the image along the film with a rotating-mirror system. Very high linear image speeds are possible by this method which has been applied extensively in a number of streak cameras for the study of sparks, explosions, and other self-luminous events. To the authors' knowledge, this method has not been employed in any camera producing an actual motion picture record, but only in cameras which yield a streak image resulting from sweeping the image of the changing event along the film. However, an interesting variant of this procedure has been applied to secure actual motion picture records at very high speeds, a system in which the aperture rather than the image is caused to sweep by the rotation of a mirror system. By way of introducing this sytem the simple multiple-camera technique will be considered.

As far back as 1877, when Muybridge made his first series of photographs of a running horse, use has been made of a series of still cameras exposed in sequence as a means of photographing motion. In recent versions use is made of an electrooptical shutter provided by the Kerr effect in a liquid such as nitrobenzene contained in a cell mounted between crossed polarizers. The optical efficiency of such shutters is low but very high speeds are possible. The most recent work with Kerr cells appears to be that of Zarem[5] in which three cameras are tripped in sequence by an electrical-pulsing arrangement giving an equivalent exposure of about $1/100,000,000$ second. In addition to shutter inefficiency (poor light transmission), this system limits the number of motion picture frames of an event to the number of complete lenses and shutters provided. Part of this limitation has been removed in the Miller and in the Bowen high-speed cameras.

In the Miller camera[6] the film and 90 identical photographic lenses are stationary, and the aperture of a primary objective is swept

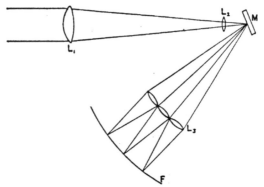

Fig. 3—Optical principle of the Miller and the Bowen cameras.
M, rotating mirror; F, film; L_1, primary camera lens; L_2, field lens; L_3, secondary camera lenses.

in succession across the 90 secondary objectives by a rotating-mirror system. In the Bowen camera[7] a somewhat similar arrangement is used to sweep an aperture across 76 separate lenses. Both the Miller and the Bowen cameras give speeds of about 400,000 frames per second, and this does not appear to be the limit attainable. The fundamental principle of these cameras is illustrated in Fig. 3. An image of the event to be photographed is formed at the surface of the rotating mirror M by a primary objective L_1. By means of a field lens L_2 an image of the aperture of the primary objective L_1 is formed upon one of the stationary secondary objectives L_3 behind each of which is an area of photographic film. The only moving part is the mirror M, the rotation of which causes the image of the aperture of L_1 to sweep successively across the apertures of the secondary objectives L_3. As the aperture of L_1 is imaged successively upon each of the lenses L_3, that lens forms an image of the mirror (and thus of the event to be photographed) upon the film F. With the arrangement shown in Fig. 3 the number of frames in the final motion picture is equal to the number of secondary objectives L_3. However, in the Miller camera by an arrangement of the multifaced mirror M it is possible to secure more frames than there are lenses L_3 in the secondary set. In spite of this the total number of frames in any run is quite limited, although sufficient to record a number of fast events.

The required film or image speed in a moving-film camera can be much reduced if the conventional shape of the motion picture frame is altered to make the dimension along the direction of film movement very small. This has been done in a camera recently described by O'Brien and Milne[8] which combines very high speed with long runs of several thousand frames in a simple mechanical system. The change of shape from the conventional motion picture frame is accomplished by a stationary optical system called the image dissector. This optical system, without moving parts, forms an image of the subject in the usual proportions of a motion picture frame, then cuts this image into a series of narrow strips, and redisposes these strips end to end to form a single long-slit image which is reimaged upon moving film with the long dimension across the direction of motion. All the information in the rectangular picture is contained in this narrow strip. It is only necessary that the negative film move the small width of this strip to record a new frame. By driving the film 400 feet per second on the inside of a shallow rotating drum, a camera speed of ten million frames per second is achieved. If a shutter opening and closing at this speed were provided, combined with image movement compensation on the film, no loss in resolving power would occur. In practice, however, the camera is used without a shutter so that the picture is blurred by the width of the slit image on the moving film. In the present model this width is only about $1/100$ mm, so the loss of resolution is comparable to the resolving power of a fine-grain negative material.

After processing, the negative is projection-printed back through an optical system similar to that which formed it, and a standard 16-mm motion picture frame thus reconstructed from $1/100$ mm of film travel. The printer is automatic and delivers a standard 16-mm motion picture print of any desired length.

In the first version of the camera reported the motion picture frame is cut into 15 or 30 strips, and the picture quality is very poor. In the second version of the camera under construction a much improved image quality is expected. The present camera permits continuous runs of 1600 motion picture frames for a 30-element image dissection, while permitting very long runs of more than 60,000 motion picture frames for a 15-element image. Because of the nature of the optical system the camera works at the high relative aperture of $f/2.0$, an aperture which has not been approached in any of the other ultraspeed cameras reported to date. If the Type II version of this camera comes up to expectations it should prove a useful addition in this field.

REFERENCES

(1) N. T. Partch and C. W. Wyckoff, Navy Department Taylor Model Basin Report R-345, June, 1948. Descriptions of many types of high-speed motion picture cameras are given in a report by John Waddell, "High-speed motion picture photography," Bell Telephone Laboratories, 1947.

(2) Brian O'Brien, "A high-speed running film camera for photographic photometry," *Phys. Rev.*, vol. 50, pp. 400–401; August, 1936.

(3) Scophony Ltd., "The Scophony high-speed camera," *Phot. J.*, vol. 86B, pp. 42–46; March-April, 1946.

(4) K. M. Baird, "High-speed camera," *Can. J. Res.*, vol. 24A, pp. 41–45; July, 1946.

(5) A. M. Zarem, Press release May 8, 1948, from Naval Ordnance Training Station, Inyokern, California; see also *Electronics*, vol. 21, p. 164; July, 1948.

(6) C. D. Miller, National Advisory Committee for Aeronautics Technical Note 140-S, June 10, 1947; see also J. *Phot. Soc. Amer.*, vol. 14, p. 669; November, 1948.

(7) Naval Ordnance Test Station Reprint 1033, Inyokern, California, "The Bowen 76-lens camera," by J. S. Stanton and M. D. Blatt.

(8) Brian O'Brien and Gordon Milne, "Motion picture photography at ten million frames per second," J. *Soc. Mot. Pict. Eng.*, vol. 52, pp. 30–41; January, 1949.

Methods of Analyzing High-Speed Photographs

By WADE S. NIVISON

RECORDAK CORPORATION, NEW YORK, NEW YORK

Summary—Qualitative, quantitative, and graphical methods of analyzing high-speed photographs are discussed and their peculiar requirements enumerated. Apparatus adapted to the several methods are described in detail.

INTRODUCTION

THE METHODS of analyzing high-speed photographs may be considered of three broad and somewhat overlapping types: qualitative, quantitative, and illustrative or graphical. Each presents peculiar problems which can be simplified by suitable apparatus.

In nearly all analyses orientation and time-scale references are desirable. In quantitative work they are, of course, essential. These references take many forms. They may be simply a linear scale, a single wire or two points of reference, and a time clock in the field. In more elaborate form, an over-all grid or checkerboard background and accurately timed spark discharge or argon glow lamps to produce pips on the edge of the film, provide orientation and timing references.

QUALITATIVE METHODS

To see the changing relationship of elements of a high-speed mechanism, the relative movement of fragments of a disintegrating subject, or to determine whether a particular phenomenon does or does not occur is readily accomplished by projecting a motion picture film taken at high speed with a conventional motion picture projector.

Such a projector should be capable of forward and reverse motion and still-picture projection. Remote controls are advantageous and a frame or footage counter is often very useful. When the exposure rate is known and the projection rate can be suitably adjusted and controlled, the duration of a cycle or a transient phenomenon can sometimes be measured with a stop watch.

Typical of projectors for qualitative analyses is the Kodascope sixteen-20, Fig. 1, which incorporates a push-button control panel, vertical-tandem pulldown claw, and forward and reverse and still-picture

49

projection. Standard equipment includes a two-inch $f/1.6$ coated lens, a 750-watt lamp, and a carrying case. A 1000-watt lamp is available for this model. The carrying case also serves as a projection stand.

Another 16-mm projector is the Bell and Howell Model 173 AD, Fig. 2. Designed especially for time and motion study it provides a two-perforation pulldown claw, forward and reverse and still-picture projection, a single-frame hand crank, and a frame counter. The electric governor-controlled motor provides a speed range from about 13 to 20 frames per second. Standard equipment includes a 2-inch $f/1.6$ coated lens, a 750-watt lamp, and a carrying case. A 1000-watt lamp may also be used in this model.

Shown in Fig. 3 is a 16-mm Keystone projector Model A-82, as modified by Professor David B. Porter of New York University to adapt it for industrial time and motion study. To reduce the inertia of the shutter and thereby minimize coasting and to reduce the flicker rate, which in all projectors becomes very objectionable at two or three frames per second, the standard shutter is replaced by a single blade of 0.010-inch Duralumin. The standard motor switch is re-

Fig. 1—Eastman 16-mm Kodascope sixteen-20.

placed by one of the more easily operated key types. A toggle switch is added for the projection lamp and one for a pilot. The pilot lamp also serves to illuminate work papers. The motor rheostat and all controls are mounted in a small box and provided with a 6-foot cable for remote operation. A further useful addition is a frame counter. With these modifications the projector can be operated at very low rates and the film can be advanced frame by frame.

Similar Kodascopes, Filmo, and Keystone projectors are available for 8-mm film.

Another useful device is the "editing viewer." Mounted between rewinds and driven by the film, it projects a small motion picture on a translucent screen. The use of a rotating prism, as in high-speed cameras, permits continuous motion of the film at low or high speeds. Usually incorporated in such a device and of particular value, is a means of marking individual frames for later detailed examination or the making of single-frame enlargements.

Fig. 2—Bell and Howell 16-mm projector Model 173 AD.

A desirable adjunct to the viewer is a footage or frame counter. A counter facilitates indexing and finding particular sections of the film and by reference to a velocity-footage curve the rate at any point in the film can be determined. One source of such counters is the Neumade Products Corporation.

The Bell and Howell filmotion viewer is shown in Fig. 4 with their

Model 136 splicer and rewind assembly. The viewer employs a 30-watt line-voltage lamp and projects an image approximately $2^1/_4$ by 3 inches. Controls are provided for framing, focus, and a marking device which produces a tiny slit in the edge of the film when it is desired to mark an individual frame. The rewinds shown accommodate 2000-foot reels. They have two-speed crank drives and are equipped with brakes.

Fig. 3—Modified Keystone 16-mm projector Model A-82.

Fig. 5 illustrates the Ciné Kodak editing viewer as part of the master editing outfit which also incorporates a splicer and 1600-foot brake-equipped rewinds. This viewer projects an image approximately $1^1/_8$ by $1^1/_2$ inches. Focus is fixed and controls are provided for framing and an edge-notching device for marking individual frames.

Similar editing viewers are also available for 8-mm film.

QUANTITATIVE ANALYSIS

Having, by previous inspection, determined that certain sequences or individual frames will yield additional data by quantitative study, other apparatus will be found especially useful.

All of the equipment hereinafter described is drawn from the seemingly unrelated field of microfilming. Film readers, as they are

Fig. 4—Bell and Howell 16-mm filmotion viewer.

Fig. 5—Eastman 16-mm Ciné Kodak editing viewer.

generally called, are, however, specifically designed to facilitate the rapid handling and critical examination of discrete images on 16-mm and 35-mm film. The very nature of their use in reading microfilm requires the production of screen images of excellent quality. High resolution, ample and uniform illumination, simplicity, and ease of operation are requisite to long periods of use without physical and visual fatigue.

By way of comparison with a probably more familiar subject, it can be stated that in current practice higher values of resolution are obtained in microfilming than in sound recording. A film commonly used for microfilming is rated at just double the resolving power of fine-grain release positive, and microfilm negatives commonly yield resolution higher in the same degree. All of these comments on microfilming are, of course, merely to qualify and recommend such film readers for the analysis of high-speed photographs.

Fig. 6—Recordak 16- and 35-mm film reader Model C.
Fig. 7—Recordak 16-mm film reader Model PD.

One of the most widely used film readers of that type is the Recordak film reader Model C, Fig. 6, which accommodates both 16- and 35-mm film. In this device images are projected on a self-contained 18- by 18-inch screen of special plastic composition which minimizes scintillation and hot spot so objectionable in ordinary ground-glass screens. Magnifications from 12 to 23 diameters are afforded merely by moving the screen outward on its hinged support. The image is automatically focused throughout the range. Variable magnification and automatic focusing facilitate adjustment of the image size to match a preselected scale or réseau.

Since at high magnification the 18-inch screen does not contain the full width of 35-mm film, a scanning mechanism is provided to move the film with respect to the projection lens and thus to bring any portion of the frame into view. The image can also be rotated on the screen through 360 degrees. By these means reference points in the image can be brought into coincidence with similar points on an overlaid screen pattern.

Illumination is obtained from a prefocused 200-watt lamp and the film is protected by a heat filter and by being clamped between glass flats. The flats are opened automatically when the film is progressed in either direction by means of the low- or high-speed hand cranks.

Fig. 8—Recordak 16- and 35-mm portable projector
Model A.

An accessory shelf is available to provide a work surface in front o the screen.

Another Recordak film reader, the Model PD shown in Fig. 7, is designed to project 16-mm film at a magnification of 23 diameters. Its self-contained translucent screen measures 14 by 15½ inches and is made of the same special material as the screen in the Model C. Other similar features are the 200-watt lamp, the heat filter, film-clamping flats, and image rotation. In addition, the Model PD can be used as an enlarger. A paper holder is incorporated in the unit and it can be readily adjusted for magnifications from 8 to 18 diameters.

In Fig. 8 is illustrated the Recordak portable projector Model A which accommodates either 16- or 35-mm film. Magnifications from

10 diameters upward can be obtained, with 30 diameters possible at $6^1/_2$ feet from the screen. Heat-filter and film-clamping flats are incorporated. The flats are manually operated. A projector of this type readily lends itself to the arrangment described by Thomas and Coles[1] wherein a mirror is placed in the beam of the projector to reflect the image back to a translucent screen near the projector so that the analyst can operate the projector and measure the screen image from the same position.

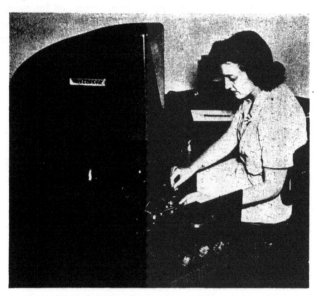

Fig. 9—Threading the Recordak transcription film reader.

Being readied for production is a new Recordak transcription film reader which incorporates many novel features. Designed to be used in making transcriptions by hand, typewriter, card punch, or posting machine its development was directed toward ease of operation, ideal placement of the screen, flexibility, and convenience of use.

The base resembles a desk-high filing cabinet with the screen mounted in a rotatable turret and the projection mechanism contained in the top drawer. Knobs on the face of the drawer control image rotation, focus, and lateral scanning of the film. A remote control for the motor-driven film transport is also included. Placed adjacent to a standard desk, as shown in Fig. 9, reel spindles are accessible and threading extremely simple and convenient. Magnification can be varied between 24 and 35 diameters merely by moving the drawer in or out.

Fig. 10 shows how the screen turret can be rotated to overhang the work surface and that the screen and work papers are equidistant from the analyst's eyes. Thus by minimizing accommodation, visual fatigue is reduced. The remote control provides three rates of film transport, both forward and reverse, and the two lower rates are

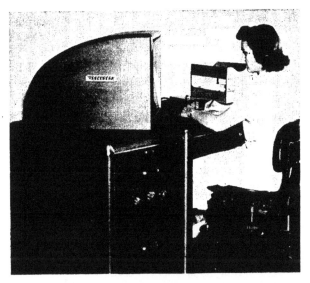

Fig. 10—Transcribing data with the Recordak transcription film reader.

adjustable. Frame-by-frame advance also is possible. Of significance is a new optical device which replaces the commonly used film-clamping flats. This device holds the film in critical focus whether stationary or in motion and without damage to the film after many thousand passes.

GRAPHICAL METHODS

Some analyses can be accomplished more readily with single-frame enlargements. Enlargements too are often desired for publication, demonstration, or record purposes. While conventional enlargers will sometimes suffice, here again microfilming equipment is eminently adapted to the task.

Fig. 11—Photostat 16- and 35-mm microfilm enlarger.

Many industrial organizations maintain photoreproduction departments and employ Photostat, Rectigraph, or similar photocopying machines. Microfilm enlargers have been designed for attachment to such machines to take advantage of the paper-handling and processing facilities which those machines incorporate. In the larger, so-called "continuous," machines completely processed and dried prints are delivered at rates of more than one per minute.

Fig. 11 shows a Photostat microfilm enlarger as mounted on the front of a typical Photostat machine. In this position it replaces the regular lens-and-prism assembly and projects an image either to a viewing screen or back onto the sensitized paper. The device comprises a projection-lamp unit, condensing lens, projection lens, prism,

Fig. 12—Haloid 16- and 35-mm microfilm enlarger.

focusing scales, and a manually operated film-winding mechanism. This enlarger is offered with one or two projection lenses and a magnification range from 5 to 30 diameters depending on the lens equipment and the model of the machine to which it is attached.

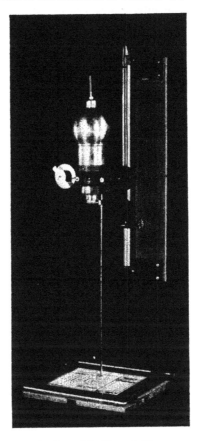

Fig. 13—Recordak 16- and 35-mm enlarger Model A.

Another form of microfilm enlarger is designed for use on the Rectigraph and other photocopying machines. It can be used independently of such machines if a suitable paper-handling mechanism is provided. Equipped with a motor-driven film transport, it incorporates a photoelectric control which advances the film, frame by frame, regardless of the length of the frame (microfilm frames often vary in size). The control operates on the difference in density of the exposed frame and unexposed frame line.

With this enlarger mounted on a Haloid Foto-Flo machine, Fig. 12, and equipped with an interlocking control, enlargements are made automatically at rates from 4 to 10 per minute depending on the size of the enlargements. Magnification range is from 8 to 26 diameters. The enlarger can be swung out of the way to permit normal photocopying operations. This enlarger, designed by the Microtronics Corporation for Haloid, is sold exclusively by the Haloid Company.

The Recordak enlarger Model A shown in Fig. 13 provides magnifications from 4 to 30 diameters with either 16- or 35-mm film. Its 63-mm Ektar lens has an extremely flat field and high resolving power. Critical focus is assured by clamping the film between glass flats and by correlating the lens scale and measuring tape which are both calibrated in diameters of magnification.

The simple and inexpensive device shown in Fig. 14 is another form of single-frame enlarger. It is no longer manufactured but still can be found in camera shops, and is extremely useful. A modified, fixed-focus, Jiffy Kodak with a 16- or 35-mm film gate provides a convenient means of making enlarged negatives on $2^{1}/_{4} \times 3^{1}/_{4}$-inch roll film.

Fig. 14—Jiffy Kodak 16-mm enlarger.

Obviously, this list is incomplete. Not all of the apparatus which might be used for the analysis of high-speed photographs have been mentioned but the author feels from his experience and that of other workers in the field the apparatus herein described can be commended to use in analysis work.

REFERENCE

(1) P. M. Thomas and C. H. Coles, "Specialized photography applied to engineering in the Army Air Forces," J. Soc. Mot. Pict. Eng., vol. 46, pp. 220–231; March, 1946.

New Developments in X-Ray Motion Pictures

By C. M. SLACK, L. F. EHRKE, C. T. ZAVALES, AND D. C. DICKSON

WESTINGHOUSE ELECTRIC CORPORATION, BLOOMFIELD, NEW JERSEY

Summary—Equipment using exposure times as short as 10 microseconds permits the radiographing of very rapidly moving objects and the use of continuously moving film without blur in a specially constructed camera without a shutter. The system may also be used in some cases to produce an image on a fluorescent screen which may be photographed with a General Radio oscilloscope camera or a synchronized motion picture camera.

INTRODUCTION

WHEN HIGH-VELOCITY ELECTRONS impinge upon matter, X rays are produced. The higher the atomic number of the element struck, the higher is the efficiency of the X-ray production; also, the higher the velocity of the electrons, the higher the efficiency and also the shorter the wavelength of the resulting X rays. These shorter-wavelength X rays are more penetrating than those generated with

Fig. 1—"Micronex" surge generator and field-emission X-ray tube making single radiographs with exposure times of $1/10$ to 1 microsecond.

61

the lower-velocity electrons.   Thus, with X rays it is possible to control both the quantity and quality of the rays produced merely by regulating the voltage and current through the high-vacuum X-ray tube.

At reasonable voltages, many times as much energy is required to produce sufficient X rays for a picture as would be needed with visible light, and since this energy must be supplied at high voltage, the equipment is, of necessity, larger and more cumbersome.   X rays

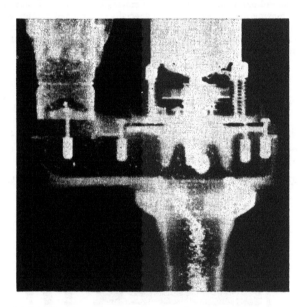

Fig. 2—Radiograph of lead dust passing through a vacuum
cleaner.   Exposure time one millionth of a second.

cannot be refracted in the usual sense of the word, so that all X-ray pictures must be shadowgraphs, with the size of the picture equal to or larger than that of the object.   If there is a sufficient quantity of X rays available, it is sometimes possible to cast an image on a fluorescent screen and then photograph this screen with a high-speed camera. Some definition is lost in this process and depending somewhat on the quality of the X rays, their intensity must be 20 to 100 times that which is required to make the radiograph directly on X-ray film with the help of the usual intensifying screens.

Since X-ray pictures are essentially shadowgraphs of the internal structure of objects, the definition is dependent upon the size of the X-ray source, which is the area of the target on which the electrons impinge. The ability of this focal-spot area to withstand the energy of the electron bombardment is a determining factor in controlling the intensity of the X rays which it is possible to produce. At any

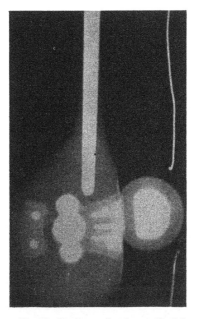

Fig. 3—Radiograph of a golf club striking a ball. Note the compressed core. Exposure time one millionth of a second.

given voltage, the quantity of X rays is proportional to the current through the X-ray tube so that the exposure time required for a picture will be inversely proportional to this current.

Recent developments have produced a new type of electron emitter called a field-emission arc, which enables currents 100,000 times larger than was possible with the hot-cathode type of X-ray tube. This has permitted X-ray pictures to be made in the extremely short time of 1 microsecond and very recent improvements have reduced this time to $1/10$ of a microsecond. This new type of equipment, shown in Fig. 1,

has been widely used for studying the action of very rapidly moving
enclosed objects and under conditions where it is not possible to use
the techniques of high-speed photography either because of the light
produced or of the inability to protect the photographic equipment

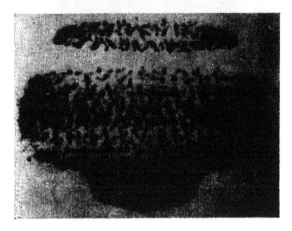

Fig. 4—Radiograph of a bursting model bomb showing
distribution pattern of fragments. Radiograph made by
the Ballistics Laboratory at the Aberdeen Proving Center
in 1 microsecond.

from flying fragments of bombs or bursting shells. With X rays, it is
feasible to use armor plate to shield the X-ray tube as well as the
photographic film. Figs. 2, 3, and 4 are typical applications of this
technique.

DEVELOPMENT OF THE HIGH-SPEED MOTION PICTURE EQUIPMENT

As in the case of photography, when it was found possible to take a single picture with a very short exposure time, the question immediately arose as to what happened just before or just after the picture was taken. In other words, what would X-ray motion pictures show? Under contract with the Navy, the development of X-ray motion picture equipment was initiated, intended primarily for the study of burning rocket propellants. Requirements to be met were 100 frames per second, each exposure to be no greater than 10 microseconds.

In order to insure the success of this project the development of two types of equipment was undertaken. The first followed more conven-

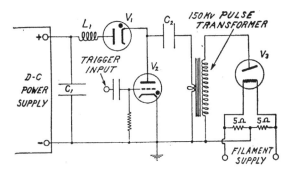

Fig. 5—Simplified choke-charging pulse-transformer circuit used to energize the X-ray tube. This circuit is similar to a radar-type line-modulator circuit with the pulse-forming network replaced by capacitor C_2.

tional circuit procedures in which the high voltage was obtained through a choke-doubling circuit which charged the final capacitor to 150 kilovolts. This was then discharged through the X-ray tube by means of a special triggering tube, the sequence being repeated 100 times a second. The second method, which proved to be the more successful, used the radar-pulsing principle in which an especially designed pulse transformer capable of stepping up the 20-kilovolt supply to 150 kilovolts was the heart of the system. This arrangement enabled the energy to be stored at relatively low voltages in capacitors and then to be discharged through the primary of the pulse transformer by means of a high-powered hydrogen thyratron similar to that used in radar techniques. Figs. 5 to 8 show the various components and assemblies of this equipment.

For certain purposes where the penetration demands are not great and where it is possible to work fairly close to the X-ray tube, it has been found feasible to utilize the technique, previously described, of photographing the image on the fluorescent screen. For many applications, including that of the rocket studies, there is not sufficient

Fig. 6—View of complete equipment; direct-recording camera on the left, oil-filled head containing the X-ray tube and pulse transformer in center, and power and control unit at the right.

X-ray intensity to employ this technique satisfactorily and it was necessary to develop a special X-ray camera for this purpose. Fig. 9 shows a photograph of the camera with the cover removed. An X-ray film $9^{1}/_{2}$ inches wide and 25 feet long with 25 feet of leader and trailer is wound on the upper spool. During the exposure, lasting 1 second, in which 100 pictures are taken, this film is wound on the lower spool, passing around the large rotating drum covered with an X-ray intensifying screen which greatly augments the intensity of the image produced. Since the exposure time for each picture is only 10 microseconds, no shutter is required and there is no blur due to the motion of the film. Fig. 10 shows a single frame taken of a simulated rocket using this technique.

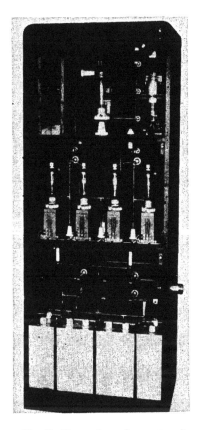

Fig. 7—Rear view of power unit. Power supply and storage capacitors below, rectifiers in center section, and triggering thyratron in the upper section.

A ¹/₄-horsepower motor is used to drive the film take-up spool and the film travels at 25 feet per second or roughly 17 miles per hour. At this speed precautions must be taken to maintain the proper tension so that the film remains in good contact with the screen and completes its run without damage.

The developed film strip, 25 feet long by $9^1/_2$ inches wide, contains 100 images.   These may be examined individually, in detail, and the

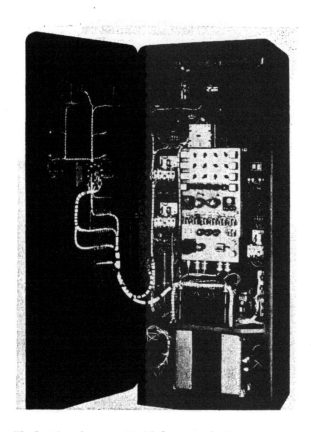

Fig. 8—View of power unit with door open showing sequencing and control components.

velocity of the observed phenomena measured with the help of a 1000-cycle-per-second time scale imprinted on the edge of the film.   If a visual impression of these phenomena is desired, the radiographs may be copied, frame by frame, on 35-mm or 16-mm motion picture film and projected in the usual manner at some suitable speed. Such a motion picture showing a Thermit reaction has been made which demonstrates the possibilities of this technique. Fig. 11 shows reproductions of several frames of this motion picture.

Fig. 9—Direct-recording camera removed from its housing.

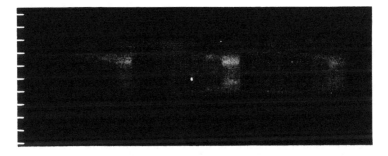

Fig. 10—Contact print from section of X-ray film showing one frame of a simulated rocket. Timing markers along the edge are recorded at a rate of 1000 per second.

CONCLUSIONS

In ordinary industrial motion picture work each new study requires a somewhat different technique, and so it is with X rays; such equipment as this can deal with only a very minute portion of the problems requiring X-ray motion pictures for their solution. It does, however, point the way and indicates that if the demand is great enough, X-ray motion pictures can be developed which will meet the requirements necessary for the solution of the industrial and ballistic problems which now cannot be completely solved by any other means.

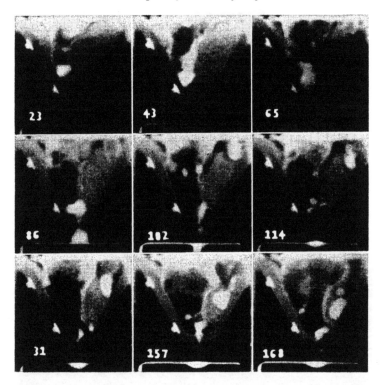

Fig. 11—Nine frames selected from a 200-frame series showing the reaction of Thermit (a mixture of iron oxide and aluminum) in a crucible. The molten iron is seen to form and run out the bottom of the crucible, below which it burns through a steel plate. The frame numbers indicate their position in the original series which was recorded at one hundred frames per second.

Some other fields in which the X-ray motion picture is expected to furnish important information include metal flow in arc welding, certain types of chemical reactions, motions of solid and liquid materials in agitators, closed conveying systems, gears and valve mechanisms, and the like.

High-Speed and Time-Lapse Photography in Industry and Research

By HENRY M. LESTER

PHOTOGRAPHER, EDITOR, PHOTO-LAB-INDEX, NEW YORK, N. Y.

Summary—High-speed and time-lapse photography are related to each other by the common factor of time mobility in photography. Similar problems exist in both types of photography, the most common of which concern lighting, timing, and interpretation. This paper discusses a number of cases in which these problems were attacked and solved by methods ordinarily considered unorthodox. A short discussion is included on the application of some of these specialized techniques to conventional cinematography.

IT APPEARS NECESSARY at the beginning of any discussion on industrial photography to distinguish between purely illustrative photography (that intended to *portray* certain processes or products), and photography as a *medium of research and investigation.* Frequently films or photographs made for investigational purposes have turned out to be fascinating subjects in themselves, wholly aside from their content of scientific data. Conversely, the occasion often arises for the use of high-speed photography as an interpretive medium, to make abstruse processes or principles intelligible to the layman. It is in these cases where the ingenuity of the industrial photographer is taxed to the utmost, and the present paper concerns itself mainly with these specialized problems.

Since we are concerned more with some of the unusual aspects of photography than with routine methods, it is well to point out here not only that unorthodox methods must frequently be used, but that these unorthodoxies often lead to improved methods of accomplishing certain photographic results in fields far removed from the original effort. In our work, however, unorthodoxy does not imply the use of highly specialized equipment in the main, though we have had occasion to devise certain units which are not generally available. More often, it involves the application of conventional apparatus in a somewhat unconventional way.

The scope of the industrial and research photographer's interests cannot be confined to any one type of work, high-speed included; even less to any one range of speeds. The fundamental utility of the cine camera in industry and research lies in the independence of camera and projector speeds. The ability to vary the rate of operation of one while maintaining constant the speed of the other gives a mobility to time, and affords the camera a function much greater than that of a mere recording medium. High-speed photography is one facet of this function; time-lapse photography is another. In a way, they may be thought of as opposite ends of a single spectrum.

When the camera operates at the same speed as the projector, the result is nothing more than a record—a mechanical notebook, so to speak. This is valuable, of course; a given action may be studied again and again, until it has delivered its full quota of information.

Obviously, there exists the possibility of varying the speed of camera or projector to effect an extension or telescoping of the time element in the film. The possibility of projector-speed variation is present but highly limited—speeds above normal by wear and tear on the film itself—speeds less than normal by discontinuity and flicker in the projected image.

But with the projector considered as the constant-speed reference, there is no necessity to limit the speed range in which the camera may be operated. Films taken at rates *above* normal and projected at normal speed are necessarily seen as slowed down; we prefer the term "time-extended" films. This would appear to be self-evident, yet much confusion exists among laymen when high-speed photography is discussed. Frequently we have been called in to discuss a proposed high-speed study, only to find that what was really desired was a speeded-up film of some very slow action; our term for this is "time-telescoped" film.

This latter form of time translation is a valuable research tool; it must be realized that many processes, particularly in chemistry and biology occur too slowly for study at their natural rate. For this reason, as industrial and scientific photographers, the scope of our work includes normal-speed, high-speed, and time-lapse photography. The value of this wide-range interest lies, as will be seen, in the frequent adaptation of techniques outside their usual fields, often a richly rewarding experience.

This may be summarized, then, by the simple statement of purpose: we are concerned with the mobility of time as it can be effected by the camera. Our problems involve means for mechanically translating the actual speed and duration of motion in time into terms that are within the range of our limited powers of visual and mental perception.

PROBLEMS OF HIGH-SPEED PHOTOGRAPHY

Strictly speaking, high-speed photography may be defined as any motion picture photographed at rates higher than that at which it will be viewed later. In our case, the term signifies a definite range of taking speeds. The common "slow-motion" camera operates at rates 4 to 8 times normal, that is from 64 to 128 frames per second. At the other extreme, there are image-dissecting cameras with speeds ranging around the rate of 1,000,000 pictures per second.

Our work mainly lies in a middle ground between the two extremes. Industrially, at least, speeds above 300 frames per second yield the most information, and we have found little to be gained by speeds above 3000 pictures per second. Most of our work has been done in even a more narrow range, using the Eastman high-speed camera, Type III, with a usable speed range of 1000 to 3000 frames per second. These speeds correspond to time magnifications of 65 to 200 times. We have made no important modifications in the camera itself, which has been fully described in the literature.[1]

The major problem in the use of this camera is strictly of a photographic nature. Operating at its peak speed of 3000 frames per second, and having an effective shutter angle of 72 degrees, the camera yields an exposure on each picture of only $1/15,000$ second, which poses a serious lighting problem.

Familiar light sources, such as reflector-photoflood (RFL-2) and reflector-spot lamps (RSP-2), are conventionally used with this camera. Their effective area coverage is measured in square inches, with even the fastest 16-mm film and large-aperture lenses. Industrial problems involving the photography of larger fields require kilowatts of light, frequently demanding special wiring and auxiliary transformers. Even when such facilities were available, the resulting films were of dubious value.

The validity of such results is doubtful because great quantities of heat accompany the enormous power dissipation of these large lamp concentrations. With industrial subjects, we were never sure that observed performance on the screen was actually inherent in the subject; there was always the suspicion that some aspects of the action may have been caused by exposure to heat. Biological subjects were frequently withered or killed before they could be photographed, and even if they were not, a definite reaction to heat was always evident.

Our eventual solution of the light-versus-heat problem was the adoption of a high-output, short-duration illuminant, the familiar still photographer's flashlamp. We found that the standard General Electric focal-plane flashlamp 31 has a light output lasting nearly 70 milliseconds, with a nearly level plateau of almost 56 milliseconds. At camera speeds of 3000 frames per second, this is enough to expose nearly 4 feet of film. Then 17 such flashlamps, fired in succession with suitable overlap, will provide one full second of continuous light, substantially uniform in level.

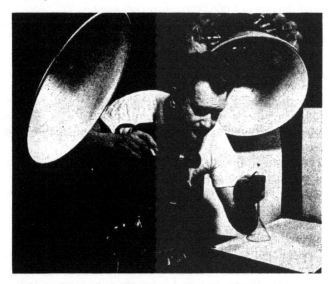

Fig. 1—The author photographing a drone fly with high-speed camera and continuous flash-lighting units. (Photo by David B. Eisendrath, Jr.)

A pair of continuous flash-lighting units (described elsewhere[2]) produce about 3,000,000 lumens for a period of one second. This is several times the brilliance of normal July sunlight. When the built-in timer on the camera is set to start the flashing after 35 to 40 feet of film have run through, the lamps in both units will continue flashing to expose the remaining 60 or 65 feet of the film as it attains the speed of 3000 frames per second.

As for exposure level, two units, each 24 inches from the subject, cover an area more than 2 feet square, with sufficient light to expose fast black-and-white film at $f/11$ to $f/16$, with the camera at peak speed of 3000 frames per second. Color film can be exposed quite adequately under the same conditions at $f/2.7$ to $f/4.5$. The heat output is trifling; what heat is radiated in the brief flashing time is blown away by the whirling lamps. Equally important is the factor of power consumption; the 2 units require only a 6-volt dry-cell battery for the flashing circuit, and less than 100 watts of 115-volt alternating current for the operation of both motors.

Even when only a small field is to be photographed, as in the drone-fly photograph shown in Fig. 1, the tremendous light output of these units is of inestimable value. The live subject was little affected by the insignificant heat output of the units, and the depth of field and sharpness obtainable with long-focus lenses stopped down to $f/12.5$

Fig. 2—Frame enlargement of drone-fly film made with setup shown in Fig. 1. Note unusual depth of field and definition possible when adequate illumination permits stopping down to $f/12.5$.

provided a new experience in the photographic quality and clarity of high-speed films. Fig. 2 is an actual enlargement of a frame of film made in this way—note the unusual sharpness and detail, particularly of the hairs covering the fly's body. This was photographed at 3000 pictures per second, on 16-mm film.

The value of such light sources for strictly scientific work cannot be overestimated; they are, however, equally important in another way. We have been able to work with as many as four such units; with the ample reserve of light available we were no longer limited to concentrating their beams frontally for the mere sake of adequate exposure. Lighting was arranged as for any normal photograph, for optimum modeling and photographic quality. The result may be said to have introduced a new realization of clarity into high-speed photography. Fig. 3 shows one such setup in which three units are visible, the fourth being behind the press for back light.

INTERPRETATIONAL PROBLEMS

Making a high-speed photographic study of a subject is only one part of the problem. The finished film must be interpreted by the

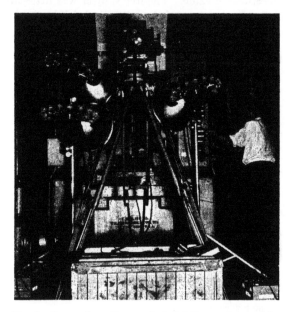

Fig. 3—Group of continuous flash-lighting units used in photographing impact extrusion process.

user, usually by projection, and frequently as well by study of single-frame enlargements, or frame-by-frame examination of the film itself.

Engineering studies are usually turned over to the client's staff for whatever use they may wish to make of them; our aid in this case is limited to providing a time base or reference where it may be required. This may take one of two forms: either a millisecond clock is included in the scene being photographed, or an argon timer built into the camera provides visible markers at intervals of $1/_{120}$ second along the edge of the film.

Films designed for nontechnical audiences pose a more difficult case. To an engineer, a slowed-down film of a machine is easily distinguishable from a normal film of the same machine running slowly. The difference, generally, lies in the presence of certain second-order effects, vibration, for example. Frequently the study of these second-order effects is the very objective of the high-speed investigation.

The layman, on the other hand, has little conception of such subtle differences, and his imagination boggs down at millisecond clocks. It is necessary in films aimed at the lay public to provide a frame of reference in more familiar terms. We have found it possible to accomplish this by careful choice of comparison objects, usually taken from nature. Thus, in a recent high-speed film made for a major industry, one in which we investigated the performance of the 750-ton high-speed power press already shown, we went far afield for comparison. Our reference object in this case was the tiny hummingbird whose wings beat 70 times per second.

Dramatic as the comparison was, it alone was not completely self-explanatory. For complete comprehension (and for those who had never seen either the hummingbird or the high-speed power press) we included several feet of normal speed (16 frames per second) photography of each. Thus, both the subject and the reference object were presented, each at its normal operating speed, and slowed down 200 times by the high-speed camera. Only in this way could we be sure that the audience would understand thoroughly not only the demonstrated process but also the method by which it was presented.

TIME COMPRESSION

While there may seem to be little apparent relation between high-speed photography and time-lapse studies, we have found it important to couple both techniques; first, because they are the logical extensions of the principle of *time mobility* and second, because some techniques applied in one field stem directly from the other.

For if rapid action may be slowed down—extended in time perhaps is a better concept—for more convenient study by the simple expedient of photographing it at rates higher than normal, then obviously the reverse procedure is equally relevant. Many industrial processes are extremely slow; either their motion is practically invisible, or a cycle extends over so long a period as to make study tedious. The same may be said for chemical processes and many natural phenomena.

For purposes of definition, rather than limitation, we may characterize time-lapse photography as covering anything photographed at rates lower than projection speed. In our work, this may mean at any rate from 8 exposures per second to a single exposure in 24 hours, though this by no means exhausts the possible range. Again, there is an optimum covering the majority of cases; our rule of thumb is based on the 100-foot roll of 16-mm film, having a projection time of slightly less than 4 minutes at 16 frames per second. Such a roll contains 4000 separate frames of film, and generally it is found convenient to divide the cycle to be studied into 4000 parts and use one such part as our time interval. This, however, is not a rigid rule; for more rapid tempo, particularly when the film is intended for lay audiences, we frequently use shorter rolls or units, perhaps 1000 or even 500 frames for a complete cycle.

Illumination, so far as quantity is concerned, is not a problem in this field; heat, however, always is. A cool light source is preferred, and we have found that the electronic condenser discharge flashtube, or so-called "strobe-light," meets every requirement of an ideal illuminant for time-lapse photography. Originally intended as a light source for high-speed photography with the shutterless type of camera, and later monopolized as a dependable light source for still photography, its remarkable uniformity, low heat radiation, and high efficiency recommend it for use as a light source for intermittent motion picture studies.[3]

The camera used for this work is a conventional Ciné-Kodak-Special, operated by a simple intermittent motor and timed by the Kodak interval timer. The only auxiliary equipment required is a simple commutator which triggers the flashtube at the open-point of the shutter cycle. The commutator is connected to the single-frame-per-turn shaft of the camera.

Photographically this is an elegant solution for a number of reasons. First, the light is not only heatless, but remarkably uniform from one exposure to the next. Second, the exceedingly short flash duration eliminates the possibility of flicker or blur caused by uneven camera speed or backlash in the shutter gears. Third, the exposure required is amazingly predictable; a simple mathematical formula[4] suffices, when the constants of the power pack are known, to determine the required exposure with a high degree of accuracy.

The problem of interpretation exists in the case of compressed-time films also. Unlike high-speed films, however, time-lapse photography is seldom studied frame by frame. The problem in this type of work is not to slow down motion so the *subject* can be studied, but to compress the time so that *movement* can be perceived. For this reason the time base, when necessary, should be included in the film; an ordinary clock or watch usually serves the purpose.

Again, for nontechnical audiences, a frame of reference is required, and again we go to nature for our comparison. In one instance the subject chosen as a comparison base for a chemical process was a talisman rose, photographed at 2-minute intervals for 5 days and 5 nights. Protected from light and air currents, the rose developed undisturbed

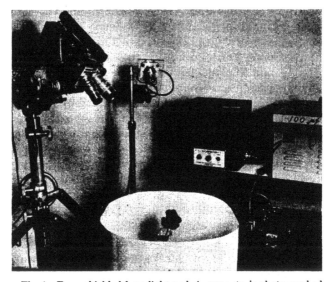

Fig. 4—Rose, shielded from light and air currents, is photographed at 2-minute intervals for 120 hours by light of bare FT-214 flashtube.

by any external influences (Fig. 4). The resulting film showed in about 3 minutes on the screen a smooth, steady progression from bud to bloom and finally a withered specter, unaccompanied by the usual gyrations of a flower subjected to alternate periods of daylight and darkness.

Filming in this case was on daylight-type Kodachrome film, which is spectrally well matched to the light of the electronic flashtube. Though this film is extremely sensitive to variations in illumination level, no visible flicker resulted from the use of an intermittent light source.

CINEMATIC BY-PRODUCTS

As already mentioned, the unorthodox use of conventional equipment, particularly illuminants, in high-speed photography, has had some interesting consequences in relation to normal cinematography. The possibility of using the electronic flashlight as a source for normal speed cinematography has been investigated, on a theoretical basis at least.[5] We have so far had several occasions to use it in this connection, with results that can be considered interesting and even promising. In general, our experience indicates that in dealing with live subjects intermittent flashlighting has a detrimental psychological effect upon organisms of a higher order. Subjects such as fish and insects seem to be affected little. People do not like it!

The energy-storage principle used in the electronic flash unit gives a deceptive picture of actual power requirements in cases other than single, widely spaced exposures. Power-pack sizes for repetitive flashing and large light outputs rapidly reach prohibitive dimensions. Our experiments have therefore centered on the use of this illuminant for small-field work, particularly in chemistry and biology, where its brilliance and low temperature permit its use close to the subject without discomfort or danger. We have also investigated its use as an illuminant for photomicrography, both by reflected and transmitted light. All these uses have been described at greater length in a previous paper.[3]

ACKNOWLEDGMENTS

To Harold Edgerton and Charles Wyckoff for design and construction of much of the equipment used in the electronic flash sequences; to John S. Carroll, Dorothy S. Gelatt, and Robert D. Olson for assistance in the preparation of the film "On Time and Light" and the gathering of material for this paper.

REFERENCES

(1) J. L. Boon, "The Eastman high-speed camera, Type III," *J. Soc. Mot. Pict. Eng.*, vol. 43, pp. 321–327; November, 1944.

(2) Henry M. Lester, "Continuous flash lighting—An improved high-intensity light source for high-speed motion picture photography," *J. Soc. Mot. Pict. Eng.*, vol. 45, pp. 358–370; November, 1945.

(3) Henry M. Lester, "Electronic flashtube illumination for specialized motion picture photography," *J. Soc. Mot. Pict. Eng.*, vol. 50, pp. 208–233; March, 1948.

(4) H. E. Edgerton, "Photographic use of electronic discharge flashtubes," *J. Opt. Soc. Amer.*, vol. 36, p. 390; July, 1946.

(5) F. E. Carlson, "Flashtubes—A potential illuminant for motion picture photography," *J. Soc. Mot. Pict. Eng.*, vol. 48, pp. 395–407; May, 1947.

Use of High-Speed Photography in the Air Forces

By E. A. ANDRES, SR.

WRIGHT FIELD, DAYTON, OHIO

Summary—As the conquest of flight continues, the use of high-speed photography by the United States Air Force is being pressed to the very limits of its uses and applications. By far the greatest and most varied application is made in the research and development center located at Wright-Patterson Air Force Base, Dayton, Ohio, and can be divided into two general categories; namely, high-speed motion pictures and high-speed still pictures.

THE ADVANCE OF HIGH-SPEED photography has been nothing less than phenomenal in view of the fact that but a few short years ago just one camera was the basis of all this type of photography. This camera, a 35-mm instrument, was capable of photographing at the rate of 240 frames per second, was tripod-mounted and hand-cranked. Today this one camera has been augmented by many additional high-speed or ultrahigh-speed cameras. From such a humble beginning, the increased use of this phase of photography has been rapid and today is recognized as a very necessary medium in connection with the research and development program. Interesting too, has been the acceptance by engineers of this type of instrument to assist them in their engineering problems to give the answer where no other medium would suffice.

Strange as it may seem, about ten years ago the use of high-speed photography in connection with research and development was a greater curiosity than a practical tool. However, the engineer quickly realized the value that this particular method of photography could have if properly applied. Certain changes in the procedure were necessary before the required answers were forthcoming. Its early application was mainly one of a visual study of the films as they were projected and the action slowed down. During the past decade high-speed photography has gathered components and accessories in its forward progress. From a very generous estimate on the part of the cameraman as to how fast he may have been cranking his camera or how accurately he may have read the camera tachometer, the methods

81

have advanced to the point of millisecond accuracies. Today high-speed photography is inseparable from timing and analysis; consequently, every high-speed camera which is utilized by the Air Force is equipped to put a time record on the film simultaneously with photographing or recording some specific phenomenon or action. The high-speed camera is no longer considered in the light of a camera; instead it is used as a photographic recording instrument to obtain a record which can later be reduced to the form of graphs or charts and more often than not the film record obtained by these instruments never is projected or sees print stage except for preservation and archival purposes.

A brief word regarding the principal methods by which time data are recorded on high-speed coverage is therefore in order.

The earliest methods of correlating time lapse with high-speed photography included the placing of a clock or disk driven by a synchronous motor in the field of the camera and simultaneously photographing the subject and the timing device. This method seriously limited the type of subject matter to that in which the clock and subject could be favorably disposed and illuminated.

This led to the logical assumption that the clock should be incorcorporated in the camera and its image projected to the film by an auxiliary optical and lighting system. The Eastman-ERPI camera, utilizing a synchronous clock driven by a portable power supply consisting of a 200-cycle tuning fork and amplifier, exemplifies one of the early practical implementations of this method. The definite advantages in facilitating analysis through direct reading of the clocks which are divided to $1/500$ of a second and readily permit interpolation to $1/1000$ of a second, still find numerous applications.

While later developments in high-speed cameras by Eastman and Bell Telephone Laboratories made available increased frame speeds, the timing problem lagged behind these developments and these cameras were locally fitted with various improvised devices for placing a time base on the film. The earliest of these devices consisted of a 60-cycle power supply or tuning fork of known frequency triggering, through a suitable electrical circuit, a spark impinging on the film. This method, while serving some purposes adequately, left much to be desired because of erratic operation. The generally poorly defined timing mark left on the film and the need for smaller and more accurate spacing and delineation of time increments, made necessary by the increasing frame speeds, proved inadequate.

The later development employed the application of gaseous-discharge lamps such as the argon lamp. The use of small argon lamps as a film-marking medium was retained as they afforded a compact concentrated light source of high actinic value capable of responding with a minimum of lag to the electrical impulses and could be readily installed in the limited spaces available in the cameras. The major problem remaining in obtaining clearer delineation of timing marks was to modify the time-voltage curve of the pulses in order that the

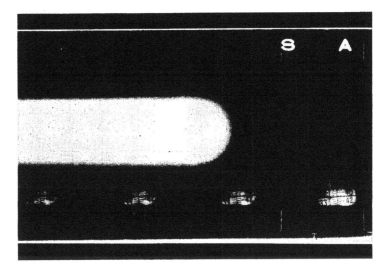

Fig. 1—Sample of synchronization mark as used in the 3-kilometer speed trials, showing clarity of synchronization point with respect to clock readings photographed on each frame. The delay or time loss from the beginning of signal to the beginning of recorded light on the film is negligible.

argon lamps would achieve full intensity or extinguish as soon as possible after each application and removal of voltage. The characteristic sine-wave time-voltage curve of alternating-current sources left images on the film which varied gradually from zero density to full density and increased the difficulty of determining the beginning and end of the timing marks. This characteristic also became increasingly pronounced and objectionable with increased frame speeds.

Taking advantage of principles developed in radar practice of World War II, Special Photographic Services Branch technicians developed a pulse amplifier to convert sine-wave forms produced by various types of frequency generators into sharp square-wave pulses. These pulses ignite and extinguish the argon timing lights in a manner which produces clearly defined timing marks. A further advantage of such amplifier has been exploited through utilizing accurate control of the duration of the pulse as an additional time scale independent of the fundamental frequency of the timing source.

One of the accepted methods of obtaining a time base for many projects, which has a wide range of application, is the 60-cycle line frequency where the frequency control is sufficiently accurate and adequate for the job. This has ground application only. Also limited to ground applications but highly accurate is the use of tuning forks of known frequency and accuracy to pulse the lamp at known intervals, thus putting a very accurate time trace on the speeding film. Obviously these two methods are entirely unsuitable for airborne use. Widely used in aircraft is the crystal-controlled chronometer of high accuracy which can be controlled with a minimum of effect from temperature and vibration. This method can be relied upon to give good performance with a high degree of accuracy. Another method fairly accurate and dependable, which can be airborne and meets space and power-supply requirements, is the electronic-oscillator circuit of the more stable designs.

Born of necessity, high-speed photographic records have proved essential and indispensable to aeronautical research. Without this type of photography endless repetition of experiments and tests, with resultant delays, would be necessary to amass the data now disclosed through the medium of high-speed film analysis.

To enumerate these applications would necessitate repeating a large portion of the research and development projects. Obviously, many of these cannot be revealed for security reasons but a few can be discussed.

HUMAN PICKUP BY AIRCRAFT IN FLIGHT

Shortly after hostilities started in World War II, it was realized that the flight of aircraft over wooded and isolated areas eventually would result in the necessity to rescue personnel who had parachuted to safety. The rescue problem in such cases was complicated by terrain unsuitable for landing a plane. The solution was to provide a means for human pickup by an aircraft in flight.

Before the apparatus developed could be proved safe for general use, it was necessary to provide a means of measuring the instantaneous forces exerted due to accelerations as functions of time and horizontal displacement when the rescue device was picked up by the plane. Here, high-speed photography and subsequent analysis of the film provided the required data.

Accurate marks were placed immediately behind the path of the rescue device and served as reference marks for horizontal displacement. Built into an Eastman high-speed camera, a timing lamp impressed known intervals of time on the edge of the film traveling at the rate of 1000 frames per second. This timing record, together with a detailed picture of the action as recorded on the film, provided a means of obtaining accurate time-displacement data. From these data accelerations and velocities could be determined.

PULSE RATE AND FLAME PROPAGATION VELOCITY DETERMINATIONS

Since operation efficiency and thrust valves for pulse jet engines are functions of pulse rate and flame propagation velocity, the determination of these quantities with precision is a vital problem of the Air Force. Again, high-speed photography proved equal to the task. A JB-2 engine, with holes 1/4 inch in diameter and five inches apart drilled in the tailpipe section along its horizontal axis, was placed on a test stand. The flame lit up successive holes as it passed through the tailpipe cone and was recorded by a Fastax camera simultaneous with the time indications. Velocities and pulse rates could then be determined.

These are commonplace applications of the use of high-speed cameras which are daily occurrences. By far the most publicized and spectacular applications of high-speed photography by this group have been in connection with the official speed trials which have been run during the past two and one-half years.

Early in 1946 the Air Force decided to make an assault on the world speed record which was held at that time by the British. Everything was progressing according to schedule except the timing. These trials must be run according to the rules and regulations set by the Fédération Aeronautique Internationale. Investigation revealed that neither a method nor suitable equipment existed to time these runs accurately according to the high standards of accuracy officially required. The problem was referred to technicians at Wright Field for solving.

Timing was required, but the British had discovered that verification of the timing equipment was also necessary. Photography was the answer, and high-speed photography the solution.

In the problem of establishing a world speed record it was necessary to produce positive data with the required accuracy as set by the Fédération Aeronautique Internationale. In general these rules required timing accuracies to at least $1/500$ of a second and comparable accuracies in location of the airplane at points of entry and exit of an accurately surveyed course.

The accurate timing was accomplished by the utilization of laboratory-type tuning forks previously calibrated by the National Bureau of Standards at Washington, D. C. These calibrations were made under temperature-control conditions to enable temperature corrections in the final data. The tuning forks used were 100-cycle-per-second accurate to $1/1000$ of a second. The tuning-fork oscillations were recorded on the film of the high-speed cameras simultaneously with photographed position of the contestant aircraft with respect to the surveyed course. A secondary timing system consisting of a synchronous clock was also impressed on the film photographing the trial plane and was used to facilitate faster analysis. These clocks were operated by a 200-cycle tuning fork with an appropriate electronic system. The time indicated by the synchronous clock was calibrated from the 100-cycle indications on each roll of film. By this method the clock indications on the film could be read in a much shorter time and still maintain the accuracy required by rules. The elapsed time for each pass could be determined by the primary timing or the 100-cycle-per-second indication but would require additional analysis time.

The method of establishing the world's record by individually operating high-speed cameras was done in the following manner. As the plane approached the entry point of the 3-kilometer course, the high-speed camera at that point was started in sufficient time to photograph the contestant aircraft as it crossed the wire locating the 3-kilometer entry point. Immediately following the entry of the course, the high-speed camera photographing the exit point was started so as to have both cameras in operation at the time the plane reached the approximate midpoint of the course. At this time a synchronizing point was established simultaneously on each photographic record by applying an electrical signal to form a pulse of light recorded on the film of each camera. After this operation had been completed, the camera at the exit point would photograph the contestant aircraft as it crossed the wire at the end of the 3-kilometer course, thus completing the operations for recording a single pass.

In order to find the elapsed time for that particular pass the time indicated on the first camera as the plane entered the course was noted. The time of the beginning of synchronization point was then noted and the difference between these two readings represented elapsed time for approximately $1/2$ of the course. The remaining portion of elapsed time was recorded by the second camera which was established by noting the time which indicated the beginning of the

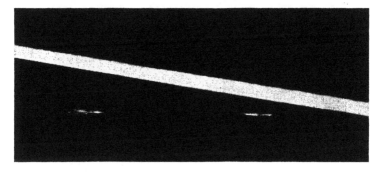

Fig. 2—A study was made of lightweight armor-plate penetration by various types of 50-caliber projectiles.

The photograph is a recording of a 50-caliber projectile trajectory just prior to deflection by armor plate. Two microflash units were triggered by the compression wave passing over two microphones and the resulting images were recorded on one plate.

The projectile was of the incendiary type. Traces between the two images are those of the tracer compound beginning to ignite.

same synchronization points on the second film and noting the time indicated upon exit of the course. The difference in these two times gave the remaining elapsed time of travel. The summation of the two elapsed times as measured by the first and second cameras was then added to give the total elapsed time for that particular pass. From this reading the speeds were calibrated to the accuracy requested. The identification of each pass with respect to camera location was made possible by slates in the field of the camera denoting pass number and station number.

In accordance with the rules, four consecutive passes must be made. The average speed of the four passes is by rule the official speed.

Keeping pace with and no less important is the application of high-speed still photography. Utilizing various gaseous-discharge tubes in combination with still cameras, remarkable results have been obtained. High-velocity projectiles have been stopped in flight with such clarity that the riflings are clearly discernible. (See Fig. 2.) Compression waves formed by fast-moving aircraft propellers are captured for analysis and study. Rupturing propellers, machine-gun malfunc-

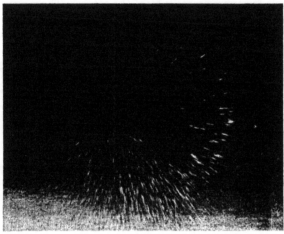

. Fig. 3—A microflash record showing disintegration of frangible ammunition striking aluminum plates similar to aircraft covering. The ammunition was made for training of aerial gunners. Gunnery practice was against standard aircraft with suitable protection. The photographic records were made for the investigation of disintegrating characteristics of training ammunition against various types of aluminum alloys.
The microflash unit was triggered by the compression wave from the projectile passing over a microphone.

tions, impact of projectiles (see Fig. 3), propeller icing, and a multitude of similar problems have been analyzed through this method.

While no sure guide exists for establishing procedures or methods for each individual problem, a careful record and file with appropriate descriptions are kept of each accomplishment. Regardless how routine or simple the process or attack may seem, many times a comparatively simple approach holds the answer to a most baffling problem. Time is too limited and qualified technicians too few to permit casting about for new methods when tried procedures have already been established and found sufficient.

This does not mean, however, that new methods are being by-passed or neglected. Thinking and planning along these lines must continue and be encouraged. Progress in aircraft and speeds of these aircraft and missiles have reached fantastic proportions. Today's camera equipment, with modifications, is being utilized to the very limits of its capabilities presenting many problems. In addition, a definite shortage of many standard and special types of photographic equipment exists.

While progress has forced the development and advancement of high-speed photography, high-speed photography has, in turn, implemented the rapid advancement of research and development programs. From the simplest application, such as timing a camera shutter, to the study of the atom, high-speed photography has played a tremendous role.

What the future holds in this respect is based to a great degree on supposition, but whatever the requirements, as in the past, high-speed photography will play an ever increasingly important function. Choose any field of endeavor and you have a field for the scientific photographer.

High-speed photography has reached its place in science and engineering. While not fully exploited, its uses and applications will continue to increase. Quality of equipment must keep in step with these advancements, a proved tool—it is necessary and indispensable.

High-Speed Photography in the Automotive Industry

By RICHARD O. PAINTER

General Motors Proving Ground, Milford, Michigan

Summary—Applications of high-speed photography at the General Motors Proving Ground since 1938 are discussed. Equipment presently in use, including cameras, lighting equipment, control equipment, and accessories are described and illustrated.

THE EXPERIENCE OF General Motors Proving Ground personnel with high-speed motion picture equipment dates back to 1938. Starting at that time with an Eastman Type II 16-mm camera, the Proving Ground has since added two more cameras and a large amount of associated equipment and now has the largest collection of high-speed photographic equipment within the General Motors Corporation. Centrally located with respect to many of the Corporation's larger Divisions, the Proving Ground is within short distances of many plants which use our facilities for high-speed photography.

The equipment necessarily is highly portable, since it must be used on the road and at many remote points on Proving Ground property, as well as being suitable for easy transportation by automobile to any Division of the Corporation. The three cameras being used are all of the so-called "rotating-prism" type. The original camera used by the Proving Ground and shown in Figs. 1 and 2 is provided with a motorclock for timing purposes, which is built into the camera base. This motorclock provides a timing image which is recorded on the film to the right of the picture area and within the projected frame area. A tuning-fork timing generator of extreme accuracy is used to supply the motorclock with a fixed frequency regardless of line-voltage and frequency fluctuations. The Western Electric 8-mm and 16-mm cameras which we acquired during the war are similar in appearance (Figs. 3 and 4). An argon-lamp flasher unit is used in these cameras to place timing marks along the edge of the film at a frequency of 120 marks per second when the unit is connected to a 60-cycle, 110-volt source. The lighting and camera-operating accessories built for this work are arranged to give portability and adaptability since this equipment must be used under a wide variety

90

of conditions. The equipment is often operated from portable motor-generator supplies and, for these cases and where limited current can be drawn from the alternating-current circuits available at the test location, it is a simple matter to divide the electrical load among several sources.

Fig. 5 shows a typical camera setup using all the camera-operating accessories. The Western Electric 16-mm camera is shown arranged to take a top view of the ejection and feeding on a 30-caliber carbine. Since a number of rolls of film were taken of this view, a camera voltage and timing-control box was used to make camera and light operation completely automatic upon pushing the remote-control-button

Fig. 1—Front view of Eastman Type II high-speed camera.

Fig. 2—Side view of Eastman Type II high-speed camera.

shown lying below the trigger of the gun. This control ties in the operation of the lights to the camera and shuts the power off after the time required for the film to pass through the camera has elapsed. This control box permits the camera operator to concentrate fully upon the subject and makes it a simple matter to take the pictures at the most desirable moment. With this device the operator can control the camera operation from any location and repeat shots for comparison purposes or other reasons take only as long as is necessary to remove the exposed film and insert a fresh roll. Fig. 6 shows a schematic diagram of this control box. The Variac permits selection of the desired camera speed by adjusting it to the corresponding voltage as read on the voltmeter. The time required for the film to

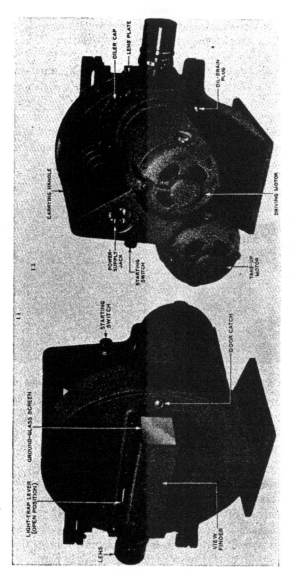

Fig. 3—Western Electric high-speed camera showing left and right side views.

pass through the camera at that voltage is set on the timer and the operation is then fully controlled and initiated by the closing of the timing-cycle contact.

Fig. 7 shows a camera set up to photograph the action of a milling machine cutting gear teeth. The box in the lower center of the picture supplies power to the lighting units on the stand above. These transformer boxes may be used in multiple to supply any number of lighting units. The number of lights used depends, of course, upon the area to be covered, the camera speed, and the type of film to be ex-

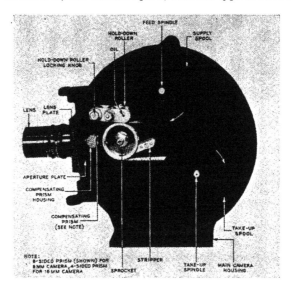

Fig. 4—Western Electric high-speed camera with door removed.

posed. These units are standard General Electric Type 150 PAR/SP projector spotlights. Having an internally silvered reflector, these units concentrate all the light into a small area when used one foot from the subject. The power-supply box contains an autotransformer which allows the lights to be operated at normal line voltage for setting up and focusing, and a high-low switch on the box permits the application of 200 to 220 volts to the lights when the pictures are taken. At this higher voltage the lights operate at extreme intensities, reaching illuminations (for three lights) of fifteen to twenty times that of sunlight under the most favorable conditions. A variety of lenses and lens extensions are used with the cameras to cover a wide range of field sizes and to allow some latitude as to the camera location.

Prior to the war, the high-speed photographic equipment was only occasionally used since there were not many Division engineers acquainted with the time and expense savings which could be achieved through the use of these techniques. With the war, the work increased so that it was necessary to acquire two additional cameras and

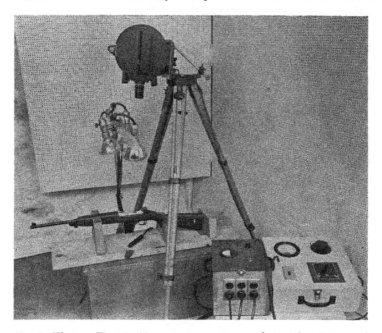

Fig. 5—Western Electric 16-mm camera setup used to take pictures of ejection on 30-caliber carbine.

for the entire period of the war the equipment was in almost continuous use with a great increase in the application of these methods to the problems of the Divisions on many different types of ordnance material and the production machinery involved in its manufacture. Although the amount of work done with the cameras dropped off considerably for about two years after the end of the war, the applications of high-speed photography have been steadily increasing since the first part of 1948 as more and more engineers and production men find that the methods are equally helpful in the solution of problems involving peacetime products and production processes. A part of this increased use of these methods is due to a report dealing with high-speed photographic equipment and its applications which was published and distributed to the Divisions of the Corporation a little over a year ago.

Unfortunately, many examples of the use of this equipment are not available for showing at this meeting because of the type of material they are concerned with and others are in the hands of the Divisions and cannot be located readily. Most of the work for which the cameras are used on the Proving Ground involves complete vehicles, with occasional tests on engines operating on engine dynamometers. Considerable work has been done on various body tests such as door-latch mechanisms and safety-glass impact tests. The action of springs, suspension parts and shock absorbers, and hydraulic brake lines has been studied through pictures taken on chassis dynamometers with cleats attached to the dynamometer rolls, and the roll speed adjusted to excite the suspension system to its highest amplitude.

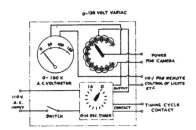

Fig. 6—Schematic of camera voltage and timing-control box.

The deflections of a tire striking an obstacle have been determined by taking 3000 pictures per second with a car speed of 40 miles per hour. Pictures taken of a tire striking a sharp-edged obstacle reveal very severe deflections.

The action of automobile engine valves has been frequently studied. At higher engine speeds the surging of the valve spring becomes quite pronounced.

A study of the flow of die-cast metal from the gate of experimental type dies was recently made. A film of this operation shows the turbulence with which the hot metal emerges from the die gate and was taken to assist in the selection of a design giving the most nearly laminar flow from the gate on the assumption that minimum cavitation exists when the flow is the smoothest.

The action of metal-cutting tools is of great interest to the production engineer. In a film showing a rather heavy cut being taken from cold-rolled steel on a milling machine using normal cutting speed and a type of tool frequently used for roughing operations, it is interesting to note the deflection of the tool and the apparent hesitation of the table as the material first strikes the tool. Higher table speeds and certain type tools lead to vibration or chatter of the tool which produces a much rougher cut than that shown here. For every type of material there is a cutting speed and tool combination which will yield the most satsifactory operation and these conditions can be very nicely compared by the use of high-speed photography.

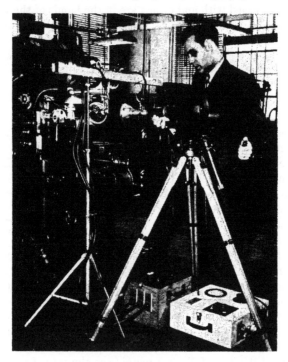

Fig. 7—Arrangement of high-speed camera equipment
for photographing a machining operation.

High-speed photography has been found to be an extremely useful tool-providing great savings in time and expense in the development of new mechanisms and in the investigation of operating faults which may exist in mechanisms and processes already in use. There is every reason to believe that use of these methods will increase steadily as engineering and production personnel become better acquainted with the advantages of high-speed photography.

Applications of High-Speed Photography

By MAX BEARD

Naval Ordnance Laboratory, White Oak, Maryland

Summary—The applications of high-speed photography at the United States Naval Ordnance Laboratory employing oscillography, streak cameras, high-speed spark and discharge illumination, and rotating-prism cameras are described.

THE USE OF PHOTOGRAPHY as a tool in research has increased with the development of new procedures for time magnification through high-speed photography. When these new procedures are introduced, they are adapted to the problem at hand, or the problem is so acute that a photographic method must be devised to study certain phases of an action.

Some of the high-speed photographic procedures for the analysis of the various problems encountered in this Laboratory's research and development activities are discussed under these general headings: 1. oscillography; 2. streak camera; 3. intermittent illumination; and 4. rotating-prism cameras.

1. Oscillography

Cathode-ray-oscillograph photography is a graphic method of obtaining permanent records in the analysis of amplitudes and frequencies of electrical and mechanical phenomena. High-speed oscillography permits the more accurate analysis of phenomena involving high frequencies.

The Naval Ordnance Laboratory has developed and built a high-speed oscillograph capable of recording six traces simultaneously on 35-mm film. · It is not only capable of recording multiple channels, but is capable of recording high-speed transients where the function of time is of considerable importance.

Basically, the instrument consists of eight RCA, Type 2BP-11, cathode-ray tubes mounted side by side, and six individual direct-current amplifiers with high gain, good high-frequency response, and excellent stability. Two of the tubes provide calibrated timing pulses.

The images of the eight tubes are transmitted to one wide-angle Baltar $f/2.3$, 25-mm lens and focused onto moving 35-mm film to form a continuous displacement-time record of the six traces and timing lines.   Up to 100 feet of film may be used for speeds from 4 inches per second to as high as 13 feet per second.   When more rapid film travel is required and the duration of the time measurement is short, 10-inch strips of film may be used around the drive drum, producing up to 38 feet per second film speed.

Fig. 1—Continuous print, enlarged 10 times from 35-mm film exposed in 6-trace oscillograph.

The Laboratory has accurately recorded signals up to 100,000 cycles per second, which is a film displacement of 0.005 inch from peak to peak at this highest film speed.   Linagraph Pan, 35-mm, recording film is used when high-speed transients are involved.

The analysis of the recording is made from a continuous photographic print enlarged ten times.   An extremely accurate 35-mm continuous printer with five or ten times magnification was designed and built for the Laboratory by Jacques Bolsey.   This instrument prints a centimeter grid on the paper, accurate to 0.001 inch, from which measurements are made and which eliminates errors due to paper shrinkage and stretching. Paper records can be made up to 200 feet in length on 30-centimeter stock. A typical record is shown in Fig. 1.

This 6-trace oscillograph has been used in the investigation of Navy guns where the function of time is of greatest importance. Information recorded is the breaking time of the primer bridge wire, the instant of appearance of flame from the primer port holes, the beginning of recoil of the gun, and pressure-time records at three positions along the shell case. The time correlation of all these phenomena, measured from the instant the firing switch closes, is required.

Other uses of this instrument have been made in the study of vibrations, acceleration, temperature and light measurements, and various types of timing measurements.

2. STREAK CAMERA

In the study of detonation phenomena, a streak camera, involving the use of stationary film, rotating mirror, defining slit, accurate synchronization, and two-lens systems, was built by the Naval Ordnance Laboratory for the direct photography of the shock waves of various sizes of self-luminous charges. This is housed in a bomb shelter with an open tunnel-way to an outside location for the explosive charge. A single photograph is obtained of phenomena on a given plane as a function of time of each detonation from which shock-wave velocity, direction, shape, duration, and intensity may be calculated. The top writing speed of the camera is 1.6 mm per microsecond.

The charge is focused onto a vertical slit, variable from $1/1000$ inch to $1/2$ inch wide and 1 inch high, by interchangeable lenses of various focal lengths. The slit image is transmitted through a 7-inch condenser lens to the rotating mirror and reflected onto a 16-inch strip of 35-mm film accurately placed on the interior surface of a portion of a cylindrical film drum. This drum is positioned to maintain a constant focal-plane distance regardless of the angle subtended by the rotating mirror. The image position on the film is controlled photoelectrically by infrared signals.

The method of synchronization of image position, firing, safety factors, and mechanical shutter involved considerable development and improvement by the scientists of the Laboratory over other models of streak cameras. Because of the fact that the instrument would be in continuous use, every factor had to be carefully considered.

With the explosive charge suspended in front of the camera so that the plane or direction to be studied is in line with the slit, the motor control for the rotating mirror is started. The camera operator presses a firing button when the rotating mirror reaches full speed.

By means of an electronic and relay system, the rest of the operation becomes automatic with the following sequence of events:

1. The ground connection to the explosive charge is broken and firing connections are made.

2. The mechanical shutter is opened, which sets up the signal in the photoelectric system for detonation.

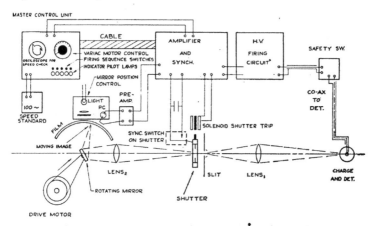

Fig. 2—Detonation camera showing optics and controls.

3. The signal from the photoelectric cell, controlled by a light reflected from the rotating mirror when in the proper photographic-image position, completes the firing operation.

The schematic diagram of this unit is shown in Fig. 2 and a typical example in Fig. 3.

Unusual detonation phenomena, hitherto unknown, have been revealed, so that the scientists working with this instrument are continuously devising new procedures for new studies of self-luminous charges.

Thus far, procedures of high-speed photography as instruments for analysis, where the phenomena are recorded graphically, have been discussed. In contrast to these methods, with the advent of intense illumination of very short duration, actual photographs are taken recording a period of action that will permit the study of the behavior of objects traveling several times the speed of sound. Such studies are vital to the development of ordnance material for high velocities.

The Naval Ordnance Laboratory now has under construction two aerodynamic ranges and a supersonic wind tunnel for the study of air flight of high-speed missiles. The ballistics ranges employ shadowgraph from a point source of light, and the wind tunnel a schlieren method of photography.

The aerodynamic ranges will have 25 photographic stations, with each station equipped to photograph on 14- × 17-inch glass plates,

Still photograph of explosive charge with double-exposed photograph of slit view superimposed.

Streak photograph of detonation as seen through slit. Time increases to right. (Total time of detonation about 50 × 10⁻⁶ second.)

Fig. 3—Photograph of detonation charge, using a streak detonation camera. Picture to the left is with mirror in stationary position. Picture to the right is the shock wave during detonation, showing the progress of detonation with time (detonation has started at bottom of view).

the vertical and horizontal planes of the missile. At each station is an open-spark discharge that casts a shadow of the projectile and its resulting shock-wave pattern on the photographic plate. The diverging light traverses the range for the vertical plate and is reflected to the horizontal plate by means of a mirror. When the projectile approaches the cross-sectional area where these two diverging beams cross, the light is triggered photoelectrically with suitable time delay to center the image in this area. Open-spark illumination is used because of the simplicity in obtaining a fairly accurate point source of light and sharp cutoff in intensity.

The value of these ballistic ranges is for the study of flight characteristics complementary to wind-tunnel models and full-scale

missiles. The data recorded on the photographic plates will be
analyzed for angular displacement from the bore line, and change in
velocity as the projectile moves down range. Deceleration is determined by an accurate timing system from station to station. This
information is used for the study of stability, yaw, and drag over a
portion of the flight, and, also, for correlation of differences in wind-
tunnel models and free-flight data of full-scale projectiles.

The study of supersonic aerodynamics by the projection of a missile
through space is confined more to the characteristics of a projectile
nearing completion in design and which has comparatively stable

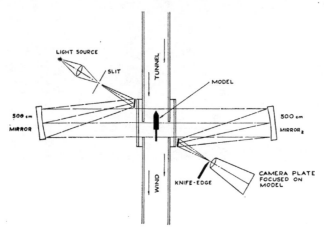

Fig. 4—Schematic diagram of optical arrangement for schlieren bench.
Supersonic wind-tunnel facilities.

flight. For velocities up to Mach number 5.18 (which is 5.18 times
the speed of sound) the Laboratory's supersonic wind tunnel is
used. This latter research facility is not limited by unstable flight
characteristics.

This tunnel was reconstructed from the famous Kochel supersonic
wind tunnel, used by the Germans for the development of the famous
V-2 rockets. Its basic principle of operation is the use of a large
sphere in which the air is reduced to a pressure of a few millimeters of
mercury by vacuum pumps. When test operations are ready, valves
are opened permitting the air to return through several tunnels back
to the sphere. Each tunnel is operated separately permitting several
tests to be performed simultaneously and independently of each other.

The air rushes by the stationary models at a Mach number which is determined by the design of the control nozzles. High-speed photographs are then taken showing the wave front of the models as if they were in free flight. The photographic procedure used is based on the principles of schlieren photography.

The general arrangement of this schlieren method is that of a light from a concentrated source focused onto a slit which then becomes a secondary source, as shown in Fig. 4. From the slit the light goes in a diverging beam to one of two 500-centimeter focal-length concave mirrors. Parallel rays from the first mirror pass through the work-

Fig. 5—40- × 40-centimeter tunnel No. 2 with schlieren bench in place.

ing area of the tunnel, where the model is placed, to the second concave mirror which reflects the light in a converging beam to a focus where a portion of the slit image is blocked out by a knife-edge. The remaining part of the slit image diverges to the lens system of the camera and onto the ground glass, which is focused on the model in the air stream of the tunnel. An exceedingly clear shock-wave pattern of the model is obtained by the use of this schlieren method.

The principle involved in the schlieren method depends upon the refraction of light due to the variation of optical densities in the vicinity of the model in the tunnel. Variations in the air density refract portions of the parallel beam farther into or away from the knife-edge, resulting in a darkening or brightening of the corresponding regions of the image at the focal plane of the camera. Dark regions in the resulting photograph correspond to deviations into the knife-edge, while bright regions in the photograph correspond to deviations away from the knife-edge. Shock waves usually have both a dark and a bright region running parallel to each other.

The final focusing lens may consist of single lenses or lens combinations which give the desired magnification in the final photograph.

The cameras, of Naval Ordnance Laboratory design, are of the reflex principle, with 5- × 7-inch plates and commercial 35-mm reflex cameras. The larger camera is wired to the lighting system for the illumination in visual observation and when the reflex mirror is displaced, the light is ready for photographing. These cameras are interchangeable and may be replaced by high-speed motion picture cameras using the continuous-light source.

The light source is a General Electric BH6 high-pressure mercury lamp rated at 900 watts. This lamp may be operated for continuous illumination or flashed intermittently. The flash duration is approximately 4 microseconds. Photographs up to $1/1000$ of a second are taken when the continuous-light source is used. The shorter exposure time is preferred for detailed study of small background disturbances, which, during longer exposures, give an integrated effect on the film. The schlieren setup is shown in Fig. 5.

An example of intermittent illumination, synchronized with high-speed motion picture cameras, is discussed under the next heading.

3. INTERMITTENT ILLUMINATION

In the study of underwater ordnance, high-speed motion pictures with intermittent flash illumination are used in the photographing of exact-scale models in glass-walled tanks. These models are fired at various speeds up to 980 feet per second into the water and are silhouetted by back lighting so that the cavitation and wave characteristics will form a definite line pattern outlining each phenomenon. A front modeling light may be used for emphasizing the shape of the model.

A paper, by Anderson and Whelan,[1] pertains to the camera and flash units used for these studies. The paper described the system used at that time, but the method has now been discarded and completely replaced by a more flexible and efficient system.

The light source for this system employs pulsed gaseous-discharge tubes and was designed to operate in synchronism with a high-speed motion picture camera. Repetition rates of as many as 3000 to 4000 intermittent flashes per second have been obtained. The average flash duration used is about 1 microsecond. Six flashlamps are mounted in one single-parabolic reflector, six feet in length and running the length of the back of the model tank. The control equipment, designed by the Naval Ordnance Laboratory, permits the intermittent flashing of each lamp independently or in any sequence or combination. This flexibility is highly desired where progressive sequence lighting is required to "follow" the action across the tank or where special lighting is needed for the elimination of unwanted shadows. A diffusion screen is used over the face of the reflector.

4. ROTATING-PRISM CAMERAS

High-speed 16-mm motion picture cameras of the revolving-prism type are used and are equipped with a new electromagnetic pickup, without any mechanical contact with moving parts of the camera, for absolute synchronization of the camera with the flashlamps. The complex synchronization and lighting system is so designed as to be used with other types of high-speed cameras and varieties of flashlamps or open-spark circuits.

The revolving-prism cameras were selected for use after comparisons were made with other types of high-speed motion picture cameras. The rotating prism satisfactorily eliminated blurring of the image which was noticeable when cameras not employing this system were tested. Image displacement for still objects, when there is no compensation for film motion, is 0.028 mm at 1 microsecond flash duration at a camera speed of 4000 frames per second on 16-mm film.

Other uses of cameras with rotating prisms are as all-purpose cameras at the Naval Ordnance Laboratory, when the magnification of time is essential. High-speed motion picture photography with these cameras has become so generally required that they are assigned to staff photographers and engineers as regular equipment. Some field tests require the use of several cameras on one assignment.

The general use of these cameras is quite similar to those applications in industry where impact, timing, or erratic operations, involving moving mechanical parts, are studied, and daylight illumination or portable floodlights are used. This type of usage, therefore, would be similar to those described by others in this symposium.

This paper has been presented to acquaint the members of the Society with a general idea of the various types of high-speed photography undertaken at the Naval Ordnance Laboratory and, therefore, the paper could not be confined to a detailed study of any one operation. The photographic methods included represent the efforts of numerous engineers and scientists from various divisions of this Laboratory, and the author is indebted to those people for the above information. Particular indebtedness is to G. E. Beyer and J. L. Jones, Mechanical Evaluation Division, Technical Evaluation Department on oscillography; S. J. Jacobs, Explosives Division, Research Department on the streak camera; members of the Mechanics Division, Research Department on aerodynamics; and members of the Hydrodynamics Subdivision, Research Department on underwater ordnance.

REFERENCE

(1) R. A. Anderson and W. T. Whelan, "High-speed motion pictures with synchronized multiflash lighting," J. Soc. Mot. Pict. Eng., vol. 50, pp. 199–208; March, 1948.

Control Unit for Operation of High-Speed Cameras

BY L. L. NEIDENBERG

INDUSTRIAL TIMER CORPORATION, NEWARK 5, NEW JERSEY

Summary—An automatic time control for the Fastax camera doubles the picture-taking speed of this camera, controls the film speed from low to high, automatically controls the camera in synchronism with an event to be photographed, allows for remote-control operation of the event and the camera, and prevents tearing of film as the camera is started. It is simple in operation, extremely accurate, and completely portable.

THE 16-MM Western Electric Fastax camera as originally developed by the Bell Telephone Laboratories could attain a maximum speed of 4000 frames per second. Later it became necessary to speed up the action of the camera so that motion picture studies could be made of detonations, vibrations of propellers, supersonic studies of airflow over wing tips, and many other ultrafast actions too numerous to mention.

Bell Laboratories determined that it was possible to double the speed of the camera by applying 280 volts directly to the camera motors. However, when starting the camera at this speed, the film would strip at the sprocket holes. Therefore, a 70-millisecond delay circuit was interposed so that the camera could start at 130 volts, run at a speed of 4000 frames per second for 70 milliseconds and, by jumping the voltage up to 280 volts, bring the camera to the full speed of 7500 frames per second. This peculiar jumping action led to the name of the "goose" for the Model J-410 control unit.

With the advent of these high speeds it became increasingly more difficult for the operator to synchronize the camera to the event which was being photographed because 100 feet of film travels through the camera in $8/10$ of a second.

The Industrial Timer Corporation was approached as to the possibility of designing a circuit which would enable the operator to synchronize, by means of time, any action-to-camera requirement which could be encountered in the laboratory or in the field. It developed the Model J-410 control unit (the "goose") which not only took care

107

of the synchronization of the camera to the event, but also included the 70-millisecond time-delay circuit previously described.

Fig. 1 is a photograph of the front panel unit showing timer 1 (camera timer), timer 2 (event timer), voltmeter (for voltage-to-speed readings), variable-transformer knob (for camera-voltage adjustment), and all the switches and receptacles necessary for the efficient operation of the high-speed camera.

The camera timer controls the period that the camera is operated and cuts off the camera voltage at the end of the preset time cycle.

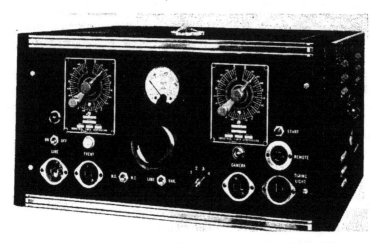

Fig. 1

This timing cutoff is important to prevent damage to the camera, which must not be run at high speeds after the film has passed through the aperture. The camera-time period is preset according to a speed-to-voltage-to-time chart based on the operating characteristics of the Fastax camera (Fig. 2).

The event timer sets up the time necessary for the event that is being studied to prestart or poststart the camera. If the event can be started or stopped by an electrical switch, the contacts on the event timer can be used efficiently for this purpose. These contacts come out to a receptacle on the front panel labeled "event." Timer 3, or the 70-millisecond delay timer is installed inside the housing.

A remote-control receptacle is provided to enable the operator to set up his equipment and operate the camera from any distant point when it is necessary to do so for safety reasons.

Because the Model J-410 is used in both tropic and arctic areas, a 100-watt heat lamp is installed within the case which acts to keep the unit dry in the tropics and warm in the arctic regions.

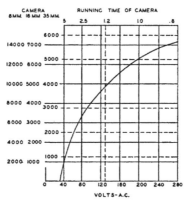

Fig. 2—Settings based on use of 100-foot film rolls only.

The "goose" can also be used efficiently with the Eastman high-speed camera by setting the voltage for the camera at 120 volts and adjusting the camera as normally used.

The housing is finished in baked black crystal enamel and measures $21 \times 10^{1}/_{2} \times 15$ inches.

Lenses for High-Speed Motion Picture Cameras

By ALAN A. COOK

WOLLENSAK OPTICAL COMPANY, ROCHESTER 5, NEW YORK

Summary—A description is given of the objective lenses which are available for use on the Western Electric Fastax cameras. The range of focal lengths is from 35 mm at an aperture of $f/2.0$ to 380 mm at $f/5.6$. A mounting of the bayonet type is provided for all lenses to make them readily interchangeable on the camera. A reflex viewfinder for the 35-mm camera is described.

THE GENERAL CHARACTERISTICS of the cameras designed for high-speed photography at the Bell Telephone Laboratories have been described in a series of papers published in this JOURNAL.[1-3]

There are three models of these cameras, one for each of the three standard size films: 8 mm, 16 mm, and 35 mm. The cameras have one important feature in common, that of a continuous film motion which in turn is compensated by a glass block or prism located between the lens and the film. This compensation is effected by rotating the prism at the exact speed required to move the image down across the aperture in synchronism with the motion of the film.

The optical effect of the prism in these cameras is an important factor in the selection of the lenses that are to be used. In the matter of shape, all the prisms are optically similar; all are glass plates with plane-parallel surfaces. The result of placing such a plate behind the lens in a camera is a familiar one in optical design.[4] In general, the effect is that of a negative lens. The back focus of the camera lens is increased. The plane-parallel plate introduces aberrations which change the pattern of image formation at the focal plane, and this change usually means a loss in image quality. The amount of the aberrations thus introduced is directly proportional to the thickness of the glass plate. It varies also with the aperture of the camera lens and the angular field covered by the lens on the film.

Therefore each type of lens used on a prism camera presents an individual problem. • The camera lens is a complex system of refracting elements designed to give a well-defined image at the film plane. It is

110

possible to modify one or more of these elements to correct for the aberrations of the plane-parallel plate and thus restore the image to its original quality. The correction is usually computed for the normal position of the prism, when the parallel faces are at right angles to the lens axis. This method of correction provides a practical solution of the rotating-prism problem. However, it is not an ideal solution; it does not take into account the complex changes in prism aberrations which are produced by rotation of the prism surfaces during the exposure cycle.[5] The results must be checked by test films exposed under actual operating conditions. This has proved to be a satisfactory and very useful method in the development of a series of lenses for the Fastax cameras.

Like all new projects involving complicated machines, the high-speed camera development proceeded gradually to its present state. In the early days the three cameras could not be considered together, with a set of drawings and a list of lens requirements attached. At that time there were only two cameras, the 8-mm and the 16-mm sizes. These two cameras have prisms of identical thickness, although the prism for the 8-mm camera is an octagon and that for the 16-mm camera is a square in cross section. A lens of 51-mm focus with an aperture ratio of $f/2$ was furnished for both these outfits and later adopted as standard equipment. The angular field for these film dimensions is small, 5.5 degrees on the 8-mm camera and 11.75 degrees on the 16-mm camera. Resolving-power tests show 28 lines per millimeter on the 16-mm camera at $f/2$, with an increase to 40 lines per millimeter at the center of the picture. At $f/3.5$, 40 lines per millimeter are resolved throughout the film area. These results apply to Super XX film and Kodachrome.

With the introduction of the 35-mm camera a few years later the lens problem became more complicated. This camera was designed to photograph an area 40 degrees in width and 12 degrees in height on a half frame of 35-mm film. The focal length required is 35 mm. An aperture of $f/2$ was specified, new prism dimensions were involved, and at the same time it was decided that lenses should be made interchangeable in all three cameras. Since the difference in prism thickness involved was only 0.008 inch, it seemed best to make a new 35-mm $f/2$ objective corrected for the larger prism and the widest angle required, assuming that any lens with definition and covering power sufficient for the wide-angle camera would be satisfactory on the smaller sizes.

Fig. 1—35-mm lens assembly.

The first results of this new design were not entirely satisfactory. Lenses from the first production lots were used on wide-angle cameras and although many useful films were produced with them, the resolution at the margin of the film was less than expected. It took more time to find the cause of the trouble—overcorrection of field curvature—and this has been eliminated within the past year. The wide-angle camera will now resolve 28 lines per millimeter, and reaches 40 lines per millimeter in the central part of the film. This 35-mm, $f/2$ camera lens is a 6-element design of the Biotar type of construction. It has been corrected for prism aberrations by the method just outlined. In respect to distortion there is still room for improvement; 1 per cent is the amount remaining at the margin of the picture area. All air-glass surfaces of the lens are fluoride-coated to minimize reflection effects, and this coating is applied on all lenses for these high-speed cameras.

The mounting and mechanical design of the 35-mm lens is shown in Figs. 1 and 2. The slotted cylindrical bearing surface of the

Fig. 2—Reflex finder and lens for 35-mm camera.

mount is 1.775 inches in diameter and 0.320 inch long. This arrangement is designed to fit the bayonet socket of the cameras, and the same lock ring is used for all the lenses of the series. The inner details of the lens mount and focusing sleeve permit a par-focal adjustment for any one of the three cameras.

Fig 3—Complete series of camera lenses.

Fig. 2 also shows a reflex finder made for the 35-mm camera. With this device the photographer can observe and check all details of his subject in setting up the camera, lights, background, and so forth. The camera lens fits on the finder for this purpose, and therefore it has two scales on the focusing ring. The red scale applies to its use as a finder lens. When the setup is complete, the finder is removed and the lens replaced on the camera. The white scale is then used, its figures giving the correct distances from the camera position.

As the high-speed cameras were developed and applied in many fields of research, the need for other types of lenses became apparent. Long-focal lengths are required to get details of distant objects, and these longer lenses were designed one by one as they were needed when the cameras were applied to various new problems in motion analysis. In considering these long-focus lenses, it is important to note that their aperature ratio is less than $f/2$. The field of view requirement, as shown in Table I, is also small. It follows that the effects of the prism surfaces on image quality are less disturbing as the focal length increases, because the angles of incidence of the separate light rays are smaller. The design of these long-focus lenses is simplified accordingly.

TABLE I

HORIZONTAL FIELD OF VIEW

Focal Length, Mm	Aperture	8-Mm Camera Degrees	16-Mm Camera Degrees	35-Mm Camera Degrees
35.0	$f/2.0$	8	16.5	40
51.0	$f/2.0$	5.5	11.75	28.5
75.0	$f/3.5$	3.6	7.25	19
101.0	$f/3.5$	2.75	·6	14
150.0	$f/4.5$	1.8	4	9.6
254.0	$f/4.5$	1.09	2.25	6
380.0	$f/5.6$	0.72	1.5	3.8

Table I gives a list of the principal dimensions of the whole series of lenses produced for high-speed cameras. For the 75-mm and 101-mm focal lengths a construction of the Tessar type has been used. The last three members of the series—150-mm, 254-mm, and 380-mm focus—are of telephoto design, with the characteristic short back focus to reduce the over-all length to reasonable limits. Fig. 3 shows some details of the mountings and the relative sizes of all lenses of the series.

The final test of any photographic lens is the result obtained on the film in the camera, under operating conditions. In general the resolution of these high-speed cameras is 28 lines per millimeter; 40 lines per millimeter are resolved in some cases. This is not a high figure for ordinary motion picture standards and it certainly will be improved in the future.

In a rotating-prism camera the requirements of synchronization at high speed are opposed to those of resolution in some respects. Synchronization demands one set of conditions, high resolving power a different one. Both optical and mechanical problems are involved here, and they must be considered together in their effect on the performance of the high-speed cameras.

REFERENCES

(1) W. Herriott, "High-speed motion picture photography applied to design of telephone apparatus," J. Soc. Mot. Pict. Eng., vol. 30, pp. 30–38; January, 1938.

(2) Howard J. Smith, "8000 pictures per second," J. Soc. Mot. Pict. Eng., vol. 45, pp. 171–184; September, 1945.

(3) John H. Waddell, "A wide angle 35-mm high-speed motion picture camera," J. Soc. Mot. Pict. Eng., vol. 46, pp. 87–103; February, 1946.

(4) I. C. Gardner, Scientific Papers of the Bureau of Standards, No. 550, p. 160; May, 1927.

(5) John Kudar, "Optical problems of the image formation in high-speed motion picture cameras," J. Soc. Mot. Pict. Eng., vol. 47, pp. 400–403; November, 1946.

High-Speed Photographic System Using Electronic Flash Lighting*

By WILLIAM T. WHELAN

NAVAL ORDNANCE LABORATORY, WHITE OAK, SILVER SPRING 19, MARYLAND

Summary—By way of summary it may be said that a system of high-speed photography has been developed which combines the desirable features of rotating prism-type motion picture cameras with those of electronic flash lighting. This has resulted in a system which delivers extremely high over-all definition and incorporates operating flexibility. Special lighting effects are made available by means of all-electronic interlacing equipment. The system can be operated in its present form at frame rates as high as 8000 frames per second without reduction of the illumination available per flash. An effective exposure time which never exceeds several microseconds is realized, and interchangeability of various commercial flashlamps is possible without modification of the electronic circuits.

INTRODUCTION

AN ELECTRONIC flash-illuminated system of high-speed photography has been developed at the Naval Ordnance Laboratory for investigation of short-duration, nonrepetitive events. The Laboratory setup has a number of features in common with other systems which have been reported previously in the literature.[1, 2] Certain specific differences exist, however, which justify this description.

The system consists of a rotating prism-type high-speed camera with which one or more electronic flashlamp units are synchronized so that the individual frame exposures are made at the instant at which the prism shutter is "wide open." Additional gating, sequencing, and interlacing equipment are included which provide picture-taking flexibility not reported heretofore.

Before entering upon a detailed description of the system, it is necessary to point out that it is not presented as an end product, but merely represents a pilot model of a larger, more complete unit which is being developed at present. However, it has been in service long enough, and a sufficient number of test scenes have been taken with the apparatus to prove that it can be used readily as a working unit in its present form.

* This work was supported by the Office of Naval Research.

General Considerations

A discussion of the system should be prefaced by a consideration of what features it makes available to the user. The most important characteristic of electronic flash illumination, that of an extremely short exposure time with its corresponding motion-arresting capabilities, is featured in the system. The light sources to be described permit an effective exposure time which varies from 1 to 3 microseconds. The exact duration of the flash is dependent upon two factors: (a) the particular type of flashlamp being used and (b) the electrical input in a specific setup. As implied by this remark, a variety of commercial flashlamps can be used interchangeably in this system, and the electrical power input can be varied to suit the needs of the operator. Any camera speed up to 8000 frames per second can be accommodated readily. Adaptation to any of the various commercial high-speed cameras, whether they are constructed with or without compensating prisms, seems readily achievable although only the Eastman Type III is in use at present. Sufficient light output is available for most scenes so that adequate exposure is not a serious problem until frame rates in excess of 4000 per second are reached. Above this speed it is necessary to increase the number of lamp units per scene over that which slower frame rates might demand in order to distribute the burden and prevent lamp overload and its resulting lamp destruction. Provision is made for the simultaneous, synchronous operation of as many as six lamps from a single camera, and the entire system is operated from a single control panel. The duration of any specific lamp's working interval, as well as the sequence of operation of the individual lamp, is adjustable and self-indicating at this panel.

An important feature is the availability of half-frame-rate operation of any specific lamp or group of lamps. If the instantaneous camera speed is, say, 3000 frames per second, then this convenience makes it possible to operate any desired lamp at a corresponding instantaneous rate of 1500 flashes per second. This method is referred to in this article as "half-frame-rate" operation and is meant to refer to the flash rate of the lamp and not the picture size. For normal projection, this half-frame-rate operation is not desirable, since, if part of the lamps are operated at half-frame rate of 1500 frames per second, and the rest are operated at the full camera speed of 3000 frames per second, the only effect one would notice on the projected motion picture would be an annoying flicker. However, for frame-by-frame analysis purposes, alternate frames which can be provided with two different types of lighting represent potentially twice as much experimental data per reel of film since such a method of photography might permit the simultaneous recording of two entirely different aspects of the action. For example, it is often necessary in the photography of objects entering water to study the event not only by means of reflected or "conventional" lighting, but also by means of transmitted or "silhouette" lighting. With two banks of lamps properly disposed, both lighting effects could be interlaced on a single reel of film.

. In addition to differential lighting effects, this film-sharing principle can be used in such a way that higher frame rates can be accommodated by a bank of flashlamps. As mentioned previously, the tolerable number of flashes per second at a given input per flash is essentially limited by the flashlamp at the present time, since under conditions of constant energy input per flash the total average power input to a lamp is directly proportional to the frequency of flashes, and a maximum average power input exists for any type of flashlamp. Because of this power-handling limitation of the flashlamps, normally one is forced either to reduce the "per flash" energy applied to the lamp or to restrict the permissible total number of flashes; i.e., the duration of the operating cycle must be reduced. Neither one of these possibilities is desirable from the photographic standpoint. This difficulty has been overcome in the present system by use of the half-frame-rate provision. Two such half-frame-rate channels operate on alternate frames and permit individual light sources to operate at one half of the instantaneous camera-frame rate. In this manner, even for camera speeds as high as 8000 frames per second, the flashlamps can be operated on a 4000-frame basis. Naturally, the two groups of lamps which are thus duplexed must be so placed that they effect the same subject illumination. The principle of film-sharing can be extended to provide for even higher camera-frame rates, but this has not been done in our system. However, a modification which permits extension of a system to 16,000 frames per second will be reported in the near future.

METHOD OF OPERATION

A block diagram of the electrical system is shown in Fig. 1. Since the individual items will be described in some detail in the following pages, only a functional description of each is given here.

The entire system is designed around the camera itself, and as it has been remarked, the system described here is adapted to the

Eastman Type III camera. Fortunately, when camera types are changed, the necessary modifications are relatively minor. The Eastman camera was chosen since it admitted of ready adaptation to a multiple-flash system, and although its maximum speed was in the neighborhood of 3000 frames per second, this speed was found entirely adequate for the specific problems under study. Referring to Fig. 1, the unit described as "master control" is a panel which contains two main switches and a group of signal lights. The first switch,

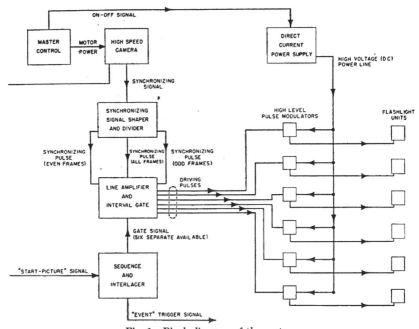

Fig. 1—Block diagram of the system.

when closed, merely permits a controlled five-minute preheat of all the electron-tube cathodes in the system. At the end of this interval, it is possible to throw the second "event" switch and thus initiate the picture-taking cycle. Under normal conditions of operation, these are the only two actions which are required to take a picture other than the usual photographic duties. These include the camera-loading, setting of the "prerun" footage to be passed through the camera while it comes up to operating speed, setting the maximum camera speed desired, placing the lights in appropriate positions, focusing, making aperture adjustments, and setting of the sequence and interlacing equipment. Normally, if a series of shots are scheduled which do not require appreciable scene changes, a run can be set up and taken in a matter of minutes.

When the event switch is thrown, the direct-current power supplies which feed the system and the camera motor are energized. The camera immediately accelerates, and a small alternating voltage, whose frequency is proportional to the optical shutter speed, begins to be fed to the "synchronizing shaper and divider" unit at the control position. In this unit, this small voltage, called the synchronizing signal, is transformed into an abrupt trigger pulse whose leading edge coincides with the instant of the wide-open shutter. A divider circuit supplies two other outputs of synchronizing pulses which correspond in time to the wide-open shutter positions of alternate frames; i.e., if the instantaneous camera speed is N frames per second, synchronizing pulses appear at one of the outputs and occur at a rate of $N/2$ frames per second. These pulses coincide with the wide-open shutter positions on, say, the even-numbered frames, while at the other output pulses also occur at $N/2$ frames per second, but these coincide in time to the open shutter on the odd-numbered frames.

The synchronizing pulses are then fed to the "line-amplifier and gate" unit where they are subsequently amplified without phase change and passed on to six channels that control specific flashlamps. No progress of the pulse through these channels would occur unless an outside influence acted at this point, since an electrical gate operates in each of the six channels and normally blocks the passage of any signals. The camera, itself, provides the outside influence to unblock the gate circuits at the proper instant and permits the synchronizing pulse to advance. This unblocking signal is generated in the Eastman camera by the closing of a microswitch at a predetermined instant. The unblocking information is relayed over the "start-picture" signal line. Normally, the unblocking "start-picture" signal is delayed until the camera approaches operating speed. When this unblocking information arrives at the "sequence and interlacer" unit, an electrical signal is developed for triggering the event to be photographed.

In the sequence and interlacer unit, in addition to the event trigger signal, the sequence and duration of the unblocking signals which close the individual electronic gates in the 6-channel line amplifier are determined. Six cascade unblocking periods whose durations are variable from zero to 3 seconds in length are available; i.e., the second timed interval commences immediately upon the cessation of the first, the third commences immediately upon the cessation of the second, and so forth, until the six timed intervals have occurred. This arrangement permits an operator to set up a sequence of lighting in which the area illuminated will change with time. In this way, progressive lighting can be preset to "follow" a relatively slow-moving object across a large field of view and so the operating time of the

Fig. 2—A portion of the equipment showing power supply (right), control panel (left), Eastman camera, and two pulse modulators with their associated flashlamps.

flashlamps is restricted to the barest minimum. Thus by proper choice of the interconnection of the three available synchronous signals, and the six variable unblocking gate signals, the operator can choose from an extremely large number of possible lighting effects.

Once the synchronous signal has passed through an appropriate "closed" line-amplifier channel, it is fed over cables to one of the six remote units referred to as the "high-level pulse modulators." These units are pulse generators that transform the direct-current energy from the main power supply into abrupt high-voltage surges of less than a microsecond duration which pass down cables to the individual gaseous flash units where they supply the electrical input energy of the light sources. The instant at which the discharge pulse, and hence the light output, takes place is precisely controlled by the arrival of the synchronizing pulse at the modulator unit. As mentioned previously, this instant is made to coincide with the wide-open position of the optical shutter in the camera.

In Fig. 2 is shown a partial setup of this multiple-flash high-speed photographic system. The rack on the left is the control panel which contains the units just described, and the rack on the right contains the direct-current power supply. At the extreme right is a pair of remote pulse modulators, and in the foreground is a pair of mounted commercial flashlamps. The Eastman camera stands behind the modulators on its own rigid tripod.

LIMITATIONS OF THE SYSTEM

Certain basic limitations exist in the system outlined above and should be noted here. (a) Only one camera can be used with it, unless one resorts to expensive and seemingly impractical "ganging" of the rotating systems of multiple cameras; (b) a single high-voltage direct-current power supply feeds six lamps, and thus any failure in the power supply removes all six lamps from service; (c) overload protection is provided in the low-voltage primary circuit of the power supply which means that a high-voltage fault at any point in the high-voltage system results in the complete failure of all lamps; (d) the high-voltage power supply requires a 3-phase primary power line of considerable capacity which restricts the number of possible locations where the equipment may be operated; and (e) the weight of the system is of the order of 2000 pounds which further restricts its flexibility.

MECHANICAL DETAILS

The pilot system described was built of conventional electronic components and hardware wherever possible. The high-voltage power supply and the control units are housed in a pair of enclosed transmitter cabinets, each of which provides seven feet of usable panel height. A third cabinet is required if operation below frame speeds of 1000 pictures per second is desired. The individual pulse-modulator units are built into relatively small steel boxes and are normally located remote from the control panel and power supplies in order to keep the line between the actual flashlamp unit and the pulse modulator short. All cables are made up in standard interconnecting lengths, and are of seven general types; (a) low-level pulse cables which connect the interlacer panel to the remote pulse-modulator units, (b) high-level pulse cables which connect modulators to lamps, (c) high-voltage direct-current power lines which carry direct-current power into the pulse-modulator units, (d) camera motor power and control lines, (e) camera-synchronizing lines, (f) conventional single-phase, low-voltage power lines, (g) heavy-duty, 3-phase power lines. All lines which carry synchronizing or high-power pulse signals are of the shielded coaxial type which prevents unwanted interaction of units. Successful photography has been accomplished with modulators, lamps, and cameras placed within an explosion chamber which was located over 150 feet away from the control panel. Normally, it is desirable to keep the cable length between the modulator and flashlamp less than 50 feet. For a problem where the modulator-to-lamp distance could be kept constant, it would be more desirable to house the pulse modulators in a single transmitter-style cabinet.

Since a variety of flashlamp light sources have been used in the past, various types of lamp housings and reflectors have been used. Perhaps the most convenient lamp units were realized when the General Electric Company's Type FT-125 lamp was used. This lamp is built into a "sealed-beam" headlight unit and thus provides its own reflector. A very simple mounting was provided by fixing the lamp unit inside a small, commercially available steel cabinet. Lamps of the long, slender geometry were usually mounted in hastily contrived cylindrical reflectors which are parabolic in cross section. For silhouette lighting effects, all of the light sources were provided with diffusers in order to provide a background of uniform illumination.

A serious problem of the electromechanical design was that of providing safe high-voltage connectors for the direct-current power cables, as well as the high-voltage, heavy-current cables which connect modulators and lamps. At the time of the development, these difficulties were eliminated by manufacturing special connectors. However, a more practical solution would be to design the cables for use with standard high-voltage X-ray fittings or their equivalent. The usual door interlocks and "dead" front panels were provided for personnel safety.

ELECTRICAL DETAILS

A thorough discussion of all of the electrical and electronic details would require an undue amount of space and would be inappropriate at this time. For this reason, the present discussion may seem sketchy, but the References should provide the interested reader with whatever information he may desire.

A common element in any electronic flash-lighting system is the one which forms and controls the short burst of electrical energy which is transferred to the gaseous-discharge lamp at the instant light output is desired. The simplest method of accomplishing this task is to charge an electrostatic capacitor to a voltage somewhat below that required to cause electrical breakdown of the gas column. Then at the appropriate instant, an external agent is brought into play which momentarily destroys the insulating property of the gas, and thus causes electrical breakdown within the gas column. In a matter of a few millionths of a second, this gas column becomes a highly conducting load on the capacitor and quickly discharges the stored charge from its plates. Normal electrodynamic-gas processes occur which result in the radiation of a large amount of energy, much of which is photographically active.

The foregoing method of triggering is usually used in small, single-flash systems. The multiple-flash system described here differs somewhat in its mode of operation from this method in that the capacitor is normally charged to a voltage considerably higher than that required to cause spontaneous breakdown of the gas column. As a consequence, the capacitor must not be connected across the lamp load until the very instant at which the light output is desired. Such a requirement demands a switch which is capable of holding off extremely high voltages until the proper instant, and which can be closed in a matter of microseconds by simple electrical means. In addition, the switch must be capable of carrying the capacitor-discharge current repeatedly without incident or excessive loss of electrical energy. A switch which meets these requirements exists in the hydrogen thyratron, a tube which was developed during the last war for certain radar problems.

These thyratrons are made in a variety of sizes, and the particular one used in the pulse modulator described here is designated as the Type 5C22. It is rated at a peak current of 325 amperes, while the normal discharge current through the flashlamp is of the order of 1500 amperes. Even though the ratings are thus exceeded each time a flash occurs, the short duration of the discharge-current pulse (about 0.75 microsecond for the usual circuit constants) and the fact that the longest picture-taking runs of less than 3 seconds in length doubtless account for the satisfactory thyratron life experienced. Ultimate failure is caused by hydrogen "cleanup" in the internal tube structure. The circuit diagram of the modulator is shown in Fig. 3. V_1 is the 5C22 hydrogen thyratron, V_2 is the flashlamp, and V_3 is a clamping diode. C_1 is the impulse energy-storage network, and L_1 is the iron-cored recharge-control inductor. The purpose of L_2 is to permit the recharge current to by-pass the flashlamp. G is a protective gap which prevents overvoltage operation in case of circuit maladjustment. L_3, L_4, and C_2 constitute a pulse-coupling network in the grid of the switching thyratron. This circuit is a conventional inductive-charging modulator, and its operation is well covered elsewhere.[3, 4] The singular features of it worthy of note are: (a) it is a

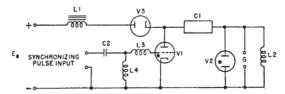

Fig. 3—Simplified schematic diagram of the pulse-modulator unit.

voltage-doubling circuit; i.e., under conditions of proper adjustment, the energy-storage capacitor will be charged every cycle to a peak voltage approximately twice that of the direct-current source, (b) the over-all charging efficiency (defined as the ratio of the capacitor energy made available for flash to the energy drawn from the direct-current source) is high, (c) it readily adjusts itself to various flash rates and hence various camera speeds.

If one has been accustomed to thinking in terms of single-flash electronic illumination, the matter of charging efficiency might appear at first to be of secondary importance, but for multiple-flash operation during which the recharge cycle may occur thousands of times per second, charging efficiency must be high in order to prevent undue loss of electrical energy. For this reason, simple resistance-controlled charging is highly undesirable, since under optimum conditions of adjustment this method results in an efficiency ratio of only 0.5. On the other hand, inductive charging approaches an efficiency ratio of

1.0 as the losses in the charge circuit are made smaller. These losses include the iron and copper losses of the choke L_1, the dielectric losses of the storage capacitor C_1, the plate-circuit losses of the clamping diode V_3, and the internal power-supply losses. This clamping diode V_3, which is indicated in Fig. 3, is a necessary evil which cannot be avoided if the recurrent flash rate of the system falls below one half the natural resonant frequency of the charging choke L_1, and the storage capacitor C_1. Its purpose is to prevent the overvoltaged storage capacitor from discharging back into the power supply. In the present system, the size of the parameters is such that this diode is not required unless the flash rate falls below 1000 per second. Housing for the six diode-clamping units requires a third transmitter-type cabinet if such low-frequency operation is desired.

It is of interest to consider the power requirements of the system under conditions when the frame speed is considerable and when the high-efficiency charging circuit is used. The average direct-current power required by a single-pulse-modulator unit is computed readily if it is assumed that the entire block of electrostatically stored energy in the capacitor is completely transferred from the capacitor to the discharge circuit during each discharge cycle, and must be re-established at the expense of the direct-current power supply. This reasoning leads to the expression for the average power given below.

$$P_{dc} = \frac{nC(V)^2}{2\eta}$$

where P = the average direct-current load demand in watts if the other terms are defined as follows:

n = number of discharge cycles per second (flash rate)
C = effective capacitance of the storage capacitor in farads
V = maximum voltage (in volts) to which capacitor is charged each cycle
η = over-all efficiency of the charging circuit expressed as a decimal.

In the system described, the nominal storage capacitor used is a 0.05-microfarad capacitor, the crest capacitor voltage is of the order of 10 kilovolts, and the charging-circuit efficiency can be expected to run about 0.9. Under these conditions, a single lamp flashing in synchronism with a high-speed camera, running 2500 frames per second, would require 3.14 kilowatts from a direct-current source of slightly more than 5000 volts. Six such lamps would require a total average power of over 18 kilowatts. Assuming a direct-current power-supply efficiency of 0.9, it is seen that the average alternating-current power-line load is more than 20 kilowatts. Such large power demands, even though they may exist for but a few seconds, require rather careful planning of the power supply itself. It would be extremely poor engineering to attempt to build a 5-kilovolt, 20-kilowatt, single-phase rectifier set which would be required to supply smooth, ripple-free direct current to the load. For this reason, a full-wave, 3-phase bridge rectifier set was used. It operates from either 220-volt or 440-volt, 3-phase circuits, and is protected against either alternating- or direct-current faults by fast-acting air-circuit breakers.

The requirement that the camera must develop an electrical synchronizing signal, which was proportional to the rotating-prism shutter speed, caused some difficulty. A scheme, reported earlier,[5]

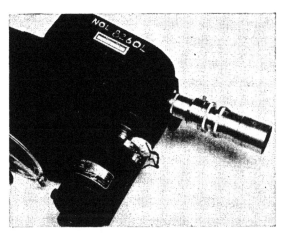

Fig. 4—Modified Eastman Type III camera showing synchronizing alternator.

whereby a mechanical commutator was attached to the driving pulley of the Eastman camera, was abandoned in favor of a drag-free electromagnetic pickup which operates on the variable-reluctance principle. A photograph of this pickup attached to the Eastman camera is shown in Fig. 4. Onto the face of the driving pulley of the Eastman-shutter system are placed nine, equispaced, thin iron wafers which rotate with the pulley. They can be seen in the photograph. Near by is mounted a small Alnico horseshoe magnet on which is wound a pickup coil. As the drive pulley rotates, the iron wafers pass in and out of the magnet's field, thereby causing reluctance variation in the magnetic circuit which results in a cyclic variation of the magnetic flux linking the coil wound on the horseshoe magnet. A voltage appears across the coil which is approximately sinusoidal in wave form, and whose instantaneous frequency corresponds to the instaneous shutter frequency. The necessity for the use of nine iron wafers on the pulley arose because the drive pulley acts through a gear train to operate the shutter at just nine times its own rotary speed. This method of obtaining a synchronizing signal is sturdy and simple to adjust. It suffers from a sensitivity to stray 60-cycle pickup from the camera-driving motor which tends to phase-modulate the final synchronizing pulses. This problem was overcome by insertion of a 60-cycle electrical filter in the output of this synchronizing system. If operation of the camera in the neighborhood of 60 frames per second is imperative, stray pickup balancing coils can be attached, and extreme shielding measures applied.

Application of this method of synchronization to cameras other than the Eastman are under way. The actual technique must be varied from camera to camera in order to meet the specific mechanical requirements of each camera. It is essential that the point in the mechanical system where the signal takeoff is placed be gear-linked to the rotating-prism shutter so that relative motion between these two points is not possible.

Before the low-level synchronizing signal from the camera can be utilized to trip the thyratron switch in the modulator, it must be shaped and amplified. At the same time, it is fed through dividing circuits which provide the interlacer flexibility mentioned previously. The divider circuit is a conventional "scale-of-two" counter which provides two outputs, each at one half the frequency of the input synchronizing signal. These two outputs are identical except that they are 180 degrees out of phase with each other, and consequently, while one of these outputs occurs on each even-numbered frame, the other occurs on each odd-numbered frame. Each of the divided output signals is exactly in phase with the wide-open position of the shutter, which assured uniform synchronization of all outputs, whether divided or not.

Both the divided and undivided signal outputs are shaped into nearly rectangular pulses. The rise time of the pulse is 0.5 microsecond, or less, while the base width is 10 microseconds. Constant-amplitude synchronizing signals of nearly 100 volts is provided from each of the three output terminals of the divider-shaper while the camera is running.

After division and shaping, the synchronizing pulses are delivered to the line amplifiers, which are six in number. Each of these amplifier stages consists of two cascade video units. The purpose of these line amplifiers is twofold: (a) to provide individual control points in the signal lines at which unblocking or interval control can be exercised, and (b) to provide electrically isolated trigger pulses at appropriate power levels for the thyratrons in the modulators. Care has been taken in the design of these line amplifiers to preserve the wave form of the synchronizing trigger pulses as they are delivered from the shaper-driver unit. Similarly, phase-delay errors have been minimized. Unblocking is provided in the line amplifier by means of electronic switching of electrode voltages in the output tube of the individual channels.

The duration of the unblocked or operating cycle of each line amplifier is set by the use of stabilized multivibrator-type timing circuits which are adjustable in fixed time steps by the operator.

A simulator unit is provided on the main control panel for "on-the-spot" investigations of trouble. Routine tests and specific symptoms have been outlined which permit relatively unskilled personnel to localize difficulties in the system. In general, it is the function of the simulator to generate artificial signals which resemble the actual operating signals, and these signals are then used as samples for sounding out the various portions of the system. This permits a step-by-step dynamic test of the various units.

Another purpose the simulator serves is that of supplying a known number of flashes. These flashes constitute a standard test burst which permit the operator to check the intensity of his subject illumination by means of a suitable exposure meter of the integrating type.[6]

REFERENCES

(1) Henry M. Lester, "Electronic flashtube illumination for specialized motion picture photography," *J. Soc. Mot. Pict. Eng.*, vol. 50, pp. 208–233; March, 1948.

(2) Robert T. Knapp, "Special cameras and flash lamps for high-speed underwater photography," *J. Soc. Mot. Pict. Eng.*, vol. 49, pp. 64–82; July, 1947.

(3) "Pulse Modulators," Book 9, MIT Radiation Laboratory Series, McGraw-Hill Publishing Company, New York, N. Y., 1948.

(4) "Principles of Radar," MIT Radar School Staff, McGraw-Hill Publishing Company, New York, N. Y., 1946.

(5) Robert A. Anderson and W. T. Whelan, "High-speed motion pictures with synchronized multiflash lighting," *J. Soc. Mot. Pict. Eng.*, vol. 50, pp. 199–208; March, 1948.

(6) An appropriate instrument has been developed and announced by H. E. Edgerton.

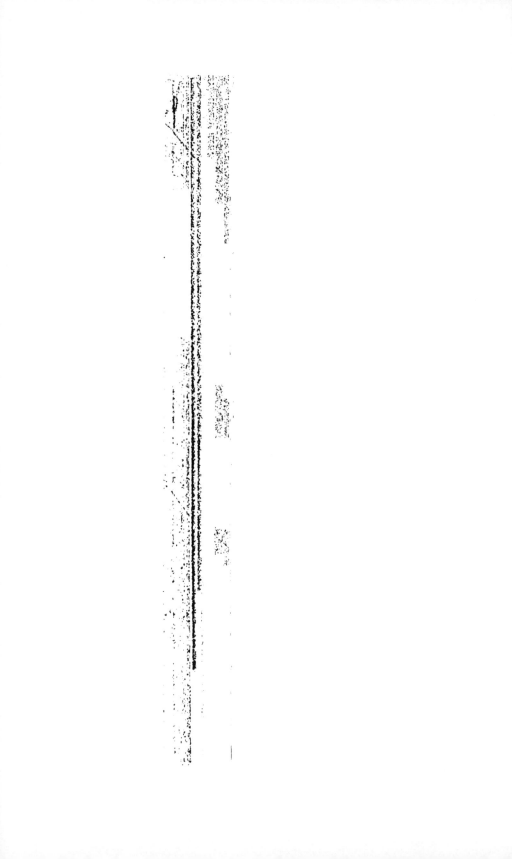

Journal of the
Society of Motion Picture Engineers

VOLUME 52   APRIL 1949   NUMBER 4

ARTHUR C. DOWNES   HELEN M. STOTE   NORWOOD L. SIMMONS
Chairman   *Editor*   *Chairman*
Board of Editors   Papers Committee

Society of
Motion Picture Engineers

342 MADISON AVENUE—NEW YORK 17, N. Y.—TEL. MU 2-2185
BOYCE NEMEC . . . EXECUTIVE SECRETARY

OFFICERS
1949–1950

PRESIDENT	PAST-PRESIDENT	SECRETARY
Earl I. Sponable	Loren L. Ryder	Robert M. Corbin
460 W. 54 St.	5451 Marathon St.	343 State St.
New York 19, N. Y.	Hollywood 38, Calif.	Rochester 4, N. Y.

EXECUTIVE VICE-PRESIDENT	EDITORIAL VICE-PRESIDENT	CONVENTION VICE-PRESIDENT
Peter Mole	Clyde R. Keith	William C. Kunzmann
941 N. Sycamore Ave.	120 Broadway	Box 6087
Hollywood 38, Calif.	New York 5, N. Y.	Cleveland 1, Ohio

1948–1949

1949

ENGINEERING VICE-PRESIDENT	TREASURER	FINANCIAL VICE-PRESIDENT
John A. Maurer	Ralph B. Austrian	David B. Joy
37-01—31 St.	1270 Avenue of The Americas	30 E. 42 St.
Long Island City 1, N. Y.	New York 20, N. Y.	New York 17, N. Y.

Governors
1949

1948–1949	James Frank, Jr. 426 Luckie St., N. W. Atlanta, Ga.	1949–1950
Alan W. Cook	William B. Lodge	Herbert Barnett
25 Thorpe St.	485 Madison Ave.	Manville Lane
Binghamton, N. Y.	New York 22, N. Y.	Pleasantville, N. Y.
Lloyd T. Goldsmith	William H. Rivers	Fred T. Bowditch
Warner Brothers	342 Madison Ave.	Box 6087
Burbank, Calif.	New York 17, N. Y.	Cleveland 1, Ohio
Paul J. Larsen	Sidney P. Solow	Kenneth F. Morgan
508 S. Tulane St.	959 Seward St.	6601 Romaine St.
Albuquerque, N. M.	Hollywood 38, Calif.	Los Angeles 38, Calif.
Gordon E. Sawyer	R. T. Van Niman	Norwood L. Simmons
857 N. Martel Ave.	4431 W. Lake St.	6706 Santa Monica Blvd.
Hollywood 46, Calif.	Chicago 24, Ill.	Hollywood 38, Calif.

Films in Television

TELEVISION FROM FILM SOURCES

MOTION PICTURE STUDIO AND "ON-THE-SPOT" FILMING FOR SUBSEQUENT TELECASTING

Cameras—At the present time films for television are being photographed with both 35-mm and 16-mm motion picture cameras at the standard speed of 24 frames per second. For production work where synchronized sound is to be used, the camera must be driven at synchronous speed. A number of television stations currently making their own newsreels use commercially available 16-mm professional cameras and associated equipment.

Composition—Data supplied by one television station indicated that because adjustment of picture size in home receivers varies greatly, all significant action and subject material be kept within a central area having $8^1/_2$ per cent top and bottom margins and 13 per cent side margins. When this is done, a large majority of commercial receivers will show all-important information.

Close-up scenes give most pleasing reproduction because viewing screens of home receivers are small and the field of action necessarily is limited. Medium shots are generally considered the outside limit and long shots rarely add anything of value to the film program. Subject matter should be kept as large as the limits and action of the scene being televised will allow without obvious crowding of action or characters.

Whenever possible checkerboard patterns with many abrupt changes of contrast should be employed as these numerous large variations in print density will reduce the horizontal-smear effect that otherwise would be caused by low-frequency defects of present systems. For the same reason, large uniform-colored or relatively dark areas and delicate or minute patterns are to be studiously avoided, particularly in the lower portion or foreground of the scene.

Subject Lighting—The limited range of picture-tube brightness requires that subject contrast be controlled wherever possible. Usually it is not necessary to resort to flat lighting in order to hold contrast within the brightness range of the television system, but *even lighting* is essential particularly over large picture areas. That is, large picture areas must have about the same average illumination. Wide variations in brightness over the scene will otherwise have to be compensated for by adjustment of the television shading controls.

Adequate foreground lighting is quite essential since the electric-energy-decay-rate characteristic of the iconoscope mosaic may cause picture degradation in the form of insufficient signal response in the lower portion of the received picture. The general intensity of illumination from scene to scene should be kept relatively constant so that the level of the television signal does not change markedly and for this reason night scenes should be avoided. For psychological reasons long fades should not be used because they interrupt program continuity and the audience may think from the long blank period that something is wrong with the receiver.

Properties—Clothing and accessories, backgrounds, furniture, and other "properties" should have definite patterns large enough to be clearly visible on the screen of the television receiver. Again, fine or delicate detail with minute changes in contrast should be avoided.

Titles—To reproduce clearly on small home receivers, the lettering of titles should be large, boldface on a textured background, and should always be located within the dimensional limits previously mentioned.

General—Action within scenes should be continuous. This, however, is not always possible, so where inanimate objects are shown for any period of time, motion of the camera by zooming, traveling, change of angle, or slow panning should be substituted to accomplish the desired effect. In the present state of the art, this type of change sometimes emphasizes the geometrical distortion in the final image. In the transition from one scene to the next, it is desirable to employ lap dissolves, quick fades, or instantaneous "cuts" timed to keep pace with the program.

FILM PROCESSING

35-Mm Negative—Normal exposure and development, as employed in motion picture negative work, should be used for pictures to be televised. Negative gamma is usually carried between 0.65 and 0.70 and the scene density is considered normal if the negative prints in the middle of the printer scale.

35-Mm Prints—Over a period of years numerous closed-circuit tests have been run in an attempt to determine optimum print density for televising.   These tests, although they were not conclusive, have shown that low-contrast prints (gamma between 1.4 and 1.6) with a general density near normal reproduce well.   When the contrast was carried to normal (gamma of 2.20 to 2.50) the print which reproduced best was at least two printer points light.

All of these tests were made from a negative exposed for motion picture theater use.   More recent tests have shown that prints of normal gamma and perhaps 1 to 2 printer points light reproduce best.   In view of the great importance of establishing proper film specifications for television this subject needs further investigating and reporting.

16-Mm Reversal—Most 16-mm film used by television stations is processed by reversal.   Current 35-mm practice shows that a negative gamma of 0.70 and a print gamma of 1.50 produce a resulting picture contrast of 1.05, while current 16-mm reversal technique produces a print gamma between 1.00 and 1.20, which has proved satisfactory and is recommended.

16-Mm Negative and Positive—A limited amount of 16-mm negative and print work is being done.   Current practice is to develop the negative in fine-grain negative developer and print normally.

TRANSFER FROM FILM TO TELEVISION SIGNAL

The translation of motion pictures into television signals is complicated by the fact that motion picture film moves at the standard rate of 24 frames per second while the rate of the television signal is 30 frames (60 fields) per second.   A simple factor can be applied to the different frame rates which satisfies the peculiar characteristics of the two systems.   Two frames of motion picture film require the same amount of time as five fields ($2^1/_2$ frames) in television scanning.   This relationship is presented graphically in Figs. 1A and 1B, which show that if one film frame is scanned for two television fields and the next film frame for three television fields the time difference of frame rate can be satisfied.   This relationship is fundamental as long as the respective frame rates are retained and applies regardless of type of camera or projector.

There are two fundamentally different types of television pickup tubes, the storage type (iconoscope—image orthicon) which stores electrical charges produced by a multitude of individual picture elements until discharged by the scanning electron beam; and the nonstorage type (image dissector—phototube) where the electrical energy

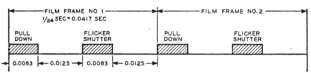

A.　35-mm motion picture projector with 72-degree shutter, 24 frames per second.

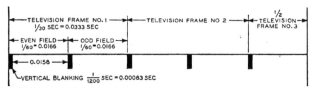

B.　Television picture with interlace scanning, 30 frames per second.

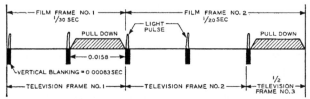

C.　35-mm television intermittent storage system of scanning motion picture film.

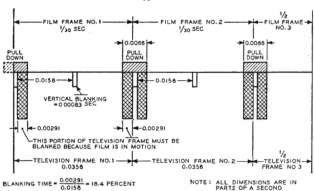

D.　Television scanning of motion picture film run at 30 frames per second, 72-degrees pulldown time.

Fig. 1

of each picture element is proportional to the incident light experienced at the instant that element is scanned. The phototube is used with the flying-spot scanner and is gaining in popularity with development engineers.

The iconoscope camera tube, however, is almost universally used for commercial-film pickup work. Because of its storage feature the iconoscope can be "pulsed" with an intense burst of light of short duration. This produces a charge picture in the tube that is then removed in the normal scanning sequence. This flash may not be applied during the actual scanning time since it would give a pulsed video signal and a noticeable black "application bar" across the receiver screen. Light is therefore applied during the vertical-blanking period and its pulse effect is further nullified by proper back lighting of the mosaic screen in the iconoscope tube and electronic gating of the beam current. Since light is applied only during vertical blanking a full scanning interval is available for pulldown of the next film frame. Fig. 1C indicates the sequence of charging the camera tube with a light pulse, scanning the resulting picture, and film pulldown in 35-mm projectors. Either mechanical or electrical means can provide the pulse. A pulldown of approximately 50 degrees and a mechanical shutter having an opening of less than 18 degrees and synchronized at 3600 revolutions per minute to open during the television vertical-blanking pulse time is practical for 16-mm projectors. Equipment is also available with an electrically timed and controlled gas-discharge tube instead of a mechanical shutter.

Control of the iconoscope camera requires adjustment of the beam current and continual monitoring of picture "shading." Beam current can be set for average light level, compromising between excessive tube noise at high beam levels and low signal with resulting amplifier noise at low beam levels. Shading, an undesirable characteristic, is a spurious signal resulting from an uneven distribution of secondary electrons on the tube mosaic and varies with picture content. Adequate correction can be obtained by properly mixing artificially generated signals, saw-tooth and parabolic, and occasionally some sine-wave forms in both the vertical- and horizontal-scanning directions, and applying the results to the camera output. Another difficulty known as edge flare, which shows up as bright areas usually on the right edge and bottom of the picture, can be improved by the adjustment of an internal edge lighter.

With proper adjustment of the controls and proper high-light illumination of the iconoscope mosaic a very satisfactory picture is obtained. Resolution usually exceeds 350 television lines and the signal-to-noise ratio is low but tone gradation is not perfectly linear. The signal-output current is approximately proportional to brightness of mosaic illumination up to about 0.1 foot-candle, but at brighter levels the signal increases less rapidly with increasing brightness. Thus in combination with the normal viewing tube both the blacks and the whites seem to be compressed.

The coating on the mosaic of the Type 1850-A iconoscope shows a preference for the blue region of the spectrum so that color films can be projected for black-and-white television pickup but tone values of various colors will not agree perfectly with those seen by the eye. Some partial correction is possible by the use of filters on the light source.

The phototube flying-spot-scanner system is now undergoing development and shows considerable promise. It has a number of very desirable advantages over the iconoscope for film pickup, namely, simplicity of components, freedom from shading and other spurious signals, no loss of stored charge during the scanning cycle, excellent contrast range, and high picture resolution.

A major difficulty of the flying-spot scanner, as in any type of non-storage television camera tube, as shown by Fig. 1B, is that no film pulldown time is available when the projector is run at 30 frames per second. Some type of nonintermittent projector would seem to be desirable but the complexity as well as unsatisfactory speed regulation of several proposed types of continuous projectors presents a serious problem.

The iconoscope camera and the flying-spot scanner are both useful with still slides or filmstrips in a standard projector. Camera switching can be accomplished by remote control and in one case, two film projectors and a slide projector can be switched into a single camera by the use of an accessory optical-mirror device. Television lends itself nicely to "fades," "dissolves," and superposition of two pictures by the simple expedient of mixing video signals at the required level before adding the standard synchronizing signal. Necessary controls are commercially available as standard studio equipment. "Wipes" are somewhat more difficult, requiring an electronic switch of a type that is not as yet commercially available.

Sound for television from film sources requires no special handling beyond equalization.

FILM FROM TELEVISION SOURCES
(CATHODE-RAY PHOTOGRAPHY)

Motion pictures photographed from a television picture tube are made as transcriptions of live-studio or remote programs for rebroadcasting and may be used at a later time by the station that presented the original program or may be syndicated with several prints from the original made for distribution to subscriber stations. Picture- and sound-quality requirements are high, demanding utmost attention on the part of station and processing laboratory personnel.

Regular record films are also made but generally at a reduced film-frame rate and have far less rigid quality requirements because they are never rebroadcast.

CAMERA REQUIREMENTS

The conversion from the 30-frame-per-second television-picture rate to the 24-frame-per-second film-picture rate presents a serious problem for television recording-camera design engineers. A currently successful solution is based on the use of successive dissimilar scanning cycles. Another proposed answer is a change of the standard film rate from 24 to 30 frames per second. The logic of this solution appears obvious but there is a serious handicap of economic inertia to consider since sound films have been made at 24 frames per second and studios and theaters have been following the present standard for over 20 years. There is also the problem of providing pulldown time if photography is on an intermittent basis. If a film rate of 30 frames per second is ever adopted, it appears that some method of continuous film motion will be desirable, if the necessary constancy of motion can be obtained.

It is possible to design cameras that use either mechanical or electrical blanking during the pulldown period. Continuously moving-film cameras are also possible but the mechanical, optical, and synchronization problems involved are most difficult.

One 16-mm television-recording camera now in use is equipped with a mechanical shutter driven by a synchronous motor from the same 60-cycle alternating-current power source as is used for the television-synchronizing generator. This shutter has a closed angle of 72 degrees and an open angle of 288 degrees. At the 24-cycle rate this represents a closed time of $^1/_{120}$ second and an open time of $^1/_{30}$ second. The latter is equivalent to one full television-frame cycle.

Fig. 2 shows the time sequence of this shutter in relation to the 30-frame (60-field) television-scanning cycle. The camera shutter remains open for exactly two television fields, closes for exactly $1/2$ field while the film is advanced, then opens again for the exact equivalent of two more television fields (actually $1/2$ plus 1 full plus $1/2$ field). It then closes for $1/2$ field while the film is advanced a second time and again opens at exactly the beginning of the next field. The two non-symmetrical cycles are then repeated.

One serious objection to the mechanical shutter for television-picture recording lies in the need for perfect synchronization between the motor that drives the shutter and the television frame-rate generator

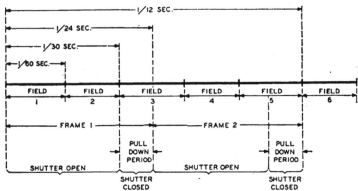

Fig. 2—Time sequence of exposure and pulldown timing of the camera in relation to the field rate of the television image.

which may not necessarily operate from the same 60-cycle alternating-current power line. The shutter action is critical in that it must rotate with extremely low flutter content since minute changes in angular velocity will result in banding, the effect of over- or under-exposure of scanning lines adjacent to the cutoff point.

With the electronic shutter now being used with some installations, this problem is minimized because the television-picture tube is electronically blanked or turned off at the end of each 525 lines (one complete television-frame cycle) and is not turned on again until the film has been pulled down and brought to rest. Also, the electronic shutter can accommodate any film-frame rate less than a given maximum determined by the practical limitations of film-pulldown time.

EQUIPMENT CONSIDERATIONS, 16-MM VERSUS 35-MM

The majority of television-film recordings are made on 16-mm rather than 35-mm film. The major reason is economic, since the cost of 35-mm film is somewhat more than three times the cost of 16-mm per unit of recording time. The current quality of television images, which undoubtedly will undergo gradual refinement, is considered to be roughly equivalent to 16-mm home motion pictures. No marked improvement, however, is to be had by recording on 35-mm rather than 16-mm film at the present time. With the use of fine-grain, high-resolution, 16-mm-film emulsions, no loss of resolution in recording the television image is noticeable.

Fire regulations covering the use of 35-mm film, which apply regardless of whether the 35-mm film is acetate safety base or the combustible nitrate base, are rigorous. The cost of providing space that meets these regulations for the use of 35-mm film is extremely high and the changes needed in existing space are difficult to accomplish. Sixteen-millimeter films are available only in acetate safety base which is classified by the Underwriters' Laboratories as having a safety factor slightly higher than that of newsprint. The use of 16-mm films, therefore, is not restricted by fire regulations. It should be noted that in New York City these restrictions apply to space in which equipment capable of operating with 35-mm film is installed, so in order to forestall trouble, all equipment should be single-purpose, 16-mm equipment rather than dual-purpose, 35-mm or 16-mm equipment.

Another factor in the choice of 16-mm film is the high cost of 35-mm projection equipment. Most television stations are providing projection facilities for 16-mm film only for this reason. In order to service these stations with syndicated programs photographed from the picture tube, 16-mm prints will be needed.

SPECTRAL CHARACTERISTICS

Film Emulsion—There are three general classifications of film emulsions in terms of their spectral characteristics and they can be matched to the phosphor spectral characteristic of the television-picture tube, for greatest actinic efficiency.

1. Panchromatic emulsions are most sensitive in the range from the ultraviolet (4000 angstrom units) through the red (7000 angstrom

units). The spectral response of these emulsions corresponds approximately to that of the eye and so they are generally used for direct photography;

2. Orthochromatic emulsions are sensitive from the ultraviole-through green (5700 angstrom units) and are used in direct photograt phy where it is desirable to reduce the red sensitivity;

3. "Ordinary," blue-sensitive emulsions, respond to the ultraviolet and blue portions of the light spectrum. This type of emulsion is used in coating films and papers generally employed in making positive prints from negatives. It is economical in comparison to the panchromatic and orthochromatic types. Another advantage is the ease of handling as relatively bright safelights may be used.

Picture-Tube Phosphors—To match these film characteristics, picture-tube phosphors are available with light output ranging from the ultraviolet through the entire visual spectrum. Three types of phosphors in common use in television techniques are as follows:

1. $P1$, green fluorescence, commonly used in oscillographic work. It is the most efficient visually, but has poor actinic efficiency.

2. $P4$, white fluorescence, used for black-and-white reproduction of television images in most home receivers. It has the advantage in picture-tube photography that picture quality is most readily judged visually. However, some $P4$ screens have undesirable decay characteristics.

3. $P5$ and $P11$; these two phosphors are blue with high ultraviolet output. Photographically, they are very efficient. There is the difficulty in using a blue phosphor in judging the quality of image visually, because of the fact that the human eye has a low response in the blue region and cannot evaluate the quality of the ultraviolet component of the image-light output at all.

Tests have indicated that for recording of television images a blue-fluorescing screen ($P5$ or $P11$) is desirable since it makes possible the use of high-resolution, low-cost, positive types of film stocks. The $P5$ screen has excellent persistence characteristics but produces a somewhat lower light level than that which can be obtained with $P11$.

FILM EXPOSURE

A method for establishing brightness range and exposure level is as follows: A plain raster is used on the tube such as would be obtained by the use of the blanking signal or pedestal without picture modulation. The brightness of this raster is varied by means of the video

gain control or picture-tube, grid-bias control. The beam current is
measured by means of a microammeter. Since the light output of the
tube is dependent upon the power input to the screen, the measure of
beam current affords a measure of the brightness of the tube. Film is
exposed to this raster with the beam currents varied in steps. The density of the film processed as a normal negative is measured and plotted
against the logarithm of the beam current. A normal negative developed to a gamma of 0.65, which has been exposed to an object with
a brightness range of 1 to 30 (in logarithmic increments, a range of
1.5) should have a density range from 0.25 in the shadows to approximately 1.4 in the high lights. The change in beam current necessary
to produce such a range on the picture tube can be read from the plot
of the log of the beam current and film density. The average brightness of the cathode-ray tube with picture then would be set by using a
beam current that produces a density in the middle of the above
range. The video signal is adjusted to a level that will put the blanking level of the composite signal just at visual cutoff of the cathode-
ray tube. A picture signal judged to have an alternating-current
axis of 50 per cent should be used for this adjustment. This method
is largely empirical, but, with experience on the part of the operator,
can be made to give consistent results.

PROCESSING AND PRINTING

A number of tests have been made in co-operation with the film
manufacturers on the processing and printing of films photographed
from a television-picture tube. Both reversal and negative processing
of the original film were tried and results show that standard processing methods result in optimum picture quality. Negatives exposed
to television images originating in iconoscope cameras are developed
to a gamma of 0.7 as determined by a standard IIb sensitometric test.
Film of orthicon pickups gives best results when processed to approximately 0.6 or 0.65. These are interim values as tests on the processing of these films have not been completed.

Printing is done according to standard motion picture laboratory
practice. Step printing in which the print stock and negative are exposed to the printing light a frame at a time is preferred over continuous printing, where the negative and print stock run past an illuminated slit at a continuous speed. There is a sufficient amount of slippage between the negative and the print stock in the continuous
printing process to degrade the resolution of the television image.

Contrary to the opinion held by many workers, the fact that the film picture of a television image is poorer in resolution than in the case of direct photography does not mean that less care is required in the handling of the film in printing and in projection. The fact is that the utmost care must be taken to maintain the original quality inherent in the film negative throughout the printing process and in the projection of the resulting print.

Films of iconoscope programs usually can be printed at one printer-light setting, that is, the densities and contrast range of the film resulting from the recording of the outputs of a number of iconoscope cameras do not change sufficiently to warrant changes in the intensity of the printing light.

In film recordings of programs picked up by orthicon cameras the picture negative often has to be timed for printing. Frequently there is some difference in the brightness range between different orthicon cameras. Much of this change can be charged to the fact that the spectral characteristics of the orthicon may vary from tube to tube. An orthicon with high infrared response has a somewhat different tonal graduation than an orthicon with lower response in this region.

In recording for retransmission through the television system a print gamma of 2.2 and a maximum density of 2.4 have been found satisfactory. Further tests may show the desirability of changing these recommendations, but to date the best results in the televising of release prints have been obtained under such conditions.

Emulsion position in the final print is of importance in television because films may be spliced with other films for special purposes. The use of a nonstandard emulsion position requires a change of focus in the film projector when interspliced with films using a standard emulsion position. This would require the constant attention of the projectionist to maintain optimum focus throughout the spliced film; therefore it is advantageous to insist upon a standard emulsion position for all film to be used in television. The American Standard for 16-mm film is emulsion "toward the screen."

In the recording of television images there are several methods of obtaining the final print:

1. The use of reversible film stock in photographing a positive cathode-ray-tube image. A dupe negative may be made of this material from which additional prints can be made. The final prints then have standard emulsion position;

2. Photography using high-contrast positive stock and a negative picture-tube image resulting in a positive print from which dupe negatives may be made if production prints are required. These prints will have standard emulsion position;

3. The use of a positive image, photographing with a negative type of film from which final prints are made, resulting in a nonstandard emulsion position. (By reversing the direction of horizontal scanning, however, the original negative may be made to have the same emulsion position as that of a dupe negative. Prints made from this negative then have standard emulsion position.)

When production prints are required Method 3 is now used almost exclusively since it eliminates the use of a dupe negative and consequently introduces less total degradation. Methods 1 and 2 do not produce production prints of suitable quality for present-day commercial television.

HISTORICAL BACKGROUND

An early system of film scanning was described by Ives (1931) as incidental to a three-channel system of television. The three channels were used to obtain the desired resolution, without increasing the frequency bandwidth beyond the technique of the art then available (40 kilocycles). The three channels were optically separated into three independent interlaced fields, giving 108 lines in all. The film was standard 35-mm, drawn continuously past a mechanically scanning Nipkow disk at 18 frames per second. No mention is made of how this was reconciled to the standard frame speed.

A later film scanner, also attributed to Ives (1938), was used to test the Bell System coaxial cable from New York to Philadelphia for television transmission. This used 240 sequentially scanned lines at the standard 24 frames per second. It also employed standard 35-mm film, drawn continuously past a Nipkow disk. The disk, however, was fitted with lenses instead of holes. Here there was no problem of frame-rate conversion.

An elaborate development was carried out in Germany by Fernseh AG (1939) on an intermediate-film quick processing device adapted to be used both for pickup of news and similar events, and for theater projection.

The pickup device used standard 35-mm film, but the exposed frame was half size in both dimensions, to save film; 16-mm film was not used because it lacked the strength necessary for the quick processing baths. In its later form it followed the German 441-line, interlaced-scanning, 25-frame-per-second, television standards. This used a multiple-spiral Nipkow disk with continuously moving film, with no frame-rate conversion problem (the film being taken at 25 instead of 24 frames per second). It was in a later form replaced by a dissector tube.

An extensive photographic investigation was made for the quick processing. A special thin emulsion was used. In the later model the processing times were

Development	5.0 seconds
Intermediate bath	2.5 seconds
Fixing	15.0 seconds
Washing	10.0 seconds
Drying	43.0 seconds

The negative film was scanned directly after drying.

In one unusually elaborate form of the apparatus the film, after using, was again washed, scraped free of emulsion, dried, coated with fresh emulsion, dried, and used again.

The whole equipment was set up in a special television truck, a series of which was built.

The intermediate-film projector was designed for the earlier 180-sequential-line scanning at 25 frames per second. It used split film, 17.5 mm wide, with an 8- × 11-mm image. This was a positive, taken from a negative image on a 12-kilovolt cathode-ray tube. The camera used intermittent film motion synchronized with the television. The processing times were

Development	24 seconds
Fixing	24 seconds
Washing	12 seconds
Drying	(not stated)

The projection was on to a 2.2- × 3-meter screen. Because of the high film cost and the rapid advances in projection tubes the German intermediate-film projector was abandoned.

A-film scanner attributed to Jensen (1941) was used by the Bell System for testing the prewar television transmission circuits over coaxial cable. This used standard size but specially printed 35-mm film. The film was drawn continuously past a gate and focused on the photosensitive cathode of an image-dissector pickup tube. Extensive study was made of the focusing and deflecting coils in the latter, to obtain improved results. The special printing of the film was used to obtain the frame-speed conversion and interlacing required, with continuous film motion and mechanical simplicity. In the specially printed film one frame is used for each television field scanning. Thus it is obtained from the original film by printing its odd-numbered frames twice in succession, and its even-numbered frames three times in succession; two successive frames of the original thus occupying five frames in the print. This enabled the television signal to follow the then current 441-interlaced-line, 30-frame-per-second standards. The blanking period was adjusted from the film standard to the television standard by a slight compensating vertical sweep in the image-dissector tube. The light source used was a 1000-watt incandescent lamp.

The sound track was printed specially on the film also, "stretched" in the direction of motion in the ratio of 2.5 to 1. Aside from this, the sound pickup was standard.

COLOR-TELEVISION SYSTEMS

An early system of color-film scanning is also attributed to Ives (1931). This made use of 16-mm Kodacolor film of that time. Film motion was continuous and the scanning mechanical with a Nipkow disk. The colors were led separately by lenses and mirrors to three phototubes (so that the system was of the simultaneous type). Because of the nature of the Kodacolor film, with its lenticular markings on the film surface, the color separation was already obtained geometrically, and no filters were necessary. The television standards were 50 lines, 18 frames per second, which previously had been used with black-and-white. The received signal was reproduced on three lamps, superposed optically on a Nipkow scanning disk, and viewed monocularly through an eyepiece by a single observer.

The Columbia Broadcasting System for some years has been intensively developing a color system. On September 27, 1946, this was proposed in a petition to the Federal Communications Commission as the basis for a commercial broadcast service in color television. After an extensive hearing on the subject, however, the petition was denied on March 18, 1947.

The Columbia system, in so far as it used film, has principally used 16-mm film because it was expected that the major available material would be in that size, but a 35-mm machine using the same principles has also been in preparation. The film in each case is standard and operates at 24 frames per second. The film is driven continuously past a special optical system using an arc lamp which focuses successive fields, as they are to be scanned, on to the photosensitive cathode of an image-dissector tube. These act in co-operation with a suitable compensating vertical sweep with this tube, to give the correct interlaced scan required. The CBS proposed standard calls for 144 fields (or scans) so that a film-frame to scanning-field conversion of 24 to 144 (or 1 to 6) is required. This is accomplished with using a special optical system which allows each film frame to be scanned six times as the frame moves past the gate aperture. There is a special adjustment for differing film shrinkage, which involves a change of magnification and a refocusing of the photosensitive cathode.

The CBS system is arranged with a set of six fixed-color filters through each of which the beam is directed in turn by the special optical system. These function in the same manner as would a rotating tricolor disk and allow successive fields to be scanned in the successive three color primaries. Six field scannings are necessary. The signal transmits the picture fields in the successive colors sequentially. However, a synchronous arrangement is also provided for adjusting the signal gain for any one color independently, to permit modifying the color balance while the apparatus is running.

The sound pickup is conventional.

Principally to obtain certain elements of flexibility not easily permitted by a sequential color system, the Radio Corporation of America has worked on a simultaneous color system, and has demonstrated experimental versions of it on various occasions.

In this case, 16-mm color film is used, which is driven past a flying-spot cathode-ray-tube scanner. The beam, after passing through the film, is separated by special mirrors via three filters to three non-storage phototubes, each of which generates one of the three simultaneous signals to be propagated.

The standards for each color have been taken by RCA to be the same as present black-and-white broadcast standards, namely, two interlaced fields per frame, at a frame rate of 30 per second. In fact the green channel is arranged to be used to reproduce a black-and-white picture, this being one of the items of flexibility desired. In the experimental demonstrations which have been given there has been no provision for frame-rate conversion, so that the action in the film is speeded up in the ratio of 30:24. Similarly no arrangement has been provided for adjusting the television to the film-blanking period (which latter is nearly zero in 16-mm film) so that a portion of the frame appears black in the reproduction.

ACKNOWLEDGMENT

This report, prepared by the Television Committee of the Society of Motion Picture Engineers, contains information on points of common interest to television and motion picture engineers.

Because of the rapidly changing state of the art, it was found impossible to present complete data on all aspects of films in television. An attempt has been made, however, to present as much information as possible concerning present practices in the hope that it will serve as a guide and aid to those seeking information in this field.

The membership of the committee is as follows:

D. R. WHITE, *Chairman*
Du Pont
R. B. AUSTRIAN
Television Consultant
F. T. BOWDITCH
National Carbon Company
F. E. CAHILL
Warner Brothers Pictures
A. W. COOK
Ansco
E. D. COOK
General Electric Company
C. E. DEAN
Hazeltine Electronics Corporation
BERNARD ERDE
Columbia Broadcasting System
R. L. GARMAN
General Precision Laboratories
FRANK GOLDBACH
International Projector Corporation
P. C. GOLDMARK
Columbia Broadcasting System
A. N. GOLDSMITH
Consultant
T. T. GOLDSMITH, JR.
Allen B. DuMont Laboratories
HERBERT GRIFFIN
International Projector Corporation
RICHARD HODGSON
Paramount Pictures
C. F. HORSTMAN
RKO Theaters

L. B. ISAAC
Loew's Theaters
P. J. LARSEN
Consultant
C. C. LARSON
Farnsworth Research Corporation
NATHAN LEVINSON
Warner Brothers Pictures
J. P. LIVADARY
Columbia Pictures
H. B. LUBCKE
Don Lee Broadcasting System
PIERRE MERTZ
Bell Telephone Laboratories
H. C. MILHOLLAND
Allen B. DuMont Laboratories
W. C. MILLER
Metro-Goldwyn-Mayer Studios
J. R. POPPELE
Bamberger Broadcasting Service
PAUL RAIBOURN
Paramount Pictures
OTTO SANDVIK
Eastman Kodak Company
G. E. SAWYER
Samuel Goldwyn Studio Corporation
R. E. SHELBY
National Broadcasting Company
E. I. SPONABLE
Movietonews
H. E. WHITE
Eastman Kodak Company

BIBLIOGRAPHY

The following is a bibliography of material dealing with the relationships between television and motion pictures. It has been divided into the following sections.

1. General   3. Film from Television Sources
2. Television from Film Sources   4. Color Television

Under "general" are classified not only discussions dealing with the broader subject, but also those covering more than one of the subsequent topics.

This bibliography has been quickly gathered from such sources as have been readily available, and does not purport to be complete.

General

(1) Alfred N. Goldsmith, "Theater television—A general analysis," *J. Soc. Mot. Pict. Eng.*, vol. 50, pp. 95–122; February, 1948.

(2) Ralph B. Austrian, "Showmanship side of television," *J. Soc. Mot. Pict. Eng.*, vol. 49, pp. 395–405; November, 1947.

(3) W. V. Wolfe, "Report of the SMPE Committee on Progress," *J. Soc. Mot. Pict. Eng.*, vol. 48, pp. 304–317; April, 1947.

(4) "Statement of SMPE on revised frequency allocations," *J. Soc. Mot. Pict. Eng.*, vol. 48, pp. 183–203; March, 1947.

(5) Lester B. Isaac, "Television and the motion picture theater," *J. Soc. Mot. Pict. Eng.*, vol. 47, pp. 482–487; December, 1946.

(6) Albert Rose, "A unified approach to the performance of photographic film, television pickup tubes, and the human eye," *J. Soc. Mot. Pict. Eng.*, vol. 47, pp. 273–295; October, 1946.

(7) Allen B. DuMont, "The relation of television to motion pictures," *J. Soc. Mot. Pict. Eng.*, vol. 47, pp. 238–248; September, 1946.

(8) P. J. Larsen, "Report of the Committee on Television Projection Practice," *J. Soc. Mot. Pict. Eng.*, vol. 47, pp. 118–120; August, 1946.

(9) Judy Dupuy, "Television Show Business" (book review), *J. Soc. Mot. Pict. Eng.*, vol. 46, p. 424; May, 1946.

(10) A. Rose, "Photographic film, television pickup tubes, and the eye," *Intern. Proj.*, May, 1946.

(11) "Technical News," *J. Soc. Mot. Pict. Eng.*, vol. 46, p. 81, January, 1946.

(12) Ralph B. Austrian, "Film—The backbone of television programming," *J. Soc. Mot. Pict. Eng.*, vol. 45, pp. 401–414; December, 1945.

(13) Ralph B. Austrian, "Some economic aspects of theater television," *J. Soc. Mot. Pict. Eng.*, vol. 44, pp. 377–386; May, 1945.

(14) Paul J. Larsen, "Statement presented before the Federal Communications Commission relating to television broadcasting," *J. Soc. Mot. Pict. Eng.*, vol. 44, pp. 123–128; February, 1945.

(15) "Technical News," *J. Soc. Mot. Pict. Eng.*, vol. 43, pp. 303–304; October, 1944.

(16) Worthington C. Miner, "Film in television: Television production as viewed by a radio broadcaster," *J. Soc. Mot. Pict. Eng.*, vol. 43, pp. 79–93; August, 1944.

(17) Wyllis Cooper, "Film in television: Television production as viewed by a motion picture producer," *J. Soc. Mot. Pict. Eng.*, vol. 43, pp. 73–79; August, 1944.

(18) "Television report, order, rules, and regulations of the Federal Communications Commission," *J. Soc. Mot. Pict. Eng.*, vol. 37, pp. 87–98; July, 1941.

(19) "Report of the Television Committee" (Flicker, Visual Fatigue, Bibliography), *J. Soc. Mot. Pict. Eng.*, vol. 35, pp. 569–584; December, 1940.

(20) M. W. Baldwin, "The subjective sharpness of simulated television images," *Bell Sys. Tech. Jour.*, vol. 19, p. 563; October, 1940.

(21) P. C. Goldmark and J. N. Dyer, "Quality in television pictures," *J. Soc. Mot. Pict. Eng.*, vol. 35, pp. 234–254; September, 1940.

(22) A. M. Skellett, "Transmission system of narrow band-width for animated line images," *J. Soc. Mot. Pict. Eng.*, vol. 33, pp. 670–677; December, 1939.

(23) G. Schubert, W. Dillenburger, and H. Zschau, "Das Zwischen Film verfahren," *Fernseh A. G. Hausmitteilungen*, vol. 1, Part I, p. 65; April, 1939; Part II, p. 162, August, 1939; Part III, p. 201, December, 1939.

(24) R. Moller and G. Schubert, "Zehn Jahre Fernsehtechnik," *Fernseh A. G. Hausmitteilungen*, vol. 1, p. 111; July, 1939.

(25) "Report of the Television Committee," *J. Soc. Mot. Pict. Eng.*, vol. 33, pp. 75–80; July, 1939.

(26) G. L. Beers, E. W. Engstrom, and I. G. Maloff, "Some television problems from the motion picture standpoint," *J. Soc. Mot. Pict. Eng.*, vol. 32, pp 121–139; February, 1939.

(27) A. D. Blumlein, C. O. Browne, N. E. Davis, and E. Green, "The Marconi-EMI television system," *J.I.E.E.* (London), p. 758; December, 1938.

(28) Herbert E. Ives, "Transmission of motion pictures over a coaxial cable," *J. Soc. Mot. Pict. Eng.*, vol. 31, pp. 256–273; September, 1938.

(29) M. E. Strieby, "Coaxial-cable system for television transmission," *Bell Sys. Tech. Jour.*, vol. 17, p. 438; July, 1938.

(30) "Television demonstration at the fall convention," *J. Soc. Mot. Pict. Eng.*, vol. 29, pp. 596–603; December, 1937.

(31) "Television from the standpoint of the motion picture producing industry," *J. Soc. Mot. Pict. Eng.*, vol. 29, pp. 144–149; August, 1937.

(32) R. R. Beal, "RCA developments in television," *J. Soc. Mot. Pict. Eng.*, vol. 29, pp. 121–144; August, 1937.

(33) Alfred N. Goldsmith, "Television and the motion picture theater," *Intern. Proj.*, May, 1935.

(34) O. H. Schade, "Electrooptical characteristics of television system," Part I, "Characteristics of vision and visual systems," *RCA Rev.*, vol. 9, p. 5; March, 1948; Part II, "Electrooptical specifications for television systems," *RCA Rev.*, vol. 9, p. 245; June, 1948.

Television from Film

(1) M. R. Boyer, "Test reel for television broadcast stations," *J. Soc. Mot. Pict. Eng.*, vol. 49, pp. 391–395; November, 1947.

(2) R. V. Little, "Film projectors for television," *Intern. Proj.*, May, 1947.

(3) Ralph V. Little, Jr., "Film projectors for television," *J. Soc. Mot. Pict. Eng.*, vol. 48, pp. 93–111; February, 1947.

(4) E. Meschter, "Television reproduction from negative films," J. Soc. Mot. Pict. Eng., vol. 47, pp. 165–182; August, 1946.

(5) Ellsworth D. Cook, "General Electric television film projector," J. Soc. Mot. Pict. Eng., vol. 41, pp. 273–292; October, 1943.

(6) R. B. Fuller and L. S. Rhodes, "Production of 16-mm motion pictures for television projection," J. Soc. Mot. Pict. Eng., vol. 39, pp. 195–202; September, 1942.

(7) Axel G. Jensen, "Film scanner for use in television transmission tests," Proc. I.R.E., vol. 29, pp. 243–250; May, 1941.

(8) Harry R. Lubcke, "Photographic aspects of television operations,".J. Soc. Mot. Pict. Eng., vol. 36, pp. 185–191; February, 1941.

(9) C. Frederick Wolcott, "Problems in television image resolution," J. Soc. Mot. Pict. Eng., vol. 36, pp. 65–82; January, 1941.

(10) R. L. Campbell, "Television control equipment for film transmission," J. Soc. Mot. Pict. Eng., vol. 33, pp. 677–690; December, 1939.

(11) Peter C. Goldmark, "Continuous type television film scanner," J. Soc. Mot. Pict. Eng., vol. 33, pp. 18–26; July, 1939.

(12) E. W. Engstrom, G. L. Beers, and A. V. Bedford, "Application of motion picture film to television," J. Soc. Mot. Pict. Eng., vol. 33, pp. 3–18; July, 1939; RCA Rev., vol. 4, p. 48; July, 1939.

(13) H. S. Bamford, "Non-intermittent projector for television film transmission," J. Soc. Mot. Pict. Eng., vol. 31, pp. 453–462; November, 1938.

(14) K. Thöm, "Neuer mechanischer Filmabtaster," Fernseh A. G. Hausmitteilungen, vol. 1, p. 24; August, 1938.

(15) H. E. Ives, "A multi-channel television apparatus," Bell Sys. Tech. Jour., vol. 10, p. 33; January, 1931.

Film from Television

(1) R. M. Fraser, "Motion picture photography of television images," RCA Rev., vol. 9, p. 202; June, 1948.

(2) C. F. White and M. R. Boyer, "A new film for photographing the television monitor tube," J. Soc. Mot. Pict. Eng., vol. 47, pp. 152–165; August, 1946.

(3) F. G. Albin, "Sensitometric aspect of television monitor-tube photography," J. Soc. Mot. Pict. Eng., vol. 51, pp. 595–613; December, 1948.

(4) J. L. Boon, W. Feldman, and J. Stoiber, "Television recording camera," J. Soc. Mot. Pict. Eng., vol. 51, pp. 117–127; August, 1948.

(5) Thomas T. Goldsmith, Jr., and Harry Milholland, "Television transcription by motion picture film," J. Soc. Mot. Pict. Eng., vol. 51, pp. 107–117; August, 1948.

Color Television

(1) W. H. Cherry, "Colorimetry in television," RCA Rev., vol. 8, pp. 427–460; September, 1947; J. Soc. Mot. Pict. Eng., vol. 51, pp. 613–643; December, 1948.

(2) R. D. Kell, "An experimental simultaneous color-television system, Part I, Introduction," Proc. I.R.E., vol. 35, pp. 861–862; September, 1947.

(3) G. C. Sziklai, R. C. Ballard, and A. C. Schroeder, "Part II, Pickup equipment," Proc. I.R.E., vol. 35, pp. 862–871; September, 1947.

(4) K. R. Wendt, G. L. Fredendall, and A. C. Schroeder, "Part III, radio-frequency and reproducing equipment," *Proc. I.R.E.*, vol. 35, pp. 871–875; September, 1947.

(5) Statements and Exhibits of CBS and RCA at FCC Hearing on Color Television, December 9, 1946.

(6) UHF Television Systems, Reports by RMA Committees, Data Bureau, RMA, November 26, 1946.

(7) Interim Report, UHF Color Television, RTPB Panel 6, RMA Television Systems Committee, Data Bureau, RMA, November 25, 1946.

(8) "Simultaneous all electronic color television," *RCA Rev.*, vol. 7, p. 459; December, 1946.

(9) R. D. Kell, G. L. Fredendall, A. C. Schroeder, and R. C. Webb, "An experimental color television system," *RCA Rev.*, vol. 7, p. 141; June, 1946.

(10) P. C. Goldmark, E. R. Piore, J. M. Hollywood, T. H. Chambers, and J. J. Reeves, "Color television—Part II," *Proc. I.R.E.*, vol. 31, pp. 465–479; September, 1943.

(11) P. C. Goldmark, J. N. Dyer, E. R. Piore, and J. M. Hollywood, "Color television—Part I," *Proc. I.R.E.*, vol. 30, pp. 162–182; April, 1942.

(12) P. C. Goldmark, J. N. Dyer, E. R. Piore, and J. M. Hollywood, "Color television," *J. Soc. Mot. Pict. Eng.*, vol. 38, pp. 311–352; April, 1942.

(13) F. W. Marchant, "New Baird color television system," *Telev. and Short Wave World*, vol. 12, p. 541; September, 1939.

(14) J. L. Baird, "Color television," *Telev. and Short Wave World*, p. 151; March, 1938.

(15) H. E. Ives, "Television in color from motion picture film," *J. Opt. Soc. Amer.*, vol. 21, p. 2; January, 1931.

(16) H. E. Ives and A. L. Johnsrud, "Television in colors by a beam scanning method," *J. Opt. Soc. Amer.*, vol. 20, p. 11; January, 1930.

(17) Bernard Erde, "Color-television film scanner," *J. Soc. Mot. Pict. Eng.*, vol. 51, pp. 351–373; October, 1948.

Possibilities of a Visible Music*

By RALPH K. POTTER†

BELL TELEPHONE LABORATORIES, MURRAY HILL, NEW JERSEY

Summary—During the past two centuries many attempts have been made to produce a visible music. If realizable, such a music combined with the existing sound music should find widespread application in screen-and-sound entertainment. Requirements for such an art are considered primarily from a scientific rather than artistic point of view, and a new development approach is proposed.

OVER 200 YEARS AGO a French mathematician and philosopher by the name of Louis Bertrand Castel proposed a visible music. He was probably the first to suggest specific possibilities of such a music and to attempt construction of an instrument. Castel thought of visible music as changing colored light and tried to associate color and musical tone. Others carried on the search in this direction, and until the late 1800's emphasis remained upon *color*. Experimental instruments built during this period were called "color organs." Then, following the color era, attention shifted to *form*. Aided by new electrical techniques, the earlier color organs became elaborate projection instruments under a variety of names. Where early experimenters with color organs had thought largely of projected color, the more recent effort was directed toward producing abstract shapes of all sorts under keyboard and other manual control.[1] While interest in form continued among those experimenting with instruments, a new trend was taking shape in the motion picture area.[2] Here emphasis was upon *movement* and for good reason; many of the experimenters, being animationists, were keenly conscious of visible and audible relationships in movement.

Summing up the results of all this background experience to date, we find the following: Efforts to produce projected light effects, that by any stretch of the imagination one might call "visible music,"

---

* Presented October 25, 1948, at the SMPE Convention in Washington.

† NOTE: For a number of years the author has been interested in efforts that have been made and are being made to produce a visible music. This interest has not been a professional one, that is, Bell Telephone Laboratories are not carrying on studies in this field.

have had very limited success. Attempts to use manually controlled projection instruments along with sound music instruments have been notably unsuccessful. Only in the film medium have encouraging results been obtained with combinations of sound music and abstract display. Here, in some notable instances at least, use of abstract color form in motion to the accompaniment of sound music has shown definite entertainment value. A familiar example is Walt Disney's classic, "Fantasia," wherein there are passages containing abstract shapes in color moving to the tune of sound music. Colored disks, cometlike points of color, waving bands of color, twisting lines, spreading beams, all move to the music. Similar passages have also appeared in some of the more recent Disney musical animations.

In compositions such as these in the Disney films and elsewhere we are certainly seeing the development of an art that deserves attention. Still, it is suggested that this development is not toward a "visible music." A better description of these performances would probably be "abstract shapes in color, moving to the rhythm of accompanying sound music" or, briefly, "dancing abstractions." It seems likely that few observers would feel inclined to think of the screen display as a visible music.

Where then must we look for a visible music? In particular, how would we know visible music if we were to see it? While this latter question seems, on first consideration, to be the kind we should prefer to leave to the philosophers, there is actually a simple answer. It is this: If we were to hear sound music and at the same time see a screen display that we feel *is* that music, the logical name for that display would be "visible music"! It would be difficult to think of it in any other terms.

When screen display and sound music blend into one the result is unison, and so in searching for a visible music the thing we should look for first is what we might call "audivisual unison." Say that we start with an exceedingly simple combination. Assume that a single tone is coming from a sound system and a single vertical line appears upon an associated screen. Assume further that the tone can be made louder, or weaker, or shifted in pitch as we please, and that the vertical line may be made larger, or smaller, or moved from side to side. Now if we hear a tone of fixed loudness and pitch and, at the same time, see the vertical line standing motionless on the screen there will probably be no feeling at all that the two are related. They simply represent two perceptual experiences. But now say that as the line moves toward us (growing larger) the accompanying tone grows louder, and when the line moves away (growing smaller) the tone grows weaker. A feeling that the two are related develops quickly. We begin to think that the line *is* the tone.

So far only one dimension of movement has been employed—that along a line toward and away from the observer. Next say that the line is moved from side to side without change in size. When it moves right the accompanying tone rises in pitch. When it moves toward the left the tone falls in pitch. Again there is clear evidence of a relationship. For this second part of the demonstration it is not essential that the line always move to the right with increasing pitch for close relationship. As a matter of fact many observers may prefer to see it turned on its side and moved upward with rising pitch. It has been my experience that a number of relationships become acceptable after brief conditioning. *The important point is this:* Once a relationship is adopted it must not be changed about, or all sense of association may disappear. If the line first moves sideways with pitch, then up and down, then twists around, the observer becomes lost. The effect seems analogous to what might result if the scale of a sound music were being changed continually.

Movement, and movement alone, is the associating factor in the above-described line-and-tone combination. To the film animationist this is no discovery; he is well aware of the power of movement in audivisual association. His art is built around these effects. But there is more to this association that the animationist has needed to know in applying his art and this less familiar part bears particularly upon the visible-music problem. Consider first the meaning of movement.

It is not generally appreciated that a tone can move just as realistically as a line can move on a screen. When we "see" an object, such as a line on a screen, a pattern of stimulation exists in the brain. When the line moves across the field of view the stimulation pattern shifts. When the line grows larger the stimulation pattern grows larger. In another part of the brain the experience of hearing a sustained tone also produces a stimulation pattern and this pattern is the same type that was produced in the visual area. Furthermore, when the pitch (or frequency) of the tone is changed the stimulation pattern shifts, and when the tone is made louder the stimulation pattern grows larger. Consequently, when line and tone in the above-described demonstration seem to move together, the observer is in reality associating two similar kinds of movements, and in this association apparently resides the basis for audivisual unison.

However, there is one fundamental difference between visible and audible movement; while we can see an object move in *three* dimensions, we can only hear a tone move in *two* dimensions. An object on a screen is able to move along one dimension toward or away from us, another up and down, and a third from side to side. The tone can become louder or weaker in one dimension, and it can shift along the pitch scale, which is another dimension. It has no third dimension of movement.

Three-dimensional seeing and two-dimensional hearing has an explanation in terms of the way eyes and ears are built. In the eye the end organs of visual perception are spread over a surface, whereas in the ear those of aural perception are arranged in a row. If our eyes were slits, so that we could only see in two dimensions, our visual experience would certainly be greatly restricted, but we might have had a visible music earlier. Relationships in audible and visible movement could then have developed without the two- versus three-dimensional conflict.

An important conclusion emerging from the foregoing discussion is that any visible display capable of very close association with sound music will have to be *two-dimensional, with one dimension in the direction of observation and the other in a lateral direction.* When we use such a display, visible movement apparently can be made to conform with audible movement, or audivisual unison is realizable, and we have what according to our specification is a visible music. Incidentally, we also have the distinction between "visible music" and "dancing abstractions." The latter is three-dimensional, and the third dimension makes it a different art with different possibilities of artistic expression.

While the simple line-and-tone combination has served to illustrate the basic principles of audivisual unison, and to indicate a fundamental requirement for visible music, a real sound-and-screen music would need to utilize many tones and many lines. Without certain technical aids it would be exceedingly difficult to demonstrate audivisual unison in the case of a real music. An aid of the required kind is the sound spectrograph that has been built experimentally in Bell Telephone Laboratories for studies of speech, noise, and other sounds.[3] This instrument translates complex sounds into two-dimensional patterns that show frequency horizontally and intensity by vertical size, as in the line-and-tone case. Patterns of sounds obtained with the sound spectrograph are sometimes made into motion picture sequences that facilitate the study of audible movements. Musical sounds have been pictured in this way and projected along with the sound music. Of particular interest in relation to this discussion is the fact that these motion pictures provide examples of audivisual unison, thus extending the line-and-tone demonstration to many lines and many tones. According to our definition these projected patterns of musical sounds are a visible music. And they seem to impress the observer as such, but it is visible music in a completely inartistic form, simply sound and pattern in unison, in black and white, presented in an arrangement that, though adequate for laboratory use, is poor artistically.

Successive frames of sound-spectrograph patterns, representing the movement in a passage by full orchestra, appear in Fig. 1. In each pattern frequency increases toward the right and the range included is approximately 3500 cycles a second. Louder tones appear as higher peaks. Each peak pictures one tone and its movements. A succession of equally spaced peaks make up the overtones of each musical note and the different notes combine to form a complex design. The designs may assume as many configurations as there are audibly different musical sounds. On the screen the action resembles that of flames, but the action is organized, not random; it is the organization of music.

Some have argued that visible movement as rapid as that in sound music is intolerable to the eye. This argument does not seem to be supported by experience with the motion pictures of musical sounds described above. Without doubt rapid visible movement *can* be exceedingly annoying. A flashbulb set off in the field of view is often extremely disturbing, but it should be borne in mind that equivalent audible effects are also disturbing. The unexpected sound of a firecracker set off near by is an illustration.

In a screen display rapid movement can be especially irritating if the visible movements are unrelated to accompanying sound. The result is audivisual bedlam. Such bedlam may be produced by throwing sound and associated sound patterns out of synchronism and its effect is similar to that of sound and picture out of step in an ordinary motion picture. Perhaps lack of relation between visible and audible movements may have been responsible for many of the difficulties that past experimenters in the visible-music field have had

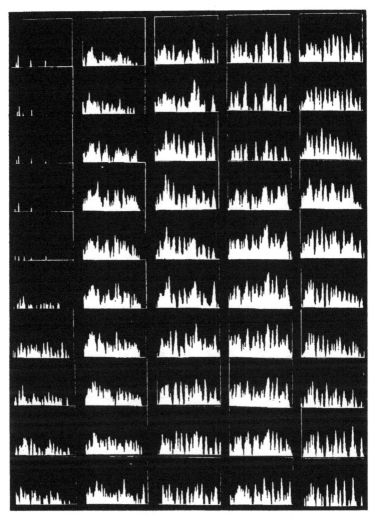

Fig. 1—Successive spectrograms of sound music (full orchestra) that in a motion picture sequence make a laboratory-type "visible music."

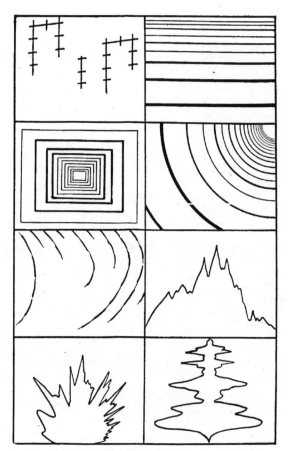

Fig. 2—A few of the two-dimensional display arrangements
that meet requirements for use as a "visible music."

with rapid visible movement in the combination of screen displays
with sound music.

Visible music evidently can assume a wide variety of forms within
the two-dimensional specification, although experience will doubtless
show that some of these are more acceptable than others. A few of
the possible arrangements are pictured by line diagrams in Fig. 2.
All these contain two distinct dimensions of movement, one toward
and away from the observer and the other in some lateral direction.

In Fig. 3 is shown a more detailed representation of one pattern form pictured in Fig. 2, specifically the first in the left-hand row. The patterns in Fig. 3 were sketched by hand, using sound spectrograms of the kind shown in Fig. 1 as a guide. First in the row at the left is a flute note. Below it is a note of the baritone horn and then one of the clarinet. At the top of the right-hand row is an organ note,

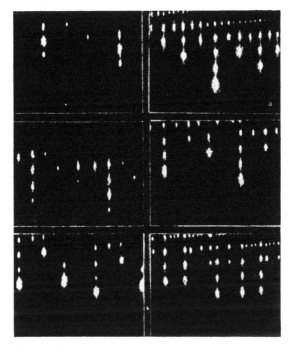

Fig. 3—Notes of several sound-music instruments shown in one type of visible-music display.

then a single violin note, and at the bottom a triad in C, E, and G. If these were displayed in unison with the corresponding sounds they would move to the right with increasing pitch, much as if a procession of patterns were gliding past a window in a slightly uphill direction. Two notes of one instrument an octave apart would produce similar patterns in exactly the same horizontal position, the higher-pitched note pattern being shifted vertically by one dot interval. Instrumental vibrato would cause the whole pattern to oscillate slightly from side to side about as rapidly as one is able to vibrate the hand. Changes in timbre would alter the pattern design. In following the musical movements such a display would have the appearance of light rippling on a surface. Unlike the ripples of nature these in the projected music would assume organized forms, and move in an organized way.

So far the discussion has centered around audivisual unison, because it was proposed initially that such unison constitutes a test for visible music. Still, unison is not the only thing we want in such a music. By far the most entertaining combinations of visible and audible music are likely to be found in visible and audible relationships that do *not* represent unison. If the visible music we have found is to be an expressive music it must be capable of showing these effects. And there is every reason to believe that it would be, for if it can show unison with accompanying sound music it should be equally capable of showing departures from unison. It should be able to reveal contrasts in movement as well as harmony, and all the elaborate interplay of audible and visible movement that would be needed to make an entertaining audivisual music.

Thus far color has hardly been mentioned, and yet color certainly will be an extremely important element in any visible music. That it does not enter into the basic formula for this music is fortunate, because this means its use will not be restricted. The indications are that it may be employed as freely as it is at present on the stage for costumes and scenery. As on the stage, color in visible music would be used in ways to support the mood of the performance. A color theme would be woven through the musical performance as it often is through a performance on the stage. Much like actors on the stage, notes of individual instruments could be dressed in distinctive hues, and the settings for their movements varied in ways to compare with change of stage scenery.

Color certainly must appear in the definition of an artistic visible music. We could, for example, call it "an abstract pattern of color in motion, with movement comparable in dimension and rate with that occurring in sound music." To visualize such a music we might think of leaping flames in color, tremulous reflections of colored lights on rippling surfaces of color, radial beams of flashing color. We could think of these displays moving in slow rhythm, or in quick, explosive movements, or wherever we please between these extremes. In any particular form of display we would need to imagine all the detailed visible movement interwoven with the movement in accompanying sound music, so that sound-and-screen performances combine in a single integrated musical pattern.

Film-animation methods would seem to provide the best means for producing audivisual music. While visible-music instruments, to project the required type of two-dimensional display under manual control, no doubt could be built, their perfection would require a great deal of time and effort. Earlier, easier, and seemingly more versatile possibilities are available in film-animation techniques, supplemented by automatic sound-to-sight translation aids.

Audivisual music based upon simple audivisual unison, and using mobile color as skillfully as it is used now in the better "dancing abstractions," should be widely acceptable—especially during the introductory period. Compositions of this character could be employed as background for title and credit lines, or in passages of musical animations. They could'start with any arrangement for a particular piece of sound music. To permit separate color treatment of notes in the visible-music display, separate pickups and sound recordings would be made for different instruments, or groups of instruments. The separate sound recordings then would be translated automatically into two-dimensional sound spectrograms of the kind shown in Fig. 1, or some other more directly adaptable form. With these as a guide, the film animationist could then picture the movements in any one of the numerous two-dimensional arrangements available. In the process he would apply to note forms and background a previously prepared color theme. Finally, the sound track would be added from one recording of all the instruments in combination.

As audivisual music develops, compositions might be written in two scores, one for the screen music and another for the sound music. The familiar notes and staves of sound music could be employed for both, perhaps with symbolized color suggestions by the composer. The screen music would be played and individual instrument, or instrument-group, recordings made as mentioned earlier. These recordings of the screen music would serve as guide material for the animation artists. The same original screen music could be arranged in an almost endless number of ways by rearrangement of the musical score, choice of display, and variation in color treatment. The sound track would be provided by recordings of the sound-score music.

CONCLUSIONS

The conclusions of the analysis so far outlined are that the possibilities of our having a visible music are excellent. The combination of such a visible music with the familiar audible type will offer the artist new opportunities for expression, and screen-and-sound audiences new and interesting entertainment. Development of this music does not seem to offer any serious difficulties—certainly it is not a venture into the unknown. Much of the way has already been blazed by a related art, the "dancing abstraction."

REFERENCES

(1) For a more extensive discussion of the early history see, "Colored Light—An Art Medium," by A. B. Klein, The Technical Press, Ltd., London, 1937, and "Light and the artist," by Thomas Wilfred, J. *Aesthetics and Art Criticism*, vol. 5, pp. 247–255; June, 1947.

(2) Early film activities are summarized in "Audivisual music," by Ralph K. Potter, *Hollywood Quarterly*, vol. 3, pp. 66–78; Fall, 1947.

(3) L. G. Kersta, "Amplitude cross-section representation with the sound spectrograph," J. *Acous. Soc. Amer.*, vol. 20, pp. 796–801; November, 1948.

Optimum Performance of High-Brightness Carbon Arcs*

By M. T. JONES and F. T. BOWDITCH

NATIONAL CARBON COMPANY, CLEVELAND, OHIO

Summary—The effects of positive-crater cooling are described, and a suitable apparatus for this purpose is illustrated. The combination of specially made high-brightness carbons with water-cooled operation permits the use of higher currents without unsteadiness, and so gives a higher brightness than has been achieved in conventional air-cooled operation. This is attributed to the fact that effective cooling of the positive carbon removes energy which would otherwise be dissipated in turbulent volatilization, so that a higher current density can be achieved in the light-producing gas ball before overload turbulence occurs. A considerable part of the more efficient crater cooling is attributable to the carbons themselves, since they will operate without water cooling at higher currents and brightnesses than other types of equal size.

Within the limit of satisfactory air-cooled operation with a given carbon, efficient water cooling always reduced the light produced at a given current; the ability to operate with higher brightness at higher currents was thus gained at the expense of a lower current efficiency. Carbons designed for efficient air-cooled operation gave no better result with water cooling; the current efficiency was sacrificed with no gain in maximum brightness.

THE HIGH-INTENSITY CARBON ARC finds extensive use in the motion picture industry because of several important attributes. First, it has a very high brightness over an area of adequate size and shape. An effective light-collecting system thus can be designed to concentrate the necessary lumens on a projector aperture within the limits of optical speed which can be utilized effectively by the projection lens. Second, the light is of excellent color quality for the faithful photography and projection of both black-and-white and colored motion pictures. Third, the carbon-arc lamp has a high degree of mechanical reliability insuring a constant trouble-free delivery of light during the period required to project one reel, or to photograph a scene.

This first attribute of a continuously maintained high brightness has been the subject of investigation by many scientists both here and abroad. In our own laboratories, we are continually searching for ways of making and operating carbon arcs which will raise the ceiling

* Presented October 26, 1948, at the SMPE Convention in Washington; Central Section, SMPE, January 13, 1949.

of brightness, although we, in common with other investigators, have at times held the opinion that certain facts of nature have determined limits beyond which we may never be able to go.

The present paper is concerned with a method of operating carbon arcs which has been found useful whenever the highest brightness and smoothest operation, particularly at high currents, is desired. This involves the use of water-cooled jaws for both the positive and negative carbons. When these jaws are properly employed in a manner to be described, they permit the effective utilization of the high-current densities required for optimum high-brightness performance. Mention has been made in previous publications of the advantages inherent in water-cooled jaw operation.[1, 2] The continued confirmation and extension of these earlier findings has made appropriate this present paper, devoted more particularly to a description of the operating methods involved.

The major source of brightness in the high-intensity carbon arc is the so-called line radiation resulting from energy exchanges between rare-earth atoms and electrons in the gas ball within the positive crater. It is apparent that the higher the current on a given-sized carbon, the higher the electron density in the crater will be. Thus a greater number of energy exchanges is to be expected, with a corresponding increase in crater brightness. However, as the current is increased beyond a rather critical value, an overload phenomenon is encountered, which is usually characterized by noise and unsteadiness. In practical operation, therefore, the user must be content with the brightness obtainable at currents below this overload point. With the present 13.6-mm super high-intensity projector carbon, for instance, the maximum recommended current is 170 amperes.

A theory of overload has been advanced in an earlier publication.[2] Briefly, it is thought to be analogous to the violent boiling of a kettle of water which accompanies a high rate of energy input from a turned-up burner. If, however, a cooling coil be inserted in the kettle (analogous to providing improved cooling of the positive-carbon crater) enough of the input energy can be absorbed so that the boiling will subside, and an even higher rate of energy input tolerated without turbulence.

So it is with the carbon arc. Effective cooling of the positive carbon dissipates peaceably energy which might otherwise produce turbulence, so that a given-sized carbon can be designed to carry more current. Finkelnburg, in an accompanying paper,[3] points out that water-cooled operation is accompanied by a lower anode drop, so that this also contributes importantly to the reduction in anode energy per ampere. By means of optimum cooling, through the use of properly constructed carbons in water-cooled jaws, the gas ball in the crater space·can thus be provided with a denser population of electrons before the limit of their peaceful absorption on the crater surface is attained. A 13.6-mm size, for instance, can be made to operate at 350 amperes and 40,000 screen lumens, instead of 170 amperes and only 20,000 lumens; while water-cooled 16-mm carbons

Fig. 1—Side view of arc-lamp mechanism incorporating water-cooled positive and negative jaws.

have been operated at currents up to 500 amperes. It is advantageous also to cool the negative carbon, particularly at high currents, as will be pointed out later.

Other workers have determined certain fundamental relationships characteristic of the carbons and methods of operation with which they were familiar, and which predict levels of operation significantly exceeded by the procedures described here. For instance, Finkelnburg[4] reports an empirical relationship between crater brightness and consumption rate, which he found characteristic of the carbons and methods of operation available to him in Germany. At 2000 millimeters per hour (79 inches per hour) for instance, this relationship

jaw, is an important feature. The guides for positive alignment of the upper jaw insure the rigid clamping of the positive carbon along a predetermined axis. The upper jaw is sloped backward along the edge nearest the arc to permit free arc-flame travel at short carbon protrusion with minimum damage to the jaw. Fig. 3 shows a front view of the mechanism with all connections, and both positive and negative carbons, in place. Special attention is directed to the negative head, which consists simply of a fixed-bore water-jacketed copper tube. An unplated negative carbon of small diameter ($5/_{16}$ inch) is employed, with a short protrusion. Current is conducted directly from this water-cooled negative head, so that the carbon carries current only along the short protrusion.

The choice of silver material for the positive head and copper for the negative is based upon the following considerations. A material for this service must combine a high electrical and thermal conductivity with freedom from excessive corrosion and rapid wear in service. Copper most economically fulfills these requirements in so far as the negative head is concerned. However, this same material fails because of excessive wear in the positive head. This is because copper is plated from the jaws onto the carbon, and this then scores the jaws as it is dragged around with carbon rotation. The reason this destructive effect is confined to the positive jaw, while the very similar usage in the negative gives no trouble, is believed to be associated with the direction of current flow, and the rectifying action of the copper-oxide and sulfide films which tend to form along the copper-carbon contacts. These films are conductive in the direction of current flow from carbon to copper in the negative holder, but they tend to block current flow from copper to carbon in the positive. Since the distance between the jaw and the carbon is so short, a contact drop of only 1 volt produces a gradient of perhaps several thousand volts per millimeter across the rectifying film. This is sufficient to rupture the film and draw copper ions across the gap to be neutralized on the carbon. Silver oxide and sulfide, on the other hand, are good conductors with no rectifying properties, and so silver is free from this difficulty. The jaws illustrated here have operated several hundreds of hours, many of them at high currents from 300 to 500 amperes, with no significant wear and every indication of prolonged satisfactory performance.

For purposes of securing comparative data, it is necessary that certain operating conditions be held constant. Factors determining the choice of these conditions in the tests to be described were as follows:

The speed of rotation of the positive carbon was chosen at 15 revolutions per minute. However, the exact speed is not critical, so long as it is above the minimum required to insure a straight crater face. In the test lamp used, the angle between the positive and negative carbon is adjustable over a wide range. This angle is not ordinarily critical over a range between about 45 and 60 degrees of the negative-carbon axis below a horizontal positive. For the tests to be described, the halfway value of 53 degrees was chosen. At shallower angles, the positive tail flame is thrown objectionably close

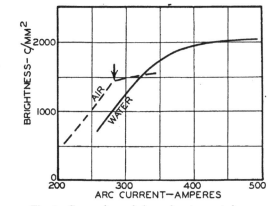

Fig. 4—Comparison of air- and water-cooled operation for new higher-current 13.6-mm "high-brightness" carbon.

to the upper jaw, the arc is less stable, and it is more difficult to hold a straight crater. At steeper angles, the negative flame tends to pass in front of (rather than into) the crater, so that the arc is more difficult to control, at least without the aid of an auxiliary magnetic field.

In order to insure optimum cooling, the protrusion of the positive carbon beyond the jaw should be held to as small a value as possible consistent with adequate jaw protection. A protrusion of $1/2$ inch was used with 16-mm carbons and only $1/4$ inch with 9-mm carbons.

The use of a small negative carbon with a short protrusion contributes importantly to a stable arc at high currents. The small carbon spindles to a sharply defined tip area, which is completely and stably filled with the negative flame at a current density of approximately 30 amperes per square millimeter. (This compares with a

positive-crater current density of between 1 and 3 amperes per square
millimeter.)   It is obvious that there is much less freedom for arc
wandering here as compared with the comparatively blunt point
formed on the much larger plated negatives conventionally employed
in heavy-current service.   The advantages of the small water-cooled
negative are more pronounced as the current is increased.

The determination of the maximum performance of a given carbon
is dependent upon the choice of a maximum operating current.   This
was chosen at a value a little below that which resulted in unstable
operation.   Over a wide range of sizes and types of positive carbons,
the same $^5/_{16}$-inch water-cooled negative was employed, giving very
satisfactory operation at all currents from 90 to 500 amperes.

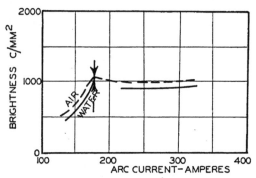

Fig. 5—Comparison of air- and water-cooled operation for 13.6-mm super high-intensity projector
carbon.

Positive carbons from 9 to 16 mm in diameter have been specially
designed to take advantage of the efficient cooling provided in the
apparatus shown in Fig. 1.   The first of these, a 13.6-mm carbon for
operation at 290 amperes, was described in an earlier paper.[1]   One of
the outstanding features of this type of carbon is its high thermal conductivity, which is essential to the efficient transfer of heat from the
floor of the crater to the water-cooled jaws.   This is an important
link in the cooling system required to postpone overload turbulence
to higher current densities, in accordance with the theories previously
expressed.   Carbon composition, as well as water cooling, are thus
involved in the achievement of crater brightness in excess of 2000
candles per square millimeter.

An interesting demonstration of this fact is given by a comparison

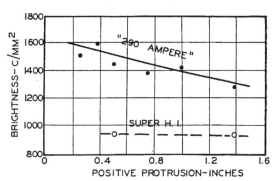

Fig. 6—Effect of positive protrusion on brightness of 13.6-mm, "290-ampere" and super high-intensity projector carbons.

of the two following figures. Fig. 4 shows the relationship between crater brightness and arc current for a new higher-current 13.6-mm carbon when operated first in water-cooled jaws at $1/2$-inch protrusion and then in conventional air-cooled jaws at $1^1/2$-inch protrusion. The outstanding feature of the water cooling, combined with the shorter protrusion which this makes possible, is the ability to carry much higher currents than with air cooling, and to attain higher brightness as a result. Within the limits of satisfactory air-cooled operation, however, the carbon reaches a higher brightness at a given current than when water-cooled, so that the current efficiency of the carbon is reduced by water cooling.

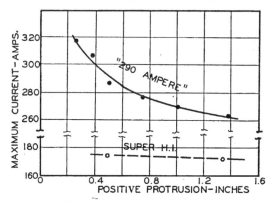

Fig. 7—Effect of positive protrusion on current capacity of 13.6-mm, "290-ampere" and super high-intensity projector carbons.

The ability to carry higher currents with water cooling is not characteristic of all carbons however. To illustrate this the performance of the 13.6-mm super high-intensity projector carbon, representative of the usual type of carbon, is shown in Fig. 5. Here water cooling in no case produces a higher brightness than can be obtained with air cooling, and the current efficiency is always less. Thus with this, as with most conventional types of carbons, water

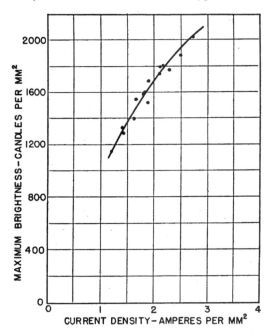

Fig. 8—Brightness variation with current density for "high-brightness" carbons.

cooling has no such advantage in increasing brightness as is exhibited by the "high-brightness" carbon of Fig. 4.

Referring again to Figs. 4 and 5, it will be noticed that sharp breaks occur in three of the four curves in the two figures, at the points indicated by the vertical arrows. These are the currents at which the carbon "overloads," with the accompanying hissing and sputtering which is familiaily encountered in such cases. At higher currents, the arc is noisy and generally unsteady, prohibiting operation under practical conditions. It is the practice, of course, to operate a carbon at a current somewhat below this "maximum" value at which overload occurs. The 13.6-mm super high-intensity pro-jector carbon, for example, overloads at about 176 amperes, whether water- or air-cooled, so that 170 amperes is the recommended maxi-mum operating current for this carbon. The high-brightness carbon (Fig. 4) reaches a similar overload condition at 282 amperes when air-cooled. However, in interesting contrast to the usual types of

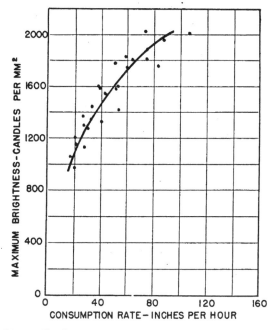

Fig. 9—Brightness variation with consumption rate for "high-brightness" carbons.

overload, this carbon does not behave in the manner just described when water-cooled, even at currents up to 500 amperes. It operates quietly up to about 325 amperes. At higher currents, the light re-mains steady, but a sort of droning noise gradually develops, which is altogether different in quality and much lower in intensity than with the conventional type of overload, and quite tolerable in many appli-cations. We have found this clear-cut difference to exist to the extent described only with carbons having relatively thin shells (less than 2 mm thick with the 13.6-mm carbon). High-brightness carbons having thicker shells (of which the 13.6-mm, "290-ampere" carbon[1] is an example) exhibit tendencies toward the hissing type of overload common to usual types of carbons, so that their "maximum" current is fairly well defined.

Another manifestation of the unique properties of the high-brightness type of carbon is the relation of brightness and of arc current to positive protrusion. Figs. 6 and 7 show these relationships for the 290-ampere, 13.6-mm carbon. As the protrusion is lessened to give improved crater cooling, the "maximum" current and the brightness increase. The usual type of carbon, exemplified again by the 13.6-mm super high-intensity positive carbon, exhibits little or no change in brightness and "maximum" current with change in protrusion.

High brightnesses have been obtained with these special carbons at significantly higher current and carbon efficiencies than have been reported by other investigators. For instance, the maximum performance predicted by Hallet[5] is exceeded by all of the 15 high-brightness carbons for which the data are plotted on Fig. 8. These carbons are from 9 to 13.6 mm in diameter and exceed the predicted performance at a given current density by as much as 10 per cent, although the general shape of Hallet's master curve is followed quite well.

Another interesting property of these carbons is their ability to produce a much higher brightness at a given consumption rate than was characteristic of the carbons which Finkelnburg[4] examined in Germany. Data on many of our high-brightness carbons ranging in size from 9 to 16 mm and burned in water-cooled jaws are plotted on Fig. 9. The brightness at a given consumption rate exceeds that reported by Finkelnburg by more than 50 per cent in all cases.

REFERENCES

(1) M. T. Jones, R. J. Zavesky, and W. W. Lozier, "A new carbon for increased light in studio and theater projection," J. Soc. Mot. Pict. Eng., vol. 45, pp. 449–459; December, 1945.

(2) F. T. Bowditch, "Light generation by the high-intensity carbon arc," J. Soc. Mot. Pict. Eng., vol. 49, pp. 209–218; September, 1947.

(3) W. Finkelnburg, "The influence of carbon cooling on the high-current carbon arc and its mechanism," J. Soc. Mot. Pict. Eng., this issue, pp. 407–417.

(4) W. Finkelnburg, "The High-Current Carbon Arc," Field Information Agency, Technical, Office of Military Government for Germany (US), Final Report 1052. (Office of Technical Services P.B. No. 81644.) Review published J. Soc. Mot. Pict. Eng., vol. 52, pp. 112–113; January, 1949.

(5) C. G. Heys Hallett, "Recent developments in carbon arc lamps," J. Brit. Kinematograph Soc., vol. 11, p. 188; December, 1947.

Effect of Carbon Cooling on High-Current Arcs*

By WOLFGANG FINKELNBURG

ENGINEER RESEARCH AND DEVELOPMENT LABORATORIES, FORT BELVOIR, VIRGINIA

Summary—The influence of carbon cooling, especially of the positive carbon, on the properties and the mechanism of the high-current carbon arc are studied systematically. An experimental super high-intensity carbon, designed by the National Carbon Company especially for this service, was used. With respect to the technical advantages of carbon cooling (increased arc steadiness, reduced positive-carbon consumption, reduced crater depth, higher possible brightness) our results agree with those presented simultaneously by the National Carbon group. Quantitative relations between current, net arc voltage, net arc wattage, gross arc wattage, crater brightness, carbon consumption, light efficiency, and crater depth for the cooled and the uncooled arc are presented in the form of graphs based on more than 1000 measurements. The considerable reduction of the net arc voltage by cooling the positive carbon indicates an influence of carbon cooling on the arc mechanism itself which is explained, on the basis of the author's anode drop theory, as a consequence of chemical and structural changes in the carbons resulting from the difference in temperatures near the burning end. These changes have been confirmed by X-ray diffraction studies.

I. INTRODUCTION

WATER COOLING of the positive as well as the negative carbon of high-intensity carbon arcs has proved to be of great advantage in cases where highest brightness, and therefore highest current density, is desired.[1-4] A systematic study of the influence of water cooling on the properties of the arc and its mechanism therefore seemed desirable and has been carried out at the Engineer Research and Development Laboratories by the author with the help of L. R. Noffsinger and C. Orr, using an excellent new super high-intensity experimental carbon of 11-mm diameter (No. 070) manufactured for this service by the National Carbon Company.

II. EXPERIMENTAL ARRANGEMENT AND METHOD OF MEASURING

The measurements were made with a Mole-Richardson lamp (see Figs. 1–3) which has a carbon angle of 52 degrees. The lamp was

---

* Presented October 26, 1948, at the SMPE Convention in Washington.

designed for automatic feeding but for these studies it was changed
to hand-controlled feeding. The carbons were cooled by copper
jackets through which water was circulated, and which enclosed the
carbons near their burning ends (see Fig. 2). The carbons protruded
from these water jackets, through holes only slightly larger than the
carbons, a distance of approximately 7 mm for the positive, and
approximately 20 mm for the negative carbon. Water jackets in the
form of semicircular jaws, pressed from both sides against the carbons, would offer certain technical advantages; however, they were

Fig. 1—Experimental lamp with probes for measuring the net arc voltage,
without water cooling.

not used for this investigation, because of the difficulty of manufacture. For comparison of measurements of water cooling with
forced-air cooling, a copper-finned head (see Fig. 3) was used. This
head was cooled with compressed air and served to prove that identical effects could be achieved by cooling the positive carbon with compressed air as with water. For the essential set of measurements,
9-mm copper-coated negative carbons were used with the 11-mm
positive carbons; the arc length of 18 mm was kept constant by
observing a greatly magnified image of the arc, with marks for the
desired position of the carbon tips. A pointer, attached to the rear
end of the positive carbon, permitted its length to be measured on a
millimeter scale during the operation of the arc. Thus, with a stop

watch, the rate of consumption of the positive carbon could be measured. This was done as soon as a stationary state of operation had been reached for each respective current.

The gross voltage of the arc, including the voltage drop in the carbon tips, was measured by connecting the voltmeter across the carbons at the water-cooled heads for the cooled arc, and at the negative clamp and the positive brush for the uncooled arc. In order to measure the net arc voltage, two carbon probes (see Figs. 1 through 3) could be made to touch the carbon tips near the burning ends by

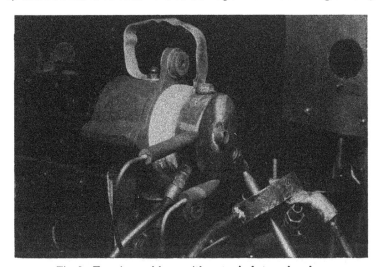

Fig. 2—Experimental lamp with water jackets and probes.

means of a magnetic relay. The crater depth was measured by means of a special gauge, while the crater diameter was measured with a standard caliper. The crater brightness was measured by projecting the crater image with a lens of known aperture on a photosensitive cell according to a method published by the author.[5, 6] While the current and arc length were kept constant with utmost care, voltage, crater brightness, and positive-carbon consumption were measured simultaneously and recorded as averaged over 2-minute runs. After each run the crater depth was measured. During the first tests, the crater diameter also was measured, but this was discontinued since it remained constant at 11 mm for the water-cooled arc, and was from 0.1 to 0.2 mm smaller for the uncooled arc.

III. General Properties of Cooled and Uncooled Arcs

Comparison of arcs without carbon cooling, with water cooling of both carbons, and with exclusive cooling of either the positive or the negative carbon revealed that water cooling increased the steadiness of the arc and of its radiation considerably. Cooling of the negative carbon alone had no effect other than that of steadying the arc, while cooling of the positive carbon increased, to a great extent, not only the steadiness of the arc, but also changed important properties of the arc, such as arc voltage and positive-carbon consumption. Further-

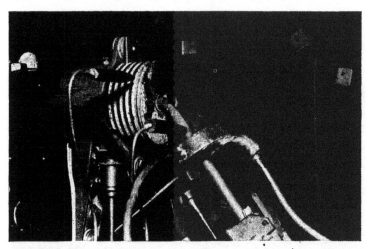

Fig. 3—Experimental lamp with probes and copper-finned head for forced-air cooling.

more, water-cooled carbons, especially if designed for this kind of operation, reached a much higher brightness than uncooled ones. With the best uncooled 11-mm carbons, sputtering and hissing of the arc began at a brightness of approximately 1500 candles per square millimeter, while steady operation of the same carbons, when water-cooled, was possible up to a brightness of 1850 candles per square millimeter. From all measurements the conclusion seems to be inevitable that cooled positive carbons behave quite differently in the high-current carbon arc than do uncooled ones, in which each part is heated to a very high temperature before the arc reaches it. A detailed study of the changes in the carbon core resulting from this heating is under way.

IV. QUANTITATIVE RESULTS

A quantitative comparison of the properties of the high-current carbon arc with cooled and uncooled 070 carbons is presented in Figs. 4 through 12, in which averaged results of a large number of measurements, carried out with many samples of 070 carbons, are plotted.

Fig. 4 shows one of the most unexpected effects of cooling the positive carbons: For all currents the net arc voltage (as measured with the probes between the carbon tips) is considerably lower than without cooling, no matter whether the negative carbon is cooled or not. As the arc stream is independent of the positive carbon, it seems safe to conclude that this decrease of the arc voltage is caused by a decrease of the anode drop, and this conclusion is in agreement with earlier investigations on the anode drop by the author.[7] According to the theory of arc radiation, developed in connection with the anode-drop work, a decrease of the anode drop always causes a decrease of the crater brightness. Fig. 5 proves that

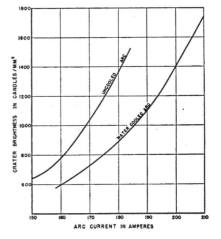

Fig. 4—Comparison of crater brightness of cooled and uncooled arcs at equal currents.

water cooling of the positive carbon actually does cause a considerable reduction of the crater brightness at the same current. In Fig. 6, the crater brightness is plotted against the arc wattage. The middle curve refers to the water-cooled arc (where the difference between the net arc wattage and the gross arc wattage falls within the limits of accuracy of our measurements because the carbon protrusions are short), while the upper and lower curves are plotted against the net arc wattage and the gross wattage of the uncooled arc. For a given net arc wattage the uncooled arc gives a higher brightness than the water-cooled arc, while a given gross wattage, actually dissipated in the arc and the carbons from the negative clamp to the positive brush, results in a higher brightness for the water-cooled arc.

With respect to the total light efficiency, measured for convenience in candles per watt gross arc wattage, the uncooled arc is always superior to the water-cooled arc, as may be seen from Figs. 7 and 8, in which the efficiency in candles per watt is plotted against the gross arc wattage and brightness, respectively. With reference to the gross arc wattage the difference is about 12 per cent, while with reference to the same crater brightness it is only approximately 6 per

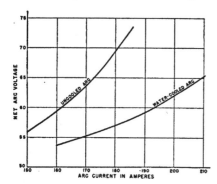

Fig. 5—Decrease of net arc voltage caused by cooling of positive carbon (same arc length).

cent. This slightly lower light efficiency of the water-cooled arc probably is caused by the fact that the water carries away part of the energy transferred to the positive carbon by the arc.

The most important feature of the water-cooled arc, next to its superior steadiness at highest brightness, is its low positive-carbon consumption as seen from Fig. 9, where consumption is plotted against crater brightness. Compared with other

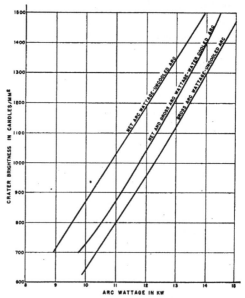

Fig. 6—Comparison of crater brightness of cooled and uncooled arcs for equal net and gross arc wattages.

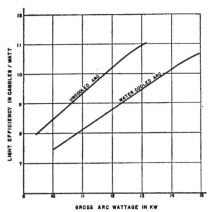

Fig. 7—Comparison of light efficiencies of cooled and uncooled arcs for equal gross arc wattages.

carbons,[8] the consumption of the experimental carbon No. 070 is very low even without water cooling. By water cooling, however, its consumption is reduced by as much as 35 per cent. In order to demonstrate the scattering of the measurements, some consumption-brightness measurements for the uncooled arc are plotted in Fig. 10.

There is also a marked influence of water cooling on the crater depth which, referred to arc wattage (Fig. 11), or brightness (Fig. 12), becomes shallower by cooling the positive carbon. This effect, although in Fig. 12 amounting to only approximately 14 per cent at higher brightness, is of technical importance, because too deep a crater is not well suited for illuminating lenses or mirrors of large apertures..

Measurements carried out with the air-cooled head (Fig. 3) gave results similar to those just described; however, the differences in the properties between the air-cooled and the uncooled arc are not quite.so large as those between the water-cooled and the uncooled arc.

All described results, attained with the carbon No. 070, were checked with two further experimental carbons, Nos. 081 and 088, and with regular 12-mm searchlight carbons. While the absolute values of all arc properties varied because of different values of core and shell diameters, the change of the arc properties as a consequence of cooling the positive carbon was similar to that for the carbon No. 070. The effect of carbon cooling

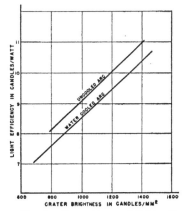

Fig. 8—Comparison of light efficiencies of cooled and uncooled arcs for equal crater brightness.

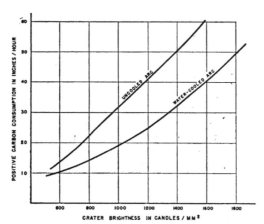

Fig. 9—Comparison of positive-carbon consumption of cooled and uncooled arcs for equal crater brightness.

on the high-current carbon arc thus seems to be a general one, though more pronounced with carbons specially designed for this purpose.

V. RESULTS AND THEORETICAL CONCLUSIONS

· From the technical point of view, water cooling of both carbons has the advantage of making possible the application of considerably higher brightness than was hitherto possible with good steadiness of the arc, with a shallower crater depth, and with greatly reduced

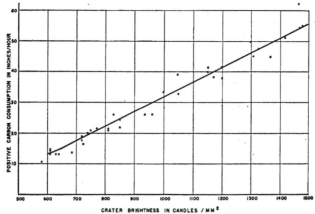

Fig. 10—Increase of positive-carbon consumption with crater brightness.

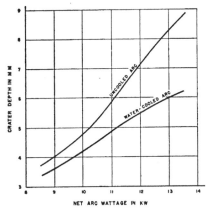

Fig. 11—Comparison of crater depth of cooled and uncooled arcs for equal net arc wattages.

carbon consumption. The only disadvantage .is that the required current is considerably higher than without cooling, while the decreased arc voltage does not form a compensating technical advantage.

From the physical point of view we have three important results:

1. The reduced net arc voltage indicates a decrease of the anode drop as a result of cooling the positive carbon.

2. No arc property was found which was *not* changed by cooling the positive carbon, leading to the conclusion that a cooled and an uncooled carbon behave like carbons of different composition or structure, these internal changes being caused by the different temperatures immediately behind their burning ends.

3. These internal changes in the carbons have been confirmed by X-ray diffraction studies, the results of which will be described in detail in a future publication.

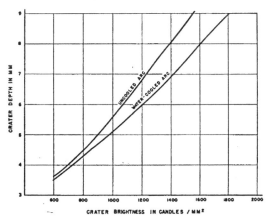

Fig. 12—Comparison of crater depth of cooled and uncooled arcs for equal crater brightness.

The mechanism of the cooling influence follows from these results in connection with earlier investigations of the arc mechanism by the author,[9] and can be indicated here only briefly.

The electrons, accelerated by the anode-drop potential, transfer an energy of 6 to 7 kilowatts to each square centimeter of the crater surface which serves to vaporize anode material and heat it to a temperature of about 6000 to 7000 degrees Kelvin. The illuminant vapor is heated furthermore by collisions with electrons in and immediately in front of the positive crater, the electrons dissipating there an additional amount of 2 to 3 kilowatts. The resulting vapor temperature of 7000 to 8000 degrees Kelvin is responsible for the high brightness of the crater vapors. The limit of load which a positive carbon can stand, and which determines the maximum brightness, is given by the transition to an unsteady, explosive vaporization instead of the desirable smooth and steady evaporation. This limit depends on the chemical and physical structure of the carbon. The possibility of using higher current density and thus attaining higher brightness by cooling the positive carbon seems to be caused by two effects: In the first place we have a reduction of the anode drop which means a reduction of the energy spent for vaporization at a given current. In the second place, the carbon, being kept cool up to a point quite near to the crater, keeps its original structure unchanged (compared with the highly heated uncooled carbon) and therefore is able to stand a higher load before beginning to evaporate unsteadily.

REFERENCES

(1) M. T. Jones, R. J. Zavesky, and W. W. Lozier, "A new carbon for increased light in studio and theater projection," J. Soc. Mot. Pict. Eng., vol. 45, pp. 449–459; December ,1945.

(2) M. A. Hankins, "Recent developments of super-high-intensity carbon-arc lamps," J. Soc. Mot. Pict. Eng., vol. 49, pp. 37–48; July, 1947.

(3) C. G. Heys Hallett, "Recent developments in carbon arc lamps," J. Brit. Kinematograph Soc., vol. 11, p. 188; December, 1947.

(4) Charles A. Hahn, "High-intensity projection arc lamps," J. Soc. Mot. Pict. Eng., vol. 50, pp. 489–502; May, 1948.

(5) W. Finkelnburg, "Leuchtdichte, Gesamtstrahlungsdichte und schwarze Temperatur von Hochstromkohlebögen," Zeit. für Phys., vol. 113, pp. 562–581; 1939.

(6) W. Finkelnburg, "The High Current Carbon Arc," Fiat Final Report 1052 (PB-81644, Office Technical Services, U. S. Dept. of Commerce, Washington, D. C.), p. 37 and Fig. 41.

(7) Pages 25 and 46 of reference 6.

(8) Page 165 and Fig. 49 of reference 6.

(9) Pages 106 through 117 of reference 6.

Disk Recorder for
Motion Picture Production*

By J. L. PETTUS

RCA Victor, Hollywood, California

Summary—A disk recording–reproducing console employing a two-speed synchronous drive is described. Improved motion is attainable by use of an oil-pressure-type thrust bearing and a compliance filter coupled to a planetary gear drive. Operation from either synchronous or interlock motor systems is possible by use of a new combination synchronous-interlock motor. Special control features incorporated in the design make its use particularly suited for all phases of disk sound recording.

D ISK RECORDING in the motion picture industry has a history parallel with that of sound motion pictures and, furthermore, it can look back upon a history of nearly sixty years in other fields of sound recording. Through these years continuous improvements have been made in the various components of this recording and reproducing method. Only lately, however, have the equipment designers been able to reach the achievement of really high-fidelity performance.[1] Concurrent with this advancement has been the ever-increasing demands of the industry for equipment having not only the best of components, but also the utmost in convenience of operation. Today, the motion picture studio finds disk recording an important tool notwithstanding that its product is released by having sound recorded with another medium. Some of the uses of disk recording today in this respect are immediate playback, protection records, and special effects, to name only a few. Phonograph recording studios, radio broadcast stations, and others make extensive use of disk recorders and are striving for the same high standards of quality.

During the past year it has been the privilege of the Radio Corporation of America to design and manufacture disk recording equipment which would meet the demands of motion picture production. Beforehand, specifications for the design were compiled and based

* Presented October 26, 1948, at the SMPE Convention in Washington.

upon obtaining the utmost in quality of performance as well as convenience of operation. To meet these specifications, the following performance requirements were established:

(1) Synchronous speed at both $33^{1}/_{3}$ and 78 revolutions per minute.

(2) Flutter not to exceed 0.05 per cent at 3000 cycles.*

(3) Frequency response flat within 2 decibels from 30 to 10,000 cycles.

(4) Mechanical noise below the surface-noise level of the best grade commercial acetate record.

(5) Simultaneous recording and playback.

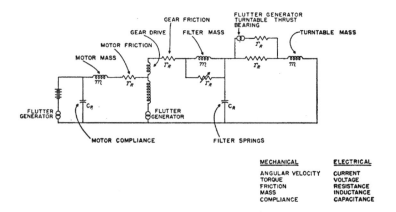

Fig. 1—Mechanical network.

(6) Cutting pitches from 50 to 156 lines per inch both "inside-out" and "outside-in."

(7) Independent synchronous or interlock motor operation.

(8) Console construction.

It is the first four of these requirements that deserve mention here since they might well be a nucleus of all disk recording-equipment designs. The first item is probably best met from the standpoint of economy and simplicity by the employment of a two-speed gear-drive mechanism. The second requires the use of a filter unit capable of removing disturbances generated by the nonuniform displacement of the driving motor and low-order oscillations produced in the gear

* As defined under "Proposed standard specifications for flutter or wow as related to sound records," J. Soc. Mot. Pict. Eng., vol. 49, pp. 147–160; August, 1947.

train.   The third is met by the choice of a well-designed transducer
system.   The fourth may be achieved after a study of all components
used in the driving system.   The aspect of these various components
can be better understood if the designer resorts to the convenience of

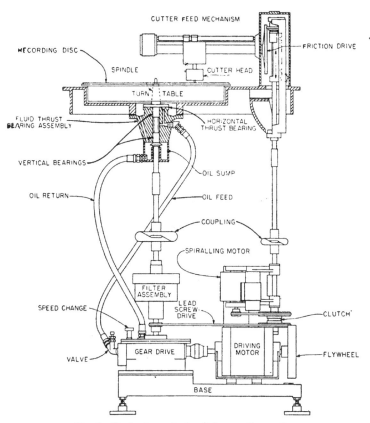

Fig. 2—Basic element of a disk recording unit.

assigning electrical analogs to the mechanical system.[2]   Fig. 1 shows
such an analogy.   From this it can be seen that elements such as
thrust, friction, compliance, and mass must be considered before
success in the end product may be achieved.   Likewise, the behavior
of each component may be predicted and its worthiness evaluated at

the initial design stage. From this point, the designer may then proceed in a creative manner to the development of the physical design.

Fig. 2 shows in outline form the physical arrangement of the essential components employed in the design which is to be described here and identified as an RCA PD-14 disk recording console. The driving motor identified at the lower portion of Fig. 2 is predicated by the design requirement of providing independent synchronous or inter-

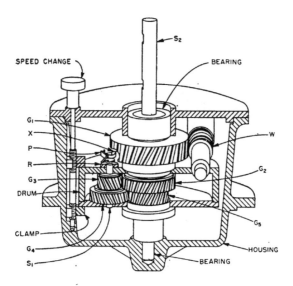

Fig. 3—Two-speed synchronous gear drive.

lock operation. This was readily met by the employment of a combination synchronous-interlock motor having the two sections common within a single frame. The torque requirement of approximately 200 mechanical watts was readily met and convenience of operation from either a 3-phase power source or a standard interlock motor system was obtained. Such a motor was conceived several years ago and has been used in many RCA designs. A flywheel is shown attached to the right-hand shaft extension and is used to provide additional inertia in the mechanical system. Attached to the opposite shaft extension is a two-speed gear-drive mechanism. This unit deserves special mention since its operation and design features are believed unique. A better understanding of the gear drive may be had by reference to Fig. 3 where the various components have been drawn in oblique fashion. From this it may be seen that the driver is worm shaft W and meshes to gear G_1 which in turn is locked to gear G_5. Gear G_2 meshes to gear G_3 which revolves about the axis of shaft S_1 along with gear G_4. The latter gear G_4 meshes with gear G_5, which in turn is keyed to shaft S_2. Gears G_3 and G_4 are located inside a drum D which provides bearings for shaft S_1. Now if the driver worm is turned at 1200 revolutions per minute by the driving motor and a step-down ratio of 36 to 1 is made by means of the worm and gear G_1, the latter will revolve about shaft S_2 at $33^1/_3$ revolutions per minute. Furthermore, if the drum D is allowed to freewheel about the axis of shaft S_2, gears G_2, G_3, and G_4 remain idle under ideal conditions. However, due to friction and resistance, some motion would result and it becomes necessary to apply a braking action to obtain the desired speed of shaft S_2. This is conveniently done by locking gears G_3 and G_4 to shaft S_1 and braking this shaft by means of a ratchet-and-pawl arrangement indicated as R, P, and X. Under this condition, the desired output speed is $33^1/_3$ revolutions per minute and synchronous with the driving motor.

In order to meet the requirement of 78-revolutions-per-minute output speed, it is only necessary to choose a step-up ratio of approximately 2.34 to 1 between gears G_2 and G_3, transmit through gears G_4 and G_5 at a ratio of 1 to 1, and lock the drum D in a stationary position by means of clamp C. Under this latter condition, the force applied by gears G_2 and G_3 are such as to unlock S_1 at the ratchet R with the result that shaft S_2 is now displaced at 78 revolutions per minute. Shifting from either speed may be made regardless of whether the unit is (or is not) in operation. A housing provides for enclosing the gears and submerging all components in an oil bath. Other features of the gear drive are a power take-off for driving the cutter-feed mechanism, and an oil pump whose function will be described subsequently.

Referring again to Fig. 2, it may be seen that a second driving mechanism is located directly behind the main driving motor. This device provides auxiliary power to the cutter-feed mechanism for spiraling purposes. An overrunning clutch located on the shaft makes it possible to increase the speed of the cutter-feed mechanism and hence to change the cutting pitch to effect a spiral in the recording medium.

Located directly above the gear drive may be seen a filter unit which provides a damping action to the disturbances generated in the driving mechanism. Such disturbances are in the form of minute oscillations due to the driving-motor pulses, gear imperfections, and uncontrollable friction. This filter mechanically performs the function of damping equivalent to that of the low-pass filter section m, r_R, and C_R of Fig. 1. Better understanding of this mechanical filter may be had by referring to Fig. 4. It may be classed as an oil-displacement device and operates just below critical damping of the resonant circuit. Its mechanical construction consists of a pair of interleaved vane assemblies viscously coupled. One of the vane assemblies is in

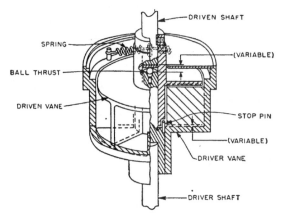

Fig. 4—Filter assembly.

the form of a cup (indicated as the driver vane) and has four equally spaced vanes attached to the inside wall of the cup extending radially toward its axis. The other vane assembly is in the form of an impeller (indicated as the driven vane) and also has four equally spaced vanes attached to the under side of a plate extending radially from its axis.

The physical dimensions of each assembly permit the centering of the driven vane within the driver vane whereby a small clearance exists between all opposite surfaces of the two sections. A spindle which is rigidly fastened inside the driver vane, serves as a support for the driven vane; likewise a second spindle is placed in a hub formed at the top of the driven vane and serves to position it with respect to the driver in the vertical plane. A ball thrust between the two shaft ends serves as a low-friction bearing. By varying the position of the driven spindle, a corresponding variation in clearance takes place between the lower edges of the driven vanes and the inside wall of the driver-vane cup. Thus a vernier damping adjustment is made and is a variable resistor r_R in the electrical analog. A neutral positioning of the vanes is accomplished by the use of a pair of two balanced springs having their ends alternately connected to the respective assemblies. These springs thus become the capacitance leg C_R of the low-pass filter section shown in Fig. 1. If the space between the opposing surfaces of the two vane assemblies is occupied by a viscous matter, a damping action takes place when the respective assemblies are axially displaced from their neutral position. Such damping may be referred to as the shunted resistor in the electrical analogy. Therefore, once having determined the values of the various electrical components of such a filter, the information may be readily referred to the mechanical design.

From the driven side of the filter unit it is then in order to transmit torque to the turntable. It is important, however, to apply a means of isolation in the connecting links of the various driving means in order that mechanical vibrations may be minimized. This is done by the use of the widely publicized vertical and angular compliance couplings.

The design of a suitable turntable is again predicated from the electrical analog since such mechanical factors as mass and bearing friction are equivalent to inductance and resistance, respectively. In this case it was determined that the mass must be great and that the bearing friction must be low. It was also evident that if no mechanical noise was to be present in the end product, special consideration to the turntable support bearing was necessary. In this respect, most designers are aware of the limitations in such devices as ball and roller bearings. With this in mind, the so-called fluid bearing appeared as a promising solution. Such a bearing may be seen in Fig. 2. Its construction is essentially a bushing and shaft with three journals lubricated under pressure. Since the shaft must be coupled to the driving mechanism below, the mass of the turntable must be supported by a horizontal thrust bearing. Likewise, support of this shaft in its vertical plane must be adequate to achieve a minimum of runout in the turntable's horizontal surface which carries the recording medium. This has been accomplished by providing a horizontal thrust collar on the shaft and two vertical bearings as shown in the illustration. The usual practice of using a bronze bearing and a hardened-steel shaft prevails in this case. For the purpose of securing adequate lubrication and also a cushion effect between the several bearing surfaces, oil is introduced at two points and with an applied force of approximately 40 pounds per square inch. The facility of oil pressure is derived from a gear pump attached to the synchronous gear drive previously mentioned and is transmitted through lines as indicated. When rotation of the turntable takes place, the oil pressure exerts a force sufficient to raise the turntable approximately 0.002 inch thereby producing a cushion effect between the horizontal-thrust-bearing surfaces. As the oil is spent from the several bearings it is directed into a sump from which it is recirculated to the gear-drive unit below. The design of fluid bearings is one of long standing, and use on such items as large celestial telescopes and other precision instruments is well known. Also the extreme low order of friction obtainable has been well established. It is noteworthy, however, that the coefficient of friction in this particular application has a factor of 10 to 1 reduction over that of a ball-bearing construction having equivalent load requirements. With this type of turntable bearing in the PD-14 recorder, flutter measurements were in accordance with the design specification and no discernible rumble could be observed on records. In tests of the oil bearing, the turntable was disconnected from the drive shaft after attaining a speed of 78 revolutions per minute and the turntable continued to rotate for 8 minutes. This is further confirmation of the extremely low friction loss in the fluid bearing.

Progressing upward in Fig. 2 from the turntable may be seen the cutter mechanism. This consists of feed screw, a carriage, a friction drive, and various novel features which make for convenience of operation. The requirement of either "inside-out" or "outside-in" cutting is met by the use of a friction drive attached to the lead screw and driven by a roller at right angles to the driving disk. Displacing the roller from the axis of the disk varies the speed as well as the direction of rotation of the lead screw. The lead screw is encased by a tube which provides support for the cutter carriage. A second tube is located directly behind and slightly upward which serves to house an automatic diameter equalizer. This is comprised of fixed resistors attached to commutator rings and connected electrically by a sliding contactor supported by the carriage assembly. A switch is provided to set the equalizer in either orthacoustic or flat frequency-response positions.

Fig. 5 shows the complete PD-14 disk recording console. Here may be seen such additional convenience items as a microscope·for groove inspection, a tone arm for reproducing either lateral or verti-

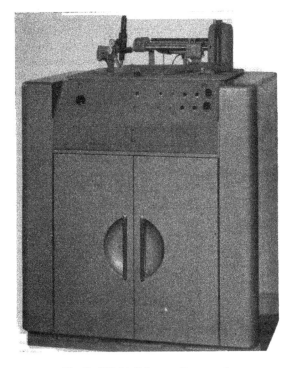

Fig. 5—PD-14 disk recording console.

cal records, and the control panel. The latter places all controls for the operation of the machine at a convenient central location. Such controls as reproducer equalization, spiraling motor, and power switches are provided. In addition, a jack row is located behind a hinged door at the lower side of the panel. All audio circuits are terminated normally at jacks. Insertion of patch cords permits easy checking for maintenance of the various audio components as well as special connections as may be required.

Where installations of disk recording equipment require adaption to existing studio facilities, it is possible to employ the components of the PD-14 without the use of the console. The mechanical driving elements may in some cases also be used to modernize existing disk recorders.

ACKNOWLEDGMENT

Grateful acknowledgment is given to Mr. George Worrall of the Worrall Camera Company, Los Angeles, who contributed the design and manufacture of the two-speed synchronous gear drive, and to the W. R. Turner Co., Los Angeles, California, for the manufacture of the synchronous-interlock motor. Specific acknowledgment is also given to Mr. H. C. Ward (RCA) whose efforts contributed to the design and construction of this device.

REFERENCES

(1) S. J. Begun, "New recording machine combining disk recording with magnetic recording with short reference to the present status of each," J. Soc. Mot. Pict. Eng., vol. 35, pp. 507–522; November, 1940.

(2) H. F. Olson, "Elements of Acoustical Engineering," second edition, D. Van Nostrand Company, New York, N. Y., 1947, chap. IV, p. 67.

Synchronous Disk Recorder Drive*

By C. C. DAVIS

WESTERN ELECTRIC COMPANY, HOLLYWOOD 38, CALIFORNIA

Summary—A new type of drive incorporates an oil-filled filter and bearing assembly which supports the turntable. Bronze bearings especially designed to prevent generation of low-frequency vibration or "rumble" are used. A mechanical filter is used to suppress flutter. This assembly is driven by a special rubber coupling which prevents the transmission of mechanical vibration from the gear box to the turntable.

INTRODUCTION

LITTLE HAS APPEARED in the literature about those mechanisms which are responsible for the constancy of tone produced by disk recorders and reproducers. Since the days of Edison, however, their importance has been recognized and elaborate means have been employed in studios to secure satisfactory disk performance. The Western Electric D85249 recorder,[1] introduced in 1926 into the infant sound-picture industry, employed the first mechanical-filter mechanism known to the art. In describing the new device, coded Western Electric RA-1388 Type drive, it may be of interest to include some of the design considerations resulting in its development.

The performance of a high-quality disk turntable is based upon its constancy of rotation as well as upon its freedom from mechanical vibration. The first of these two qualities is expressed in terms of flutter measurements and the second in terms of signal-to-noise or "rumble" meaurements, the latter indicating the relative amount of recorded background noise resulting from vibration in the turntable.

To meet rigid requirements both as to flutter and rumble, two considerations appear which at first seem incompatible. A close association of driving source and turntable is desirable to minimize flutter but the reverse is true in reducing rumble. The driving mechanism must not only drive the turntable at a velocity constant within a small fraction of 1 per cent in spite of stylus-load variations of several per cent, [2,3] but at the same time it must introduce no mechanical vibration in appreciable amount. Thus the driving source, such as the

---

* Presented October 26, 1948, at the SMPE Convention in Washington.

motor and gear box, must be isolated from the turntable as far as induced vibration is concerned but it must deliver rotary motion to the
turntable in an otherwise direct manner.

The necessary separation of drive and turntable introduces a certain
amount of flexibility or compliance in the transmission system. The
combination of this compliance and the mass of the turntable will
cause amplification of steady-state-drive disturbances near their
resonance frequency and may actually generate oscillatory disturbances as the result of transients unless damping is applied. When
this is done the combination may be referred to as a mechanical filter.
Therefore the basic design of synchronous disk recorder drives in
general consists of a motor-and-gear-box combination, capable of de-
livering the desired turntable speeds, connected to the turntable by
means of a flexible coupling and filter device. The mechanical filter
serves the further important function of removing gear and other rota-
tional discrepancies such as motor misalignment, which might otherwise reach the recording medium in the form of high-frequency flutter.

The RA-1388 Type Disk Recorder Drive

The design of the Western Electric RA-1388 Type drive, shown in
Fig. 1, is based on the foregoing general principles. Every effort has
been made to proportion the elements for optimum performance under
average recording conditions as well as to produce a simple, easily
maintained mechanism. The materials and tolerances correspond to
those of the highest quality precision devices.

Filter and Bearing Assembly

A feature of this drive is the combination of the filter and turntable
supporting shaft and associated bearings into a single oil-filled assembly, also shown in Fig. 1. It is a modernized version of a previous
design[4] but it makes use of a heavy oil-film type of turntable-thrust
bearing in place of the usual ball-thrust bearing in the interest of
greater signal-to-noise ratio. This unit which is mainly responsible for
the performance of the whole drive may be readily checked or disassembled for inspection.

The assembly consists of three concentric members. The outer
portion consists of a large aluminum-bronze centrifugally cast member
terminating in a housing at its lower end which is directly driven by
the vertical drive shaft connected to the gear box. This housing contains the mechanical filter which drives the turntable through the

inner or turntable supporting shaft. It is oil-sealed with an "O" ring and is filled to the mid-point of a small reservoir at the top, which is covered by a dust cover. The third member is a heavy walled steel tube which is rigidly attached to the body of the recording machine, and constitutes the stationary support upon which the other members revolve.

A view of the assembly with cover removed is shown in Fig. 2. The fact that the filter unit is oil-filled offers the opportunity to use a new type of vertical thrust bearing to support the turntable. The rotating portion of this bearing which bears the weight of the turntable consists of a flat bronze surface which revolves upon a stationary steel surface. This steel portion of the bearing contains radial oil grooves sloping gradually into the bearing surface. The amount and direction of this slope are such that the turntable is lifted on a thick film of oil by the hydraulic wedging action of the oil when rotating in normal direction. This avoids metal-to-metal con-

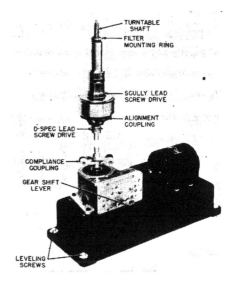

Fig. 1—RA-1388 Type synchronous disk recorder drive.

tact between the two halves of the bearing at operating speeds and since metal-to-metal contact is eliminated, the bearing is incapable of creating vibration or "rumble." Such a bearing is similar to the "Kingsbury" type which, in applications such as large marine installations, has individual self-aligning segments or shoes capable of bearing loads reaching one ton per square inch of bearing surface. The oil grooves of the filter unit are of such dimensions that a turntable weighing twenty pounds is lifted on a thick oil film when its rate of revolution exceeds a few revolutions per minute. Thus the turntable reaches an operating condition in less than $1/2$ revolution after starting when driven by a synchronous motor.

The upper bearing of the filter unit appearing immediately below the turntable, has been manufactured to extremely close tolerances. The bearing proper is a self-lubricating type and the revolving steel half, which is a section of the turntable drive shaft, is lapped to a close fit in the bearing. This is done to avoid sidewise play in the turntable which may cause characteristic "grouping" of recorded grooves due to wandering of the recording medium. In view of the fact that the manufacturer could meet the necessary tolerances, the above construction was selected in preference to an adjustable type because wear is minimized and the construction is simplified.

The damping element of the mechanical filter consists of two identical dashpots and plungers. These provide viscous damping without

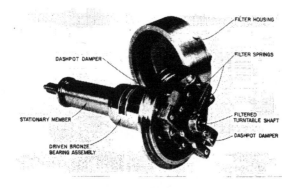

Fig. 2—Filter-and-bearing assembly with filter housing removed.

static friction, the amount of which s controlled by their relative diameters and by the viscosity of the oil contained in the filter unit. Self-alignment between plungers and pots is accomplished by the use of piston rings fitted in deep grooves, capable of generous lateral freedom. Backlash has been eliminated by the plunger design. This eliminates the tendency to oscillate in an undamped manner within the limits of the free play at frequencies close to the natural period of the filter system. Microscopic examination of a small target on the turntable shows it may be deflected from rest by minute forces applied to its periphery and further, that these small deflections are damped.

MOTOR-AND-GEAR-BOX ASSEMBLY

The gear box and motor are mounted on a large rigid casting which is fastened to the floor beneath the recording machine. This casting is equipped with four adjustable leveling feet to assist in aligning the drive shaft in correct relationship with the recorder. Both the interlock and synchronous type of motors used with the RA-1388 drive are rubber-mounted as well as the gear box which prevents direct metallic contact between these units and the base casting. The remaining vibration in the base casting is prevented from reaching the recording machine by its attachment to the floor and by the special coupling connecting the gear box to the filter and the bearing unit.

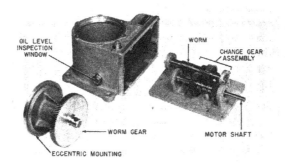

Fig. 3—Gear-box assembly.

The gear box shown in Fig. 1 is conventional in that it contains no filtering and produces speeds of 78.26 and $33^{1}/_{3}$ revolutions per minute by means of a gear shift. A disassembled view is shown in Fig. 3. It contains three ball-bearing-equipped shafts: the worm shaft, the worm-gear shaft, and the motor or change-gear shaft. Mating gears are helical Bakelite and steel combinations. A unit-type construction is used in which two shafts are mounted in pillow blocks, then aligned and doweled on a front face plate. After installation of this plate, the worm-gear assembly is adjusted in proper mesh by means of an eccentric mounting with which this unit is equipped. The gear box is oil-filled and provided with an oil-inspection window and is built for either 1200- or 1000-revolution-per-minute-type driving motors.

The design of the gear box is based on two theoretical factors. First, the limitation of turntable mass necessary for practical considerations, such as synchronous operation and satisfactory thrust-bearing performance, forbids the choice of filter elements which might otherwise be effective at the once-per-revolution rate of the turntable. Therefore the worm gear which revolves at this rate must be manufactured to tolerances which, when weighted for stylus disturbances, do not exceed anticipated performance. The tolerances of lead errors and eccentricity for the worm gear are such as to hold the once-per-revolution flutter to 0.03 per cent or less.

The second design factor recognizes the fact that all gear-box flutter velocities greater than the once-per-revolution rate must be removed by the mechanical filter. Therefore it is desirable to make these frequencies as high as practicable in order to fall within the high-attenuation range of the filter. For this reason the worm and associated shaft and gears are designed to revolve at relatively high speeds, generating a minimum frequency of $13^1/_3$ cycles.

FILTER DESIGN

Selection of filter constants for optimum flutter performance resulted from many actual recording and flutter-measuring tests under varied conditions as well as from transmission tests of electrical elements analogous to the mechanical elements of the drive system. A curve representing the attenuation characteristic of the RA-1388 Type drive is shown in Fig. 4. Fortunately, the controlling elements of the filter and bearing unit, namely, the piston rings and coil springs, may be readily changed for the purpose of experimentation. Practical use of an electrically equivalent circuit was made to remove a slightly disturbing amount of $13^1/_3$-cycle flutter originating in the worm-shaft assembly in spite of its rigid manufacturing tolerances. By means of the electrical circuit a value of torsional compliance was selected for the rubber coupling of the vertical drive shaft which aids attenuation of flutter at $13^1/_3$ cycles without serious effects at lower frequency disturbances. The difference is indicated in the two curves of Fig. 4. Transient conditions were also investigated by an electrically analogous circuit and by this means, for example, it was found advisable to reduce motor compliance by avoiding the use of interlock motors of small types.

PERFORMANCE

The design objectives of the RA-1388 Type drive originally included an unweighted signal-to-noise ratio of 50 decibels and flutter performance not exceeding 0.05 per cent at any significant rate and positive

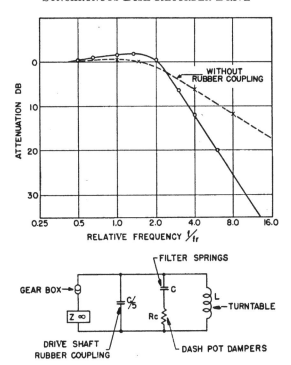

Fig. 4—Filter-attenuation characteristics and analogous
electrical circuit.

gear-driven action. Signal-to-noise ratios of 60 decibels have consistently been measured with normal playback equalization. Reference is made here to a 300-cycle recording of ±2-mil amplitude representing 9.5-centimeter-per-second velocity. Recordings have been made wherein all flutter bands measured considerably below 0.05 per cent and in some cases the so-called "total" flutter measured below 0.05 per cent.

REFERENCES

(1) L. A. Elmer and D. G. Blattner, "Machine for cutting master disc records," *Trans. Soc. Mot. Pict. Eng.*, vol. 13, pp. 227–247; May, 1929.

(2) S. J. Begun, "Some problems of disk recording," *Proc. I.R.E.*, vol. 28, pp. 389–399; September, 1940.

(3) H. E. Roys, "Force at the stylus tip while cutting lacquer disk-recording blanks," *Proc. I.R.E.*, vol. 35, pp. 1360–1364; November, 1947.

(4) U. S. Patent No. 2,337,976 (12-28-44, C. C. Davis).

Test-Film Calibration— Proposed Standards*

By F. J. PFEIFF and E. S. SEELEY

ALTEC SERVICE CORPORATION, NEW YORK 13, N. Y.

Summary—Despite a decade and a half of widespread use of multifrequency test reels throughout the motion picture industry, the method of calibration has not been standardized. One result has been nonuniformity of calibration by various agents and another is lack of clear understanding of the significance of the calibration data. This paper discusses prevailing diversity of calibration results, describes a means of securing an absolute calibration of relative level at various frequencies, and proposes a standard definition of level of recorded signal and methods for measuring it.

IN MANY MEASUREMENTS with calibrated multifrequency test film a considerable degree of accuracy is required. For example, in the service field we have striven for good accuracy in frequency-response measurement since its inception. It was found that vigilance in all phases of the measurement, from test-film calibration to the final adding up of figures, is essential if acceptable over-all tolerances are not to be exceeded. With hundreds of calibrated test films in constant use by the field forces, any disagreement between results obtained with several films on the same sound system would be soon discovered and faith in the test jeopardized.

Unfortunately, it must be reported that 15 years or so of calibration and use of test film has not produced uniformity of results as between calibrating agencies. The purpose of this paper is to present differences of calibration accuracy obtained by the several agencies and to urge that the Society of Motion Picture Engineers and other interested organizations adopt as a matter of substantial importance the project of resolving these differences.

PART I—CALIBRATION OF RELATIVE LEVEL AT VARIOUS FREQUENCIES

The procedure that has always been and still is followed for calibrating a considerable number of multifrequency test films is first to calibrate a standard with great care and using special equipment developed for the purpose; then the lot of films are calibrated by comparison with the standard through the use of a suitable reproducing

* Presented October 26, 1948, at the SMPE Convention in Washington.

system. In our earliest calibrations, the standard was calibrated by the Moll microdensitometer.[1] In about 1935 the Moll instrument was no longer available for this purpose and a new type of calibrating equipment was constructed based on the novel principle of inverse speed. In this system, the film was reproduced at speeds inversely proportional to the recorded frequency so that the electrical signal produced was of the same frequency for all frequency sections of the film. This procedure eliminated the frequency response of all electrical components of the reproducing system from the phototube to the output meter, thus leaving the scanning loss of the optical system as the only source of frequency discrimination. The scanning loss was minimized by employing a projection-type (rear view) scanning system, Fig. 1, in which the slit was adjusted to 1.5 mils. This in combination with a magnification of 8 times, resulted in a computed scanning width of about 0.19 thousandths of an inch at the film.

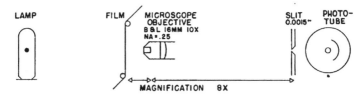

Fig. 1—Optical system of inverse-speed calibrator.

Films previously calibrated with the Moll microdensitometer were found to have a greater high-frequency content when measured by the inverse-speed equipment. The differences, shown by the lower curve in Fig. 2, increase to 4.5 decibels at 8000 cycles per second. Microdensitometers were soon constructed by others and these equipments provided calibrations which differed from each other and from the inverse-speed results. The amounts of difference varied with the calibrating equipment but the differences were consistent in one respect: they always showed less relative high-frequency level on the sound track than the inverse-speed calibrator found. Since the only error remaining in the inverse-speed instrument was the scanning loss, and since this loss always discriminates against high frequencies and never in their favor, it was argued that the inverse-speed calibrator, by finding a greater relative level at high frequencies than other calibrators, was the most accurate of the calibrators compared. This argument, although not refuted, did not apparently impress other

owners of calibrating equipment. Each of these had a carefully designed equipment with a frequency discrimination of negligible value
as calculated from the known physical constants. No direct method
of calibrating the calibrators was available.

Comparisons made in 1938 and 1939, Fig. 2, show the differences
between the calibrations provided by other calibrators and that given
by the inverse-speed instrument.

Our own film calibrations, as previously pointed out, were originally
Moll calibrations, and the new inverse-speed calibrator provided new
calibrations for the same films. Thus a reproducing-system frequency

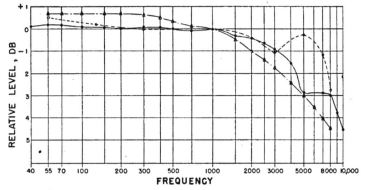

Fig. 2—Comparison of 4 calibrators, inverse-speed calibrator taken as zero.

response measured with the film could be either of two curves depending upon which calibration one preferred to believe. Field-
servicing information was originally based on the Moll film calibrations, and in view of the disagreement of results produced by the
various film-calibrating agencies, the change-over of the norms from
the Moll basis was postponed until general agreement could be
reached. When the Moll equipment was retired, the inverse-speed
calibrator came into use, but the resulting calibrations were modified
arithmetically to convert them to Moll calibrations.

During the first half of 1947, arrangements were made with the
Motion Picture Research Council to provide us with test-film prints
according to a specification dictated by the particular requirements
of theater sound service. Calibrated samples were supplied and,
happily, excellent agreement was found with our inverse-speed calibrations. It therefore appeared that the differences between our

calibrations and those of the Research Council no longer existed. We accordingly revised the servicing information to place all equipment performance data on the basis of the inverse-speed calibration.

Unfortunately, recalibration of 338 calibrated prints received after the sample again revealed some differences in average calibration and inconsistencies of calibration between prints. One of the prints was then sent to the West Coast for calibration on a modern microdensitometer there. The various results are summarized in Table I.

Line 1 shows the excellent agreement already mentioned. Calibrations made in October, 1946. Line 2 shows how closely inverse-speed calibration of a given film was repeated after 7 months. Lines 3 and 4 present differences between three calibrators: the inverse-speed system, the Research Council calibration, and the microdensitometer in Hollywood.

Line 5. The close agreement which existed seven months earlier between the calibrations by the Research Council and by the inverse-speed equipment has not continued, although the inconsistencies listed could not be considered to be of serious magnitude unless the change in 7 months be taken as cause for alarm. Question: How great a difference will be found after another 7 months?

Lines 6, 7, and 8. These data preclude any satisfaction with the results based on averages. Transmission-test tolerances do not allow for errors so great as these in film calibration. The area of uncertainty (line 8) may be compared with a specification under the heading "Tolerance" in the Research Council publication Standard Electrical Characteristics for Theater Sound Systems:[2] "A tolerance of ±1 decibel up to 3000 cycles, increasing progressively to ±2 decibels at 7000 cycles, is permitted for any of the following electrical characteristics. This tolerance is necessary as it is often extremely difficult, or impossible, to adjust the electrical response of the amplifier system to the exact characteristic. Whenever such conditions exist that the particular characteristic recommended does not give satisfactory results, it is recommended that the acoustic characteristics of the auditorium be corrected." Further evidence of the inadequacy of this calibration accuracy is revealed by a proposed standard prepared by Section Z22 of the American Standards Association covering 35-mm multifrequency test film which specifies a calibration tolerance of ±0.25 decibel.

The reported differences in calibration could be caused by an error on the part of either calibrating agency or both. The authors do not

TABLE I
RECENT CALIBRATION COMPARISONS

					Cycles					
40	70	130	300	500	1000	2000	3000	5000	7000	8000
1. Sample ED-35: Inverse-Speed Calibration Minus Research Council Calibration 10/46										
+0.2	+0.3	0.0	+0.1	-0.2	0.0	+0.2	0.0	-0.4	+0.2	0.0
2. Sample ED-35: Inverse-Speed Calibration 10/46 Minus Inverse-Speed Calibration 5/47										
0.0	-0.1	0.0	+0.1	+0.2	0.0	+0.1	+0.2	0.0	-0.1	0.0
3. Print #2 ED-35: Inverse-Speed Calibration 5/47 Minus Research Council Calibration										
+1.3	+0.6	+0.3	+0.1	-0.2	0.0	-0.2	+0.1	+0.6	+0.4	+1.7
4. Print #2 ED-35: Inverse-Speed Calibration 5/47 Minus Microdensitometer Calibration										
+1.8	+1.0	+1.1	+0.7	+0.8	0.0	+0.5	+0.9	+1.2	+1.6	+1.7
5. 338 Prints ED-35: Average Differences Between Inverse-Speed and Research Council Calibration										
+0.8	+0.2	+0.1	-0.2	-0.4	0.0	-0.2	+0.1	+0.5	+0.7	+0.8
6. 338 Prints ED-35: Extreme Differences Between Inverse-Speed and Research Council Calibration										
+1.4	+0.6	+0.5	+0.2	0.0	0.0	+0.3	+0.8	+1.1	+1.9	+2.0
7. +0.2	-0.2	-0.4	-0.6	-0.9	0.0	-0.7	-0.4	0.0	-0.1	0.0
8. 338 Prints ED-35: Spread Between Extreme Differences (IS−RC)										
1.2	0.8	0.9	0.8	0.9	0.0	1.0	1.2	1.1	2.0	2.0

claim complete accuracy, but serious efforts have been made to minimize the possibility of error in transferring calibration of a "standard"
to all the films of a production lot (see Appendix), and a study has been
made to evaluate errors in the calibration of a "standard." Since, as
previously pointed out, frequency-discrimination errors cannot be
introduced into the calibration by any electrical element in the system, the latter problem reduces to one of determining the scanning
loss by direct means.

A method was devised for obtaining an absolute calibration of the
scanning loss of the calibrator. A piece of fine wire was attached
transversely to clear film to bridge the normally scanned area, as
illustrated in Fig. 3, and a sound track was thus produced consisting of
a single striation 2.0 thousandths
of an inch wide with sharply
defined edges. An optical system having zero scanning loss
would reproduce such a track
as a single rectangular pulse
with vertical sides. The effect
of scanning loss would be to
slope the sides, and the difference between this slope and
the vertical could be taken as
a measure of the scanning loss.

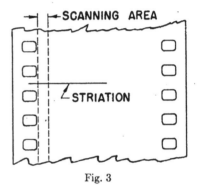

Fig. 3

The output voltage of the
phototube was connected to the vertical deflection plates of a
cathode-ray tube through a direct-current amplifier. A single-sweep
time-axis circuit having negligible nonlinearity was connected to the
horizontal deflection plates.

To carry out the calibration, the synthetic sound track was run
through the calibrator at a speed of 4 thousandths of an inch per
second. Just before the striation passed across the slit the sweep was
triggered and the trace on the cathode-ray tube was photographed.
Fig. 4 (a) is the unretouched photograph so obtained and Fig. 4 (b) is
the same photograph with lines drawn to facilitate numerical
evaluation.

The effective width of the scanning beam is revealed by the duration
of the fall from the light to dark phototube condition. Since the
striation width is known to be 2.0 thousandths of an inch, projections
of the slopes on the axis can be evaluated in units of length referred to

the film. This evaluation is complicated somewhat by several features of the trace:

1. The two slopes are not quite equal. This property was unrelated to the direction of film travel, and was not due to inexactness of the striation since it was found with all the striations tried. In the evaluation, the mean of the two slopes was taken.

2. The corners of the pulse are not sharp but exhibit toe and shoulder curvatures. These curvatures occupy so small a part of the pattern that they were neglected in the evaluation.

3. The small humps before and after the pulse were not definitely explained but were presumed to be caused by reflections off the edge of the wire and the film surface. They are neglected in the evaluation.

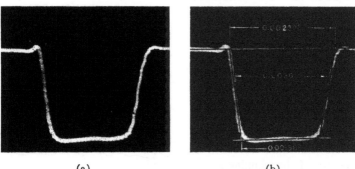

(a)　　　　　　　　　　　　　　　(b)

Fig. 4—Average length of slope = 0.00032.

As shown in Fig. 4 (b) the effective scanning width is evaluated as 0.32 thousandth of an inch. The actual slit opening and the magnification had been measured by obvious and reliable methods and the scanning width computed to be 0.2 thousandth inch. The difference between expected and measured results might be explained by the interference fringe which extends into the shadow of the striation. From the numerical aperture of the objective and the geometry of the incident light, the primary fringes are computed to extend approximately 0.2 thousandth of an inch into the shadow area at the wavelength of light for which an $S1$ phototube cathode has its greatest response to light from a tungsten source. Exact evaluation of this effect is complicated by the wide range of wavelengths present, but it is felt that this method of calibrating the scanning system portrays the true relationships as they apply when calibrating film.

The scanning width evaluated in Fig. 4 (b) would produce an 8000-cycle scanning loss of 0.34 decibel. Personal errors—the only kind known which could produce negative scanning-loss errors in calibrating film with the inverse-speed system—are minimized by carrying out the calibration with care, repetition, and reference to films previously calibrated. (See Appendix.)

Since all other calibrators compared reveal lower high-frequency level, it is concluded tentatively that the inverse-speed calibrator is the most nearly accurate. In the interest of industry standardization of test-film calibration, it is urged that this conclusion be challenged by others who possess means for calibrating multifrequency film.

Part II—Film-Level Calibration

One important difference between film-reproducing systems is manifested by differences in output power developed when a particular film is reproduced. The fundamental property involved here has been called over-all system gain or, for brevity, system gain. However, any given system will develop various output-power levels (up to the limit of its power capacity) depending upon the particular sound track being reproduced. The term film level has long been used to describe this pertinent property of the sound track.

Specific definitions were established for system gain and film level in the earliest days of film calibration and these have been in continuous use ever since by one part of the industry. The same definition for film level was adopted elsewhere in the industry a decade ago, but it appears that the principles have not been well understood since we have found some of these film-level figures as much as 10 decibels apart from those measured by our methods for the same films.

With the thought that others in the industry will also find valuable use for the film-level calibration of multifrequency test film and for the measurement or specification of system gain, definitions for these quantities and methods for measuring them will be proposed for consideration by the Standards Committee.

The basic relationship between these quantities is shown in the following: A test film is reproduced and a particular output-power level (less than the power capacity of the system) is developed. Then

$$\text{output-power level} = \text{over-all system gain} + \text{film level}. \qquad (1)$$

Since gain of an amplifier system has long been clearly defined, it is convenient to divide over-all system gain into two parts:

$$\text{over-all system gain} = \text{amplifier gain} + \text{relative pick-up sensitivity}. \qquad (2)$$

In this division of the reproducing system, the amplifier system is considered to include all elements other than the speaker system which follow the phototube terminals, and the pickup system includes all elements except the film which are ahead of the phototube terminals. Gain of a communication system is defined as

$$\text{gain} = 10 \log \frac{P_2}{P_1}$$

where (Fig. 5) P_1 is signal power produced in a load of specified value (usually matching) terminating a signal generator of internal impedance selected as the specified sending impedance of the system under

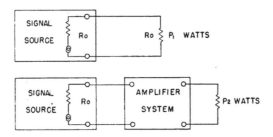

Proposed Standard:

    (A) For amplifiers to be coupled to phototubes, $R_0 = 10$ megohms.

    (B) Phototube-coupling resistor, transformer, or network considered as part of amplifier system. Thus, amplifier system comprises all elements following phototube terminals.

Fig. 5—Gain of amplifier system $= 10 \log \dfrac{P_2}{P_1}$.

measurement, and P_2 is the power produced in the system load when the generator load is replaced by the system. In simple words, gain of an amplifier system is the increase in power obtained when the amplifier is used, over that obtained when the amplifier is not used.

Because the amplifier in a reproducing system is connected to a phototube, the sending impedance R_0 should be the phototube impedance. Phototube impedance was standardized years ago in our work at 10 megohms, and this selection will be discussed later. Thus amplifier gain is readily evaluated from the equation in Fig. 6.

The pickup system is a part of all film-reproducing systems; since the sensitivity of pickup systems varies substantially, its sensitivity must be included in the over-all system gain if the latter is to have any meaning. Combining (1) and (2),

output-power level − film level = amplifier gain
+ relative pickup sensitivity.   (3)

Since power level is an observation and amplifier gain is already defined, (3) establishes the relation between film level and relative pickup sensitivity.

In order to define film level in terms of power, a particular pickup system is borrowed, and film level will be defined as the power level developed in a load on that pickup system. The reference pickup chosen years ago consisted of the elements used in a sound system which was very widely distributed at the time. The phototube impedance was taken as 10 megohms; consequently the reference load for the reference pickup was set at 10 megohms. It was found that

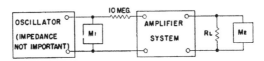

M_1 and M_2, similar decibel meters.

Gain = M_2 decibels − M_1 decibels + $\log \dfrac{500}{R_L}$ + 49 decibels.

Fig. 6—Measurement of gain of phototube amplifier.

an idealized film recording, having 50 per cent average transmission and 100 per cent modulation (this being the maximum sinusoidal film level theoretically possible), would produce a power in the 10-megohm load on the reference pickup of just 1.0 microwatt or a power level of −37.8 decibels relative to 6 milliwatts. The level which any film would develop in the load on the reference pickup is the film level, which is therefore defined as

$$\text{film level} = 20 \log 2TM - 37.8 \text{ decibels} \qquad (4)$$

where T is the average transmission and M is the modulation expressed as a decimal fraction. If the densitometric constants of the recording are known, the film level may be computed from (4).

With film level completely defined, the related definition for relative pickup sensitivity is established. The reference pickup is assigned a relative pickup sensitivity of 0.0 decibel. The idealized recording with 50 per cent average transmission and 100 per cent modulation, will cause the phototube current to swing from I_0, the phototube current with no film, to zero, or a total peak-to-peak swing of I_0

(see Fig. 7). Since this swing produces a power of 1 microwatt in the 10-megohm load on the reference pickup, the current through 10 megohms is computed to be 0.895 microampere. Taking the cell impedance to be 10 megohms, the current swing becomes twice as great when the phototube load is short-circuited and this factor leads to the useful conclusion that the phototube current I_{00}, measured with zero load resistance and with no film, is 1.79 microamperes in the case of the reference pickup. The relative pickup sensitivity of any pickup system is greater or less than 0.0 decibel as its no-film, no-load phototube current is greater or less than 1.79 microamperes. Thus, for any pickup,

$$\text{relative pickup sensitivity} = 20 \log \frac{I_{00}}{1.79} \qquad (5)$$

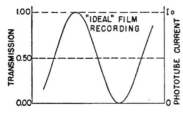

With reference pickup and 10-megohm load:
Power $= 1 \times 10^{-6}$ watt $=$
$$\left(\frac{I_0}{2\sqrt{2}}\right)^2 \times 10^7.$$

Therefore $I_0 = 0.895$ microampere.
With reference pickup and zero load resistance:
$I_{00} = 2I_0 = 1.79$ microamperes.
For any pickup, pickup sensitivity $=$
$$20 \log \frac{I_{00}}{1.79} \text{ decibels.}$$

Fig. 7

where I_{00} is measured in microamperes. This equation provides a very convenient means for evaluating the relative pickup sensitivity for any pickup system.

Having established a means for measuring relative pickup sensitivity and amplifier system gain without the use of a test film, we now have two methods for determining film level: first, by measuring the densitometric constants of the film and applying (4); and second, by measuring relative pickup sensitivity and amplifier gain, and subtracting their sum from output-power level. Measurements were made using both methods and the results differed by only 0.05 decibel. This difference is less than · was warranted by the inaccuracies of the instruments used and by the errors involved in the assumption that phototube impedance is 10 megohms, but the measurements were made carefully and verify that the two methods may be expected to produce equivalent results.

In connection with the 10-megohm choice of phototube impedance, note that pickup sensitivity is measured with zero impedance connected to the phototube, yet system gain depends in part on a finite phototube load impedance. If the phototube impedance were 10 megohms as assumed, the foregoing relationships would apply without error. However, phototube impedance varies considerably, a range of 7 megohms to 20 megohms being commonly found. For phototube impedance as low as 5 megohms or as high as infinity, the disagreement between the two methods for measuring film level would be ±0.2 decibel, respectively, for amplifiers with 0.25-megohm input resistance and ±0.4 decibel for amplifiers with 0.5-megohm input resistance.

These errors are not believed to be great enough to constitute a serious handicap in the use of the proposed definitions. In service, phototubes of a given type operated under identical conditions vary in sensitivity by as much as ±5 decibels and more, and much greater tolerance for the sensitivity of the phototube must be allowed than for the entire system excluding the phototube. For this reason, it is naïve to attempt to achieve the ultimate in accuracy when the sensitivity of the phototube is part of the picture.

The foregoing definitions for film level and relative pickup sensitivity are recommended for standardization for the following reasons:

1. They are consistent with present practice.

2. The several terms are clearly defined and simply related. Clarity of definition should lead to widespread employment of the film-level calibration to the benefit of the industry.

3. Methods are described for measuring film level and for evaluating relative pickup sensitivity which are not difficult to carry out and offer opportunity for cross check. A clear understanding of these methods should help reduce the diversity of results obtained for the same film by different calibrating agencies.

4. Over-all system gain may be predicted from knowledge of the two significant factors; namely, relative pickup sensitivity and amplifier gain.

ACKNOWLEDGMENT

The methods of calibration of multifrequency film and the development of the coherent system of relationships that have been presented were the product of the effort and thought of several people. The authors wish to acknowledge particularly the pioneering work of Mr. O. C. Johnson on the film-level problem and the contribution of Mr. C. R. Keith who first suggested the inverse-speed system of calibration.

APPENDIX

CALIBRATION PROCEDURE

When a substantial number of films are to be calibrated by the inverse speed equipment, procedure is as follows:

1. The equipment is stabilized and carefully adjusted.

2. One or more films calibrated at an earlier date are recalibrated. The earlier results should be closely duplicated. . The film is driven at such speeds that all frequency sections from 40 to 130 cycles, inclusive, are reproduced at 50 cycles, 130 to 1000 cycles, inclusive, are reproduced at 100 cycles, and 1000 to 8000 cycles, inclusive, are reproduced at 1000 cycles.

3. Two films are chosen from the new lot and marked Standard and Substandard. These are examined under a microprojector for the presence of azimuth (at 8000 cycles) and for correctness of soundtrack position with respect to the guided edge of the film. They are then calibrated by the inverse-speed process and, finally, recalibrated for check.

4. The equipment is set to operate at constant speed and the Standard and Substandard are reproduced to calibrate the system.

5. All films of the lot are reproduced at constant speed, the Substandard being repeated after each four films or less to recheck the system calibration. If a change occurs, it is verified by reproducing the Standard and after restabilization, films as required are recalibrated.

6. The final calibration data are scrutinized to discover possible arithmetic errors or other anomalies.

REFERENCES

(1) W. R. Goehner, "The microdensitometer as a laboratory measuring tool," J. Soc. Mot. Pict. Eng., vol. 23, pp. 318–328; December, 1934.

(2) "Standard Electrical Characteristics for Theater Sound Systems," 1948, page 4, Motion Picture Research Council, Hollywood, California.

Proposed American Standard

THE PROPOSED AMERICAN STANDARD for a common type 35-mm film perforation for both positive and negative film which appears in the following pages is published here for 90-day trial and criticism. It is now being proposed for adoption as an American Standard in order to solve the registration problems that exist in the printing of certain types of color release prints. It is possible to meet the problems of exact registration needed for color prints by the use of cine negative perforations in the release prints. However, many people are reluctant to do this because they fear that they cannot in this way make release prints which will have satisfactory projection lives. One answer to this problem is the use of combination positive and negative perforations proposed herewith. While the problem of preparing satisfactory color release prints is the reason for the presentation of this proposal, the whole question of 35-mm standards is involved. This problem is an old one and dates back to 1916. The following comments are intended to furnish a background for a proper consideration of the standards.

The first informative paper on perforating motion picture film, by D. J. Bell, was published in October, 1916. Since most of the projector manufacturers at that time had fairly well-established intermittent sprocket diameters, it remained for the film manufacturers to establish standard film dimensions. Mr. Bell proposed a perforation having a width of 0.110 inch and a height of 0.073 inch. This "Bell and Howell" perforation has become the standard for negative films and is described in detail in American Standard Z22.34-1949.

By 1918, Mr. Bell's proposed perforating standard had grown in favor until it was almost universally accepted. (Formal approval by the Society did not take place until May, 1922.) In light of this film standardization, A. C. Roebuck published a paper in the JOURNAL in November, 1918, which recommended redesign of sprocket teeth to provide greater picture steadiness with the accepted perforations. This sprocket study took into consideration film shrinkages to be expected from base materials of that day.

J. G. Jones became much interested in this problem about 1920 and did a great deal of experimenting. He became chairman of a subcommittee to report on this subject. This committee proposed a new perforation because the standard Bell and Howell perforation gave evidence of fracturing when run through projection equipment. Mr. Jones published a paper in this JOURNAL in 1923 (volume 17, pages 55 to 76) which gave rise to much discussion. In this paper he proposed the rectangular perforation now used in cine positive release film.

In order to determine what changes should be made to give positive film greater projection life, punches and dies were made up and film-life tests were run on several types of perforations having rounded corners. Of those tested, a rectangular perforation, 0.110 inch by 0.078 inch, having filleted corners gave the best projection life. It was also pointed out that increasing the height from 0.073 to 0.078 inch enabled the new perforation to accommodate existing sprockets which had given interference with the former perforation. It was also reported that no difficulty had been experienced when printing from Bell and Howell perforated negative on to positive stock with the proposed new perforation. Since this new perforation might have given trouble in some cameras then in use, it was recommended that it not be used for negative films.

This whole question of positive and negative perforating was thoroughly discussed at the next meeting of the Standards Committee in May, 1924. While the Bell and Howell perforation was accepted for negative films without objection, the new rectangular perforation for positive stock was strongly opposed. Many believed the laboratory tests which had been conducted were not entirely indicative of what could be expected in the field and, in any event, sufficient time had not elapsed to assure its superiority. Consequently, even though several million feet of positive film had been perforated in this manner during the preceding year, it was recommended the whole matter of positive perforation be referred back to the Film Perforating Committee. In particular, it was recommended that the committee coordinate its work internationally to insure world-wide acceptance of any standard that was adopted in the United States.

Consequently, in July, 1925, the American proposal for the positive perforation was submitted to the Sixth International Congress of Photography held in Paris. The results of this meeting indicated that the foreign film manufacturers favored the rectangular perforation for positive stock. Therefore, in October, 1925, the Standards Committee recommended that the Society adopt 0.110- by 0.078-inch rectangular perforation as the American Standard for positive film.

(continued on page 451)

Proposed American Standard Cutting and Perforating Dimensions for **35-Millimeter Motion Picture** Combination Positive-Negative Raw Stock	**Z22.1-** **April 1949**

Page 1 of 2 pages

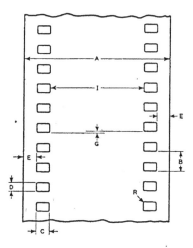

Dimensions	Inches	Millimeters
A	1.377 ± 0.001	34.98 ± 0.025
B	0.1870 ± 0.0005	4.750 ± 0.013
C	0.1100 ± 0.0004	2.794 ± 0.01
D	0.073 ± 0.0004	1.85 ± 0.01
E	0.079 ± 0.002	2.01 ± 0.05
G	Not > 0.001	Not > 0.025
I	0.999 ± 0.002	25.37 ± 0.05
L*	18.70 ± 0.015	474.98 ± 0.38
R	0.013 ± 0.001	0.33 ± 0.03

These dimensions and tolerances apply to the material immediately after cutting and perforating.

* This dimension represents the length of any 100 consecutive perforation intervals.

NOT APPROVED

Proposed American Standard
Cutting and Perforating Dimensions for
35-Millimeter Motion Picture
Combination Positive-Negative Raw Stock

**Z22.1-
April 1949**

Page 2 of 2 pages

Appendix

The dimensions given in this standard represent the practice of film manufacturers in that the dimensions and tolerances are for film immediately after perforation. The punches and dies themselves are made to tolerances considerably smaller than those given, but owing to the fact that film is a plastic material, the dimensions of the slit and perforated film never agree exactly with the dimensions of the punches and dies. Shrinkage of the film, due to change in moisture content or loss of residual solvents, invariably results in a change in these dimensions during the life of the film. This change is generally uniform throughout the roll.

The uniformity of perforation is one of the most important of the variables affecting steadiness of projection.

Variations in pitch from roll to roll are of little significance compared to variations from one sprocket hole to the next. Actually, it is the maximum variation from one sprocket hole to the next within any small group that is important.

Perforations of this size and shape were first described in the Journal of the SMPE in 1932 by Dubray and Howell. In 1937 a subcommittee report reviewed the work to date. The main interest in the perforation at that time was in its use as a universal perforation for both positive and negative film. The perforation has been adopted as a standard at this time largely because it has a projection life comparable to that of the perforation used for ordinary cine positive film, and the same over-all dimensions as the perforations used in the negative film.

NOT APPROVED

(continued from page 448)

Following the establishment of the two standards in the mid-twenties, no further action by the Society has been recorded up until April, 1932, when Messrs. Howell and Dubray proposed a combined positive and negative perforation. They believed that the rectangular style of perforation had advantages which were desirable to retain for use on release prints. Their proposal combined the advantages of both the current styles of perforations and film perforated according to this proposal could still be used on all existing equipment without alteration. It was a perforation similar in shape to the existing positive perforation but it had the negative perforation height of 0.073 inch.

In 1933, however, it was decided to adopt the then existing rectangular perforation as the universal standard for negative as well as positive film. Following this action, the whole matter remained at a standstill. It was never actually adopted in practice mainly because of the difficulty of inducing owners of equipment designed for the negative perforation to alter their apparatus. Cameras in particular presented the greatest obstacle because of their widespread use throughout the world and the difficulty of reaching all owners.

In 1937, the Subcommittee on Film Perforating Standards published a complete report proposing that the 1933 standard be withdrawn and the universal perforation proposed by Dubray and Howell be accepted for both negative and positive film. In this report it was concluded that this perforation would operate satisfactorily in equipment designed for either positive or negative perforations. In spite of the favorable report from the Film Dimensions Subcommittee, the universal perforation was turned down by the Standards Committee. It was felt at that time that the large amount of background film accumulated in the libraries would prevent the universal perforation from being used.

When it became apparent that it was going to be extremely difficult to get the industry to use a single perforation, it appeared desirable to re-establish the old negative perforation (the Bell and Howell perforation) as a standard for negative film. This was finally accepted by the American Standards Association in 1944. With the close of the war, however, Sectional Committee Z22 referred all the existing American standards for film dimensions to the Society of Motion Picture Engineers for revision. In due course all were revised. However, the Motion Picture Research Council did not approve the revised negative perforating standard. Instead they suggested that the whole question be reinvestigated and suggested that Dubray and Howell perforation be studied.

Consequently, during the years 1947 and 1948, another thorough study of the whole question was carried on by a committee under the chairmanship of Dr. E. K. Carver. During this time, various modifications were proposed for a universal positive-negative perforation. Extensive life tests on films having these perforations were conducted. It was determined in parallel experiments that while film having the Dubray-Howell perforation, when run to destruction, had slightly shorter projection life than film having positive perforations, it was markedly superior in this respect to film with negative perforations. It was also shown that the Dubray-Howell perforation would operate satisfactorily in most equipment designed either for negative or positive perforations and produce films of satisfactory steadiness. Meanwhile, one producer used film having this perforation for a number of color releases and obtained very satisfactory results when printing from a standard negative perforated film.

Therefore, in view of favorable results obtained, the proposed standard appearing in the preceding pages is being published for your trial and comment. In addition, the former negative perforating standard, Z22.34, has been submitted to the ASA for reapproval for the use of those not faced with the exacting registration requirements of color work.

Your Society
Report of the Executive Secretary

A T THE PRESENT TIME the Society's prestige in the industry is at an
all-time high. Over thirty-two years of effort in the direction
originally set by the Society's founders have made the JOURNAL a
standard reference for motion picture and television engineers everywhere; Society membership has grown to nearly 3000 without a
prolonged or intense effort in this direction; income from Sustaining
members for each of the last three years has been in excess of $20,000;
and the Society has established itself as a clearing house for industry
standards as well as a major producer of 16-mm and 35-mm test films.

PUBLICATIONS

1948 was a banner year for Society publications. We published
1,376 JOURNAL pages, an increase of nearly 10 per cent over the previous year. In addition, the appearance of JOURNAL reprints was improved, and their sale to authors and others increased to the extent that
the dollar volume of such sales was nearly two and one-half times the
equivalent for 1947. We published a Membership Directory which
was distributed to all members. We also published a 155-page Ten-
Year Index to the Journal that was mailed free of charge to all members. Through a very simple promotion program, we pointed out to a
selected list of librarians and nonmember engineers in the industry
that these indexes as well as back copies of the JOURNAL were available. Resulting sales exceeded our estimate by about 40 per cent.

The major part of our work on the Society's new book "The Motion
Picture Theater—Planning and Upkeep" was accomplished in 1948,
but printing and binding were delayed so long by production difficulties that we were unable to begin our sales campaign until the first
quarter of 1949. However, promotion plans made early in 1948 give us
an orderly sales program which is now under way.

We published a group of papers on High-Speed Photography for distribution at no extra cost to all members and subscribers as a separate
supplement to the March issue of the JOURNAL. Because this is the
first publication of its kind on the subject and because of its obvious
technical merit, we printed 1000 additional copies for sale to nonmembers.

Our publications operations for 1948 underwent some profound changes which have accrued to the benefit of the entire Society. The break with tradition, represented primarily by the JOURNAL's "New

In order that the Society of Motion Picture Engineers may serve its membership to greater advantage, a survey is being made to determine what type of articles the members wish published in the Journal and to obtain definite suggestions on possible papers and authors.

(Please Print or Typewrite)

I should like to read more papers on_____

_____

_____

I am willing to prepare a paper on_____

_____

Mr._____

_____
(address)

might write a paper on_____

_____

RETURN THIS CARD PROMPTLY

Fig. 1.

Look" in January, 1948, was considerably less painful than we had anticipated. It was a substantial forward step but there is still plenty of room for improvement with little or no increase in the costs of production. Without increasing the page size or changing the general format of the JOURNAL, we could make some typographical improvements, which would give us a JOURNAL outstanding in itself as well as in content.

To continue serving the needs of our current members, and attract new members in related fields, we must co-ordinate the policies of the Papers Committee, Board of Editors, Section Program Committees, and the basic policies of our public relations program. As a guide to forming this policy, we follow the replies to the papers program inquiry cards that were distributed with the membership dues bills in January. (Fig. 1.) Another guide being followed is the "Editorial Policy of the SMPE JOURNAL," a joint statement of the Papers Committee and the Board of Editors.

The current average of 120 pages per JOURNAL issue are set by budgetary limitations and will not be increased unless we are forced to do so because the Papers Committee has procured a large quantity of really valuable material

A subscription price increase, from $10.00 to $12.50 per year, became effective March 15, 1949, and was caused by continued increases in the costs of publication.

TEST FILMS

Test film sales in 1948 were 55 per cent greater than we had estimated at the beginning of the year in spite of the fact that two new 16-mm test films, one Sound and one Visual, that we had planned to have available, were not completed. This represents an increase of nearly 60 per cent over test-film business for 1947 and although it is roughly 20 per cent less than sales in 1946, which were primarily a few large Government orders, it was a good year's business.

Sales of test films have been promoted in two ways. Technical editors of the motion picture trade magazines were given copies of the last catalog which nearly all of them used in their publications. Mimeographed sales letters were sent to the chief engineers of all television stations then operating or for which permits had been requested, and we also wrote to universities that operate their own film libraries. The cost was small but the returns have been high.

Test films sold by the Society represent a substantial portion of the Society's annual income and we may be justly proud of this program. It is a valuable service to the industry. If we are to expand these operations and develop new test films as need for them arises, we must have an aggressive sales program aimed in the right direction. To this end, new test-film catalogs have been printed for distribution by both the Society and the Motion Picture Research Council.

About two years ago the Society joined with the Research Council in a reciprocal test-film sales program which has been an additional worth-while service to the industry.

A 16-mm Sound Service Test Film, one of two new 16-mm test films which have been under development for quite a while, has now been completed and is described in the new test-film catalog which is available to everyone.

TECHNICAL ACTIVITIES

The work of our eighteen engineering committees suffered some interruption during 1948 but recent changes have been extremely beneficial, as is indicated by the increase in the number of standards developed and engineering reports completed. Changes in organization of the Standards Committee have simplified the processing of standards because at the present time only the chairmen of other engineering committees, plus a few "old timers," serve. They are all familiar with standards procedures with the result that the time required for processing standards has been substantially reduced. Another valuable benefit is that the Engineering Vice-President now uses the Standards Committee as an advisory body to guide his decisions regarding new proposals for standardization.

The year 1949 promises to be the most active in history as well as the most efficiently managed so far as our engineering committees are concerned. It is unfortunate though that there, has been a notable lack of enthusiastic interest from many quarters, and it is the earnest hope of the Engineering Vice-President and all engineering committees that all members will take a more active interest in the engineering matters for which the Society was originally formed.

ADMINISTRATIVE

The past year and a half have seen many major changes in the methods and results of the Society's business administration.

We moved from the Pennsylvania Hotel to an office with about double the floor space. We have added four additional people to the staff, two of whom are concerned primarily with the JOURNAL, one other is a bookkeeper, and the fourth, an order clerk and secretary to the office manager.

As outmoded membership record system as well as cumbersome

systems for processing new applications and purchase orders have been replaced within the past nine months by simplified and more suitable systems. The conversion costs were nominal since one extra girl working for about two weeks was all the extraordinary labor expense involved.

The most important change from the fiscal standpoint is the adoption of an accrual system of accounting developed by the SMPE office manager, which will now allow the year-end financial statements to give a more accurate picture of the financial results of the Society's activities. In addition, it will be possible to draw worth-while comparisons between operating periods.

The year 1948 was the best in Society history. A great deal of active interest on the part of Society members has made the JOURNAL, the technical programs, and Society conventions tremendously successful. A great deal of effort was applied in the proper places and if Society members continue their interest, 1949 can be even better.

Most of the administrative changes which seemed necessary as a result of recent expansion have been made and are working well.

BOYCE NEMEC
Executive Secretary
March 20, 1949

"The Motion Picture Theater, Planning and Upkeep" is a proud presentation of the SMPE, based on a project directed by James Frank, Jr., who also has written the Foreword. The book is attractively bound with a hard cover of green buckram. There are 428 pages, generously illustrated and distinctively designed with modern typography.

The book is available at $5.00 per copy. Add 2 per cent sales tax for New York City deliveries. Price per copy, postage prepaid, outside the continental United States $5.50. SMPE, 342 Madison Avenue, New York 17, N. Y.

Officers of the Society

EARL I. SPONABLE
President
1949–1950

PETER MOLE
Executive Vice-President
1949–1950

LOREN L. RYDER
Past-President
1949–1950

JOHN A. MAURER
Engineering Vice-President
1949–1950

Officers of the Society

CLYDE R. KEITH
Editorial Vice-President
1949–1950

DAVID B. JOY
Financial Vice-President
1949

WILLIAM C. KUNZMANN
Convention Vice-President
1949–1950

ROBERT M. CORBIN
Secretary
1949–1950

RALPH B. AUSTRIAN
Treasurer
1948–1949

Governors of the Society

PAUL J. LARSEN
Governor
1948–1949

LLOYD T. GOLDSMITH
Governor
1948–1949

ALAN W. COOK
Governor
1948–1949

FRED T. BOWDITCH
Governor
1949–1950

GORDON E. SAWYER
Governor
1949–1950

Governors of the Society

HERBERT BARNETT
Governor
1949–1950

NORWOOD L. SIMMONS
Governor
1949–1950

JAMES FRANK, JR.
Governor
1949

WILLIAM B. LODGE
Governor
1949

KENNETH F. MORGAN
Governor
1949–1950

461

Officers and Managers of Sections

R. T. VAN NIMAN W. H. RIVERS S. P. SOLOW

ATLANTIC COAST

W. H. RIVERS, *Chairman*
EDWARD SCHMIDT, *Secretary-Treasurer*

Managers

E. A. BERTRAM	PIERRE MERTZ
H. B. BRAUN	E. S. SEELEY
W. F. JORDAN	R. E. SHELBY

CENTRAL

R. T. VAN NIMAN, *Chairman*
G. W. COLBURN, *Secretary-Treasurer*

Managers

E. E. BICKEL	A. SHAPIRO
F. E. CARLSON	M. G. TOWNSLEY
R. E. LEWIS	H. A. WITT

PACIFIC COAST

S. P. SOLOW, *Chairman*
WATSON JONES, *Secretary-Treasurer*

Managers

G. M. BEST	J. P. LIVADARY
P. E. BRIGANDI	G. C. MISENER
J. P. CORCORAN	F. R. WILSON

STUDENT CHAPTER

●

.- UNIVERSITY
OF SOUTHERN
CALIFORNIA

○

ALGERNON G. WALKER

○

ALGERNON G. WALKER
President
JAMES L. WILKINSON
Secretary
NORTON D. SOKOLOW
Treasurer

○

Constitution and Bylaws of the Society of Motion Picture Engineers

CONSTITUTION

Article I

NAME

The name of this association shall be SOCIETY OF MOTION PICTURE ENGINEERS.

Article II

OBJECTS

Its objects shall be: Advancement in the theory and practice of motion picture engineering and the allied arts and sciences, the standardization of the equipment, mechanisms, and practices employed therein, the maintenance of a high professional standing among its members, and the dissemination of scientific knowledge by publication.

Article III

ELIGIBILITY

Any person of good character may be a member in any grade for which he is eligible.

Article IV

OFFICERS

The officers of the Society shall be a President, a Past-President, an Executive Vice-President, an Engineering Vice-President, an Editorial Vice-President, a Financial Vice-President, a Convention Vice-President, a Secretary, and a Treasurer.

The term of office of all elected officers shall be for a period of two years. Of the Engineering, Editorial, Financial, and Convention Vice-Presidents, the Secretary, and _the Treasurer, three shall be elected alternately each year, or until their successors are chosen. The President shall not be immediately eligible to succeed himself in office. Under such conditions as set forth in the Bylaws, the office of Executive Vice-President may be vacated before the expiration of his term.

Article V

BOARD OF GOVERNORS

The Board of Governors shall consist of the President, the Past-President, the five Vice-Presidents, the Secretary, the Treasurer, the Section Chairmen, and ten elected governors. Five of these governors shall be resident in the area operating under Pacific and Mountain time, and five of the governors shall be resident in the area operating under Central and Eastern time. Two of the governors from the Pacific area and three of the governors from the Eastern area shall be elected in the odd-numbered years, and three of the governors in the Pacific area and two of the governors in the Eastern area shall be elected in the even-numbered years. The term of office of all elected governors shall be for a period of two years.

Article VI

MEETINGS

There shall be an annual meeting, and such other meetings as stated in the Bylaws.

Article VII

AMENDMENTS

This Constitution may be amended as follows: Amendments shall be approved by the Board of Governors, and shall be submitted for discussion at any regular members' meeting. The proposed amendment and complete discussion then shall be submitted to the entire Active, Fellow, and Honorary membership, together with letter ballot, as soon as possible after the meeting. Two thirds of the vote cast within sixty days after mailing shall be required to carry the amendment.

BYLAWS

Bylaw I

MEMBERSHIP

SEC. 1—The membership of the Society shall consist of Honorary members, Fellows, Active members, Associate members, Student members, and Sustaining members.

An **Honorary member** is one who has performed eminent services in the advancement of motion picture engineering or in the allied arts. An Honorary member shall be entitled to vote and to hold any office in the Society.

A **Fellow** is one who shall not be less than thirty years of age and who shall comply with the requirements of either (a) or (b) for Active members and, in addition, shall by his proficiency and contributions have attained to an outstanding rank among engineers or executives of the motion picture industry. A Fellow shall be entitled to vote and to hold any office in the Society.

An **Active member** is one who shall be not less than 25 years of age, and shall be (a) a motion picture engineer by profession. He shall have been engaged in the practice of his profession for a period of at least three years, and shall have taken responsibility for the design, installation, or operation of systems or apparatus pertaining to the motion picture industry; (b) a person regularly employed in motion picture or closely allied work, who, by his inventions or proficiency in motion picture science or as an executive of a motion picture enterprise of large scope, has attained to a recognized standing in the motion picture industry. In case of such an executive, the applicant must be qualified to take full charge of the broader features of motion picture engineering involved in the work under his direction.

An Active member is privileged to vote and to hold any office in the Society.

An **Associate member** is one who shall be not less than 18 years of age, and shall be a person who is interested in or connected with the study of motion picture technical problems or the application of them. An Associate member is not privileged to vote, to hold office, or to act as chairman of any committee, although he may serve upon any committee to which he may be appointed; and, when so appointed, shall be entitled to the full voting privileges of a committee member.

A **Student member** is any person registered as a student, graduate, or undergraduate, in a college, university, or educational institution, pursuing a course of studies in science or engineering that evidences interest in motion picture technology. Membership in this grade shall not extend more than one year beyond the termination of the student status described above. A Student member shall have the same privileges as an Associate member of the Society.

A **Sustaining member** is an individual, a firm, or corporation contributing substantially to the financial support of the Society.

SEC. 2—All applications for membership or transfer, except for Honorary or Fellow membership, shall be made on blank forms provided for the purpose, and shall give a complete record of the applicant's education and experience. Honorary and Fellow membership may not be applied for.

SEC. 3—(a) **Honorary membership** may be granted upon recommendation of the Board of Governors when confirmed by a four-fifths majority vote of the Honorary members, Fellows, and Active members present at any regular meeting of the Society. An Honorary member shall be exempt from all dues.

(b) **Fellow membership** may be granted upon .recommendation of the Fellow Award Committee, when confirmed by a three-fourths majority vote of the Board of Governors. Nominations for Fellow shall be made from the Active membership.

(c) Applicants for **Active membership** shall give as references at least one member of Active or of higher grade in good standing. Applicants shall be elected to membership by the unanimous approval of the entire membership of the appropriate Admissions Committee. In the event of a single dissenting vote or failure of any member of the Admissions Committee to vote, this application shall be referred to the Board of Governors, in which case approval of at least three fourths of the Board of Governors shall be required.

(d) Applicants for **Associate membership** shall give as references one member of the Society in good standing, or two persons not members of the Society who are associated with the industry. Applicants shall be elected to membership by approval of a majority of the appropriate Admissions Committee.

(e) Applicants for **Student membership** shall give as reference the head of the department of the institution he is attending, this faculty member not necessarily being a member of the Society.

Bylaw II
OFFICERS

SEC. 1—An officer or governor shall be an Honorary, a Fellow, or an Active member.

SEC. 2—Vacancies in the Board of Governors shall be filled by the Board of Governors until the annual meeting of the Society.

Bylaw III
BOARD OF GOVERNORS

SEC. 1—The Board of Governors shall transact the business of the Society between members' meetings, and shall meet at the call of the President, with the proviso that no meeting shall be called without at least seven · (7) days' prior notice, stating the purpose of the meeting, to all members of the Board by letter or by telegram.

SEC. 2—Nine members of the Board of Governors shall constitute a quorum at all meetings.

SEC. 3—When voting by letter ballot, a majority affirmative vote of the total membership of the Board of Governors shall carry approval, except as otherwise provided.

SEC. 4—The Board of Governors, when making nominations to fill vacancies in offices or on the Board, shall endeavor to nominate persons who in the aggregate are representative of the various branches or organizations of the motion picture industry to the end that there shall be no substantial predominance upon the Board, as the result of its own action, of representatives of any one or more branches or organizations of the industry.

Bylaw IV
Committees

Sec. 1—All committees, except as otherwise specified, shall be appointed by the President.

Sec. 2—All committees shall be appointed to act for the term served by the officer who shall appoint the committees, unless their appointment is sooner terminated by the appointing officer.

Sec. 3—Chairmen of the committees shall not be eligible to serve in such capacity for more than two consecutive terms.

Sec. 4—Standing committees of the Society shall be as follows to be appointed as designated:

(a) *Appointed by the President and confirmed by the Board of Governors—*
Progress Medal Award Committee
Journal Award Committee
Honorary Membership Committee
Fellow Award Committee
Admissions Committees
  (Atlantic Coast Section)
  (Pacific Coast Section)
European Advisory Committee

(b) *Appointed by the Engineering Vice-President—*
Sound Committee
Standards Committee
Studio Lighting Committee
Color Committee
Theater Engineering Committee
Exchange Practice Committee
Nontheatrical Equipment Committee
Television Committee
Test Film Quality Committee
Laboratory Practice Committee
Cinematography Committee
Process Photography Committee
Preservation of Film Committee

(c) *Appointed by the Editorial Vice-President—*
Board of Editors
Papers Committee
Progress Committee
Historical Committee
Museum Committee

(d) *Appointed by the Convention Vice-President—*
Publicity Committee
Convention Arrangements Committee
Apparatus Exhibit Committee

(e) *Appointed by the Financial Vice-President—*
Membership and Subscription Committee

Sec. 5—Two Admissions Committees, one for the Atlantic Coast Section and one for the Pacific Coast Section, shall be appointed. The former Committee shall consist of a Chairman and six Fellow or Active members of the Society residing in the metropolitan area of New York, of whom at least four shall be members of the Board of Governors.

The latter Committee shall consist of a Chairman and four Fellow or Active members of the Society residing in the Pacific Coast area, of whom at least three shall be members of the Board of Governors.

Bylaw V
Meetings

Sec. 1—The location of each meeting of the Society shall be determined by the Board of Governors.

SEC. 2—Only Honorary members, Fellows, and Active members shall be entitled to vote.

SEC. 3—A quorum of the Society shall consist in number of one fifteenth of the total number of Honorary members, Fellows, and Active members as listed in the Society's records at the close of the last fiscal year.

SEC. 4—The fall convention shall be the annual meeting.

SEC. 5—Special meetings may be called by the President and upon the request of any three members of the Board of Governors not including the President.

SEC. 6—All members of the Society in any grade shall have the privilege of discussing technical material presented before the Society or its Sections.

Bylaw VI
DUTIES OF OFFICERS

SEC. 1—The **President** shall preside at all business meetings of the Society and shall perform the duties pertaining to that office. As such he shall be the chief executive of the Society, to whom all other officers shall report.

SEC. 2—In the absence of the President, the officer next in order as listed in Article IV of the Constitution shall preside at meetings and perform the duties of the President.

SEC. 3—The five Vice-Presidents shall perform the duties separately enumerated below for each office, or as defined by the President:

(a) The **Executive Vice-President** shall represent the President in such geographical areas of the United States as shall be determined by the Board of Governors and shall be responsible for the supervision of the general affairs of the Society in such areas, as directed by the President of the Society. Should the President or Executive Vice-President remove his residence from the geographical area (Atlantic Coast or Pacific Coast) of the United States in which he resided at the time of his election, the office of Executive Vice-President shall immediately become vacant and a new Executive Vice-President elected by the Board of Governors for the unexpired portion of the term, the new Executive Vice-President to be a resident of that part of the United States from which the President or Executive Vice-President has just moved.

(b) The **Engineering Vice-President** shall appoint all technical committees. He shall be responsible for the general initiation, supervision, and co-ordination of the work in and among these committees. He may act as Chairman of any committee or otherwise be a member ex-officio.

(c) The **Editorial Vice-President** shall be responsible for the publication of the Society's JOURNAL and all other technical publications. He shall pass upon the suitability of the material for publication, and shall cause material suitable for publication to be solicited as may be needed. He shall appoint a Papers Committee and an Editorial Committee. He may act as Chairman of any committee or otherwise be a member ex-officio.

(d) The **Financial Vice-President** shall be responsible for the financial operations of the Society, and shall conduct them in accordance with budgets approved by the Board of Governors. He shall study the costs of operation and the income possibilities to the end that the greatest service may be rendered to the members of the Society within the available funds. He shall submit proposed budgets to the Board. He shall appoint at his discretion a Ways and Means Committee, a Membership Committee, a Commercial Advertising Committee, and such other committees within the scope of his work as may be needed. He may act as Chairman of any of these committees or otherwise be a member ex-officio.

(e) The **Convention Vice-President** shall be responsible for the national conventions of the Society. He shall appoint a Convention Arrangements Committee, an Apparatus Exhibit Committee, and a Publicity Committee. He may act as Chairman of any committee, or otherwise be a member ex-officio.

SEC. 4—The **Secretary** shall keep a record of all meetings; he shall conduct the correspondence relating to his office, and shall have the care and custody of records, and the seal of the Society.

SEC. 5—The **Treasurer** shall have charge of the funds of the Society and disburse them as and when authorized by the Financial Vice-President. He shall make an annual report, duly audited, to the Society, and a report at such other times as may be requested. He shall be bonded in an amount to be determined by the Board of Governors and his bond filed with the Secretary.

SEC. 6—Each officer of the Society, upon the expiration of his term of office, shall transmit to his successor a memorandum outlining the duties and policies of his office.

Bylaw VII

ELECTIONS

SEC. 1—All officers and governors shall be elected to their respective offices by a majority of ballots cast by the Active, Fellow, and Honorary members in the following manner:

Not less than three months prior to the annual fall convention, the Board of Governors shall nominate for each vacancy several suitable candidates.

Nominations shall first be presented by a Nominating Committee appointed by the President, consisting of nine members, including a Chairman. The committee shall be made up of two Past-Presidents, three members of the Board of Governors not up for election, and four other Active, Fellow, or Honorary members, not currently officers or governors of the Society. Nominations shall be made by three-quarters affirmative vote of the total Nominating Committee. Such nominations shall be final unless any nominee is rejected by a three-quarters vote of the Board of Governors present and voting.

The Secretary shall then notify these candidates of their nomination. From the list of acceptances, not more than two names for each vacancy shall be selected by the Board of Governors and placed on a letter ballot. A blank space shall be provided on this letter ballot under each office, in which space the names of any Active, Fellow, or Honorary members other than those suggested by the Board of Governors may be voted for. The balloting shall then take place.

The ballot shall be enclosed in a blank envelope which is enclosed in an outer envelope bearing the Secretary's address and a space for the member's name and address. One of these shall be mailed to each Active, Fellow, and Honorary member of the Society, not less than forty days in advance of the annual fall convention.

The voter shall then indicate on the ballot one choice for each office, seal the ballot in the blank envelope, place this in the envelope addressed to the Secretary, sign his name and address on the latter, and mail it in accordance with the instructions printed on the ballot. No marks of any kind except those above prescribed shall be placed upon the ballots or envelopes. Voting shall close seven days before the opening session of the annual fall convention.

The sealed envelope shall be delivered by the Secretary to a Committee of Tellers appointed by the President at the annual fall convention. This committee shall then examine the return envelopes, open and count the ballots, and announce the results of the election.

The newly elected officers and governors of the general Society shall take office on January 1st following their election.

Bylaw VIII

DUES AND INDEBTEDNESS

SEC. 1—The annual dues shall be fifteen dollars ($15) for Fellows and Active members, ten dollars ($10) for Associate members, and five dollars ($5) for Student members, payable on or before January 1st of each year. Current or first year's dues for new members in any calendar year shall be at the full annual rate for those notified of acceptance in the Society on or before June 30th; one half the annual rate for those notified of acceptance in the Society on or after July 1st.

SEC. 2—(a) Transfer of membership to a higher grade may be made at any time. If the transfer is made on or before June 30th the annual dues of the higher grade are required. If the transfer is made on or after July 1st and the member's dues for the full year have been paid, one half of the annual dues of the higher grade is payable less one half the annual dues of the lower grade.

(b) No credit shall be given for annual dues in a membership transfer from a higher to a lower grade, and such transfers shall take place on January 1st of each year.

(c) The Board of Governors upon their own initiative and without a transfer application may elect, by the approval of at least three fourths of the Board, any Associate or Active member for transfer to any higher grade of membership.

SEC. 3—Annual dues shall be paid in advance. A *new* member who has not paid dues in advance shall be notified of admittance but shall not receive the JOURNAL and is not in good standing until initial dues are paid. All Honorary members, Fellows, and Active members in good standing, as defined in SECTION 5, may vote or otherwise participate in the meetings.

SEC. 4—Members shall be considered delinquent whose annual dues for the year remain unpaid on February 1st. The first notice of delinquency shall be mailed February 1st. The second notice of delinquency shall be mailed, if necessary, on March 1st, and shall include a statement that the member's name will be removed from the mailing list for the JOURNAL and other publications of the Society before the mailing of the April issue of the JOURNAL. Members who are in arrears of dues on June 1st, after two notices of such delinquency have been mailed to their last address of record, shall be notified their names have been removed from the mailing list and shall be warned unless remittance is received on or before August 1st, their names shall be submitted to the Board of Governors for action at the next meeting. Back issues of the JOURNAL shall be sent, if available, to members whose dues have been paid prior to August 1st.

SEC. 5—(a) Members whose dues remain unpaid on October 1st may be dropped from the rolls of the Society by majority vote and action of the Board, or the Board may take such action as it sees fit.

(b) Anyone who has been dropped from the rolls of the Society for non-payment of dues shall, in the event of his application for reinstatement, be considered as a new member.

(c) Any member may be suspended or expelled for cause by a majority vote of the entire Board of Governors; provided he shall be given notice and a copy in writing of the charges preferred against him, and shall be afforded opportunity to be heard ten days prior to such action.

SEC. 6—The provisions of SECTIONS 1 to 4, inclusive, of this Bylaw VIII given above may be modified or rescinded by action of the Board of Governors.

Bylaw IX
EMBLEM

SEC. 1—The emblem of the Society shall be a facsimile of a four-hole film reel with the letter S in the upper center opening, and the letters M, P, and E, in the three lower openings, respectively. The Society's emblem may be worn by members only.

Bylaw X
PUBLICATIONS

SEC. 1—Papers read at meetings or submitted at other times, and all material of general interest shall be submitted to the Editorial Board, and those deemed worthy of permanent record shall be printed in the JOURNAL. A copy of each issue shall be mailed to each member in good standing to his last address of record. Extra copies of_the JOURNAL shall be printed for general distribution and may be obtained from the General Office on payment of a fee fixed by the Board of Governors.

Bylaw XI
LOCAL SECTIONS

SEC. 1—Sections of the Society may be authorized in any state or locality where the Active, Fellow, and Honorary membership exceeds 20. The geographic boundaries of each Section shall be determined by the Board of Governors.

Upon written petition, signed by 20 or more Active members, Fellows, and Honorary members, for the authorization of a Section of the Society, the Board of Governors may grant such authorization.

SECTION MEMBERSHIP

SEC. 2—All members of the Society of Motion Picture Engineers in good standing residing in that portion of any country set apart by the Board of Governors tributary to any local Section shall be eligible for membership in that Section, and when so enrolled they shall be entitled to all privileges that such local Section may, under the General Society's Constitution and Bylaws, provide.

Any member of the Society in good standing shall be eligible for nonresident affiliated membership of any Section under conditions and obligations prescribed for the Section. An affiliated member shall receive all notices and publications of the Section but he shall not be entitled to vote at sectional meetings.

SEC. 3—Should the enrolled Active, Fellow, and Honorary membership of a Section fall below 20, or should the technical quality of the presented papers fall below an acceptable level, or the average attendance at meetings not warrant the expense of maintaining the organization, the Board of Governors may cancel its authorization.

SECTION OFFICERS

SEC. 4—The officers of each Section shall be a Chairman and a Secretary-Treasurer. The Section chairmen shall automatically become members of the Board of Governors of the General Society, and continue in such positions for the duration of their terms as chairmen of the local Sections. Each Section officer shall hold office for one year, or until his successor is chosen.

SECTION BOARD OF MANAGERS

SEC. 5—The Board of Managers shall consist of the Section Chairman, the Section Past-Chairman, the Section Secretary-Treasurer, and six Active, Fellow, or Honorary members. Each manager of a Section shall hold office for two years, or until his successor is chosen.

SECTION ELECTIONS

SEC. 6—The officers and managers of a Section shall be Active, Fellow, or Honorary members of the General Society. All officers and managers shall be elected to their respective offices by a majority of ballots cast by the Active, Fellow, and Honorary members residing in the geographical area covered by the Section.

Not less than three months prior to the annual fall convention of the Society, nominations shall be presented to the Board of Managers of the Section by a Nominating Committee appointed by the Chairman of the Section, consisting of seven members, including a chairman. The Committee shall be composed of the present Chairman, the Past-Chairman, two other members of the Board of Managers not up for election, and three other Active, Fellow, or Honorary members of the Section not currently officers or managers of the Section. Nominations shall be made by a three-quarters affirmative vote of the total Nominating Committee. Such nominations shall be final, unless any nominee is rejected by a three-quarters vote of the Board of Managers, and in the event of such rejection the Board of Managers will make its own nomination.

The Chairman of the Section shall then notify these candidates of their nomination. From the list of acceptances, not more than two names for each vacancy shall be selected by the Board of Managers and placed on a letter ballot. A blank space shall be provided on this letter ballot under each office, in which space the names of any Active, Fellow, or Honorary members other than those suggested by the Board of Managers may be voted for. The balloting shall then take place.

The ballot shall be enclosed in a blank envelope which is enclosed in an outer envelope bearing the local Secretary-Treasurer's address and a space for the member's name and address. One of these shall be mailed to each Active, Fellow, and Honorary member of the Society residing in the geographical area covered by the Section, not less than forty days in advance of the annual fall convention.

The voter shall then indicate on the ballot one choice for each office, seal the ballot in the blank envelope, place this in the envelope addressed to the Secretary-Treasurer, sign his name and address on the latter, and mail it in accordance with the instructions printed on the ballot. No marks of any kind except those above prescribed shall be placed upon the ballots or envelopes. Voting shall close seven days before the opening session of the annual fall convention.

The sealed envelopes shall be delivered by the Secretary-Treasurer to his Board of Managers at a duly called meeting. The Board of Managers shall then examine the return envelopes, open and count the ballots, and announce the results of the election.

The newly elected officers and managers shall take office on January 1st following their election.

SECTION BUSINESS

SEC. 7—The business of a Section shall be conducted by the Board of Managers.

SECTION EXPENSES

SEC. 8—(a) As early as possible in the fiscal year, the Secretary-Treasurer of each Section shall submit to the Board of Governors of the Society a budget of expenses for the year.

(b) The Treasurer of the General Society may deposit with each Section Secretary-Treasurer a sum of money, the amount to be fixed by the Board of Governors, for current expenses.

(c) The Secretary-Treasurer of each Section shall send to the Treasurer of the General Society, quarterly or on demand, an itemized account of all expenditures incurred during the preceding interval.

(d) Expenses other than those enumerated in the budget, as approved by the Board of Governors of the General Society, shall not be payable from the general funds of the Society without express permission from the Board of Governors.

(e) A Section Board of Managers shall defray all expenses of the Section not provided for by the Board of Governors, from funds raised locally by donation, or fixed annual dues, or by both.

(f) The Secretary of the General Society shall, unless otherwise arranged, supply to each Section all stationery and printing necessary for the conduct of its business.

SECTION MEETINGS

SEC. 9—The regular meetings of a Section shall be held in such places and at such hours as the Board of Managers may designate.

The Secretary-Treasurer of each Section shall forward to the Secretary of the General Society, not later than five days after a meeting of a Section, a statement of the attendance and of the business transacted.

SECTION PAPERS

SEC. 10—Papers shall be approved by the Section's Papers Committee previously to their being presented before a Section. Manuscripts of papers presented before a Section, together with a report of the discussions and the proceedings of the Section meetings, shall be forwarded promptly by the Section Secretary-Treasurer to the Secretary of the General Society. Such material may, at the discretion of the Board of Editors of the General Society, be printed in the Society's publications.

CONSTITUTION AND BYLAWS

SEC. 11—Sections shall abide by the Constitution and Bylaws of the Society and conform to the regulations of the Board of Governors. The conduct of Sections shall always be in conformity with the general policy of the Society as fixed by the Board of Governors.

Bylaw XII
AMENDMENTS

SEC. 1—These Bylaws may be amended at any regular meeting of the Society by the affirmative vote of two thirds of the members present at a meeting who are eligible to vote thereon, a quorum being present, either on the recommendation of the Board of Governors or by a recommendation to the Board of Governors signed by any ten members of Active or higher grade, provided that the proposed amendment or amendments shall have been published in the JOURNAL of the Society, in the issue next preceding the date of the stated business meeting of the Society at which the amendment or amendments are to be acted upon.

SEC. 2—In the event that no quorum of the voting members is present at the time of the meeting referred to in SECTION 1, the amendment or amendments shall be referred for action to the Board of Governors. The proposed amendment or amendments then become a part of the Bylaws upon receiving the affirmative vote of three quarters of the Board of Governors.

Bylaw XIII
STUDENT CHAPTERS

SEC. 1—Student Chapters of the Society may be authorized in any college, university, or technical institute of collegiate standing.

Upon written petition, signed by twelve or more Society members, or applicants for Society membership, and the Faculty Adviser, for the authorization of a Student Chapter, the Board of Governors may grant such authorization.

CHAPTER MEMBERSHIP

SEC. 2—All members of the Society of Motion Picture Engineers in good standing who are attending the designated educational institution shall be eligible for membership in the Student Chapter, and when so enrolled they shall be entitled to all privileges that such Student Chapter may, under the General Society's Constitution and Bylaws, provide.

SEC. 3—Should the membership of the Student Chapter fall below ten, or should the technical quality of the presented papers fall below an acceptable level, or the average attendance at meetings not warrant the expense of

maintaining the organization, the Board of Governors may cancel its authorization.

CHAPTER OFFICERS

SEC. 4—The officers of each Student Chapter shall be a Chairman and a Secretary-Treasurer. Each Chapter officer shall hold office for one year, or until his successor is chosen. Officers shall be chosen in May to take office at the beginning of the following school year. The procedure for holding elections shall be prescribed in Administrative Practices.

FACULTY ADVISER

SEC. 5—A member of the faculty of the same educational institution shall be designated by the Board of Governors as Faculty Adviser. It shall be his duty to advise the officers on the conduct of the Chapter and to approve all reports to the Secretary and the Treasurer of the Society.

CHAPTER EXPENSES

SEC. 6—The Treasurer of the General Society may deposit with each Chapter Secretary-Treasurer a sum of money, the amount to be fixed by the Board of Governors. The Secretary-Treasurer shall send to the Treasurer of the General Society at the end of each school year an itemized account of all expenditures incurred during that period.

CHAPTER MEETINGS

SEC. 7—The Chapter shall hold at least four meetings per year. The Secretary-Treasurer shall forward to the Secretary of the General Society at the end of each school year a report of the meetings for that year, giving the subject, speaker, and approximate attendance for each meeting.

Awards

In accordance with the provisions of Administrative Practices of the Society, the regulations for procedure in granting the Journal Award, the Progress Medal Award, and the Samuel L. Warner Memorial Award, a list of the names of previous recipients, and the reasons therefore, are published annually in the JOURNAL as follows:

JOURNAL AWARD

The Journal Award Committee shall consist of five Fellows or Active members of the Society, appointed by the President and confirmed by the Board of Governors. The Chairman of the Committee shall be designated by the President.

At the fall convention of the Society a Journal Award Certificate shall be presented to the author or to each of the authors of the most outstanding paper originally published in the JOURNAL of the Society during the preceding calendar year.

Other papers published in the JOURNAL of the Society may be cited for Honorable Mention at the option of the Committee, but in any case should not exceed five in number.

The Journal Award shall be made on the basis of the following qualifications:

(1) The paper must deal with some technical phase of motion picture engineering.

(2) No paper given in connection with the receipt of any other Award of the Society shall be eligible.

(3) In judging of the merits of the paper, three qualities shall be considered, with the weights here indicated:

(a)	Technical merit and importance of material......	45 per cent.
(b)	Originality and breadth of interest...............	35 per cent.
(c)	Excellence of presentation of the material........	20 per cent.

A majority vote of the entire Committee shall be required for the election to the Award. Absent members may vote in writing.

The report of the Committee shall be presented to the Board of Governors at their July meeting for ratification.

These regulations, a list of the names of those who have previously received the Journal Award, the year of each Award, and the titles of the papers shall be published annually in the April issue of the JOURNAL of the Society. In addition, the list of papers selected for Honorable Mention shall be published in the JOURNAL of the Society during the year current with the Award.

-The Awards in previous years have been as follows:

1934—P. A. Snell, for his paper entitled "An Introduction to the Experimental Study of Visual Fatigue." (Published May, 1933.)

1935—L. A. Jones and J. H. Webb, for their paper entitled "Reciprocity Law Failure in Photographic Exposure." (Published September, 1934.)

1936—E. W. Kellogg, for his paper entitled "A Comparison of Variable-Density and Variable-Width Systems." (Published September, 1935.)

1937—D. B. Judd, for his paper entitled "Color Blindness and Anomalies of Vision." (Published June, 1936.)

1938—K. S. Gibson, for his paper entitled "The Analysis and Specification of Color." (Published April, 1937.)
1939—H. T. Kalmus, for his paper entitled "Technicolor Adventures in Cinemaland." (Published December, 1938.)
1940—R. R. McNath, for his paper entitled "The Surface of the Nearest Star." (Published March, 1939.)
1941—J. G. Frayne and Vincent Pagliarulo, for their paper entitled "The Effects of Ultraviolet Light on Variable-Density Recording and Printing." (Published June, 1940.)
1942—W. J. Albersheim and Donald MacKenzie, for their paper entitled "Analysis of Sound-Film Drives." (Published July, 1941.)
1943—R. R. Scoville and W. L. Bell, for their paper entitled "Design and Use of Noise-Reduction Bias Systems." (Published February, 1942; Award made April, 1944.)
1944—J. I. Crabtree, G. T. Eaton, and M. E. Muehler, for their paper entitled "Removal of Hypo and Silver Salts from Photographic Materials as Affected by the Composition of the Processing Solutions." (Published July, 1943.)
1945—C. J. Kunz, H. E. Goldberg, and C. E. Ives, for their paper entitled "Improvement in Illumination Efficiency of Motion Picture Printers." (Published May, 1944.)
1946—R. H. Talbot, for his paper entitled "The Projection Life of Film." (Published August, 1945.)
1947—Albert Rose, for his paper entitled "A Unified Approach to the Performance of Photographic Film, Television Pickup Tubes, and the Human Eye." (Published October, 1946.)
1948—J. S. Chandler, D. F. Lyman, and L. R. Martin, for their paper entitled "Proposals for 16-Mm and 8-Mm Sprocket Standards." (Published June, 1947.)

The present Chairman of the Journal Award Committee is C. R. Daily.

PROGRESS MEDAL AWARD

The Progress Medal Award Committee shall consist of five Fellows or Active members of the Society, appointed by the President and confirmed by the Board of Governors. The Chairman of the Committee shall be designated by the President.

The Progress Medal may be awarded each year to an individual in recognition of any invention, research, or development which, in the opinion of the Committee, shall have resulted in a significant advance in the development of motion picture technology.

Any member of the Society may recommend persons deemed worthy of the Award. The recommendation in each case shall be in writing and in detail as to the accomplishments which are thought to justify consideration. The recommendation shall be seconded in writing by any two Fellows or Active members of the Society, who shall set forth their knowledge of the accomplishments of the candidate which, in their opinion, justify consideration.

A majority vote of the entire Committee shall be required to constitute an Award of the Progress Medal. Absent members may vote in writing.

The report of the Committee shall be presented to the Board of Governors at their July meeting for ratification.

The recipient of the Progress Medal shall be asked to present a photograph of himself to the Society and, at the discretion of the Committee, may be asked to prepare a paper for publication in the JOURNAL of the Society.

These regulations, a list of the names of those who have previously received the Medal, the year of each Award, and a statement of the reason for the Award shall be published annually in the April issue of the JOURNAL of the Society.

Previous Awards have been as follows:

The 1935 Award was made to E. C. Wente, for his work in the field of sound recording and reproduction. (Citation published December, 1935.)

The 1936 Award was made to C. E. K. Mees, for his work in photography. (Citation published December, 1936.)

The 1937 Award was made to E. W. Kellogg, for his work in the field of sound reproduction. (Citation published December, 1937.)

The 1938 Award was made to H. T. Kalmus, for his work in developing color motion pictures. (Citation published December, 1938.)

The 1939 Award was made to L. A. Jones, for his scientific researches in the field of photography. (Citation published December, 1939.)

The 1940 Award was made to Walt Disney, for his contributions to motion picture photography and sound recording of feature and short cartoon films. (Citation published December, 1940.)

The 1941 Award was made to G. L. Dimmick, for his development activities in motion picture sound recording. (Citation published December, 1941.)

No Awards were made in 1942 and 1943.

The 1944 Award was made to J. G. Capstaff, for his research and development of films and apparatus used in amateur cinematography. (Citation published January, 1945.)

No Awards were made in 1945 and 1946.

The 1947 Award was made to J. G. Frayne for his technical achievements and the documenting of his work in addition to his contributions to the field of education and his inspiration to his fellow engineers. (Citation published January, 1948.)

The Award for 1948 was made to Peter Mole for his outstanding achievements in the field of motion picture studio lighting which set a pattern for lighting techniques and equipment for the American Motion Picture Industry. (Citation published January, 1949.)

The present Chairman of the Progress Medal Award Committee is J. G. Frayne.

PROGRESS MEDAL AWARDED FOR ACHIEVEMENT IN MOTION PICTURE TECHNOLOGY

SAMUEL L. WARNER MEMORIAL AWARD

Each year the President shall appoint a Samuel L. Warner Memorial Award Committee consisting of a chairman and four members. The chairman and committee members must be Active Members or Fellows of the Society. In considering candidates for the Award, the committee shall give preference to inventions or developments occurring in the last five years. Preference should also be given to the invention or development likely to have the widest and most beneficial effect on the quality of the reproduced sound and picture. A description of the method or apparatus must be available for publication in sufficient detail so that it may be followed by anyone skilled in the art. Since the Award is made to an individual, a development in which a group participates should be considered only if one person has contributed the basic idea and also has contributed substantially to the practical working out of the idea. If, in any year, the committee does not consider any recent development to be more than the logical working out of details along well-known lines, no recommendation for the Award shall be made. The recommendation of the committee shall be presented to the Board of Governors at the July meeting.

The purpose of this Award is to encourage the development of new and improved methods or apparatus designed for sound-on-film motion pictures, including any step in the process.

Any person, whether or not a member of the Society of Motion Picture Engineers, is eligible to receive the Award.

The Award shall consist of a gold medal suitably engraved for each recipient. It shall be presented at the Fall Convention of the Society, together with a bronze replica.

These regulations, a list of those who previously have received the Award, and a statement of the reason for the Award shall be published annually in the April issue of the JOURNAL of the Society.

SAMUEL L. WARNER MEMORIAL AWARD PRESENTED ANNUALLY FOR MOST OUTSTANDING WORK IN SOUND MOTION PICTURE ENGINEERING

The 1947 Award was made to J. A. Maurer, for his outstanding contributions to the field of high-quality 16-mm sound recording and reproduction, film processing, development of 16-mm sound test films, and for his inspired leadership in industry standardization.

The 1948 Award was presented to Nathan Levinson for his outstanding work in the field of motion picture sound recording, the intercutting of variable-area and variable-density sound tracks, the commercial use of control track for extending volume range, and the use of the first soundproof camera blimps. (Citation published January, 1949.)

The present Chairman of the Samuel L. Warner Memorial Award Committee is W. V. Wolfe.

Society Announcement

Progress Medal Award

The SMPE Progress Medal Award is presented to an individual in recognition of his technical contributions to the motion picture industry. This is an annual award; however, it need not be presented in any given year if the Progress Medal Award Committee feels that there is no qualified candidate. Candidates may be proposed by any member of the Society as outlined in the formal committee procedure on page 475 of this issue of the Journal.

Proposals for consideration by the committee may be addressed to any member of the committee which is listed below, but must be received prior to May 15, 1949.

R. M. Corbin, Eastman Kodak Company, 343 State St., Rochester 4, N. Y.

J. G. Frayne, Electrical Research Products, 6601 Romaine St., Los Angeles 38, Calif.

R. L. Garman, General Precision Laboratories, 68 Bedford Rd., Pleasantville 1, N. Y.

Barton Kreuzer, RCA Victor Division, Camden, N. J.

T. T. Moulton, 20th Century-Fox Films, Beverly Hills, Calif.

The Progress Medal was inaugurated during the term of office of President J. I. Crabtree but much credit is due Mr. G. E. Matthews, the then chairman of the Historical Committee, for his efforts in obtaining an outstanding design. Sketches for the proposed medal were submitted by some of the better-known artists in New York City but these were mostly conventional featuring the laurel wreath. Fortunately Mr. Alexander Murray, a co-worker with Messrs. Crabtree and Matthews in the Research Laboratories of the Eastman Kodak Company, became interested in the problem and submitted a unique design incorporating many symbols peculiar to the photographic and motion picture art and donated his work to the Society. A picture of the medal is shown on page 476.

The design was approved unanimously by the Board of Governors, precision dies made by the Metal Arts Company, Rochester, N. Y., and the first gold medal struck in the year 1935 which, on recommendation of the Progress Award Committee, was awarded to Dr. E. C. Wente of the Bell Telephone Laboratories.

A complete description of the Medal appeared on pages 414–415 of the April, 1948, issue of the JOURNAL.

Society of Motion Picture Engineers

REPORT OF THE TREASURER

January 1—December 31, 1948

Members' Equity, January 1, 1948		$90,759

Receipts, Jan.–Dec., 1948:

Membership Dues	$51,402	
Test Films	42,549	
Publications	14,584	
Standards and Certificates	839	
Other (Interest, etc.)	2,108	
Total Receipts		$111,482

Disbursements, Jan.–Dec., 1948:

Test Films	$34,194	
Publications	34,448	
Standards and Certificates	699	
Engineering (Salaries and Committees)	8,229	
Nonengineering Committees	1,331	
General Office	35,996	
Officers Expenses	295	
Sections	2,150	
SMPE Affiliations	888	
Conventions (Net)	1,976	
Total Disbursements		120,206

Excess Disbursements over Receipts		$ 8,724
Loss on Sale of U. S. Treasury Bond		158
Members' Equity, December 31, 1948, Cash Basis		$81,877

Adjustments necessary to change to accrual basis:

Add: Other Assets (Net)	$ 19,641	
Deduct: Other Liabilities and Reserves	4,670	
Net Adjustments		14,971
Members' Equity, December 31, 1948—Accrual Basis		$96,848

Respectfully submitted,
R. B. AUSTRIAN, *Treasurer*

The cash records of the Treasurer were audited for the year ended December 31, 1948, by Sparrow, Waymouth and Company, Certified Public Accountants, New York, and are in conformity with the above report.

D. B. JOY
Financial Vice-President

Society of Motion Picture Engineers

MEMBERSHIP CHANGES*

Year Ended December 31, 1948

	Hon.	Sust.	Fel.	Act.	Asso.	Stu.	Total
Membership, Jan. 1, 1948	6	42	165	675	1636	127	2651
New Members		31		97	336	77	541
Reinstatements			1	4	8		13
	6	73	166	776	1980	204	3205
Delinquent		−2		−31	−96	−19	−148
Resignations		−1	−3	−14	−33	−3	−54
Deaths	−2		−1	−2	−5		−10
	4	70	162	729	1846	182	2993
Changes in Grade:							
Active to Fellow			16	−16			
Associate to Active				24	−24		
Active to Associate				−2	2		
Student to Associate					2	−2	
Associate to Student					−1	1	
Student to Active				1		−1	
	4	70	178	736	1825	180	2993
Corrections per physical count—December 31, 1948				−2	−65	−26	−93
Membership—December 31, 1948 as per physical count	4	70	178	734	1860	154	2900

NONMEMBER SUBSCRIPTIONS TO JOURNAL
As of Dec. 31, 1948

Subscriptions, January 1, 1948	926
New Subscriptions	250
	1176
Expirations and Cutoffs	409
Subscriptions, Dec. 31, 1948—as per physical count	767

* Grades: Honorary, Sustaining, Fellow, Active, Associate, and Student.

Committees of the Society
(CORRECT TO APRIL 1, 1949)

ADMISSIONS

To pass upon all applications for membership, applications for transfer, and to review the Student and Associate membership list periodically for possible transfer to the Associate and Active grades, respectively. The duties of each committee are limited to applications and transfers originating in the geographic area covered.

E. A. BERTRAM, *Chairman, East*
DeLuxe Laboratories
850 Tenth Ave.
New York 19, N. Y.

R. B. AUSTRIAN	H. D. BRADBURY	C. R. KEITH
HERBERT BARNETT	RICHARD HODGSON	W. H. RIVERS

G. E. SAWYER, *Chairman, West*
Samuel Goldwyn Studio Corporation
1041 N. Formosa Ave.
Hollywood 46, Calif.

GEORGE FRIEDL, JR.	L. T. GOLDSMITH	S. P. SOLOW
	H. W. MOYSE	

BOARD OF EDITORS

To pass upon the suitability of all material submitted for publication, or for presentation at conventions, and publish the JOURNAL.

A. C. DOWNES, *Chairman*
2181 Niagara Dr.
Lakewood 7, Ohio

L. F. BROWN	A. M. GUNDELFINGER	G. E. MATTHEWS
A. W. COOK	C. W. HANDLEY	PIERRE MERTZ
J. G. FRAYNE	A. C. HARDY	J. H. WADDELL
	P. J. LARSEN	

CINEMATOGRAPHY

To make recommendations and prepare specifications for the operation, maintenance, and servicing of motion picture cameras, accessory equipment, studio and outdoor-set lighting arrangements, camera technique, and the varied uses of motion picture negative films for general photography.

C. G. CLARKE, *Chairman*
Twentieth Century-Fox Film Corporation
Beverly Hills, Calif.

J. W. BOYLE	A. J. MILLER	ARTHUR REEVES
KARL FREUND		JOSEPH RUTTENBERG

COLOR ·

To make recommendations and prepare specifications for the operation, maintenance, and servicing of color motion picture processes, accessory equipment, studio lighting, selection of studio set colors, color cameras, color motion picture films, and general color photography.

H. H. DUERR, *Chairman*
Ansco
Binghamton, N. Y.

J. A. BALL	L. E. CLARK	A. M. GUNDELFINGER
R. H. BINGHAM	R. O. DREW	A. J. MILLER
M. R. BOYER	ALBERT DURYEA	C. F. J. OVERHAGE
H. E. BRAGG	R. M. EVANS	G F. RACKETT
O. O. CECCARINI	J. G. FRAYNE	L. E. VARDEN
	L. T. GOLDSMITH	

CONVENTION

To assist the Convention Vice-President in the responsibilities pertaining to arrangements and details of the Society's technical conventions.

W. C. KUNZMANN, *Chairman*
National Carbon Company
Box 6087
Cleveland, Ohio

E. R. GEIB	L. B. ISAAC	F. B. ROGERS, JR.
C. W. HANDLEY	O. F. NEU	N. L. SIMMONS
H. F. HEIDEGGER	W. H. RIVERS	S. P. SOLOW

EUROPEAN ADVISORY COMMITTEE

To act as liaison between the general Society and European firms, individuals, and organizations interested in motion picture engineering. To report to the Society on general motion picture affairs in Europe, on new technical developments, and to assist the Papers Committee in soliciting papers for publication in the JOURNAL.

I. D. WRATTEN, *Chairman (British Division)*
Kodak, Ltd.
Kingsway, London, England

R. H. CRICKS	L. KNOPP	A. W. WATKINS
W. M. HARCOURT		A. G. D. WEST

L. DIDIÉE, *Chairman (Continental Division)*
Association Française des Ingenieurs et Techniciens du Cinéma
92, Champs-Elysees
Paris (8e), France

EXCHANGE PRACTICE

To make recommendations and prepare specifications on the engineering or technical methods and equipment that contribute to efficiency in handling and storage of motion picture prints, so far as can be obtained by proper design, construction, and operation of film-handling equipment, air-conditioning systems, and exchange office buildings.

(Under Organization)

FELLOW AWARD

To consider qualifications of Active members as candidates for elevation to Fellow, and to submit such nominations to the Board of Governors.

L. L. RYDER, *Chairman*
Paramount Pictures
5451 Marathon St.
Hollywood 38, Calif.

R. B. AUSTRIAN	W. C. KUNZMANN	W. H. RIVERS
R. M. CORBIN	J. A. MAURER	S. P. SOLOW
D. B. JOY	PETER MOLE	E. I. SPONABLE
C. R. KEITH		R. T. VAN NIMAN

FILM DIMENSIONS

To make recommendations and prepare specifications on those film dimensions which affect performance and interchangeability, and to investigate new methods of cutting and perforating motion picture film in addition to the study of its physical properties.

E. K. CARVER, *Chairman*
Eastman Kodak Company
Kodak Park
Rochester 4, N. Y.

E. A. BERTRAM	E. FEHNDERS	M. G. TOWNSLEY
A. W. COOK	A. M. GUNDELFINGER	FRED WALLER
A. F. EDOUART	A. J. MILLER	D. R. WHITE
	W. E. POHL	

FILM-PROJECTION PRACTICE

To make recommendations and prepare specifications for the operation, maintenance, and servicing of motion picture projection equipment, projection rooms, film-storage facilities, stage arrangement, screen dimensions and placement, and maintenance of loudspeakers to improve the quality of reproduced sound and the quality of the projected picture in the theater.

(Under Organization)
L. W. DAVEE, *Chairman*
Century Projector Corporation
729 Seventh Ave.
New York 19, N. Y.

HIGH-SPEED PHOTOGRAPHY

To make recommendations and prepare specifications for the construction, installation, operation, and servicing of equipment for photographing and projecting pictures taken at high repetition rates or with extremely short exposure times.

J. H. WADDELL, *Chairman*
Bell Telephone Laboratories
463 West St.
New York 14, N. Y.

H. E. EDGERTON, *Vice-Chairman*
Massachusetts Institute of Technology
Cambridge 38, Mass.

E. A. ANDRES, SR.	W. R. FRASER	W. S. NIVISON
K. M. BAIRD	H. M. LESTER*	BRIAN O'BRIEN
D. M. BEARD	L. R. MARTIN	EARL QUINN
A. A. COOK	J. J. McDEVITT	M. L. SANDELL
H. W. CROUCH	C. D. MILLER	KENNETH SHAFTON
D. S. L. DURIE†	A. P. NEYHART	N. F. OAKLEY
R. E. FARNHAM		CHARLES SLACK

† Alternate.   * Representing Photographic Engineering Society.

HISTORICAL AND MUSEUM

To collect facts and assemble data relating to the historical development of the motion picture industry to encourage pioneers to place their work on record in the form of papers for publication in the JOURNAL, and to place in suitable depositories equipment pertaining to the industry.

(Under Organization)

EDWARD F. KERNS, *Chairman*
Museum of Modern Art
11 W. 53 St.
New York 19, N. Y.

HONORARY MEMBERSHIP

To search diligently for candidates who through their basic inventions or outstanding accomplishments have contributed to the advancement of the motion picture industry and are thus worthy of becoming Honorary members of the Society.

G. A. CHAMBERS, *Chairman*
Eastman Kodak Company
343 State St.
Rochester 4, N. Y.

HERBERT GRIFFIN	W. C. MILLER	R. O. STROCK
	TERRY RAMSAYE	

JOURNAL AWARD

To recommend to the Board of Governors the author or authors of the most outstanding paper originally published in the JOURNAL during the preceding calendar year to receive the Society's JOURNAL Award.

C. R. DAILY, *Chairman*
Paramount Pictures
5451 Marathon St.
Hollywood 38, Calif.

OTTO SANDVIK   FRED SCHMID   J. R. VOLKMANN
                 M. G. TOWNSLEY

LABORATORY PRACTICE

To make recommendations and prepare specifications for the operation, maintenance, and servicing of motion picture printers, processing machines, inspection projectors, splicing machines, film-cleaning and treating equipment, rewinding equipment, any type of film-handling accessories, methods, and processes which offer increased efficiency and improvements in the photographic quality of the final print.

(Under Organization)

J. G. STOTT, *Chairman*
Eastman Kodak Company
342 Madison Ave.
New York 17, N. Y.

MEMBERSHIP AND SUBSCRIPTION

To solicit new members, obtain nonmember subscriptions for the JOURNAL, and to arouse general interest in the activities of the Society and its publications.

L. E. JONES, *General Chairman*
Neumade Products Corporation
427 W. 42 St.
New York 18, N. Y.

A. G. SMITH, *Chairman, Atlantic Coast*
National Theater Supply
356 W. 44 St.
New York 18, N. Y.

BERTIL CARLSON   C. F. HORSTMAN   P. D. RIES
A. R. GALLO   W. C. KUNZMANN   C. W. SEAGER
T. J. GASKI   O. F. NEU   HARRY SHERMAN
N. D. GOLDEN   C. R. WOOD, SR.

MEMBERSHIP AND SUBSCRIPTION (*continued*)

CARRINGTON H. STONE, *Chairman, Central*
Suite 2020
205 W. Wacker Dr.
Chicago 6, Ill.

B. W. DEPUE	JOHN POWERS	LLOYD THOMPSON
R. E. FARNHAM	T. I. RESS	ELMER VOTZ
C. E. HEPPEBERGER	JOHN SPINNEWEBER	JOHN ZUBER

G. C. MISENER, *Chaïrman, Pacific Coast*
Ansco
6424 Santa Monica Blvd.
Hollywood 38, Calif.

L. W. CHASE, JR.	L. T. GOLDSMITH	H. W. MOYSE
J. P. CORCORAN	HERBERT GRIFFIN	H. W. REMERSCHEID
C. R. DAILY	WILLIAM HARRIS	G. E. SAWYER
J. G. FRAYNE	EMERY HUSE	W. V. WOLFE
	WATSON JONES	

A. G. PETRASEK, *Chairman, 16-Mm*
RCA Victor Division
Harrison, N. J.

W. C. BARRY, JR.	WILSON LEAHY	W. H. OFFENHAUSER, JR.
G. A. CHAMBERS	C. L. LOOTENS	F. B. ROGERS, JR.
A. W. COOK	L. R. MARTIN	R. J. SHERRY
W. F. KRUSE		LLOYD THOMPSON

R. O. STROCK, *Chairman, Foreign*
Westrex Corporation
111 Eighth Ave.
New York 11, N. Y.

A. F. BALDWIN	VERNON T. DICKINS	H. R. HOLM
WALTER BIRD	R. J. ENGLER	H. S. WALKER
	Y. A. FAZALBHOY	

R. B. AUSTRIAN, *Chairman, Television*
Television Consultant
Room 701
1270 Avenue of the Americas
New York 20, N. Y.

NOMINATIONS

To recommend nominations to the Board of Governors for annual election of officers and governors.

D. E. HYNDMAN, *Chairman*
Eastman Kodak Company
342 Madison Ave.
New York 17, N. Y.

HERBERT BARNETT	R. E. FARNHAM	A. N. GOLDSMITH
F. T. BOWDITCH	GEORGE GIROUX	T. T. GOLDSMITH
F. E. CAHILL		K. F. MORGAN

PAPERS

To solicit papers and provide the program for semiannual conventions, and make available to local sections for their meetings papers presented at national conventions.

N. L. SIMMONS, *Chairman*
6706 Santa Monica Blvd.
Hollywood, Calif.

JOSEPH E. AÏKEN, *Vice-Chairman*   E. S. SEELEY, *Vice-Chairman*
116 N. Galveston St.   Altec Service Corporation
Arlington, Va.   161 Sixth Ave.
                                              New York 13, N. Y.

L. D. GRIGNON, *Vice-Chairman*   R. T. VAN NIMAN, *Vice-Chairman*
20th Century-Fox Films Corporation   4331 W. Lake St.
Beverly Hills, Calif.   Chicago 24, Ill.

H. S. WALKER, *Vice-Chairman*
1620 Notre Dame St., W.
Montreal, Que., Canada

F. G. ALBIN	J. P. CORCORAN	W. J. MORLOCK
JOHN ARNOLD	G. R. CRANE	O. W. MURRAY
G. M. BEST	C. R. DAILY	EDWARD SCHMIDT
P. E. BRIGANDI	W. P. DUTTON	V. C. SHANER
G. A. BURNS	J. L. FORREST	W. L. TESCH
PHILLIP CALDWELL	L. R. MARTIN	J. W. THATCHER
	G. E. MATTHEWS	

PRESERVATION OF FILM

To make recommendations and prepare specifications on methods of treating and storage of motion picture film for active, archival, and permanent record purposes, so far as can be prepared within both the economic and historical value of the films.

PRESERVATION OF FILM (*continued*)

J. W. CUMMINGS, *Chairman*
National Archives
Washington, D. C.

HENRY ANDERSON	J. I. CRABTREE	J. E. GIBSON
J. G. BRADLEY	RAYMOND DAVIS	TERRY RAMSAYE
H. T. COWLING	J. L. FORREST	V. B. SEASE

PROCESS PHOTOGRAPHY

To make recommendations and prepare specifications on motion picture optical printers, process projectors (background process), matte processes, special process lighting technique, special processing machines, miniature-set requirements, special-effects devices, and the like, that will lead to improvement in this phase of the production art.

(Under Organization)

LINWOOD DUNN, *Chairman*
RKO Radio Pictures
780 Gower St.,
Los Angeles 3, Calif.

PROGRESS

To prepare an annual report on progress in the motion picture industry.

C. R. SAWYER, *Chairman*
Western Electric Company
167 Chambers St.
New York 7, N. Y.

J. E. AIKEN	R. E. LEWIS	J. W. THATCHER
C. W. HANDLEY	W. A. MUELLER	W. V. WOLFE
	W. L. TESCH	

PROGRESS MEDAL AWARD

To recommend to the Board of Governors a candidate who by his inventions, research, or development has contributed in a significant manner to the advancement of motion picture technology, and is deemed worthy of receiving the Progress Medal Award of the Society.

J. G. FRAYNE, *Chairman*
Electrical Research Products
6601 Romaine St.
Los Angeles 38, Calif.

R. M. CORBIN	R. L. GARMAN	T. T. MOULTON
	BARTON KREUZER	

PUBLICITY

To assist the Convention Vice-President in the release of publicity material concerning the Society's semiannual technical conventions.

HAROLD DESFOR, *Chairman*
RCA Victor
Camden, N. J.

LEONARD BIDWELL	HARRY SHERMAN	R. T. VAN NIMAN
GEORGE DANIEL	N. L. SIMMONS, JR.	HAROLD WENGLER

PUBLIC RELATIONS

To assist the President at all times in improving the Society's public relations.

IRVING KAHN, *Chairman*

RALPH AUSTRIAN	HAROLD DESFOR	R. T. VAN NIMAN
	PETER MOLE	

SAMUEL L. WARNER AWARD

To recommend to the Board of Governors a candidate who has done the most outstanding work in the field of sound motion picture engineering, in the development of new and improved methods or apparatus designed for sound motion pictures, including any steps in the process, and who, whether or not a Member of the Society of Motion Picture Engineers, is deemed eligible to receive the Samuel L. Warner Memorial Award of the Society.

W. V. WOLFE, *Chairman*
Motion Picture Research Council
1421 N. Western Ave. ·
Hollywood 27, Calif.

E. M. HONAN	J. P. LIVADARY	E. A. WILLIFORD
	W. W. LOZIER	

SCREEN BRIGHTNESS

To make recommendations, prepare specifications, and test methods for determining and standardizing the brightness of the motion picture screen image at various parts of the screen, and for special means or devices in the projection room adapted to the control or improvement of screen brightness.

E. R. GEIB, *Chairman*
Box 6087
Cleveland 1, Ohio

HERBERT BARNETT	W. F. LITTLE	ALLEN STIMSON
F. E. CARLSON	W. W. LOZIER	C. W. TUTTLE
GORDON EDWARDS	G. M. RENTOUMIS*	C. R. UNDERHILL, JR.
L. D. GRIGNON	N. L. SIMMONS	H. E. WHITE
ARTHUR HATCH, JR.		A. T. WILLIAMS

* Alternate.

16-MM AND 8-MM MOTION PICTURES

To make recommendations and prepare specifications for 16-mm and 8-mm cameras, 16-mm sound recorders and sound-recording practices, 16-mm and 8-mm printers and other film laboratory equipment and practices, 16-mm and 8-mm projectors, splicing machines, screen dimensions and placement, loudspeaker output and placement, preview or theater arrangements, test films, and the like, which will improve the quality of 16-mm and 8-mm motion pictures.

H. J. HOOD, *Chairman*
Eastman Kodak Company
343 State St.
Rochester 4, N. Y.

W. C. BOWEN	C. R. FORDYCE	A. G. PETRASEK
F. L. BRETHAUER	R. C. HOLSLAG	L. T. SACHTLEBEN
F. E. BROOKER	RUDOLF KINGSLAKE	R. SPOTTISWOOD
F. E. CARLSON	W. W. LOZIER	H. H. STRONG
S. L. CHERTOK	D. F. LYMAN	A. L. TERLOUW
E. W. D'ARCY	W. C. MILLER	LLOYD THOMPSON
J. W. EVANS	J. R. MONTGOMERY	M. G. TOWNSLEY
	W. H. OFFENHAUSER, JR.	

SOUND

To make recommendations and prepare specifications for the operation, maintenance, and servicing of motion picture film, sound recorders, re-recorders, and reproducing equipment, methods of recording sound, sound-film processing, and the like, to obtain means of standardizing procedures that will result in the production of better uniform quality sound in the theater.

L. T. GOLDSMITH, *Chairman*
Warner Brothers Pictures, Inc.
Burbank, Calif.

G. L. DIMMICK, *Vice-Chairman*
RCA Victor Division
Camden, N. J.

A. C. BLANEY	ROBERT HERR	OTTO SANDVIK
D. J. BLOOMBERG	J. K. HILLIARD	G. E. SAWYER
F. E. CAHILL, JR.	L. B. ISAAC	R. R. SCOVILLE
E. W. D'ARCY	E. W. KELLOGG	W. L. THAYER
R. J. ENGLER	J. P. LIVADARY	M. G. TOWNSLEY
J. G. FRAYNE	W. C. MILLER	R. T. VAN NIMAN
L. D. GRIGNON		D. R. WHITE

STANDARDS

To survey constantly all engineering phases of motion picture production, distribution, and exhibition, to make recommendations and prepare specifications that may become proposals for American Standards. This Committee should follow carefully the work of all other committees on engineering and may request any committee to investigate and prepare a report on the phase of motion picture engineering to which it is assigned.

F. E. CARLSON, *Chairman*
General Electric Company
Nela Park
Cleveland 12, Ohio

Chairmen of Engineering Committees

F. S. BERMAN	E. R. GEIB	LEONARD SATZ
CHARLES CLARKE	L. T. GOLDSMITH	J. G. STOTT
J. W. CUMMINGS	M. A. HANKINS	J. H. WADDELL
H. H. DUERR	H. J. HOOD	D. R. WHITE
	D. E. HYNDMAN	

Members-at-Large

E. K. CARVER	E. W. KELLOGG	G. T. LORANCE
GORDON EDWARDS	RUDOLF KINGSLAKE	D. F. LYMAN
C. R. KEITH		OTTO SANDVIK

Members Ex-Officio

F. T. BOWDITCH	V. O. KNUDSEN	GEORGE NIXON
L. A. JONES	J. A. MAURER	F. W. SEARS

STUDIO LIGHTING

To make recommendations and prepare specifications for the operation, maintenance, and servicing of all types of studio and outdoor auxiliary lighting equipment, tungsten light and carbon-arc sources, lighting-effect devices, diffusers, special light screens, etc., to increase the general engineering knowledge of the art.

M. A. HANKINS, *Chairman*
Mole-Richardson Company
937 N. Sycamore Ave.
Hollywood 38, Calif.

W. E. BLACKBURN	KARL FREUND	C. R. LONG
RICHARD BLOUNT	C. W. HANDLEY	W. W. LOZIER
J. W. BOYLE		D. W. PRIDEAUX

SUSTAINING MEMBERSHIP

To solicit new sustaining members and thereby obtain adequate financial support required by the Society to carry on its technical and engineering activities.

L. L. RYDER, *Chairman*
Paramount Pictures
5451 Marathon St.
Hollywood 38, Calif.

D. E. HYNDMAN   PETER MOLE

TELEVISION

To study television art with special reference to the technical interrelationships of the television and motion picture industries, and to make recommendations and prepare specifications for equipment, methods, and nomenclature designed to meet the special problems encountered at the junction of the two industries.

D. R. WHITE, *Chairman*
E. I. du Pont de Nemours and Company
Parlin, N. J.

R. B. AUSTRIAN	A. N. GOLDSMITH	PIERRE MERTZ
F. T. BOWDITCH	T. T. GOLDSMITH, JR.	H. C. MILHOLLAND
F. E. CAHILL*	HERBERT GRIFFIN	W. C. MILLER
A. W. COOK	RICHARD HODGSON*	J. R. POPPELE
E. D. COOK	C. F. HORSTMAN	PAUL RAIBOURN
C. E. DEAN	L. B. ISAAC	L. L. RYDER
BERNARD ERDE	P. J. LARSEN	OTTO SANDVIK
R. L. GARMAN	C. C. LARSON	G. E. SAWYER
FRANK GOLDBACH	NATHAN LEVINSON	R. E. SHELBY
P. C. GOLDMARK	J. P. LIVADARY	E. I. SPONABLE
	H. B. LUBCKE	H. E. WHITE

* Alternate.

TEST-FILM QUALITY

To supervise, inspect, and approve all print quality control of sound and picture test films prepared by any committee on engineering before the prints are released by the Society for general practical use.

F. S. BERMAN, *Chairman*
Movielab Film Laboratory
1600 Broadway
New York 19, N. Y.

C. F. HORSTMAN   F. R. WILSON

THEATER ENGINEERING

To make recommendations and prepare specifications of engineering methods and equipment of motion picture theaters in relation to their contribution to the physical comfort and safety of patrons, so far as can be enhanced by correct theater design, construction, and operation of equipment.

LEONARD SATZ, *Chairman*
Century Theaters
132 W. 43 St.
New York 18, N. Y.

F. W. ALEXA	E. J. CONTENT	E. H. PERKINS
HENRY ANDERSON	C. M. CUTLER	BEN SCHLANGER
A. G. ASHCROFT	JAMES FRANK, JR.	SEYMOUR SEIDER
CHARLES BACHMAN	AARON NADELL	EMIL WANDELMAIER

THEATER TELEVISION

To make recommendations and prepare specifications for the construction, installation, operation, maintenance, and servicing of equipment for projecting television pictures in the motion picture theater, as well as projection-room arrangements necessary for such equipment, and such picture-dimensional and screen-characteristic matters as may be involved in high-quality theater-television presentations.

D. E. HYNDMAN, *Chairman*
Eastman Kodak Company
342 Madison Ave.
New York 17, N. Y.

G. L. BEERS	T. T. GOLDSMITH, JR.	HARRY RUBIN
F. E. CAHILL, JR.	C. F. HORSTMAN	L. L. RYDER
A. W. COOK	L. B. ISAAC	OTTO SANDVIK
JAMES FRANK, JR.	A. G. JENSEN	EDWARD SCHMIDT
R. L. GARMAN	P. J. LARSEN	A. G. SMITH
E. P. GENOCK	NATHAN LEVINSON	E. I. SPONABLE
A. N. GOLDSMITH		J. E. VOLKMANN

SMPE REPRESENTATIVES TO OTHER ORGANIZATIONS

AMERICAN STANDARDS ASSOCIATION

Standards Council
D. E. Hyndman

Sectional Committees

Standardization of Letter Symbols and Abbreviations for Science and Engineering, Z10
S. L. Chertok

Motion Pictures, Z22
F. T. Bowditch, *Chairman*
F. E. Carlson　　　D. F. Lyman*
Pierre Mertz

Acoustical Measurements and Terminology, Z24
H. F. Olson

Photography, Z38
J. I. Crabtree

* Alternate.

INTER-SOCIETY COLOR COUNCIL
R. M. Evans, *Chairman*
J. A. Ball　　L. E. Clark
F. T. Bowditch　A. M. Gundelfinger
M. R. Boyer　　H. C. Harsh
H. E. Bragg　　W. H. Ryan

AMERICAN DOCUMENTATION INSTITUTE
J. E. Abbott

UNITED STATES NATIONAL COMMITTEE OF THE INTERNATIONAL COMMISSION ON ILLUMINATION

R. E. Farnham, *Chairman*

Herbert Barnett　　　H. E. White

Section Meetings

Atlantic Coast

At the meeting of the Atlantic Coast Section on February 16, 1949, the principal topic was "Direct Positive Variable-Area Recording with the Light Valve." A paper on this subject by L. B. Browder of Western Electric Company, Hollywood, was read by C. R. Keith. It was shown how direct positive sound tracks are obtained by reflecting light from the ribbons rather than by transmitting light between them. A 16-mm recording of speech and music made by this method was played.

"A New 16-Mm Studio Re-Recording Machine," by G. R. Crane of Western Electric Company, Hollywood, was read by C. R. Sawyer. In addition, two examples of variable-density sound tracks on Kodachrome were played. One showed the result which may be obtained by printing from a suitable black-and-white positive and the other was the result of recording directly on the Kodachrome film.

Central

Four papers previously presented at the Washington Convention were given at the January 13, 1949, meeting of the Central Section.

M. J. Yahr of RCA's Theater Equipment Division read the first two papers. "Theater Installation, Instantaneous Large-Screen Television," by Roy Wilcox and H. J. Schlafly, reported installation problems, programming methods, and audience reactions to the installation made in June, 1948, at the Fox-Philadelphia theater for the Louis-Walcott fight. "Equipment for Television Photography," by Ralph V. Little, Jr., reviewed the art of television kinescope photography for record purposes, or for rapid processing, and subsequent rebroadcast or theater projection uses.

C. E. Heppberger delivered a paper on "Optimum Performance of High-Brightness Carbon Arcs," by M. T. Jones and F. T. Bowditch. The first draft of this paper was presented at the June, 1948, Cleveland meeting; the completed manuscript and slides covered subsequent work with water-cooled jaws and special carbons of high heat conductivity developed to meet the demand for more and more light.

I. F. Jacobsen presented "Influence of Carbon Cooling on the High-Current Carbon Arc and Its Mechanism," by Wolfgang Finkelnburg. This paper reported the results of an extensive independent investigation of this subject.

General discussion of future possibilities for brighter screens for drive-in theaters and cooler, more dependable operating arc lamps followed.

At the February 17, 1949, meeting of the Central Section Reid H. Ray presented "The Use of 35-Mm Ansco Color for 16-Mm Release Prints." Since the requirements of his company's clients called for 35-mm color release prints as well as 16-mm color releases within a short space of time, the following procedure was worked out:

Section Meetings

Original photography is 35-mm on Ansco Type 735 camera film with appropriate filters. This original is processed at the Houston Laboratory in Los Angeles and a "daily" is printed on all footage using an average printer light and average color filter balance on 732 print stock. After editing the final picture a scene-by-scene color-corrected and density-balanced print is made. When this is approved the 35-mm release prints are made.

A selected 35-mm release print is used for a master to make 16-mm release prints on Kodachrome duplicating film in a standard Depue reduction printer. At the time of making the final re-recording of the sound track for the 35-mm prints, a second track is made with special equalization for reduction printing to the 16-mm Kodachrome releases.

Demonstration reels of identical scenes in the 35-mm releases and the 16-mm reductions were presented. Although the sound in both cases was excellent and the color in the 16-mm matched the 35-mm exceedingly well, the 16-mm suffered some loss of definition in the reduction printing process.

The next paper was one previously presented to the National Society in Washington on "Recent Advances in Densitometry" by Monroe H. Sweet. In addition to showing the newly developed densitometer using a photomultiplier tube and sharp cutoff filters, several additional pieces of equipment were described: (a) an attachment for measuring liquid color densities and (b) an automatic strip reading device which traces the readings directly on a paper record.

The next meeting of the Central Section will be held on May 12, 1949, at 8:00 P.M.—82 East Randolph St., Chicago. Samuel R. Todd, of the Bureau of Electrical Inspection, Department of Buildings, City of Chicago, will present a paper on "Potential Trends for Projection-Room Booth Specifications Due to Advent of Acetate Film." "Film-Distribution Considerations" will be given by Thomas McConnell, Attorney.

Book Review

Friese-Greene: Close-up of an Inventor, by Ray Allister

Published (1948) by Marsland Publications, Ltd., 122 Wardour St., London, W.˙1, England. Distributed by The Falcon Press, 6 and 7 Crown Passage, Pall Mall, London, S.W. 1, England. 176 pages, + XIII pages, + 6-page index. 19 illustrations, $5^3/_4 \times 8^3/_4$ inches. Price, 12s. 6d. net.

We have here an oddment of the lore of the motion pictures' yesterdays, restating the curious, contradictory, and erratic tradition of William Friese-Greene, one of the more nebulously connected of the many claimant fathers of the motion picture.

496

Book Review

The telling has heart-appeal in its picturing of the dreaming photographer's apprentice in the warm terms of an author who appears to have based his writing mostly upon the testimonies of relatives of his hero. None of them has been more scientifically or even technologically informed than was the romantic Mr. Greene or the author. It becomes a sort of literary adventure for persons who would study the manner of origin and growth of tradition. Despite its well-meant prefatory insistence that the work is predicated on "two years' careful research," including conversations with the family, and that "every statement can be authenticated," its findings are not supported by existing record, or by the researches of litigants and others over a period of forty years. Had Mr. Friese-Greene's claims been supportable much of the history of the motion picture would have been very different.

In this volume Mr. Greene is credited with original concept of principles and devices long antecedent to his advent, including the work of Baron Franz von Uchatius, Louis Ducos du Hauron, Coleman Sellers, and many another. His alleged initial and so-called prior presentations of the motion picture and the screen, when examined, even on the showings of this volume, are not reductions to practice of anything beyond devices and methods of years before. The existing documented record on all this is clear, and we have not now occasion to take the whole intricate history of the industry apart for answer to this prejudiced little volume. It is prejudiced alike both to the American origins of the motion picture and to the constructive work of such distinguished and modestly able scientists as Britain's late Robert W. Paul and Louis and Auguste Lumiere of France.

This book does at long last make clear how William Green came to be Friese-Greene. He married a Swiss lady, hyphenated her name to his and added an "e" for euphony. He was a graceful fellow, one gathers, and as a portrait photographer had a skillful way of making customers, including babies, smile prettily. He was ever on the eve of vast triumph, but mishaps were always arriving, and he made excursions to the pawnshops between investments by speculators on his wonders-to-come. Indubitably he believed in himself, and always considered tomorrow at the golden end of the rainbow.

The tale of his life, so friendly told, falls with considerable exactness into the cliché pattern so dear to Sunday supplement journalism keynoted with the idea of "neglected inventor, ragged, hungry, is true father of million-dollar industry." There are tales like it in all the arts and industries and every now and then they get printed. Then the dramatic conclusion in which Mr. Friese-Greene dropped dead at a dinner in his honor, given by some British showmen, did give accent to his tragic story. That, however, had no relevancy to the fact that the motion picture was conceived and brought to birth without him.

TERRY RAMSAYE
Motion Picture Herald
New York 20, N. Y.

497

Current Literature

THE EDITORS present for convenient reference a list of articles dealing with subjects cognate to motion picture engineering published in a number of selected journals. Photostatic or microfilm copies of articles in magazines that are available may be obtained from The Library of Congress, Washington, D. C., or from the New York Public Library, New York, N. Y., at prevailing rates.

American Cinematographer

30, 1, January, 1949
Photographing Films for Television (p. 9) W. STRENGE
Changing Trends in Cinematography (p. 10) H. A. LIGHTMAN
Modern Title Making (p. 12) N. KEANE
Color and Color Reproduction (p. 13) H. MEYER
A Synchronous Magnetic Recorder (p. 14) R. LAWTON

30, 2, February, 1949
Mercury Cadmium Lamps for Studio Set Lighting (p. 47) R. B. FARNHAM
The Use of Films in Television (p. 50) P. H. DORTE

British Kinematography

13, 6, December, 1948
Development of Theatre Television in England (p. 183) A. G. D. WEST

14, 1, January, 1949
Film Production Technique (p. 1) A. HITCHCOCK
Nitrate and Safety Film Base Characteristics (p. 7) G. J. CRAIG
Coloured and Directional Lighting as Applied to the Stage (p. 17) L. G. APPLEBEE

International Photographer

21, 1, January, 1949
Cameras of Yesteryear (p. 16) W. W. CLENDENIN

21, 2, February, 1949
Cameras of Yesteryear. Pt. II (p. 12) W. W. CLENDENIN
Fallacy of the "Persistence" Theory (p. 18) G. H. SEWELL

International Projectionist

24, 1, January, 1949
Projected Light and the Curved Screen (p. 10)
Expanding Use of Infra-Red Film (p. 16) A. STOUT

24, 2, February, 1949
Sound System Components (p. 12) R. A. MITCHELL

Kinematograph Weekly

382, December 30, 1948
Kine Television Is Here (p. 6)

Philips Technical Review

10, 7, January, 1949
A Demonstration Studio for Sound Recording and Reproduction and for Sound Film Projection (p. 196) Electro-Acoustics Department

RCA Review

9, 4, December, 1948
Electro-Optical Characteristics of Television Systems (p. 653) O. H. SCHADE

498

Journal of the
Society of Motion Picture Engineers

VOLUME 52   MAY 1949   NUMBER 5

Subscription to nonmembers, $10.00 per annum; to members, $6.25 per annum, included in their annual membership dues; single copies, $1.25. Order from the Society's General Office. A discount of ten per cent is allowed to accredited agencies on orders for subscriptions and single copies. Published monthly at Easton, Pa., by the Society of Motion Picture Engineers, Inc. Publication Office, 20th & Northampton Sts., Easton, Pa. General and Editorial Office, 342 Madison Ave., New York 17, N. Y. Entered as second-class matter January 15, 1930, at the Post Office at Easton, Pa., under the Act of March 3, 1879.

Society of Motion Picture Engineers

342 MADISON AVENUE—NEW YORK 17, N. Y.—TEL. MU 2-2185

BOYCE NEMEC . . . EXECUTIVE SECRETARY

OFFICERS
1949-1950

PRESIDENT	PAST-PRESIDENT	SECRETARY
Earl I. Sponable	Loren L. Ryder	Robert M. Corbin
460 W. 54 St.	5451 Marathon St.	343 State St.
New York 19, N. Y.	Hollywood 38, Calif.	Rochester 4, N. Y.

EXECUTIVE VICE-PRESIDENT	EDITORIAL VICE-PRESIDENT	CONVENTION VICE-PRESIDENT
Peter Mole	Clyde R. Keith	William C. Kunzmann
941 N. Sycamore Ave.	120 Broadway	Box 6087
Hollywood 38, Calif.	New York 5, N. Y.	Cleveland 1, Ohio

1948-1949   1949

ENGINEERING VICE-PRESIDENT	TREASURER	FINANCIAL VICE-PRESIDENT
John A. Maurer	Ralph B. Austrian	David B. Joy
37-01—31 St.	1270 Avenue of The Americas	30 E. 42 St.
Long Island City 1, N. Y.	New York 20, N. Y.	New York 17, N. Y.

Governors
1949

1948-1949	James Frank, Jr. 426 Luckie St., N. W. Atlanta, Ga.	1949-1950
Alan W. Cook 25 Thorpe St. Binghamton, N. Y.	William B. Lodge 485 Madison Ave. New York 22, N. Y.	Herbert Barnett Manville Lane Pleasantville, N. Y.
Lloyd T. Goldsmith Warner Brothers Burbank, Calif.	William H. Rivers 342 Madison Ave. New York 17, N. Y.	Fred T. Bowditch Box 6087 Cleveland 1, Ohio
Paul J. Larsen 508 S. Tulane St. Albuquerque, N. M.	Sidney P. Solow 959 Seward St. Hollywood 38, Calif.	Kenneth F. Morgan 6601 Romaine St. Los Angeles 38, Calif.
Gordon E. Sawyer 857 N. Martel Ave. Hollywood 46, Calif.	R. T. Van Niman 4431 W. Lake St. Chicago 24, Ill.	Norwood L. Simmons 6706 Santa Monica Blvd. Hollywood 38, Calif.

New Series of Lenses for Professional 16-Mm Projection*

By A. E. NEUMER

BAUSCH AND LOMB OPTICAL COMPANY, ROCHESTER 2, NEW YORK

Summary—This paper describes the technical details covering the mounting, optical performance, and optical design of a new series of 16-mm projection lenses for professional use. Focal lengths of 2 to 4 inches, in half-inch steps, with speeds of $f/1.6$ are provided. Measurements of resolving power, optical corrections, and vignetting are included.

THE LENSES currently used in practically all 16-mm projectors traditionally have been of the Petzval type, designed originally by Joseph Petzval in Vienna around 1840, primarily as a portrait lens. The basic design has remained the same although numerous detailed changes and improvements have been made over the years. Essentially it consists of two sets of lenses, each set separately achromatized, and with a comparatively large separation between them. The front half is usually cemented while the rear lenses are separated by a small air space as shown in Fig. 1. The basic form was notably improved by Dallmeyer who in 1866 reconstructed the back combination by reversing the elements and changing their shape. Further changes have been made by uncementing the front combination, thereby giving an extra degree of freedom in the design. This step has the obvious disadvantage, however, of adding two extra air glass surfaces which tend to reduce the over-all transmission and contrast in the image.

The Petzval construction readily lends itself to lenses of very large aperture ratios with excellent center definition, but the covering power is seriously limited by heavy curvature of field which becomes noticeable a few degrees off the lens axis. Many attempts have been made to flatten this field both artificially, in the parlance of the lens designer, by introducing large amounts of astigmatism or by adding a field-flattening element close to the film plane. In spite of all these efforts, a good state of correction of the field aberrations has never been achieved. Nevertheless, considering the factors which in the

* Presented October 26, 1948, at the SMPE Convention in Washington.

past have motivated the 16-mm projector design, the choice of the Petzval lens was seemingly justified in that speed and cost were factors of more importance than screen definition.

Because of the limitations imposed by the Petzval design, mainly on resolving power,and because of the demand for a better lens, about fifteen years ago designers abandoned it for 35-mm projection in favor of an anastigmat design. Anastigmat, as used here, means a lens corrected simultaneously for astigmatism and curvature of field and, at the same time, being fully color-corrected. The latter requirement is equally important for black-and-white as well as color film. Such lenses were in rather common use in photography, but were of relatively low speed and it was, therefore, apparent at the outset that a more complicated and therefore more expensive lens form would

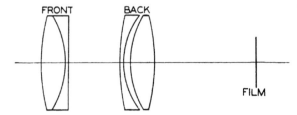

Fig. 1—Typical projection objective of the Petzval type.

have to be developed. As far back as 1921, Lee in England had succeeded in developing a 6-element anastigmat type which worked reasonably well at an aperture of $f/2.0$. This lens, which can be traced all the way back to the original Gauss telescope objective,[1] consists of two single outside elements with two cemented doublets in between. This basic design formed the background for the $f/2.3$ Baltar and later the $f/2.0$ Super Cinephor. Because of the unusual success of these lenses, and after a careful survey of other possible approaches, it was decided to use the same basic form for the Super Cinephor 16, in focal lengths from 2 to 4 inches in steps of $1/2$ inch, all rated at $f/1.6$. (See Figs. 2 and 3.)

One of the big differences between 16-mm and 35-mm projectors is in the inherent need for faster lenses in order to achieve sufficient screen illumination. The main reason for this is that because of the desire to make most 16-mm equipment portable, the use of carbon arcs with their attendant power supply is not feasible. The only alternative is to use a tungsten filament or, as has more recently been suggested, a concentrated arc. However, the inherent brightness per unit area of a projection-type tungsten filament is something of the order of $1/7$ that of a low-intensity carbon, and about $1/30$ that of a high-intensity carbon. Similarly, the concentrated arc has an inherent brightness per unit area about $1/1.75$ that of low-intensity carbons and $1/8$ that of high-intensity carbons. Furthermore, even if carbon arcs are used in the 16-mm projector, since the magnification is usually much greater than with 35-mm, the brightness per unit area of the screen will be much less.

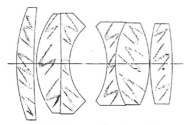

Fig. 2—Super Cinephor 16.

The result of these factors is that for 16-mm projection, a fast lens is imperative. To the lens designer this imposes an almost formidable problem since aberrations increase rapidly with lens speed. Spherical aberration, for example, increases as the square of the aperture. Thus, an $f/1.6$ lens identical to

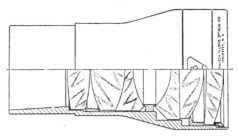

Fig. 3—Cross section showing the mounting details of the 4-inch Super Cinephor 16.

and of the same focal length as an $f/2.0$ lens would have more than twice the spherical aberration. Furthermore, since image detail on the 16-mm frame is smaller than on the 35-mm frame, the 16-mm projection lens should be capable of resolving powers considerably in excess of its big brother if it is expected to do the same type of job.

Considering all these factors, the design of the Super Cinephor 16 was not easy. It was completed only after about two years of painstaking effort and was greatly abetted by the use of new dense barium crown glasses which up to the present have not been generally available. Each focal length in the series was designed individually, but they are all of the same basic form. The final results on paper looked extremely promising, but frequently it happens that a particular pattern of aberrations which looks well according to computations will add up to yield a poor actual result. Therefore, final specifications were not released until sample lenses of each focal length were made and thoroughly tested. The results of these tests exceeded our fondest hopes. For example, the 2-inch lens, which can be considered as the standard focal length, at full aperture has a measured spherical aberration of about 0.1 per cent of the focal length or 50 microns. At the extreme corner of the 16-mm frame there is no measurable astigmatism, but approximately 0.2 per cent or 100 microns inward curvature of field. A more understandable picture of what these corrections mean can be obtained from the fact that any lens of the series will resolve visually more than 90 lines per millimeter anywhere in the 16-mm frame. This is not only about double what the average Petzval lens will resolve in the corners of the frame, but in addition the quality of resolution, which is the one hidden factor in any statement of resolving power, is excellent. While the Super Cinephor 16 is classified as a projection lens actually, in every respect, it is a high-quality photographic lens.

No lens is any better than it is mounted, and for that reason a considerable amount of thought was given to this case keeping in mind the particular application. We have experimented and actually used for some time a one-piece barrel type of mount, the inside of which is a true hollow cylinder. The lens components are individually mounted into cells which are accurately turned to fit the inside diameter of the barrel. Assembly is accomplished by stacking the cells inside the barrel with the addition of spacer rings, with the addition of a threaded retainer at one end to complete the job. (See Fig. 4.) This method has been quite successful, but it does not eliminate some of the troubles which have always been a problem in lens-mounting. First of all, the lens cells must be accurately turned to fit the barrel with no more than about 0.001 inch clearance. Second, the degree of centering possibly depends on how accurately the lens elements can be edged and how well they can be fitted to their respective cells. In actual manufacture, edging is a difficult operation, particularly on weak lenses, with the result that the finished diameter of a lens is not always concentric with its axis (the axis being the line through the centers of curvature of the two surfaces). All these sources of error frequently build up causing objectionable decentering and therefore rejection. An entirely new technique has been developed in which the lenses are not centered by means of their edged diameter. Instead, spacer rings which are turned to fit the bore of the barrel contact the lens surfaces near the periphery. The elements are actually edged to a smaller diameter than the bore and are therefore free to seek their own center between any two spacer rings. In other words, centering is effected by means of the differential thickness of the lenses and obviously eliminates the need for accurate edging as well as the expensive operation of fitting the elements to

Fig. 4—Exploded view showing the optics and spacer rings.

individual cells. When the spacer rings are designed correctly, this method has proved highly successful. (Figs. 4 and 5.)

Because of the weight factor all metal parts are made of aluminum with a dull black anodized finish inside and a satin anodized finish outside. This eliminates the necessity of using any lacquer inside the mount and avoids the trouble encountered with lacquer eventually flaking off and sticking to the inner surfaces of the lens elements. Finally, the lenses are sealed at both ends against dust and moisture.

The external dimensions of all the lens mounts have been made in accordance with standard Z52.1-1944 of the American Standards Association, which in this respect is identical with more recent Joint Army-Navy specifications. The diameter of the mounting section which fits into the projector is 2.062 inches. In addition the 2-inch lens is being offered in the semistandard $1^3/_{16}$-inch diameter rolled thread focusing type of mount which has long been used on both 8-mm and 16-mm projectors. Because of size limitations, it is not possible to fit any of the longer focal lengths in this style of mounting without reducing the speed.

As regards screen illumination with the subject lenses, every effort has been made to take full advantage of the $f/1.6$ speed. All

Fig. 5—Photograph of the series showing focal lengths from 2 to 4 inches. The 2-inch rolled thread mount is also included.

air glass surfaces are "Balcoted." However, so far as illumination is concerned, any lens is no better than the condenser and the light source behind it. In the absence of no other interference in the projection train, it is the combination of these three elements which determines the total amount of lumens reaching the screen, and also the degree of uniformity of illumination from the center to the edge of the screen. While it is not the main purpose of this paper to describe condensing systems, a few words concerning them is in order.

Based on the published[2] brightness values of a 750-watt, 25-hour tungsten projection filament, it has been computed that with a coated $f/1.6$ projection lens of the Super Cinephor 16 type, a perfect condenser, a mirror behind the lamp, and with no shutter or film in the gate, theoretically it should be possible to deliver approximately 605 lumens to the screen. Assuming a shutter efficiency of 50 per cent this would reduce to 302 lumens. The previously mentioned ASA specification (also JAN-P-49) requires under the same conditions 275 lumens with 65 per cent average corner-to-center distribution. A Navy specification CS-P 41A requires the same lumen output, but with 75 per cent average corner-to-center distribution.

Obviously, then, in order to reach the required total lumen output, practically all of the $f/1.6$ speed of the entire system must be utilized. Inherently, any lens, as is well known, will transmit less light the farther off the axis we go. In the case of the 2-inch Super Cinephor 16, which will vignette the most since it is the shortest focal length in the series, there is about a 30 per cent loss of light in the extreme corner of the 16-mm frame. A Petzval-type lens, because of its simpler construction, vignettes under the same conditions anywhere between 20 and 30 per cent, depending on the lens and whether or not it has a field-flattening element. Therefore, the requirement of 75 per cent corner-to-center uniformity demands not only a 100 per cent efficient condenser, but actually in most cases a deliberate reduction of light in the center of the field. This can be done in the condenser design, but obviously it will reduce the total lumen output and, as already stated, there is practically no room to move in this direction. Therefore, even at best to meet the illumination requirements as outlined above, it requires a very delicate balance between total output and uniformity.

Coupled with the fact that it is almost impossible to design a perfect condenser, and with the many other variables in the system such as variation in light sources and misalignment of the optical system, it is felt that the above specifications are not realistic for practical projector performance. However, under carefully controlled conditions, and with a well-designed condenser, 65 per cent corner-to-center uniformity with approximately 550 total lumens (with no shutter) can be achieved with an $f/1.6$ Super Cinephor 16 lens, and a 750-watt, 25-hour lamp. (NOTE: All illumination measurements made in accordance with ASA specification Z52.1-1944.) This represents about the limits that can be obtained without resorting to faster lenses or different light sources.

Finally, in keeping with the design requirements of projection lenses, so that they be in every way comparable to larger lenses, the Super Cinephor 16's are held in manufacturing to a focal length tolerance of ±1 per cent. This eliminates the necessity of matching when the lenses are used in pairs in the usual manner in continuous projection.

Acknowledgment

The author wishes to acknowledge those who did most of the actual optical and mechanical design referred to in this paper, namely, the late Dr. W. B. Rayton, Miss Lena M. Hudson, Dr. K. Pestrecov, Mr. C. DeGrave, and Mr. D. Gottschalk.

Bibliography

(1) A. E. Murray, "The Baltar series of lenses," *Internat. Photog.*, vol. 19, p. 12; June, 1947.

(2) F. E. Carlson, "Light source requirements for picture projection," *J. Soc. Mot. Pict. Eng.*, vol. 24, pp. 189–206; March, 1945.

(3) W. B. Rayton, "A new series of lenses for 16-mm cinematography," *J. Soc. Mot. Pict. Eng.*, vol. 48, pp. 211–217; March, 1947.

Discussion

Mr. George Lewin: The Signal Corps projector developed 230 to 250 lumens of light with the application you mentioned.

Mr. A. E. Neumer: With what uniformity?

Mr. Lewin: The uniformity as specified.

New Series of Lenses for 16-Mm Cameras*

By RUDOLF KINGSLAKE

EASTMAN KODAK COMPANY, ROCHESTER, NEW YORK

Summary—The Eastman Kodak Company has recently announced a matched series of high-grade interchangeable lenses for 16-mm motion picture cameras, to be known as Kodak Ciné Ektar lenses. There are six focal lengths, from 15 to 152 mm, in geometrical progression with a common ratio of about 1.6. Relative apertures are unusually high, from $f/1.4$ for the 25-mm lens to $f/4.0$ for the 152-mm lens. The mechanical back focus in all cases has been designed to be greater than 12.7 mm to allow space for a reflex finder and a camera turret.

The reasons underlying the choice of formula for each lens are discussed, with particular reference to aperture and angular field. The definition in all cases is remarkably good over the entire frame, even at the maximum aperture. The lens barrels are large for convenience but light in weight, and the diaphragm scales are uniformly spaced. Unusually complete and equally spaced focus scales have been adopted for increased convenience in use. All glass-air surfaces are hard-coated, and the lens barrels are baffled to give the maximum image contrast and freedom from flare. The lenses can be fitted to any 16-mm camera by means of suitable adapters.

INTRODUCTION

FOR MANY YEARS the Kodak line of lenses for 16-mm ciné cameras has grown in a somewhat irregular fashion. The lens on the original hand-cranked Ciné-Kodak Model A of 1923 was a 25-mm $f/3.5$ lens, of the air-spaced quadruplet type. In 1926 this was replaced by a 1-inch $f/1.9$ modified Petzval lens which was particularly suitable for the lenticular "Kodacolor" film on account of its low vignetting. In spite of some shortcomings, this lens was very successful, and it has remained the standard Ciné-Kodak camera lens ever since. In 1926 also, a 78-mm $f/4.5$ lens of the well-known 4-element type with cemented rear component was sold as an interchangeable lens of long focus. The model B camera (1925) was fitted at first with a 20-mm $f/6.5$ rapid rectilinear lens, but this was shortly afterwards replaced by the same 25-mm $f/3.5$ quadruplet lens as on the Model A camera. In 1929, a 20-mm $f/3.5$ triplet lens was introduced, which has been manufactured more or less regularly since.

* Presented October 28, 1948, at the SMPE Convention in Washington.

The current wide-angle 15.8-mm $f/2.7$ lens was first made in 1931, and a 50-mm $f/3.5$ was added in 1932, both lenses being also simple triplets. Three $f/4.5$ telephoto lenses of 76-mm, 114-mm, and 152-

63-mm $f/2$
102-mm $f/2.7$ 40-mm $f/1.6$ 25-mm $f/1.9$
15-mm $f/2.5$ 25-mm $f/1.4$ 152-mm $f/4$

Fig. 1—The new line of Kodak Ciné Ektar lenses.

mm focal length were introduced between 1931 and 1933. In 1936 the 63-mm and 102-mm $f/2.7$ lenses of the Petzval type were added, and shortly afterwards (1940) the list was augmented by the addition of the 2-inch $f/1.6$ lens of the same type. Since the war, seven of these lenses have been retained in production, namely, the 15-mm $f/2.7$, the 25-mm $f/1.9$, the 50-mm $f/1.6$, the 50-mm $f/3.5$, the 63-mm and 102-mm $f/2.7$, and the 152-mm $f/4.5$.

As many of these formulas were designed years ago and no longer represent the best modern standards in motion picture lenses, the Company has introduced an entirely new unified series of "Ciné Ektar" lenses. These are lenses of the highest quality, incorporating new types of optical glass and the latest developments in lens design and manufacturing methods. The designers have not confined themselves to the use of only three or four elements as was the case in our previous ciné lenses. The exterior appearance of all the lenses has been planned with an eye to utility, convenience, and beauty of line, and the whole series is to be as uniform in appearance as possible.

This is the first time that an integrated series of professional-type 16-mm camera lenses of this high aperture and quality has been manufactured in this country.

Choice of Focal Lengths

The principal reason for using interchangeable lenses of various focal lengths on a camera is to change the scale of the picture without the necessity of going closer to or farther away from the subject. For this purpose, it is mathematically logical that all the focal lengths in a given series of lenses should be in geometrical progression, each focal length bearing a constant ratio to the next below it or above it in the series.

With this basic principle established, the first question was to determine what value should be adopted for this common ratio. After much discussion, it was decided to retain the 25.4-mm (1-inch) and 102-mm (4-inch) sizes, and to insert two other focal lengths between them. This leads to a common ratio of $\sqrt[3]{4} = 1.59$, and the two intermediate focal lengths become 40 and 63 mm, respectively. The series has been carried one step each way beyond the 1-inch and 4-inch sizes, giving a wide-angle lens of 15 mm and an extreme telephoto lens of 160 mm, thus making six lenses in all. Actually, the focal length of this last lens has been reduced to 152 mm for convenience since the existing adjustable viewfinders can accommodate lenses with focal lengths up to only 6 inches.

Choice of Relative Apertures

In general, because 16-mm motion picture cameras have a fixed exposure time of about $1/32$ of a second at the standard speed of 16 frames per second, the user frequently requires a high relative aperture. There are, however several factors that set a practical upper limit to the aperture that can be used in each focal length. The chief problem is that of designing high-aperture lenses that will give sufficiently sharp definition over the required field. With the *shorter* focal lengths, central definition is not a problem, but the wide angular field makes the design difficult. In the *longer* focal lengths where the angular field is small, the zonal spherical aberration and spherochromatism tend to become large, and it then becomes difficult to obtain sufficiently good central definition. However, it is then possible to select types of construction, such as the Petzval lens, that are noted for their excellent central definition but narrow covering power. Long-focus lenses of the telephoto type have been avoided because they give inherently poorer definition than normal lenses of the same focal length and aperture.

Mechanically, lenses of high aperture are likely to be bulky and heavy and to have a short back focus which causes interference with the front of the camera, the shutter, and the lens turret (if any). All the lenses in the new Ektar series have an optical back focus of at least 13.4 mm, and a mechanical back clearance of at least 12.7 mm to allow for the reflex finder on the Ciné-Kodak Special camera. Moreover, at high apertures the depth of focus is small, and it becomes a problem to hold the relative positions of lens and film with sufficient accuracy.

Another factor which must be considered when long-focus lenses are used is depth of field. In the case of a camera equipped with a series of interchangeable lenses, where the pictures taken by all lenses are to be viewed from the same distance, we can tolerate a constant circle of confusion in the image. On this basis, it can be shown that the depth of field will be approximately proportional to s^2/fd, where s is the object distance, f is the focal length of the camera lens, and d is the diameter of the entrance pupil of the lens. Consequently, in order to keep the depth of field approximately the same for all lenses when photographing objects at a constant distance s, the relative aperture (f/d) of the long-focus lenses must be very much less than that of the short-focus lenses. Indeed, for a group of objects at any given distance from the camera, we shall find the same depth of field for a 1-inch $f/1.4$ lens at full aperture that we have with a 6-inch lens if the latter is used at about $f/50$!

Wherever possible, it was felt desirable to make the maximum aperture of all lenses one of the standard series, namely, $f/1$, 1.4, 2,

2.8, and so forth, and this has been done in several of the new Ciné Ektar lenses. The series of lenses finally adopted is listed in Table I.

TABLE I

Focal Length, mm	f Number	Angular Semifield,* Degrees
15.8	2.5	21.4
25.5	1.4	13.6
25.4	1.9	13.7
40.1	1.6	8.8
63.8	2.0	5.6
101.7	2.7	3.5
152.4	4.0	2.3

* Computed for the corners of the standard camera gate, with a 12.4-mm diagonal.

DESIGN OF THE MOUNTS

In order that the new lenses should be as uniform in appearance as possible, the greatest care has been given to the design of their mounts. For instance, it is very desirable to keep the focusing ring and the diaphragm ring in the same relative positions on all the lenses, although this could not be done on the 25-mm lenses because of mechanical restrictions. Similarly, every effort was made to have a nonrotating barrel with a differential screw-focusing device, so that the index marks for both the diaphragm and focusing scales would remain in a fixed position at the top of the mount. This was realized for all the members of the series except the 15-mm wide-angle.

The lenses are interchangeably attached to the camera by the pin and threaded-ring coupling already familiar in Ciné-Kodak interchangeable lens adapters. However, in the Ciné-Kodak Special II camera the threaded-ring adapter is incorporated as an integral part of the camera turret.

Every lens is equipped in front with a suitable screw thread for attaching standard series VI filters, attachments, and lens hoods. The mounts are carefully baffled on the inside to eliminate any possible loss of contrast due to light reflected from the interior of the barrel, and all glass-air surfaces are "lumenized" to reduce surface reflections.

A considerable reduction in weight has been achieved by the extensive use of aluminum in the mounts, and of high-strength aluminum alloys for the moving parts. The exterior portions of the mounts are clear anodized without lacquer, to improve the durability and to increase the resistance to fungus and moisture attack. Wherever a dead black is not required, black anodizing has been extensively employed on internal surfaces, to eliminate the flaking which commonly occurs with paint.

IRIS DIAPHRAGM

A new type of iris diaphragm has been used on these lenses. This employs special L-shaped leaves designed so as to give a uniformly spaced diaphragm scale. Thus, equal angular rotations of the diaphragm ring alter the image brightness by the same proportion in all parts of the scale. As is well known, the normal type of semicircular leaf gives a scale which is very crowded at the smaller end and expanded at the larger end. All lenses can be stopped down to $f/22$.

FOCUS SCALES

In accordance with the recommendations of Specification No. Z52.51-1946 of the American Standards Association, all object distances are measured from the *film plane*, and not from such an indefinite point as the front of a camera or the front of a lens barrel. The nearest distance on the focusing scale varies from 6 inches for the 15-mm lens to 6 feet for the 152-mm lens. Distance markings are chosen to give an approximately uniformly spaced scale in all cases, and the spacing of adjacent marks is seldom greater than the total depth of field at $f/8$. The shorter distances are marked directly in inches with red figures to distinguish inches from feet.

Because of random variations in lens radii, thicknesses, air spaces, and the refractive indexes of the glasses, the focal lengths of the lenses in a production run may vary over a range of perhaps ± 1 per cent. As this is too great a variation to be ignored, some procedure must be established to handle the problem. Two alternatives are available. We can adjust each lens to bring the focal length back to its nominal value, which is a troublesome and expensive process and may lead to some deterioration in the optical quality, or we can make several focusing scales for each lens, computed for a series of focal lengths differing slightly from the nominal value, and select the correct one in each case. The latter procedure has been adopted for the Ciné Ektar lenses, and any adjustment that may be made on individual lenses during inspection is done to improve the definition and not to adjust the focal length.

Depth-of-Field Scale

A depth-of-field scale (colored yellow to prevent confusion) is engraved on each lens adjacent to the focusing-scale index. For the computation of the depth-of-field scale, a limiting circle of confusion of $1/1000$ inch ($1/40$ mm) was assumed. This represents about the limit of visual resolution for an observer situated at a distance of $2^1/_2$ screen widths from the projected image, assuming that he can resolve an angle of 1 in 1000 ($3^1/_2$ minutes of arc). It is of interest to note that each division on the depth-of-field scale is a direct measure of the shift of focus in thousandths of an inch, since, for example, at $f/8$ the depth of focus is ± 0.008 inch.

Photographic Performance Tests

A typical sample of each lens was tested photographically in two ways:

(a) A distant point source was imaged by the lens, and the image so formed in the focal plane was magnified by a microscope objective and photographed on ordinary 16-mm panchromatic film at each of a succession of obliquities from the center of the field out to the corner of the 16-mm picture.

(b) A large high-contrast resolution chart was photographed directly with each lens at full aperture, at a reduction of 50 power in all cases. The focus position was changed in steps of $1/1000$ inch to ensure that the best focus position was properly covered. The resolving powers at the center, sides, ends, and corners of the frame were read from the film under a 50-power microscope. The lens was then stopped down without change in focus to determine the resolution at lower apertures. It should be noted that in some cases a slight over-all improvement can be obtained by refocusing at the lower aperture.

As the properties of the film play a very prominent part in the ultimate resolution obtainable with any lens-film combination, it is important that the resolving power of the film itself be known. Measurements at the Kodak Research Laboratories indicate that standard 16-mm films have the maximum resolving powers shown in Table II, assuming that the image density, processing conditions,

TABLE II

RESOLVING-POWER DATA FOR KODAK 16-MM CINÉ FILMS
(MEASURED SINCE OCTOBER, 1946)

Code	Film	Gamma	Density for Maximum Resolving Power	Resolving Power (Lines/ mm) 30:1
1230	Background X Ciné Neg. Pan.*	0.7	1.0	95
1232	Super XX Ciné Neg. Pan.†	0.8	0.9	90
5256	Super X Ciné-Kodak Pan (Reversal)	1.2	0.5	80
5261	Super XX Ciné-Kodak Pan (Reversal)	1.7	0.6	70

* Developed in SD-21 for 7 minutes.
† Developed in SD-21 for 12 minutes.
Other films given regular processing.

gamma, and so forth, are all maintained at their optimum values. An object contrast of 30:1 has been assumed. In all ordinary photography where the conditions are not so carefully controlled, film-resolution values decidedly inferior to these must be expected, especially for ordinary objects of low contrast.

TABLE III

ILLUMINATION RATIOS

Lens	Maximum Aperture (Top)	(Corner)	At $f/5.6$ (Top)	(Corner)
15-mm $f/2.5$	0.75	0.20	0.90	0.62
25-mm $f/1.9$	0.90	0.72	0.97	0.95
25-mm $f/1.4$	0.85	0.63	0.97	0.92
40-mm $f/1.6$	0.90	0.78	0.98	0.95
63-mm $f/2.0$	0.91	0.80	0.99	0.98
102-mm $f/2.7$	0.89	0.68	0.99	0.97
152-mm $f/4.0$	0.91	0.79	0.99	0.93

VIGNETTING

The illumination at the top and corner of the standard camera frame relative to that at the center of the field was measured photoelectrically for each lens in the series, at both the full aperture and at $f/5.6$, with the lens hood or series VI adapter in place. The measured ratios given in Table III show that except for the 15-mm wide-angle lens at the extreme corners of the frame, vignetting is in all cases very small and practically negligible.

DESCRIPTION OF THE SEPARATE LENSES

The construction and properties of each lens will now be considered separately in some detail.

(a) *The 15.8-millimeter f/2.5 wide-angle lens.* For this lens, a special design was necessary. Although the aperture (f/2.5) and

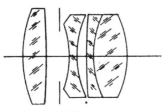

Fig. 2

field angle (21 degrees semifield) are moderate, it was necessary to provide a back focus greater than 13 mm in order to clear the viewfinder mirror on the Ciné-Kodak Special camera. To secure this exceptionally long back focus with lens elements that are thick enough to be easily manufactured presented a severe problem to the lens designer. The design finally adopted is shown in Fig. 2, and is covered by United States Patent 2,308,007. The star images formed by this lens are only a few microns in diameter except at 20 degrees obliquity, where they are about 0.02 mm long. The measured resolving power of the lens at full aperture,

on Ciné-Kodak Eight Panchromatic Safety Film* with the ordinary reversal processing, was found to be 80 lines per millimeter in the center of the frame, dropping gradually to 38 lines per millimeter at the corners. At f/5.6, the corner resolution rises to 48 lines per millimeter.

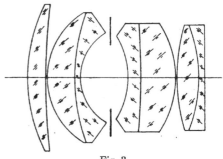

Fig. 3

This wide-angle lens can be focused down to 6 inches, the general arrangement of the mount being shown in Fig. 1. It will be noticed

* This emulsion is rather slow and is now obsolete. It was chosen for these tests because it has somewhat finer grain and a little higher resolution than current reversal films.

that the whole lens rotates for focusing, and that the diaphragm ring is in front of the focusing ring.

(b) *The 25.5-millimeter f/1.4 normal lens.* This lens is of the 4-component meniscus type, illustrated in Fig. 3. The outer components are positive and the inner are negative. It is a matter of great difficulty to design a lens with an aperture as high as $f/1.4$ to give acceptable definition over a 14-degree semifield, especially with the long back-focus requirement, but it has been found possible to do so with the aid of the new high-index glasses that are now available. The manufacture of this lens presents many problems, as the thicknesses, air spaces, refractive indexes, surface curvatures, and centering of the various elements must be controlled to an exceptionally high degree of accuracy. This lens is covered by United States Patent 2,350,035.

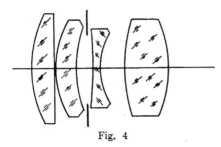

Fig. 4

The star images formed by this lens are only a few microns in diameter out to an obliquity of about 9 degrees, after which they begin to expand to a diameter of about 0.025 mm at the corner of the frame. The measured resolving power at $f/1.4$ with Ciné-Kodak Eight Panchromatic Safety film was found to be about 50 lines per millimeter in the center of the frame, and 38 lines at the corners. At $f/2.0$ the central resolution rises to 65 lines per millimeter and at $f/5.6$ it becomes 86 lines.

The mount is shown in Fig. 1. The lens is exceptionally light in weight, and its relatively large size gives adequate space for the various scales. Furthermore, the long, open barrel in front of the lens itself acts as an effective hood or sunshade. The lens can be focused from infinity to one foot.

(c) *The 25.4-millimeter f/1.9 lens.* This lens has been added to the line as a lower-priced substitute for the $f/1.4$ lens. It contains only four elements, but it represents a considerable improvement over the old four-element Petzval-type lens. The construction is shown in Fig. 4. It is found that the star images expand continuously from 0.009 mm on the axis, to 0.015 mm at 5 degrees, and finally reach 0.06 mm at 14 degrees from the axis. They are thus somewhat larger than for the $f/1.4$ lens, which is reasonable since there are four elements instead of seven. However, at full aperture the measured photographic resolving power on Ciné-Kodak Eight Pan Safety film is 60 lines per millimeter in the center, dropping to 40 lines at the corners. This corner definition is better than we should expect from the size of the star image, because of the distribution of light within the star image itself. When stopped down to $f/5.6$, these figures be-

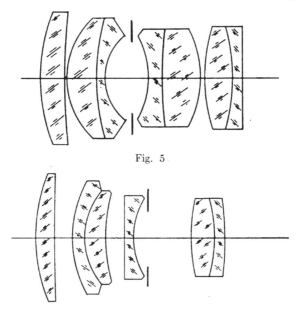

Fig. 5

Fig. 6

come 80 and 53, respectively. The design is covered by United States Patent 2,432,387.

The exterior appearance of the mount is similar to that of the $f/1.4$ lens, as can be seen in Fig. 1.

(d) *The 40-millimeter $f/1.6$ lens.* This lens is very similar to the 25-mm $f/1.4$, except that the field has been reduced to 9 degrees since that is all that the lens is required to cover. The aperture was reduced to $f/1.6$ because of the mechanical difficulty of fitting such a large lens into the opening in the front of the camera. The actual lens construction is illustrated in Fig. 5, the design being covered by United States Patent 2,262,998.

The definition given by this lens is exceptionally sharp. The star image is a few microns in diameter from the axis out to 7 degrees, and it rises to only 0.01 mm at the corners of the frame. The measured resolving power at $f/1.6$, on Ciné-Kodak Eight Panchromatic Safety Pan film, was found to be 55 lines per millimeter at the center of the frame and 45 lines at the corners. At $f/5.6$ these figures become 75 and 68 lines, respectively.

Details of the mount are clearly shown in Fig. 1. The focus scale runs from infinity to 2 feet.

(e) *The 63-millimeter $f/2$ lens.* For this lens, the semifield to be covered is only $5^1/_2$ degrees, but as the focal length is rather long, the

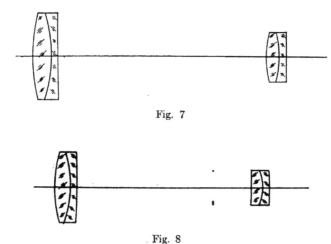

Fig. 7

Fig. 8

designer's attention had to be directed more toward securing good central definition than to covering a wide angular field. Consequently, a different type of construction was chosen, comprising two front positive components, a single negative lens, and a rear positive component (Fig. 6). The image given by this lens is unusually clean and free from aberration haze.

The diameter of the star image increases to 0.01 mm at 4 degrees from the axis, and reaches about 0.018 mm in the corners of the field. At full aperture, the measured photographic resolving power on Ciné-Kodak Eight Panchromatic Safety film at full aperture is 54 lines per millimeter in the center and 36 lines in the extreme corners. At $f/5.6$, these figures become 80 and 60 lines, respectively.

The details of the mount are shown in Fig. 1. The focusing range extends from infinity to 2 feet.

(f) *The 102-millimeter f/2.7 lens.* The optical system of this lens has been familiar to Ciné-Kodak users for several years. It is a very long lens of the Petzval or projection type of construction (Fig. 7), which is known to give clean, sharp images over the required narrow field of only $3^1/_2$ degrees semiangle. The star-image diameter increases slowly from the axis to 0.01 mm at about 2 degrees, and it reaches about 0.017 mm at the corners of the frame. The measured photographic resolving power on Ciné-Kodak Eight Panchromatic Safety film at full aperture is 60 lines per millimeter in the center, dropping to 50 lines at the corners. At $f/5.6$ these values become, respectively, 80 and 57 lines per millimeter.

The mounting, shown in Fig. 1, is a great improvement over the previous style of mount for this lens. The focusing scale runs from infinity to 3 feet.

(g) *The 152-millimeter f/4 lens.* This lens is of novel construction (Fig. 8), and it is intended to replace the previous 152-mm $f/4.5$ telephoto. Optically, it consists essentially of a telescope objective with a specially designed zero-power achromatic field flattener fairly close to the image plane. For the low covering power of only $2^1/_2$ degrees half angle, this type of construction is entirely satisfactory and the images are unusually sharp and clean. The star-image diameter increases to 0.01 mm at about $1^1/_2$ degrees obliquity, and it reaches about 0.014 mm at the corners of the frame. The measured photographic resolving power on Ciné-Kodak Eight Panchromatic Safety film at full aperture is 63 lines per millimeter, dropping to 40 lines at the corners. At $f/8.0$ these figures improve slightly.

As the over-all length and linear aperture of this lens closely resemble the corresponding dimensions of the 102-mm $f/2.7$, the two lenses are mounted in similar barrels (see Fig. 1), but the sunshade is made longer on the 152-mm to distinguish it from the other. The focusing scale runs from infinity to 6 feet.

Zero-Shift Test for Determining Optimum Density in Variable-Width Sound Recording*

By C. H. EVANS and R. C. LOVICK

EASTMAN KODAK COMPANY, ROCHESTER, NEW YORK

Summary—In variable-width sound recording, the fidelity of the recorded wave form depends in part upon the sensitometric conditions to which the photosensitive materials are subjected. The well-known cross-modulation test is available for determining the optimum sensitometric conditions which result in minimum distortion.

Another method of determining the optimum density under any given processing conditions, the zero-shift test, has been used for several years in the Kodak Research Laboratories. In this test, the average transmission of a high-frequency sine-wave track is compared with that of an unmodulated, unbiased track at the same image density, by means of a simple physical densitometer. At optimum density, where photographic distortion is minimum, these two transmissions are equal. Measurements at a single image density show directly whether that density lies above, below, or at the optimum. Optimum density thus determined is in good agreement with that found by the cross-modulation test. The equipment required is simple and only a small amount of film is needed. The zero-shift test is recommended for use in cases where the more extensive cross-modulation test is not believed to be justified.

INTRODUCTION

IN THE PRODUCTION of high-quality photographic sound records, it is necessary to control sensitometric conditions carefully. The *zero-shift test* is useful for this purpose in variable-width recording. As a preface to the description of this method, there follows a brief review of the properties of photographic emulsions which may cause the developed photographic image to differ in size from the geometrical pattern of light by which the film was exposed.

The turbidity of an emulsion tends to increase the size of the developed image. A ray of light incident on the surface of a photographic film does not, in general, follow a straight path within the emulsion layer. Reflection and refraction at the boundaries between the halide grains and the surrounding gelatin, together with diffraction effects, cause the light to be scattered. Some of this scattered light produces latent image in grains lying outside the boundaries defined by the beam of light impinging on the surface. As the intensity

---

* Presented October 28, 1948, at the SMPE Convention in Washington. Communication No. 1216 from the Kodak Research Laboratories.

⌐ of the exposing light is increased, other factors remaining constant, latent image is produced in grains lying farther and farther outside these boundaries. In other words, as the density of the developed image increases, its size also tends to increase.

Fortunately, there is also an effect which tends to decrease the size of the developed image. This is called the gelatin effect. It is produced by developer oxidation products which tan the gelatin in the regions where silver is developed. These regions dry more rapidly after processing than do the untanned regions which surround them, because they contain less water. As the gelatin shrinks down in the drying

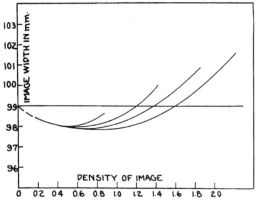

Fig. 1—Curves showing the change in size of an image as a function of the image density. The family of curves represents 2, 4, 8, and 12 minutes' development, respectively, from left to right.

process, this differential drying sets up forces at the edge of the image which are directed in toward the area containing developed silver. The wet gelatin is deformed by these forces, and, as a result, when drying is complete the image is somewhat smaller than it was immediately after development. Like the turbidity effect, the tanning effect increases with increasing density of the image.

Experimentally it is found that under fixed development conditions there is a density at which these two opposite effects cancel, with the result that the photographic image corresponds in size with the area of the film surface which was exposed. Below this density, the shrinkage effect predominates, while above it, the spreading effect predominates. Fig. 1, which was taken from a paper by Jones and Sandvik,[1] shows a family of curves relating the width of the image to its density. The emulsion was exposed through a slit 0.99 mm wide, placed in contact with the surface. Four curves are shown, for four different times of development. The cancellation point shifts toward higher densities as development is increased.

In variable-width sound recording, factors external to the emulsion itself may also contribute to image spreading, but even so a density exists at which the spreading is offset by contraction of the gelatin. Examination of a variable-width sound-track negative exposed to produce a series of negative densities covering a wide range will reveal that the wave form is seriously distorted at low and at high densities. At low densities, where the shrinkage effect is predominant, the peaks of the developed image are narrower, and the valleys between the peaks are broader, than they should be. At high densities, the reverse is true. The peaks are broadened, while the valleys are narrowed by filling in from the sides. At some intermediate density, the mutual cancellation of the spreading and shrinkage effects results in minimum distortion of the wave form. This is the proper density for a recording which is to be used as a "direct-playback" record. In preparing a negative from which prints are to be made, however, it is better to allow some spreading in the negative, which is later compensated for by a complementary spreading of the image in the print. By reason of the higher negative and print densities involved, the signal-to-noise ratio obtainable in this way is superior to that of a low-distortion print made from a low-distortion negative.

It is common practice to use fixed development conditions for the sound negative, and to print to a fixed print density, usually about 1.35 for release prints. Development conditions for the print are also fixed. This means that the density of the negative must be chosen to fit these fixed conditions. That negative density which results in the lowest distortion in the print is termed the optimal negative density. The cross-modulation test described in 1938 by Baker and Robinson[2] has become a standard method of determining this density. In this test, a signal is recorded which consists of a high-frequency sine wave whose amplitude is modulated at a low frequency. Recordings are made at several different negative densities, and a print is then made under standard conditions. The print is examined by playing it on a standard sound-film reproducer, passing the output into a filter which transmits only a narrow band of frequencies centered about the modulating frequency, and measuring the signal transmitted by the filter. Nonlinearity in the photographic process, caused by image spreading or contraction, demodulates the recorded signal, and the magnitude of the signal appearing at the output of the filter is related to the degree of wave-form distortion. By plotting the level of this signal for each section of the print against the corresponding negative density, the optimum negative density may be found, at which demodulation is minimum. In a recording of actual complex sounds, such as speech or music, the cross-modulation type of distortion is more objectionable than the accompanying simple wave form or harmonic distortion.

The chief purpose of the present paper is to describe another method of detecting image spreading or contraction. It is called the zero-shift test and has been used in these Laboratories since 1938. It was originally employed for comparing the relative amounts of image spreading which occurred in various emulsions. Recently, however, we have more thoroughly explored its use as a control test for practical sound recording, and have found it possible to obtain results in good agreement with those of the cross-modulation test. The following description of the zero-shift test assumes its application to the problem of finding the optimal negative density for use in making prints. Negative development, print density, and print development are assumed to be fixed. It will be obvious, however, that the test can also be used for other purposes, such as finding the correct density for a recording which is to be played back directly, studying the effect on this density of processing variations, and so on.

THE ZERO-SHIFT TEST

The sound-recording negative material whose optimal density is to be found is exposed on a standard variable-width recorder. The test signal is a high-frequency sine wave of constant amplitude which produces a modulation of 80 per cent on the sound track. There are advantages in recording a Class A push-pull track for this test, but with certain precautions a regular duplex track can be used, as will appear later. A frequency of 9500 cycles per second is used for 35-mm films, and 4000 cycles per second for 16-mm films. The signal is periodically removed from the galvanometer by means of vibrating contacts which short-circuit the signal during approximately one-half cycle of their vibration. This results in an exposure on the film of an alternate series of modulated and unmodulated sound-track sections of about equal length. A 60-cycle vibrator is used for 35-mm film, so the individual sections are each about 0.15 inch long. Exposures are made at several different lamp currents to produce a range of densities in the developed negative. Only a few inches of track need be exposed at each lamp current. It is convenient to remove the signal from the galvanometer during each change in lamp current, and simultaneously to bias the galvanometer with direct current to produce a track whose density can be measured readily. This can be accomplished with a suitably wired push-button switch. Lamp-current changes are made with the switch closed, and then the button is released to record the

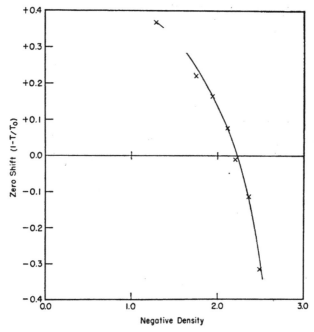

Fig. 2—Zero-shift curve, measured on the print.

signal. Track densities of the developed film are read immediately adjacent to the modulated sections, where the lamp-current change has been completed.

After development, the negative is printed to standard print conditions. Densitometric measurements made on the print enable the optimal negative density to be determined. In each part of the print, the average transmission T of the modulated track, and the transmission T_O of the unmodulated track, are measured by means of a physical densitometer. The aperture of this densitometer is rectangular. Its width can be equal to the width of sound track scanned on a standard reproducer, and its length should be sufficient to include many wavelengths of the recorded frequency on the film. The average transmission T of a modulated portion of the track is the ratio of the amount of light transmitted by this aperture when covered by modulated track to that transmitted when the film is removed. The transmission T_O of unmodulated track is found similarly. Several readings of T and T_O should be made in each part of the print in order to get good representative values. The quantity $(1 - T/T_O)$ for each part of the print is then plotted against the corresponding negative density. The re-

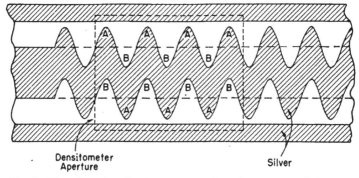

Densitometer
Aperture   Silver

Fig. 3—Class A push-pull print, superposed on the aperture of the zero-
shift densitometer.

sult is a zero-shift curve, such as that shown in Fig. 2. Optimal negative density is read from this curve at the point where $(1 - T/T_O)$ is equal to zero. Fig. 3 is a drawing of a push-pull modulated track placed over the densitometer aperture. If the wave form of this track is not distorted by the photographic process, the clear portion of the track has the same area as the clear portion of an equal length of unmodulated track, because for each opaque area A introduced by the modulation there is also introduced an exactly equal clear area B on the opposite edge of the track. In this case, T must equal T_O. If distortion is introduced by the photographic process, then T no longer will be equal to T_O. For example, when the negative density is higher than optimal, spreading in the negative will reduce the size of the clear areas through which the print is exposed to such an extent that the spreading of the exposed areas of the print is insufficient to establish the correct wave form. If T is measured on this part of the print, it will be greater than T_O, and the zero shift $(1 - T/T_O)$ will be negative.

On the other hand, if the negative density is too low, then the exposed areas of the print will be so large that the spreading which takes place will extend the image past the correct point, resulting in positive zero shift. Any physical densitometer which can be provided with an aperture of correct size may be used for determining T and T_o, provided that the illumination is uniform over the aperture. This is a very important condition, as may be seen by again referring to Fig. 3, which depicts an undistorted track placed over the aperture. Suppose that on the densitometer the center of the track receives more light than the edges. It is clear that the image areas A will then cut off less light than the clear areas B will transmit. The measured transmission

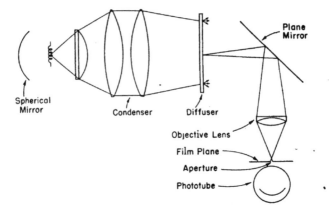

Fig. 4—Optical schematic of the zero-shift densitometer.

of the modulated track, therefore, will be higher than that of unmodulated track, which would have all of its clear area in the region of low illumination. This would indicate image contraction in the overall photographic process, corresponding to excess spreading in the negative for the problem which we have assumed.

In the construction of a new densitometer for use in these Laboratories, great care was taken to obtain uniform illumination over the aperture. Fig. 4 is a diagram of the optical system used. The light source is a projection lamp with four vertical coiled filaments. Ground glass is placed beyond the condensing-lens system to diffuse the light, and the lens tube is blackened to reduce reflections. An objective lens forms an image of the ground glass slightly above the aperture. If measurements are to be made on tracks having selective spectral absorption, such as those employed on various color films, then it is important to match the quality of radiation falling on the film in the densitometer with that encountered in the sound reproducer. Also, the photocell of the densitometer must have the same type of sensitive surface as that employed in the reproducer. The reason for this is that such a track may have a different amount of image spread when examined by light of one color than it does when examined by light of a different color. The ground glass which was first used in our densitometer was found to absorb an undue amount of infrared radiation, and hence had to be replaced by a different material.

Fig. 5 is a schematic diagram of the electrical part of the densitometer, used to measure the light passing through the aperture. The photocell which collects this light forms one arm of an alternating-current bridge.[3] The adjacent arm of the bridge is an identical photocell which is kept dark. Light falling on the first photocell alters its impedance and so unbalances the bridge, resulting in the flow of alternating current through resistor R_4. The voltage developed across this resistor is amplified and applied to the input of an external vacuum-tube voltmeter. Readings of this meter are proportional to the amount of light falling on the photocell. A potentiometer R_2 is used for balancing the bridge when both cells are dark. It is also necessary to balance the capacitances of the two arms containing the photocells. This may require a small shunt capacitance across one tube or the other, or it may be possible to obtain balance merely by changing the position of the wiring. Connections to the phototubes are soldered directly to the pins, and heavy wire is used to avoid microphonics. The potentiometer R_7 is the only operating control, and is used for setting the full-scale deflection of the voltmeter when there is no film over the aperture. If the alternating supply voltage is regulated, this control requires but little adjustment after a brief warm-up period.

Before closing this section on apparatus and procedure, it is advisable to point out the precautions which should be observed when zero-shift measurements are to be made on duplex rather than on Class A push-pull tracks. The first point of difference is that the transmission of the unmodulated portion of a duplex track is dependent upon the rest position assumed by the recording galvanometer mirror when the signal is short-circuited. Frictional forces at the pivot cause this rest position to be somewhat erratic, and, with the triangular duplex mask, this causes variations in width of the clear portions of successive unmodulated sections of the negative. With the parallel-sided push-pull mask, however, the width of the clear portion stays fixed, and only its position is affected. This difficulty with duplex track can be avoided by placing a resistance in series with the signal-short-circuiting contacts of the vibrator, such that the modulation is reduced about 20 decibels when the contacts close, instead of being removed entirely. This is sufficient to keep the galvanometer mirror in motion, without adversely affecting the results. With duplex track, the length of the densitometer aperture should, strictly speaking, be chosen equal to an integral number of wavelengths on the film; otherwise, the longitudinal placement of the modulated track over the aperture will affect the value T. With Class

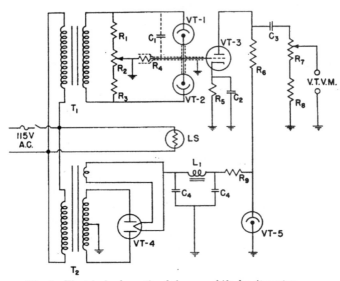

Fig. 5—Electrical schematic of the zero-shift densitometer.

PARTS

T_1	General Radio Type 578-A bridge transformer	R_9	3000-ohm, 20-watt
T_2	600-volt, 40-milliampere power transformer	L_1	10-henry, 40-milliampere
		VT-1	925 phototube
R_1	10,000 ohms	VT-2	925 phototube
R_2	2000-ohm potentiometer	VT-3	6SF5
R_3	10,000 ohms	VT-4	5Y3-GT
R_4	0.5-megohm Continental Type X	VT-5	VR-150
		C_1	Bridge-balancing capacitor
R_5	3500-ohm, 1-watt	C_2	15-microfarad, 50-volt
R_6	100,000 ohms	C_3	0.1-microfarad
R_7	50,000-ohm potentiometer	C_4	20–20-microfarad, 450-volt
R_8	50,000 ohms	LS	Light-source Kodaslide projector Model 2A (modified)

A push-pull track this effect does not occur. However, the greater the number of wavelengths included by the aperture, the less important the effect becomes with duplex track. An aperture 75 mils long includes about 40 wavelengths of 9500 cycles on 35-mm film, and the error under these conditions would be negligible.

CORRELATION BETWEEN ZERO-SHIFT AND CROSS-MODULATION TESTS

In 1939, Daily and Chambers[4] published a densitometric method of measuring cross-modulation prints. Their method was similar to the zero-shift method, except that a regular cross-modulation track was used instead of a track recorded at constant amplitude. The agreement between this densitometric method and the conventional dynamic method of analysis was excellent. They reported, however, that use of a constant-amplitude high-frequency track in place of the cross-modulation track resulted in optimum print densities that were about 0.35 lower.

Initially, we found that the zero-shift test indicated optimal negative densities lower by 0.2 to 0.3 than those indicated by the cross-modulation test. The cause of the discrepancy was finally traced to the densitometer used in measuring the average transmission of the sound track. The illumination falling on the aperture was not sufficiently uniform. In the construction of a new densitometer this defect was remedied, and at the same time operating convenience was improved by changing from a direct- to an alternating-current light-measuring circuit which had much better stability. Optimum densities indicated by the zero-shift and cross-modulation tests then showed good agreement. Typical results are shown in Table I. The zero-shift tests were made at 9500 cycles, and the cross-modulation tests at 9500 cycles, modulated at 400 cycles.

It is hoped that because of its simplicity the zero-shift test will find application in cases where use of the more extensive cross-modulation test is not believed to be justified. In many instances, the vibrator and the galvanometer-biasing unit are not needed. For instance, at the end of a normal recording, there could be recorded at the same lamp current a short section of high-amplitude, high-frequency sinusoidal track, followed by a section of track recorded at the same frequency, but with the amplitude reduced about 20 decibels. When the negative has been printed, comparison of the average transmissions of these two sections of the print will show whether or not the optimal negative density has been attained. If it has not, then the measurements show directly whether the negative density is too high or too low for the print density employed. In such an application, where only one use of the zero-shift curve is determined, it is especially important to make several readings in each section, to ensure that a good average is obtained.

TABLE I

CORRELATION OF ZERO-SHIFT AND CROSS-MODULATION TESTS

Test Method	Film Measured	Optimum Negative Density
Cross-modulation	Direct-playback	1.08
Zero-shift	Direct-playback	1.10
Cross-modulation	Print	2.20
Zero-shift	Print	2.21

ACKNOWLEDGMENT

For the help which they have given us, we thank Dr. O. Sandvik and Mr. W. K. Grimwood, who initiated the use of the zero-shift test for sound-recording films.

REFERENCES

(1) Loyd A. Jones and Otto Sandvik, "Photographic characteristics of sound recording film," J. Soc. Mot. Pict. Eng., vol. 14, pp. 180–204; February, 1930.

(2) J. O. Baker and D. H. Robinson, "Modulated high-frequency recording as a means of determining conditions for optimal processing," J. Soc. Mot. Pict. Eng., vol. 30, pp. 3–18; January, 1938.

(3) C. Butt and R. S. Alexander, "A method of recording low-intensity flashes of light," Rev. Sci. Instr., vol. 13, pp. 151–153; April, 1942.

(4) C. R. Daily and I. M. Chambers, "A densitometric method of checking the quality of variable-area prints," J. Soc. Mot. Pict. Eng., vol. 33, pp. 398–403; October, 1939.

DISCUSSION

MR. S. P. SOLOW: Have you found the field of the Western Electric RA-1100 sufficiently uniform?

MR. C. H. EVANS: We made one test and it appeared not sufficiently uniform.

DR. J. G. FRAYNE: This zero-shift test, of course, was not considered in the design of that densitometer, so the degree of uniformity was not considered too important. However, if this type of test should be considered comparable to the cross-modulation test, I am sure it would not be difficult to modernize the optical system to obtain the correct uniformity.

From an operating standpoint, is this a more difficult test to make than the cross-modulation test or is it easier?

MR. EVANS: Our thought in bringing this to your attention is that it can be made in places where the equipment is rather simple. In addition to the special densitometer, it requires only a simple oscillator. All other operations could be taken care of by push-button switches. We are not suggesting that it should replace the cross-modulation test, although there are places where it may do as well.

DR. FRAYNE: Have you considered the application of intermodulation tests to variable area? I do not believe cross modulation tells the whole story.

MR. EVANS: We have not done any recent work on it. We took one set of intermodulation data, but it was not too interesting. This is to be expected, as the amplitude of the high-frequency component of the intermodulation-test signal is constant, and therefore, in the variable-width system film distortion will not vary at the low-frequency rate.

MR. REVESTOCK: Since the aperture in your densitometer must be even in illumination, do you have the same problem in case the illumination is not absolutely even all the way across in the recorder or printer?

MR. EVANS: It does not cause the same type of difficulties. The same thing would occur in cross modulation.

MR. REVESTOCK: Would it not be advantageous to introduce just enough light in a feedback circuit to compensate for the light-source fluctuations that exist?

MR. EVANS: We want to keep this a very simple circuit. Yes, it is desirable to stabilize the light source. We do this by means of a voltage-regulating transformer in the line which serves the densitometer.

MR. M. H. SWEET: Are you measuring primarily a ratio of areas? You are really integrating an area rather than measuring the density of the photographic deposit.

MR. EVANS: We are dealing with the ratio of the areas.

MR. LEWIN: In some tests we made at the Signal Corps the high frequency selected made a difference in the optimum density. In other words, if you are using 9500 cycles, you might find that the optimum print density turns out to be 1.6; but if you make exactly the same test using the same magnitude of density but using the high frequency of 7000 cycles, then the optimum density comes out to be 1.4. Did you notice the same thing with respect to this test?

MR. EVANS: We find that there are differences in optimum negative density according to the frequency, both with cross modulation and with zero shift. They stay together.

MR. LEWIN: You made the point that the question of field uniformity is rather important. It is difficult to maintain uniformity in the average projector. That is pretty well recognized. Would there not be a discrepancy in the result you obtain by this method as compared with the dynamic test which you read on the actual projector that will reproduce from the sound track?

MR. EVANS: In obtaining these correlations, we were also careful to keep our projector scanning beam uniform. I believe you will find that it is more important to have uniform illumination on the zero-shift densitometer than it is in the projector.

Standard Quality of Photographic Chemicals*

By CARROLL V. OTIS

EASTMAN KODAK COMPANY, ROCHESTER, NEW YORK

PHOTOGRAPHIC PROCESSING inherently involves chemical operations which require adequate purity of the chemicals to attain a high standard of photographic quality. As a result, it has been found desirable to set up standards of purity for "photographic grade" chemicals to meet the requirements of the various sensitive materials and processes. These chemicals fall into groups according to their use or chemical nature: developing agents, alkalies, sulfites, restrainers and antifoggants, fixing agents, acids, hardeners, and a group of miscellaneous chemicals.

Standards for these "photographic grade" chemicals take into consideration the fact that in some instances the photographic criteria may permit more of certain impurities than could be allowed for pharmaceutical chemicals or analytical reagents whereas in other cases even small traces of chemical impurities that would be unimportant in other uses may be extremely harmful in photographic processing.

The ASA Committee on Standardization in the Field of Photography, Z38, soon after its organization recognized this need for specifications covering chemicals used in processing photographic materials. In 1940 it asked the Subcommittee on Processing and Processing Equipment, No. 8, to consider this project. The late Dr. S. E. Sheppard of the Kodak Research Laboratories, a member of Subcommittee 8 and representative of the American Chemical Society on Sectional Committee Z38, was asked to undertake the preparation of preliminary drafts of specifications for the chemicals commonly listed in published photographic processing formulas.

* A report of the ASA Subgroup on Photographic Grade Chemicals which developed the new American Standards establishing criteria of purity for chemicals used for processing photographic materials. Reprinted in part from *Standardization*, vol. 20, pp. 10–14; January, 1949.

Standards for 52 chemicals are now listed by the committee having been prepared with the aim of establishing minimum standards of purity for photographic grade chemicals. All are not available in approved form at the present time, however. Twenty-four standards received final approval on July 12, 1948, and are now available in published form. It is anticipated that 24 more will be available in approved and published form early in 1949. Of the 52 chemicals listed at present, approximately 25 have fairly wide usage in processing formulas while the remaining 27 have limited or specialized use. As the need arises, specifications defining photographic grade will be considered for additional chemicals.

PHOTOGRAPHIC GRADE DEFINED

The term *photographic grade* refers to chemicals in which impurities known to be photographically harmful are limited to a safe quantity, while inert impurities are restricted to amounts that will not reduce the required assay strength. In general, the chemical tests and analyses of the specifications are intended to indicate the presence of photographically detrimental impurities which might occur in the production and supply of the chemicals.

Instances of harmful types of substances which are limited in appropriate chemicals by these specifications are the following: (1) chemical reducing and sulfiding agents which can initiate fog or stain when present in developers, (2) chemical reducing and sulfiding agents which can produce fog or stain when present in fixing baths, (3) powerful restraining agents, (4) excessive silver-halide solvents in developers, (5) catalysts capable of accelerating the deterioration of photographic solutions, and (6) precipitate-forming compounds and insoluble matter.

In some cases no reference has been made in the specifications to substances known to be harmful. Such references have been omitted to eliminate unnecessary testing on the part of those using the specifications. This was decided upon after testing currently supplied material and consulting with manufacturers on present methods of manufacture. Thus arsenic which even in traces is known to cause deterioration of acid fixing baths was not specifically limited in several of the specifications after manufacturers and suppliers assured the committee that the chemicals as produced and supplied for photographic use at the present time are free from this impurity.

Specifications Based on Existing Data and Research

The specifications take into account existing data in the United States Government Specifications (comprising U. S. Federal Specifications and U. S. Treasury Department Procurement Division Specifications), in the Analytical Reagent Specifications of the American Chemical Society, and in the U. S. Pharmacopoeia. Considerable use has been made of the facilities and testing methods of the leading American manufacturers of chemicals and photographic film and paper. Reference has been made to manufacturing and purchase specifications which have been made available by producers and suppliers of photographic chemicals.

The specifications for the organic restrainers and the developing agents (with the exception of four for which there were existing U. S. Government Specifications) were developed as a result of research carried on by members of the Subgroup on Photographic Grade Chemicals which Dr. Sheppard organized to draft the specifications. It is hoped eventually to add assay procedures to the specifications for the developing agents but it was felt that, since the present specifications are generally agreed to be quite satisfactory, they should not be held up until the necessary research and testing could be completed.

It has been necessary for the present to omit reference to any specific technique for determining melting points in the American Standard specifications for the developing agents and organic restrainers due to the lack of a widely used, generally accepted, and reproducibly accurate method. As a result, melting-point ranges are wider than would otherwise be desirable. In the past, certain Federal Specifications have referred to the method described in the U. S. Pharmacopoeia for the determination of melting points for Class I solids using American Society of Testing Materials calibrated thermometer E 1 (2C-39). Attention of the U.S.P. Committee on Revision was called early in 1946 to the experience of members of the Subgroup on Photographic Grade Chemicals indicating that this method, when followed strictly in accordance with the instructions, does not give satisfactorily reproducible results when employed by different operators or when different equipment is involved. The National Bureau of Standards concurred that this U.S.P. technique was unsatisfactory. We understand that they are working on the design for a standard melting-point apparatus incorporating the use of a copper block with precision controls. When such an apparatus has been designed and approved and is commercially available, reference can be made to it for melting-point determinations in the American Standard specifications. At that time it may be possible to narrow the melting-point ranges specified.

SPECIFICATIONS IN TERMS OF CHEMICAL TESTS

The chemical tests and analyses have been assembled from the above and other sources, modified when required, tested, and correlated with photographic requirements. While it is considered that the chemical methods of testing are adequate in most cases, in view of present knowledge, it is recognized that the ultimate criterion of satisfactory quality is the successful performance of a given chemical in a use test, that is, in the recommended formula with a given sensitive material and under the prescribed conditions of use.

The methods of analysis, as given in the specifications, in general do not require the use of special apparatus or equipment. If an equally accurate and generally recognized method exists, its use is allowed as an alternate. An exception to this should, however, be noted where the tests bear the requirement "to pass test." In these instances alternate methods should not be employed.

Reagents used in making the tests should be recognized reagent grade chemicals normally used for careful analytical work. These chemicals should generally meet the Recommended Specifications for Analytical Reagent Chemicals of the American Chemical Society.

The new American Standards for photographic grade chemicals are part of a large program of standards for photography being developed by the ASA Sectional Committee on Standardization in the Field Photography, Z38. One hundred and thirteen standards have already been completed and approximately 43 are now under consideration. This committee works under the sponsorship of the Optical Society of America, and has a membership that is broadly representative of manufacturers, distributors, and users of photographic materials and equipment, and technical experts in the photographic field.

Comments on the new American Standards for photographic grade chemicals, and suggestions for additions, are invited. Please address the American Standards Association, 70 E. 45 St., New York 17, N. Y.

AMERICAN STANDARD AND PROPOSED AMERICAN STANDARD SPECIFICATIONS FOR
PHOTOGRAPHIC GRADE CHEMICALS

The standards in this list which have year dates following their designation numbers are approved American Standards and copies may be obtained for 25 cents per copy from the American Standards Association, 70 E. 45 St., New York 17, N. Y. Those standards lacking year dates are proposed American Standards, and are not yet available.

I. DEVELOPING AGENTS
   1. Mono-Methyl-Para-Aminophenol Sulfate   Z38.8.125-1948
      (Armol, Elon, Genol, Graphol, Metol, Photol, Pic-
      tol, Rhodol, Veritol)   Z38.8.126
   2. Hydroquinone
      (Para-Dihydroxybenzene, Quinol, Hydrochinone,
      Hydroquinol)
   3. 2,4-Diaminophenol Hydrochloride   Z38.8.127-1948
      (Acrol, Amidol, Dianol, Dolmi)
   4. Para-Hydroxyphenylglycin   Z38.8.128
      (Athenon, Glycin, Iconyl, Monazol)
   5. Para-Aminophenol Hydrochloride   Z38.8.129-1948
      (Kodelon, P.A.P.)
   6. Pyrogallic Acid   Z38.8.130-1948
      (1,2,3-Trihydroxybenzene, Pyro, Pyrogallol)
   7. Catechol   Z38.8.131-1948
      (Ortho-Dihydroxybenzene, Pyrocatechin, Pyrocat-
      echol)
   8. Para-Phenylenediamine   Z38.8.132-1948
      (1,4-Diaminobenzene)
   9. Para-Phenylenediamine Dihydrochloride   Z38.8.133-1948
      (1,4-Diaminobenzene Dihydrochloride)
  10. Chlorhydroquinone   Z38.8.134-1948
      (2-Chlor-1,4-Dihydroxybenzene, Adurol, C.H.Q.)

II. ALKALIES
   1. Sodium Hydroxide   Z38.8.225-1948
   2. Potassium Hydroxide   Z38.8.226-1948
   3. Sodium Carbonate, Monohydrate   Z38.8.227-1948
   4. Sodium Carbonate, Anhydrous   Z38.8.228-1948
   5. Potassium Carbonate   Z38.8.229-1948
   6. Sodium Tetraborate   Z38.8.230-1948
   7. Sodium Metaborate   Z38.8.231-1948
   8. Ammonium Hydroxide   Z38.8.232-1948

III. SULFITES
   1. Sodium Sulfite   Z38.8.275-1948
   2. Sodium Bisulfite   Z38.8.276
   3. Potassium Metabisulfite   Z38.8.277-1948

IV. RESTRAINERS AND ANTIFOGGANTS
   A. INORGANIC
      1. Potassium Bromide   Z38.8.200
      2. Potassium Iodide   Z38.8.201-1948
      3. Potassium Chloride   Z38.8.202-1948
      4. Sodium Chloride   Z38.8.203-1948
   B. ORGANIC
      1. Benzotriazole   Z38.8.204-1948
      2. 5-Methylbenzotriazole   Z38.8.205-1948
      3. 6-Nitrobenzimidazole Nitrate   Z38.8.206-1948

V. FIXING AGENTS
  1. Sodium Thiosulfate, Anhydrous   Z38.8.250
  2. Sodium Thiosulfate, Crystalline   Z38.8.251
  3. Ammonium Thiosulfate   Z38.8.252

VI. ACIDS
  1. Acetic Acid, Glacial   Z38.8.100
  2. Sulfuric Acid   Z38.8.101
  3. Citric Acid   Z38.8.102
  4. Boric Acid, Crystalline   Z38.8.103
  5. Hydrochloric Acid   Z38.8.104
  6. Sodium Acid Sulfate, Fused   Z38.8.105
  7. Acetic Acid, 28 Per Cent   Z38.8.106

VII. HARDENERS
  1. Aluminum Potassium Sulfate, Crystalline   Z38.8.150
  2. Chromium Potassium Sulfate, Crystalline   Z38.8.151
  3. Formaldehyde Solution   Z38.8.152
  4. Paraformaldehyde   Z38.8.153

VIII. MISCELLANEOUS
  1. Sodium Sulfate, Anhydrous   Z38.8.175
  2. Sodium Acetate, Anhydrous   Z38.8.176
  3. Potassium Dichromate   Z38.8.177
  4. Potassium Permanganate   Z38.8.178 .
  5. Potassium Ferricyanide   Z38.8.179
  6. Copper Sulfate   Z38.8.180
  7. Potassium Persulfate   Z38.8.181
  8. Sodium Sulfide, Fused   Z38.8.182
  9. Ammonium Chloride   Z38.8.183
  10. Ammonium Sulfate   Z38.8.184

-------- ◆ --------

Correction

In the December, 1948, issue of the JOURNAL, there was published an article by Lorin D. Grignon.

The reprint permission for Fig. 1, page 556, of the paper "Flicker in Motion Pictures: Further Studies," should read, "Fig. 1 reprinted from the *Journal of General Physiology*, July, 1936, with the permission of the copyright owners." Under Fig. 2, page 557, of the same paper should be, "Reprinted with permission from M. Luckiesh and F. Moss, "Science of Seeing," D. Van Nostrand Company.

Theater Television System*

By RICHARD HODGSON

Paramount Pictures, New York 18, New York

Summary—The paper reviews the development and performance of the theater television system developed and now being used by Paramount Pictures. Paramount has chosen the intermediate film method because of the flexible manner in which it can be integrated into theater operations, and its technical excellence.

ONE OF THE PRIME QUESTIONS up for discussion wherever motion picture people gather today is what effect will television have on the industry and how can the activities of the two industries be made to complement each other. How can television be used to the best advantage? There are three obvious roads open to the motion picture companies: (1) operate television broadcast stations; (2) produce motion pictures for television; and (3) bring television to theater audiences by means of theater television.

Paramount Pictures has been actively exploring two of these alternatives through the operation of stations in Hollywood and Chicago for the past seven years and by the development and operation of its own theater television system. This paper describes its method of bringing television to a theater audience.

Following several preview demonstrations of its theater television system before unapprised audiences in the Paramount Theater in New York early this year, the incorporation of full-screen theater television was announced in July as a regular entertainment policy. There have been twelve showings up to date. Fig. 1 shows a program being picked up from Central Park by a Paramount mobile television unit, and in Fig. 2 is shown the microwave relay equipment used for transmitting the picture to the theater.

Theater television, as developed and shown by Paramount, is an intermediate film system consisting of three units: a receiver, including recording cathode-ray picture tube; a 35-mm sound-recording camera, operating without a shutter, which is unique to this system; and a high-speed processing machine. However, they function as one and can be operated by a single skilled technician.

The intermediate film method was chosen because it provides the five following definite advantages:

* Presented October 25, 1948, at the SMPE Convention in Washington.

1. The film recording allows the theater to hold to its schedule of regular showing and to present the television program at an interval between other portions of the show, while a direct projection showing of the television projection system by its nature necessitates the immediate showing of the television program, interrupting, if necessary, the feature film.

2. Only an extremely small percentage of all theaters in the United States have sufficient space either behind the screen or in an orchestra

Fig. 1

pit to accommodate direct-projection equipment. To install direct-projection television would require eliminating a substantial number of seats which would otherwise be available for use by the theater's patrons. While difficulties might also be encountered in making room for an intermediate recorder in or near the projection room of the theater, less than 200 square feet is now required and less than 50 square feet will finally suffice.

3. The intermediate recording system allows opportunity for cutting and editing of the program before it is presented on the

theater screen, an asset which is most desirable from an entertainment point of view.

6. It can be used in the largest theater with the largest screen with regular illumination of approximately 14 foot-candles of incident light. No special projectors are required. No special directional screens are required.

5. There is no possibility of eye irritation from a projected intermediate film picture which is not always true of an interlaced elec-

Fig. 2

tronic picture of the same subject scanned with the same number of lines projected on a large screen.

RECEIVER

Any good quality commercial television receiver can be used.

The receiver is fed a stabilized signal whose levels have been properly balanced through a stabilizing amplifier. In an over-all theater television system the picture resolution may be limited by numerous factors, but horizontal resolutions as high as 600 lines are possible with existing television pickup and relay equipment, so that full advantage should be taken of these potentials. From the receiver the signal passes through auxiliary equipment for the separation of synchronizing pulses, and for amplification. A negative picture is produced on the cathode-ray tube by polarity reversal of the video signal. Use of the negative picture allows direct photography to give as the end result a positive picture for projection. It is desirable to introduce electrical gamma corrections by means of nonlinear amplifiers in order to produce accurate contrast gradations in the final picture. This subject was discussed in detail in a recent paper.[1] In addition, the usual circuits are included for the control of picture size, linearity, position, focus, and brightness. However, these circuits necessarily have been designed for optimum picture quality and reliability.

Included with the synchronizing circuits is an electronic shutter which converts the 30 television frames per second to 24 frames per second, corresponding to the camera running rate. The intermittent action of the camera is electrically phased with the electronic shutter so that at all times, including periods of line-current frequency drift, a complete 525-line television frame is displayed on the face of the cathode-ray tube at the rate of 24 frames per second.

FRAME-RATE RELATIONSHIPS

The problem of reducing the television image frequency of 30 frames per second on the cathode-ray tube to one of 24 frames per second on film is well known. Since the ratio of these two frequencies is 5 to 4, the transition from television frequency to film frequency can be accomplished by omitting one half a television field after every two fields. During this half field, which is scanned in $1/120$ second, the film in the camera is pulled down from frame to frame.

To reproduce one full frame or two fields of a television picture, the film remains stationary for the exposure period of $1/30$ second. If this exposure period is greater or less than $1/30$ second, more or less than 525 lines will scan the film, resulting in an underlap or overlap of exposure. This results in a band of underexposure or overexposure whose width is proportional to the degree in error of exposure time. If the frame frequencies of the two systems were equal, the exposure error could be hidden in the vertical blanking, but this would require that the film be pulled down during the vertical blanking period of $1/800$ second, which is considered an impossible requirement under present camera design and film-strength limitations. It would also mean that the film could not be run in standard projectors.

With the standard motion picture 24-frame rate, after one television frame of 525 lines has scanned the film in $1/30$ second there remains a period of $1/120$ second before the next film picture cycle begins. During this interval the film is pulled down. Also, during this period, the scanning, which is a continuous process, has progressed one-half field which takes it to the middle of the picture. At this point the film is exposed to the next television picture and remains

Fig. 3

so for $1/30$ second. It can be seen that for this film frame the television scanning has started and stopped in the middle of a field. The cycle now repeats with alternate frames having a join-up in the middle. The precision required for a perfect join-up is not easily attained with a conventional camera mechanical shutter. This led to the development of the particular electronic shutter by which the exposure time can be controlled to and phased with the camera action to within a microsecond.

CATHODE-RAY TUBE

The conversion of the incoming television signal to a reproduction on film of the original scene, depends greatly on the characteristics of the cathode-ray tube used to display the negative picture to be photographed. In the course of development of this system, cathode-ray tubes varying in diameter from 5 to 20 inches were tried and a 12-inch flat-faced tube selected as the most satisfactory. A picture size approximately 6 × 8 inches is used for photographing. This tube must have other characteristics to provide the high picture quality and satisfactory brightness levels; such as extremely fine, uniform grain size, short-persistence phosphor coating, spectral characteristics

Fig. 4

consistent with the spectral sensitivity of the film used to photograph the tube, metal backing of the fluorescent screen of the tube for maximum brightness and contrast, and small spot size for good definition.

Both P-5 (blue-glowing calcium tungstate) and P-11 (blue-glowing zinc sulfide) screen materials (these two being standard for cathode-ray-oscilloscope photography) have been tried. A special version of P-5, having a higher efficiency than the regular P-5 tungstate phosphor with a grain size approximately 5 microns and a decay time of 5 microseconds to 10 per cent of its light output, has been selected. The standard P-11 screen has also been extensively used. Its grain size, however, is approximately three times as large as the special P-5 and has a decay time of approximately 100 microseconds. The tubes are operated at 25,000 volts to obtain high brightness levels and minimum spot size. Projection-type gun structures have been incorporated and magnetic deflection and focus of the tube is used.

CAMERA

The photographing of the picture on the cathode-ray tube is done with a special 35-mm single-system, sound-recording camera. This

Fig. 5

camera has two unique features essential for this application. It employs no shutter to block off the light from the film during film pulldown, and it contains a special intermittent mechanism which pulls the film down in less than $1/120$ second. Sound track is recorded by the modulator in the camera at the standard displacement between picture and sound, $19^1/_2$ frames.

Cameras of this type, built by the Akeley Camera Company, are currently being used. Fig. 3 shows the camera in position. Other designs are being studied and model tests conducted, including one recently built by the Mitchell Camera Company.

The camera is equipped with a specially designed film magazine which can be loaded with 12,000 feet of 35-mm film, permitting continuous recording of over two hours. A Cooke $f/1.3$ coated lens is used at a normal aperture, $f/2.3$.

Du Pont, Type 228, fine-grain, master, positive film, or Eastman 5302 is used for the recording of either positive or negative pictures.

Fig. 6

The film has a peak sensitivity at 4600 angstroms. It has an emulsion which is capable of standing up under the high temperatures and high solution concentrations used in the rapid processing of the film as it leaves the camera, and is continuously fed to the theater projector. Fig. 4 shows the complete equipment layout in the Paramount Theater in New York, and Fig. 5 shows the chute through which the film is fed continuously to a standard projector.

PROCESSING EQUIPMENT

The tube-type processing equipment, designed by Paramount, shown in Fig. 5, is being superseded, in other installations, by high-speed developing machines built by us in collaboration with the Eastman Kodak Company (Fig. 6). ·

This machine processes film traveling at the required rate of 90 feet per minute, in 40 seconds. It includes three vertical stainless-steel cabinets for developing, fixing, and washing, respectively, and a drying drum. Because of the high drying temperature, safety-base stock must be used.

The hot developer, fixer, and wash water are brought into contact with the traveling film by sprays from nine nozzles emitting a solid cone of solution or water supplied by pumps. The developer is highly caustic and is maintained at a temperature of 117 degrees Fahrenheit. The film passes through in 5 seconds. The fixing solution is a high-speed, commercial-type solution maintained at 120 degrees Fahrenheit, with the film passing through two loops in 10 seconds. The 5-second wash is at 100 degrees Fahrenheit. Drying time is 5 seconds.

The drive sprocket above the drying drum is a 64-tooth sprocket which draws the film through the unit to the projector or take-up. The drum itself is constructed from an ordinary free-running bicycle wheel, with an aluminum rim extension. A concentric exterior heater surface of nichrome strip, one-half inch from the surface of the film supplies the heat which bears directly on the emulsion surface. The strip has a temperature of approximately 1000 degrees Fahrenheit.

New air from a 30-pound compressor source is blown through No. 60 drill size holes in edge-positioned tubing, operating separately on each quarter section of the drying wheel. Approximately 7 kilowatts, at 50 volts, is required for drying.

Tests completed recently indicate that the film drying time can be reduced to less than a second. With total processing time in the order of 10 seconds, these modifications will be incorporated into future units. By changing sprockets, 16-mm film also can be processed on the same machine.

REFERENCE

(1) E. Meschter, "Television reproduction from negative films," *J. Soc. Mot. Pict. Eng.*, vol. 47, pp. 165–182; August, 1946.

DISCUSSION

MR. WILLIAM PRAEGER: Why do you use safety stock?

MR. RICHARD HODGSON: At the present time we are not using safety stock. The film you saw projected is nitrate. However, the last machine that was worked out with the Eastman Kodak Company has drying temperatures so high that nitrate stocks are dangerous to use. That is the only reason.

Demonstration of Large-Screen Television at Philadelphia*

BY ROY WILCOX

RCA VICTOR DIVISION, CAMDEN, NEW JERSEY

AND

H. J. SCHLAFLY

TWENTIETH CENTURY-FOX FILM CORPORATION, NEW YORK, NEW YORK

Summary—In an effort to determine some of the problems and to learn some of the limitations and potentialities of the instantaneous projection method of large-screen theater television, Twentieth Century-Fox, with the assistance of the Radio Corporation of America, made an installation in the Fox-Philadelphia Theater, June, 1948. It is the purpose of this paper to give a factual report on the problems encountered in locating, installing, programming, and using the equipment in a typical operating downtown theater, and to comment on the reaction of the audience to a showing of the Louis-Walcott world's championship fight.

LARGE-SCREEN THEATER TELEVISION has, until recently, shown an increasing similarity to the weather in Mark Twain's famous commentary—everyone talks about it but nobody does anything about it. At the present time there are several groups in the motion picture industry who are "doing something about it." The leading exhibitors are showing increasing interest—even to the point of spending money. Research and engineering laboratories have been organized by several film corporations, which can investigate and perhaps even improve upon the equipment and techniques now employed to produce a television picture on the large-size screens used in motion picture theaters.

It is now well known that there are two basic methods being considered—the instantaneous projection method wherein the television picture formed on the phosphor screen of a cathode-ray tube is immediately projected by an optical system onto the viewing screen for the theater audience; and the film-storage method wherein the television picture formed on the end of the cathode-ray tube is

* Presented October 25, 1948, at the SMPE Convention in Washington.

actually photographed by a motion picture camera, the film processed either by standard laboratory techniques or, more likely, one of the new rapid processing methods, and then projected onto the theater screen using standard theater equipment. Each system has certain advantages and disadvantages. It is difficult to predict at this time whether either method will be generally adopted by the industry or whether both or perhaps even a third method will be required for commercialization of theater television.

As in the case of all involved problems, the unknown quantities must be approached one by one. In an effort to determine some of the limitations and potentialities of the instantaneous projection method of large-screen theater television, Twentieth Century-Fox, with the assistance of the Radio Corporation of America, made an installation in the Fox-Philadelphia Theater in June of 1948. This installation was not a "stunt." It was made primarily for the contribution that it could make to the growing fund of knowledge that theater men must have before they can properly evaluate their new entertainment tool. It is the purpose of this paper to give a factual report on the problems encountered in locating, installing, programming, and using the equipment in a typical, operating, downtown theater, and to comment on the reaction of the audience to a showing of the Louis-Walcott World Championship Fight. Technical details of the instantaneous projection equipment involved have been presented by Little and Maloff.[1,2] Briefly, however, the equipment is composed of two major units: the *projector* which houses the cathode-ray tube, driving circuits, high-voltage supply, and the correcting lens and 42-inch mirror of the Schmidt optical system; and the *control rack* which contains all operational controls, power supplies, and input circuits.

Projector Location

The choice of projector location was seriously restricted because of the size, weight, and short throw distance of the unit. Rear projection sacrifices too much of the precious light in transmission loss; locations in the orchestra floor and in the balcony sacrifice view from too many of the best seats in the house; a platform suspended from the ceiling presents structural and servicing problems and a large projection angle brings into consideration keystoning and top-to-bottom focus. An investigation of the geometry of the Fox-Philadelphia Theater showed that it was particularly well suited

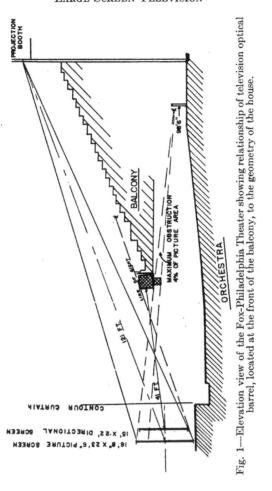

Fig. 1—Elevation view of the Fox-Philadelphia Theater showing relationship of television optical barrel, located at the front of the balcony, to the geometry of the house.

for the purpose. The front edge of the balcony loge in this theater is at just about the correct throw distance, nominally 40 feet, to the theater screen, and the load-bearing structure of the loge was quite adequate to permit a hanger to be installed in front of the railing. Fig. 1 is a profile view of this theater. It shows that the chosen location not only permits the correct throw but also nestles the optical barrel in the audience "blind spot" in front of the balcony, obstructing the least possible number of seats. To do this, •the projector had to be stripped down—the carriage and trim which normally enclosed the lower portion of the optical barrel were removed. The circuit chassis housed in this carriage was relocated in a small rack placed on the hanger immediately adjacent to and on a level with the optical barrel. The housing for the high-voltage supply hangs below the platform, but its projected area is negligible and caused no audience complaint. Fig. 2 is a photograph of the actual installation showing the hanger, optical barrel, and high-voltage housing, and the adjacent circuit rack. Total weight of this equipment, including hanger, was 1600 pounds.

Fig. 2—Television projector and associated circuit rack in position on hanger in front of balcony railing.

It should be pointed out that very few theaters would be so geometrically ideal for this type of installation. It is recognized that each theater presents a problem of its own. Meanwhile, any steps that can be made in equipment design which will allow a reduction in size or weight or an increase in throw distance should certainly be encouraged. The most recent model of the RCA projector, publicly demonstrated for the first time during the September, 1948, Theater Equipment and Supply Manufacturers Association Convention in St. Louis, is certainly a step toward a more practical size.

Fig. 3—Control rack (with front door removed)
for television projector as installed in projection
booth. The partially filled rack on the right contains portions of test and termination equipment.

CONTROL-RACK LOCATION

The control rack for the television projector used in Philadelphia can be removed as far as 100 feet from the optical-barrel assembly. Additional footage probably can be accommodated with slight circuit changes. Although several locations in the theater were considered for the control, the projection booth was selected. Several reasons influenced this choice:

1. The operator had a full and undistorted view of the screen through a projection-booth port.

2. The control rack, voltage regulator, picture and sound line terminations, and test equipment were completely removed from public access.

3. Electric power and the input to sound-amplifying equipment were immediately available. Communication channels with other parts of the theater were already established.

4. Test and operation of this portion of the equipment were possible while the house was in normal operation.

Only one objection to this location was voiced—the controls were too far away from the screen (about 120 feet) for the operator to set accurate focus without the assistance of field glasses or directions from some one in the auditorium, but this is not an uncommon practice for film projection. Fig. 3 shows the control rack as it was installed in the projection booth. The front door has been removed to permit a view of the circuit panels. The auxiliary rack on the right contains a portion of the terminal and test chassis.

Three rubber-covered multiconductor cables and one video cable connect the projector and the control rack. These cables were run directly from the circuit rack adjacent to the optical barrel, under the balcony, up through the floor of the projection booth, and into the control rack. Needless to say, the job of making this cable run presented its own unique problems but it did permit the desired location of the control rack without the necessity of increasing the 100-foot length of interconnecting cable.

Again, we were favored by the geometry of this particular theater. The projection booth was large enough to accommodate the additional equipment and the cable run was possible without any major alterations. Those interested in new theater construction might well consider the possibility of making the projection room a bit oversize and allowing a clear cable channel to the front of the balcony.

INSTALLATION

Drawings were made and the angle-iron hanger was fabricated. Professional riggers were engaged to make the physical installation. Moving the control rack into the projection booth and securing the hanger in place on the front of the loge railing was routine. Handling the optical barrel, because of its bulk and because of the precautions observed to prevent any possible damage, proved rather difficult.

Unloading this unit from the truck and flying it to its position on the hanger caused some anxious moments for the entire crew.

SCREEN

One of the serious limitations of the instantaneous projection method of large-screen television is low screen brightness. All of the light on the screen is generated, not by a carbon arc, but by electron excitation of a phosphor screen on the inside of the cathode-ray tube. It is one problem to obtain the required light level for direct viewing of a 50- or even 150-square-inch surface on the cathode-ray-tube face, and quite a different problem to project and spread that light, at the required level, over the 300 or 400 square feet of a large-size theater screen. Furthermore, motion picture technique permits each picture element to receive full illumination approximately 50 per cent of the time; television-scanning technique permits full picture-element brilliance only about 4 millionths of a second out of every second and a value greater than half brilliance less than 10 per cent of the time. Thus, every known device must be used to conserve or utilize the light that is available. Extremely high voltages are used on the cathode-ray tube to increase the energy content of the electron beam. For this particular projector, an 80,000-volt second-anode potential is used. The Schmidt system is provided because it is optically "fast," the approximate equivalent of an $f/0.8$ lens. And, finally, the theater screen itself offers an opportunity for light conservation by directing most of the incident light back to the audience. For the Philadelphia demonstration, an experimental aluminized screen having controlled directional characteristics was installed. This screen had a gain of approximately 2.5 over a perfectly diffuse screen. Its size was 15×22 feet, the television picture being pushed off into the top and bottom masking to retain the proper 3-to-4 aspect ratio. The screen was perforated for normal sound transmission.

Even with these devices for increasing light, the apparent high-light brightness was in the order of 3 to 4 foot-lamberts. This measurement was made in the white area of a monoscope test pattern and probably should not be directly compared with figures given for motion picture projectors where measurements are made of incident light on the screen when there is no film in the projector.

While it cannot be denied that additional light is required, particularly to improve fine detail and contrast in the television picture, it should also be noted that only the most critical observer objected to the light level during the actual demonstration.

SOUND

The television projector has no provisions for audio portions of the program. Sound can be treated as an entirely independent problem and, as such, has the advantage of tried-and-tested methods and techniques. The use of the special directional screen for the television picture did provide an interesting audio problem. The house speakers were attached to and could be flown with the regular picture sheet. Thus, if they were used for the television picture, the sound would have to be transmitted through two screens, the regular sheet and the directional sheet. It was thought that such an arrangement would attenuate the high frequencies very noticeably. Therefore, although the theater had not used its public-address speakers for several years, these units were dusted off and mounted on a dolly which could be rolled into position directly behind the television screen. Comparative tests showed that high-frequency attenuation caused by the second screen was considerably less than had been expected and that the sound from the regular speakers, even with the two screens, was superior to the older cone-type public-address units.

PROGRAMMING

One of the objectives of this installation was to observe audience reaction in terms of program interest. The proper question to ask was not, "Does the audience enjoy this program?" but, "Is the reaction of the audience viewing the program at the theater comparable to the reaction of the audience at the actual event?" It was to our benefit, therefore, to obtain an event for the demonstration which had possibilities of high audience interest. The Louis-Walcott fight presented an ideal vehicle for these observations.

The problem then seemed simply one of securing from the interested parties the necessary rights to carry the telecast of the fight in the Philadelphia theater. Negotiations were started many weeks in advance. The final arrangements were made and the final permissions were secured—almost a day and a half before the scheduled day of the fight.

Programming is undeniably one of the major concerns of those who are interested in theater television. Television rights have been and probably will continue to be guarded jealously until a sufficient number of well-thought-out precedents have been established. It will be interesting to follow the trend.

PROGRAM DELIVERY

Yankee Stadium, the location of the fight ring, was almost 100 miles away from 16th and Market Streets in Philadelphia. Arrangements made with the National Broadcasting Station, WNBT, included utilization of their camera pickup of the event, and delivery,

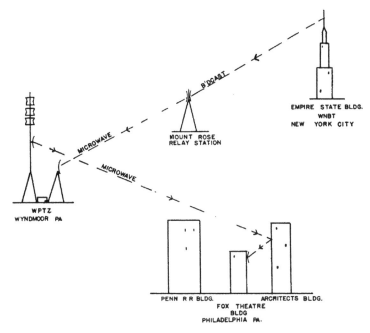

Fig. 4—Transmission path employed to bring Louis-Walcott fight from New York City to the Fox-Philadelphia Theater.

either via the Bell System coaxial cable, or via the WPTZ microwave relay, to Philadelphia. If the cable were used, the remaining jump between the Philadelphia terminal and the theater building would require installation of a special equalized video line. An order for this line was issued several weeks in advance of the fight date but it was canceled during one of the periods when completion of the program negotiations appeared impossible. When the final arrangements were made, insufficient time remained to make the cable installations. Program transmission from the Philadelphia terminal of the radio-relay circuit, a distance of $9^{1}/_{2}$ miles north of the theater building, was made via a microwave relay link installed and operated by RCA. An interesting feature of this installation was the fact that line of sight between the roof of the theater building and the WPTZ transmitter tower was cut off by a corner of the Pennsylvania Railroad Suburban Building. As you know, microwave frequencies do not bend around corners. With an unusual amount of good luck, we were able to pick up the transmitted signal on the first bounce, reflecting it from the side of one of the large buildings in the downtown Philadelphia area. Instead of having the relay transmitting antenna and receiving antenna facing each other, they were actually within a few degrees of being pointed in the same direction. The reflected signal was quite strong and of good quality; it was not affected by fading or by secondary reflections.

In order to avoid an excessively long run of a multiconductor cable, the power supply, monitor, and control for the micowave relay were located near the receiver on the roof of the theater building. Fortunately, this equipment was protected from the weather. Only one coaxial cable and a communication line made the run from the roof to the theater projection booth eleven stories below.

Program sound was carried in a routine manner by the Bell System Long-Lines Department. An equalized line terminating in the theater projection booth was installed in a matter of two days from the time the order was placed. This termination was connected to the input selector switch of the house amplifier through a level setting attenuator pad.

TEST AND AUXILIARY EQUIPMENT

At the present state of the television art, test equipment is essential. In addition to a portable oscilloscope, vacuum-tube voltmeter, and circuit analyzer, it was necessary to include a monoscope camera and synchronizing generator. This latter item provided a standard television test pattern of consistent high quality and was used for placing the projector in proper optical and electrical alignment. Of course, all test work involving the projector had to be done after closing hours and, at that time, no broadcast signals were available. Auxiliary equipment included a line amplifier, stabilizing amplifier, and video switching and monitoring facilities.

AUDIENCE REACTION

We considered ourselves fortunate when at three o'clock on the afternoon before the fight, June 23, we were informed that the event was postponed to the next day. We had last-minute details that needed attention. The next day the postponement notice did not reach us until after 8:00 P.M. That, too, was received with relief since the same storm that postponed the fight had misaligned our relay transmitter antenna. The third evening, June 25, 1948, we carried the fight as scheduled.

With the exception of one announcement placed near the box office at 8:00 P.M. the day the fight was actually presented, there had been no advance publicity. Even so, the theater, a 2400-seat house, was nearly filled at fight time.

The audience followed the fight with mild enthusiasm during the first few rounds. The picture was good; close-ups were better than medium shots. Occasionally, a critical observer would notice signs of camera overloading in a particular bright reflection from the shiny head or shoulders of the two contestants. As the fight progressed, the viewers became completely absorbed. During the action preceding the eleventh-round knockout, the entire audience was on its feet, cheering as wildly as the ringside crowd. When Walcott was counted out in New York, he was counted out in the Fox Theater. A veteran showman, watching from the rear of the theater, commented, "This is the most spontaneous reaction I have ever seen in a theater."

REMARKS

Criticism of picture quality is certainly a function of program interest. Where the interest is high a surprisingly poor picture will be accepted without comment; when interest lags, a surprisingly good television picture may bring complaints.

Theater surroundings are, of course, designed to emphasize the picture and to subjugate all distractions. For this reason, it may be easier for the viewer to become absorbed in the action on the screen in a theater than in some other public place or even in his own home. Television as a medium of picture transmission has a powerful psychological ally. The viewer knows that he is seeing an event that is happening *now* and that neither he nor anyone else knows exactly what is going to happen next. It is possible that a delay in the order of a few minutes might not detract from this feature of simultaneous presentation—particularly in surroundings where an accurate sense of time is lost.

Is the operation and programming of large-screen television by the instantaneous projection method, or any other method, sufficiently practical and is the picture quality sufficiently good to present this new medium of entertainment to the theater public? This vital question is not answered by one demonstration nor by one paper before the Society of Motion Picture Engineers.

Admittedly the equipment used for the Philadelphia demonstration was large and television techniques are new and involved. Admittedly the television picture quality does not now and probably in the near future will not equal that of 35-mm film. Improvements in present equipment undoubtedly will be made. Improvements are desired not only in screen brightness but in linearity, interlace, picture resolution, and particularly in the reproduction of the tone scale. Many of these present limitations are not problems affecting the projector alone but apply as well to the pickup and transmission of the television signal.

The answer to our question still requires thought and work and the co-operation of the equipment manufacturers, laboratories, engineers, showmen, promoters, in short, the industry.

REFERENCES

(1) Ralph V. Little, Jr., "Developments in large-screen television," *J. Soc. Mot. Pict. Eng.*, vol. 51, pp. 37–47; July, 1948.

(2) I. G. Maloff, "Optical problems in large-screen television," *J. Soc. Mot. Pict. Eng.*, vol. 51, pp. 30–37; July, 1948.

DISCUSSION

CHAIRMAN N. D. GOLDEN: Is it not true that the equipment now used for theater large-screen television is much more compact than the equipment used at the theater in Philadelphia?

MR. H. J. SCHLAFLY: I think there has been a definite step forward in reduction of size; the equipment is not so small as I should like to see it. The latest equipment which RCA has built and demonstrated uses a 20-inch mirror compared to the 42-inch mirror that was used in the Philadelphia demonstration.

MR. C. J. STAUD: We have developed audience-reaction studies and we should be very glad to make that information available and that might be an important tem in connection with your work.

We have flooded our auditorium with infrared light and taken pictures of people viewing a picture or listening to a lecture. I must confess that the program material was not so interesting as yours and the pictures were not complimentary, but the facts are there.

Precision Speed Control*

By A. L. HOLCOMB

WESTERN ELECTRIC COMPANY, HOLLYWOOD 38, CALIFORNIA

Summary—A precision speed control is capable of maintaining motor speed within 1 part in 25,000 under voltage variations of plus or minus 20 per cent and load variations from zero to full load. For motion picture production this order of regulation provides the equivalent of synchronous operation without physical connection between such motors.

THE TERM "PRECISION" in speed control, as in all things, is a matter of definition. For most industrial use speed regulation of 3 or 4 per cent is considered excellent, and regulation of synchronous motors operated from the supply lines of a large public utility is considered as precision control. For sound recording the speed regulation of synchronous motors is satisfactory for most applications even though a critical examination of line frequency and the resultant synchronous motor speed will show variations of $^1/_4$ per cent as common on the best-regulated supply lines; however, the inertia of large power networks is high and the rate of change is consequently slow, so that resultant speed variations are not objectionable. Thus, in the field of sound motion pictures, line-frequency speed control can be defined as satisfactory but not as a precision control. In some other fields such as the recording of vibration in geophysical prospecting, and other research and development projects, the required order of regulation may be as much as 50 times better than that which is satisfactory for motion picture work.

An electronic speed control developed by the Bell Telephone Laboratories[1] is capable of providing this latter high order of precision control in a very compact form. Regulation of 1 part in 25,000, or $^4/_{1000}$ of 1 per cent, can be obtained from this unit. This degree of regulation normally would serve no useful purpose in motion pictures when used as the speed control element of an interlocked motor system since $^1/_4$ of 1 per cent has proved to be satisfactory. However, it may prove valuable for some special applications.

Because of the high order of regulation, it is possible to operate two or more separate motors each with its individual speed control

* Presented October 26, 1948, at the SMPE Convention in Washington.

The crystal CR-1 is a special duplex type developed by the Bell Telephone Laboratories[2-4] and is mounted in a vacuum-tube envelope. The resonant frequency of this particular crystal is 720 cycles, the Q is 25,000 or better, and the temperature coefficient is relatively very good, about 2 parts in 1 million per degree centigrade. The crystal operates effectively as a motor generator, the motor voltage being derived from the bridge and the generator section being connected across the grid resistor R_3. Since the Q of the crystal is so high its sensitivity is very low to frequencies off resonance; consequently the crystal is inoperable as a control without the bridge to maintain the speed and pilot frequency close to crystal resonance. The action of the bridge and crystal complement each other inasmuch as the crystal output voltage is maximum at the resonant point where the voltage across the bridge is at a minimum. It will be apparent that the phase shift applied to the grid must be implemented by sufficient voltage amplitude or the phase-shift function loses its effect. Thus the sharpness of regulation that can be obtained from the bridge alone is limited by the fact that the higher the bridge Q, the lower the voltage at resonance, and motor speed thus will drift back and forth across the ideal. Since the output voltage of the crystal is highest at resonance, the crystal becomes most effective in the region of lowest bridge output. It should be noted that the circuit is completely operable and capable of regulation of the order of $1/10$ of 1 per cent or better with the bridge alone, and the crystal may be removed from the circuit at will. The total phase shift of the grid of VT_1 with respect to the plate supply is approximately 90 degrees at normal frequency and a shift of approximately 90 degrees each side of this mean value is available. The average plate current of VT_1 is of the order of $1/10$ of a milliampere and, as previously noted, the current which is passed by the tube consists of one-half wave of the pilot frequency. A capacitor C_4, back of the load resistor R_8, provides sufficient filtering to maintain essentially a direct-current bias on the grid of the beam-power tube VT_2.

This beam-power tube VT_2 operates as a direct-current valve controlling the current through the regulating field in the motor and thus the speed of the motor over a considerable range. The space current is controlled by the grid voltage which is a function of the phase relation of grid and plate in VT_1. In order to increase the sensitivity of VT_2 to changes in plate current of the phase-comparator tube, the negative voltage generated across R_8 is made high and a positive bias is introduced to cancel approximately $1/2$ of it. This positive voltage is supplied by the resistor R_6 which is a portion of a bleeder circuit R_{5-6-7} across the direct-current supply. This provides an additional feature since the positive canceling bias across R_6 varies directly with the input voltage and increases the average value of regulating field current so as to correct partially the tendency of the motor to speed up with increased supply voltage. The resistor R_5 in the bleeder circuit serves to provide a small negative bias on the grid of VT_1 and also serves to provide a positive feedback action since the regulating current returning to the cathode of VT_2 also passes through this resistor, thus increasing the negative bias on VT_1 as the regulating current increases. This positive feedback

Fig. 2—Lightweight motor and control.

tends to intensify regulating action for both load and voltage changes. The capacitor C_5 and resistor R_9 connected to the grid of VT_2 and ground constitutes an antihunt circuit since grid voltage can only be changed as the capacitor C_5 is changed through resistor R_9. Thus the steady-state sensitivity of the grid may be exceedingly high whereas the effective amplification of cyclic variations drops off sharply. The resistor R_{10} in series with the grid serves only to damp high-frequency oscillations to which beam-power tubes are subject.

Fig. 2 is a photograph of the control chassis and motor just described. With respect to the control chassis the bridge network is at the extreme left, the crystal next, then the input transformer T_1, power tube VT_2, and the phase-comparator tube VT_1. The chassis as shown weighs 45 ounces.

The motor in the photograph weighs 57 ounces and delivers a maximum power of 50 watts. This is sufficient to drive the majority of 16-mm cameras but would not be adequate for 35-mm use. A motor for 35-mm cameras can be built weighing about 15 pounds although such a motor probably would have to operate at higher than normal speeds for the required power output and thus might be somewhat noisier than conventional units. The control circuit shown would be adequate for the control of a 35-mm camera motor without major changes.

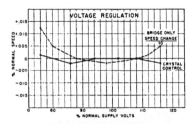

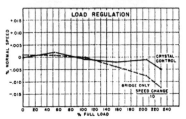

Fig. 3—Regulation characteristics.

Fig. 3 shows the characteristic regulation curves for both the bridge circuit alone and for the circuit complete with crystal. The regulation shown for the crystal control is plotted in solid lines between X's and refers to the ordinates as shown. The bridge-only regulation, shown by dots and dotted line, is reduced by a factor of 10; or in other words, the regulation shown for this mode of operation is only $^1/_{10}$ as good as it would appear relative to crystal control. In measuring speed which varies by such small values it is necessary to use a strobolamp excited from a frequency standard oscillator which is stable to at least 1 part in 100,000. In addition, it is necessary to count the pattern departure from a given point for two minutes or more. It will be noted that under voltage regulation, the speed change for the crystal control circuit was $^2/_{1000}$ of 1 per cent for voltage changes from 20 per cent below normal to 15 per cent above normal. For the same set of conditions, the regulation with the crystal removed was $^7/_{100}$ of 1 per cent. The shape of the regulation curve for the bridge only is due to the direct-voltage compensation previously noted. This compensation is more effective at low values of regulating current and low-supply voltage, and thus the low-voltage condition is overcompensated while the high-voltage condition is undercompensated, and is unable to prevent some speed increase with voltages above normal. In the load-regulation characteristics shown, the crystal-control circuit is within $^4/_{1000}$ of 1 per cent from no load to 200 per cent of rated full load. The last point shown at 230 per cent was on the edge of pullout. In the case of the bridge circuit without crystal, the regulation is $^8/_{100}$ of 1 per cent from no load up to 200 per cent of full load. It should be noted that this latter figure is still about three times better than line-frequency regulation.

Fig. 4 shows an adaptation of the above-described circuit that has been made to provide speed control for existing $^1/_5$- and $^1/_2$-horsepower distributor motors. The electronic-type speed control is useful where line-frequency regulation is too unstable for a synchronous-motor drive, or on location work where the alternating current supplied to the distributor system is from small engine-driven generators, or inverters. Another useful application is found where the desired system speed, such as 1440 revolutions per minute, cannot be derived directly from line frequency. The control circuit used is basically similar to the circuit described except that no provision is made for the addition of the crystal and the bridge is made externally tunable. Control speeds of 1000, 1200, and 1440 revolutions per minute are standard. The vacuum tubes used are quick-heating filamentary types and the maximum power output is approximately twice that of the smaller unit. A pentode tube V_2 is used as a phase comparator which increases the ratio of output-current change to phase shift, and the screen voltage of this tube is maintained constant by means of the voltage regulator tube V_1. The antihunt network is made adjustable and the capacitor C_{10} which forms a portion of this network is transferred by a relay. The input transformer T_2 may be connected for either 220- or 110-volt operation; tube filaments are operated on alternating current; plate supply is derived from a rectifier V_4; and both alternating- and direct-current circuits are fused and provided with disconnect switches.

The relays S_1, S_2, and S_3 are used to provide the following features: to prevent the motor from being started until the vacuum tubes have heated sufficiently to function; to prevent damage to the power tube during the heating period; to prevent arcing at the motor brushes by maintaining the starting series field in circuit until running speed is reached; to ensure reaching any one of the three operating speeds under heavy loads and low-voltage; and to prevent overshooting particularly for 1000-revolution-per-minute operation which is usually associated with 50-cycle stator-supply frequency. The coil

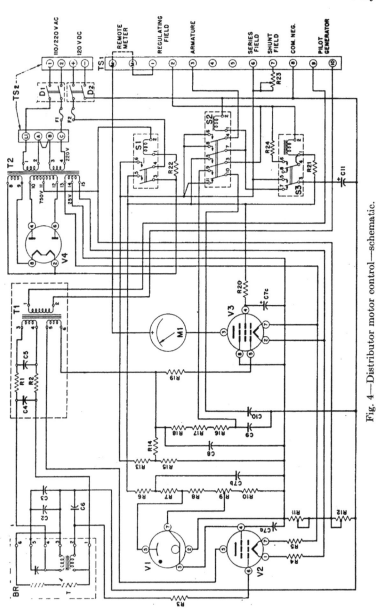

Fig. 4—Distributor motor control—schematic.

of the relay S_1 is connected across the rectifier tube V_4 and will not operate the relay S_1 until the rectifier is functioning. One pole of the relay is in series with the supply circuit to the motor and the other pole is in series with the plate supply to the power tube V_3. Thus, the motor cannot start until the rectifier tube has heated and the only load on the rectifier during the heating period is the few mils re-

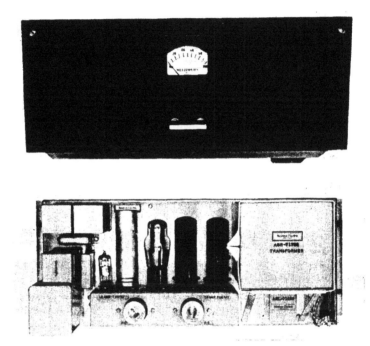

Fig. 5—Distributor motor control—front and back view.

quired to operate the coil of S_1. Also, no plate voltage appears on the power tube V_3 during the heating time of the filament. This heating time is normally not more than 2 seconds but under very cold conditions where the unit has not operated for some time this may be increased considerably. The coil of relay S_3 and its series resistance R_{21} are initially in series with the regulating field which limits the regulating field current to about 25 milliamperes during start, and also operates this relay as soon as S_1 pulls up. The capacitor C_{11} is transferred to the direct-current supply line by the operation of S_3

and is thus charged when motor speed reaches the beginning of the control range. At this point the grid of the power tube is driven practically to cutoff which allows the relay S_3 to drop back due to the reduced current in the coil and thus transfers the charged capacitor C_{11} to the coil circuit of S_2 which is thereby operated. The 4-pole relay S_2 then is locked up by means of one pole, a second pole short-circuits the series field in the motor, a third pole short-circuits the coil of S_3 which puts the regulating field into normal operation, and a fourth pole transfers a positively charged capacitor C_{10} to the antihunt circuit where it remains during operation. This capacitor, being positively charged when transferred, introduces a positive voltage on the grid of V_3, momentarily increasing the regulating current to maximum to prevent overshooting. This positive charge then leaks off through the series resistance and the grid of the power tube thereafter is normally controlled.

The regulation of this control unit is essentially the same as shown in Fig. 3 for bridge-only control; namely, 1 part in 1000 or better. A crystal can be used in conjunction with this unit, but there appears to be no reason why this should be done since the regulation is already better than is required.

Fig. 5 shows both a front and back view of the control cabinet. In the front view will be noted a meter which shows the value of regulating current and the two disconnect switches D_1 and D_2 which are barred together. In the back view the bridge network appears in the lower left-hand corner. Immediately above it is the hedgehog-type capacitor used to tune the bridge. The miniature tube is the phase-comparator pentode; to the right is a 3-section electrolytic capacitor, the voltage-regulator tube, the power tube, rectifier tube, and the power transformer. Immediately under the power transformer may be seen the pilot-frequency transformer. The unit is mounted on a 7-inch panel and weighs approximately 30 pounds. It is intended as a replacement unit for some of the older-type speed controls such as the 700-A cabinet which, by comparison, weighs 144 pounds.

REFERENCES

(1) H. M. Stoller, United States Patent 2,395,517, February 26, 1946.

(2) C. E. Lane, "Duplex crystals," *Bell Labs. Rec.*, vol. 14, p. 59; February, 1946.

(3) A. W. Ziegler, "Wire supported crystals," *Bell Labs. Rec.*, vol. 13, p. 140; April, 1945.

(4) C. E. Lane, United States Patent 2,410,825, November 12, 1946.

Automatic Tempo Indicator

By BRUCE H. DENNEY and GEORGE TALLIAN

PARAMOUNT.PICTURES, HOLLYWOOD, CALIFORNIA

Summary—The automatic tempo indicator provides a visual cuing system to replace, or augment, the stage playback loudspeaker system used in the production of musical sequences.

A FAMILIAR SCENE in a motion picture shows a ballroom filled with dancing couples and with the principal characters in the scene's center. Orchestral music is heard; the dancers are dancing to the music's rhythm and as the principals of the scene approach the camera their dialog lines are spoken in low confidential voices. The tempo of the dancers and of the orchestra, if seen, does not vary.

Several methods have been used to record the intimate dialog of the actors in such a scene without recording too much sound from the playback loudspeakers that supply the prescored music. One,method is to reduce the loudspeaker volume as much as possible and direct the sound away from the microphone   Another method is to cut off the music during the dialog sequence and rely upon the rhythmic memory of the actors and the actor-musicians to keep them in tempo. At times a music director wearing headphones may give manual off-stage directions to assist in keeping strict tempo. With patience and practice a satisfactory scene may be photographed and recorded, but the editing problem may be involved if the recorded dialog has an accompanying background of low-level music which must match a continuous music recording, the two to be combined in re-recording.

This problem is repeated and varied by the requirements of many pictures. It results in a loss of "shooting time," editing time, and re-recording time. Many ideas have been tried and discarded in favor of the low-volume side-line loudspeaker assisted by manual direction.

A solution for this problem is the use of a group of small flashing lights located about the scene but outside of the camera's view. With enough lights to avoid the concentration of the actors in any certain direction, the flashing lights provide strict tempo signals for the dancers and actor-musicians. The playback loudspeakers are turned off during dialog recording. Editing and re-recording are simplified.

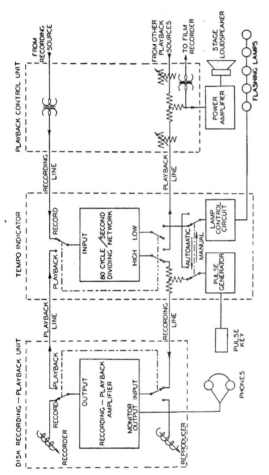

Fig. 1—Block diagram of automatic tempo indicator.

There are, however, several exacting requirements; the lights must flash in exact synchronism with the music beat, the duration of each flash must be constant, and the flashes must be controlled from information recorded upon a playback disk; the information should be controlled or keyed by a music or dance director in tempo with the music being recorded. This dual-purpose playback disk would supply not only the music signals for rehearsal loudspeakers but also

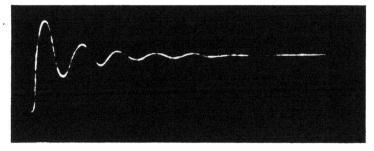

A. 45-cycle pulse voltage across capacitor of inductance-capacitance circuit. Note direct-current component.

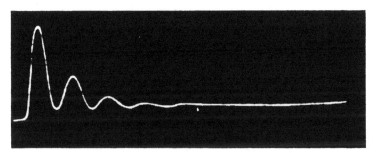

B. 45-cycle pulse voltage at output of disk-recording amplifier. The direct-current component has been attenuated by the circuit amplifiers and transformers.

Fig. 2

provide the signals that energize the flashing-light control circuit. In this manner the relationship between the music and the light flashes, once established and checked, would not change. The loudspeakers should be controlled, or turned off, without affecting the flashing lights.

The following described system is connected with a portable disk re-recording and playback unit, a loudspeaker system, and a playback control panel. The tempo-indicating device is not only the manually controlled source of *pulses of 45-cycle-per-second* tone that are recorded upon a disk with the music but also the unit that automatically operates the flashing tempo lights, from the recorded pulses of 45-cycle-per-second signal, when the disk is played back.

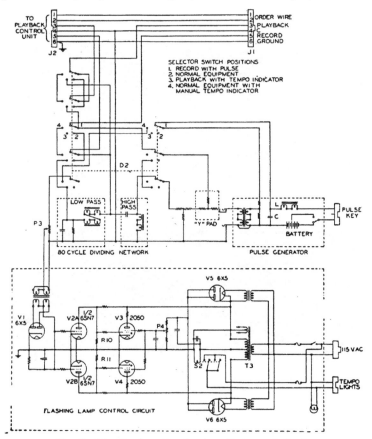

Fig. 3—Circuit schematic of automatic tempo indicator.

Fig. 1 is a block diagram showing the basic components of the system. The recorder playback unit consists of a $33^1/_3$-revolution-per-minute disk turntable (usually driven from an interlocked motor system) and a recorder-playback amplifier with necessary controls. Through monitor phones the operator hears the signal being recorded and operates the push switch that inaugurates each pulse of 45-cycle-per-second signal; each depression of the switch, in time with the music, starting another pulse; and each pulse causing a flash of the tempo lights when the disk is played back. These pulses are generated by the closing of a battery into a high-Q inductance-capacitance circuit. Damped oscillations, Fig. 2, at 45 cycles per second are generated, ceasing when the capacitor is fully charged. This frequency was chosen because it was within the lower limits of the recording system, and because its separation from the music by means of an 80-cycle-per-second dividing network would not cause the music quality to suffer unduly.

The 80-cycle-per-second dividing network is used to remove frequencies below 80 cycles per second from the music being recorded. This prevents the overlapping of low-frequency music and the control pulses. The same network is used in the playback circuit to divide the music into the playback loudspeaker circuit and the 45-cycle-per-second pulses into the flashing-lamp control circuit.

The circuit, Fig. 3, may be divided into four sections; a multiple-pole four-position switch, the 80-cycle-per-second dividing network, the pulse generator, and the flashing-lamp control unit. The switch connects the pulse generator into the recording circuit, the playback circuit into the flashing-light control, the pulse generator directly into the flashing-lamp control unit for manual nonautomatic operation, and an interconnecting circuit that removes the tempo-indicator equipment from the recorder-playback equipment.

Recording Circuit—A signal from the music source via the playback control panel appears on leads *5* and *4* (common) of *J2* and is connected through the selector switch *D2* to the 80-cycle-per-second dividing network. All low-frequency components of the music are attenuated. The high-passed components are mixed with the 45-cycle-per-second pulses in the "Y" pad and connected through the selector switch to the recording circuit through leads *5* and *4* of *J1*.

Automatic Playback Circuit—The signals from the disk appear on leads *3* and *4* (common) and are connected through the selector switch to the 80-cycle-per-second dividing network. The high-frequency components pass on through the selector switch to the playback control unit and eventually to the power-amplifier loudspeaker system. The low-frequency components are chiefly 45-cycle-per-second pulses inasmuch as the music recorded on the disk was previously high-passed. The pulses are attenuated by *P3*, the sensitivity control, and are adjusted to a level sufficient to operate the electronic control circuit.

The voltage from the rectified pulses is applied to the grids of the

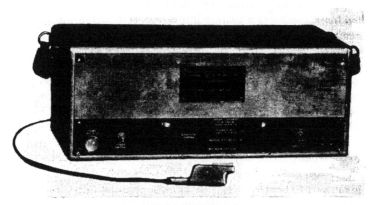

A. Unit and pulse key.

B. Chassis.

Fig. 4—Tempo indicator.

twin triode $V2A$-$V2B$. With no signal the conducting-cycle plate-cathode resistances of the triodes are low. These resistances in parallel with $R10$ and $R11$ reduce the voltages appearing on the control grids of the Type 2050 Thyratrons, $V3$ and $V4$, to less than 10 peak volts on positive cycles from each half of the power transformer *T3*. With the screen grids of *V3* and *V4* negatively biased, by *V5* and *V6* via *P4*, to approximately 40 volts the Thyratrons will not conduct on any part of the positive (550 peak volts) cycle. As the grids of the triodes *V2A-V2B* are biased negatively by the rectified pulse, the *V2A-V2B* conducting-cycle plate-cathode resistances increase. At an instant when the resistances have increased to over 30,000 ohms the Thyratron whose plate voltage is positive and over 300 volts will "fire" and conduct until the positive cycle ends. If the triode plate-cathode resistances are still high the opposite Thyratron will "fire" after its positive cycle, in about 33 degrees, has reached 300 volts. The Thyratrons alternately conduct until the rectified pulse voltage decreases below "firing" requirements. The alternating-current relay *S2*, in series with the *V3* and *V4* cathodes operates without chattering from the pulsating direct current through the Thyratrons. The relay contacts connect 115 volts alternating current into a multiple of Mazda S-14 lamps. These commercial 10-watt lamps have a low thermal inertia and are satisfactory as flashing lights.

Manual Tempo Indicator—Both recording and playback circuits are normalized through the selector switch and both the recording and playback lines are free of all attachments. However, through selections of the selector switch, the pulse-generator circuit and the tempo-indicator control circuit are connected. The pulses can be used to operate the flashing-lamp control circuit directly when testing or when it is desirable to eliminate or change the automatic control. The pulse inaugurating key is a microswitch in a casing designed to fit comfortably in the hand of the operator.

The unit, Fig. 4, is compact and self-contained. It is easily connected with the stage playback and recording equipment and within a few moments a tempo-indicating disk can be recorded and be ready for the control of the automatically flashing tempo lights. Its use has reduced the complexity and cost of many previously difficult scenes.

Editorial Policy of the Journal

THE BOARD OF EDITORS has asked for a statement of editorial policy to guide them in considering papers submitted for publication in the JOURNAL. The following brief statement may also be of interest to prospective authors as a guide in preparing papers for presentation at Society meetings and for publication.

As stated in the Constitution, the objects of the Society include "dissemination of scientific knowledge by publication." Thus, the main purpose of the JOURNAL is to provide members with up-to-date, reliable information on engineering and scientific developments in motion picture and allied fields. Its scope includes new developments in materials, processes, and equipment from the raw film to projection in theater or home.

The chief requirements of an engineering paper are that it be clear, concise, accurate, and, above all, that it describe the results of genuine engineering investigation.

To elaborate on the above points, clarity in writing means not only good English which is not ambiguous, but it means that the author should have a clear picture of what he wants to say. Generally it is well to state the objective which the paper is intended to cover, to include experimental or calculated data relating to the subject of the paper, and to draw logical conclusions from a study of the data. Clarity of expression is often achieved by first outlining the main points and then filling in the details. The author must assume some knowledge of the subject on the part of the reader since a full explanation of the fundamentals would make the paper longer than could be published. References to previously published material often serve in place of long explanations which would be required for a person having little or no familiarity with the subject.

Suitably drawn curves and diagrams are essential to most engineering papers. Standard electrical and mechanical symbols should, of course, be used in all drawings. It is important that lines and lettering be of such a size that they will be clear and legible when reduced to fit in a JOURNAL page.

The question of how long a paper should be can be answered only by saying that it should be no longer than is necessary to express clearly the information and conclusions applicable to the subject. In particular, a paper should not contain a long historical review, minor details, or information which properly belongs only in a supplier's catalog. The high cost of printing also makes it necessary to ask authors not to use any more illustrations than necessary, particularly avoiding multiple views of commercial equipment.

Accuracy is as essential in writing a paper as in performing an experiment. This means not only accurate statements but the use of accurate and reliable data, taken under carefully controlled condition and expressed in standard terms. Accuracy of statement includes implications as well as facts. A statement which implies more than the facts warrant has no place in an engineering paper. Quotations or statements attributed to others should, of course, not be used without permission. For the salesman a recital of the accomplishments of a new piece of equipment may be satisfactory, but an engineer wants to know how it works and why. The explanation may not be mathematically exact but it should be logical and should be based on sufficient actual test data so that it may reasonably represent the results to be expected by other users of the same apparatus and methods. A single test or observation should never be the basis of a scientific paper.

In conclusion it may be stated that the aim of the JOURNAL is to present scientific and engineering papers of the highest grade. To this end authors should keep in mind that papers should be written with the object of imparting engineering information and not for the sake of advertising a product. It is the duty of the Papers Committee and the Board of Editors to select papers which in their opinion most nearly fill the above requirements.

CLYDE R. KEITH
Editorial Vice-President

Report of SMPE
Progress Committee*

INTRODUCTION

A NY REVIEW OF PROGRESS within the motion picture industry during 1948 must take into account the much greater progress of
television throughout that period. The phenomenal rise of the new
medium of entertainment is a social change, which affects not only
motion pictures but all other forms of entertainment and education.
The year was marked by a series of dire predictions as to the closing
of thousands of motion picture theaters, along with estimates of the
millions of feet of film required for presenting television programs by
means of motion pictures. It was therefore natural that these
changes at home, together with limitations on the world-wide exchange of films, created a condition of economic uncertainty which
influenced technical developments, installations, and operations.

This summary of progress is presented under the headings of Photography, Sound Recording, Picture and Sound Reproduction, Television, and Standards, with further divisions relating to film widths of
35, 16, and 8 mm. However, these classifications are closely interrelated and there is increasing similarity in the processes and equipment used in the preparation of films for various purposes and of different dimensions. A large number of earlier developments were
carried forward during 1948, some of which are not discussed herein
because of having been covered adequately in previous reports.[1]

A. PHOTOGRAPHY

1. 35-MM

a. General

.- Progress in the fields of motion picture film, cameras, and studio
lighting during the past year was largely a consolidation of techniques and equipment previously announced. Considerable emphasis was placed on more efficient production methods, and numerous
time- and cost-saving procedures were developed. For example,
several of the Hollywood studios made widespread use of walkie-

* Original manuscript received by the Society, April 25, 1949.

talkie radio equipment for the control of boats, vehicles, trains, and so forth, while shooting on location.

In England, new silent cameras, improved lenses and lighting equipment were manufactured and put into extended use since the war.[2,3] The assimilation of these developments enabled excellent photography to be realized along with economical studio operations. A new form of production technique was introduced, called the "Independent Frame," which involved complete, mobile sets and background projection.[2] Many other countries throughout the world made definite advances in their motion picture production and photographic quality, in some cases with the encouragment or sponsorship of the government.

b. Color Processes

The feature picture "Eiffel Tower" was being produced in Ansco color in Hollywood, but the method of making prints was not announced.[4] Other preliminary or well-advanced work was carried on toward changing two-color processes into three-color, the adoption of one or more negative-positive processes, and other approaches toward the solution of the color bottleneck. The use of color in newsreels was introduced by Warner Brothers in a monthly release of a color sequence in the Pathé Newsreel. Applications of the principles of latensification, involving an additional exposure to the latent image for increased sensitivity and latitude, were used by Cinecolor as well as by others previously mentioned.[5]

c. Lighting Equipment and Techniques

Carbon Arcs—The use of synchronized venetian-blind-type shutters for controlling the light output of carbon-arc or incandescent lamps has increased.[6] These enable the studio to create many lighting effects which require accurate dimming, or changes in light level from a number of lamps operating in unison. There was increased use of motor-driven, telescoping parallels for adjusting the elevation of carbon arcs used in lighting color sets.

Incandescent Bulbs—A 750-watt R40 Type photographic bulb was developed for high-speed photography providing 75,000-beam candle power.[7] On film productions of a documentary nature such as "Naked City," overvolted standard lamps were used and equipment voltage control was worked out. These units are generally carried on location where the extreme in portability is indicated.[8]

Cadmium-Mercury Lamps—Further tests were conducted on compact, cadmium-mercury sources for studio-set lighting, but they are still considered in the experimental phase and have not been used in production in west coast studios.[9, 10] However, these sources, developed and first manufactured in England, were used in British studios last year in 2- and 5-kilowatt sizes.[2]

d. Camera Accessories

A new electrically operated camera crane, designed by the Research Council, was delivered to Hollywood studios late in the year. The Research Council also designed a new type of camera gear head and a dolly permitting the use of the above crane on location. Servo-operated repeat mechanisms were used quite extensively for the exact duplication of camera movements on various types of composite photography. General availability of such facilities enabled freedom of movement for cameras, even for takes of three[11] to ten minutes' duration.

e. Set Construction

There was increased use of extremely large photographic enlargements of exterior backgrounds for large sets, both in color and black and white. Twentieth Century-Fox developed a process for preserving and flameproofing foliage used on sets. The color of the foliage is not preserved but can be restored if necessary for color photography by the application of flameproof paints. Paramount developed a material called "Paralite" that can be used in place of ordinary plaster. It is made up of plaster and fiberglass and is applied with a special gun.

2. 16-Mm Photography

Motion picture photography directly on 16-mm film has been employed over a period of twenty-five years and it is estimated that at least 325,000 cameras for such film are now in active use in the United States.[12] This broad field includes large numbers of cameras used for home motion pictures as well as many professional applications, both with and without color and sound. Recent developments in equipment for the amateur field are characterized by improved performance and ease of operation at moderate cost.

There was a marked increase during 1948 in the production of films for various professional applications such as education, entertainment, business, and science. This work was stimulated by the requirements of television with many producers specializing in the making of animations, commercials, and short subjects for this field. Improved films and photographic equipment became available in greater quantities, having operational and performance characteristics similar to their 35-mm counterparts. Another significant change was the transfer to 16-mm work of many skilled personnel having 35-mm production experience. These trends enabled the production of many 16-mm educational and documentary films that were truly professional in their technical aspects. One outstanding example was a film produced by Jerry Fairbanks for the Dole Pineapple Company, which, with the aid of Kodachrome film and sound, depicted the growing and processing of the product. The success of wartime Government training films and other factors gave impetus to the production of more and better training films not only by the Services but for schools and colleges.

Remarkable progress has also been made in high-speed motion picture photography, and a Committee of the Society of Motion Picture Engineers was formed early in the year to further the knowledge in this field and assist in making the equipment more usable through portability and better performance.[13] The subject was covered in a symposium at the Washington Convention in October, and fourteen of the papers were published together as a JOURNAL supplement. They covered work done in recent years by the United States Army Air Forces, Naval Ordnance Laboratory, General Motors Proving Ground, and other organizations. In general, cameras employing a rotating prism operate over the range of 250 to 10,000 frames per second and the ultrahigh speed rates between 10,000 and 15,000,000 frames per second are usually provided by a form of strip camera operating only an instant.

A series of high-grade lenses for 16-mm cameras was announced by Eastman Kodak Company. These seven Ektar lenses cover the range of focal lengths from 15 to 152 mm in a common geometrical ratio of 1.6, with relative apertures of $f/1.4$ to $f/4.0$, respectively.[14]

3. 8-MM PHOTOGRAPHY

The first 8-mm motion picture films and cameras were introduced in 1932, and it is estimated that approximately 900,000 such cameras have been made.[12] These facilities, including 274,122 cameras manufactured in 1947,[15] are used primarily in the home on a highly personal basis, with initial and operating costs being important considerations. Recent progress in 8-mm cameras was generally directed toward simplifying consumer operations in exposure, threading, handling, or aids in cutting.

The Bell and Howell Filmo-Auto-8, in introducing the lift-out turret, has advanced a solution to enable reduction of the back clearance necessary for the lenses. The viewfinder visible footage indicator aids in judging scene length while shooting, and is superior to the audible-click system. Judging exposure while shooting is a similar

Fig. 1—DeJur Amsco D-300 8-mm camera.

problem, and in the Eumig C-3, 8-mm camera a solution was presented in the form of a built-in photoelectric exposure meter, semiautomatically coupled which enables the diaphragm to be varied continuously during a take.

Magazine loading generally has taken the lead over the more complex automatic-threading approach. The types of magazines currently in use prevent loss of frames on loading, but force the camera to rely on the precision of the magazine for accuracy of film focus. There are recurrent questions as to whether this is satisfactory for lens speeds faster than $f/2.5$ due to depth of focus problems.

The DeJur Amsco 8-mm camera (Fig. 1) is notable for the automatic scene-fade system operating on the trigger. This avoids trick printing never done by the amateur and removes the necessity for bulky special-purpose attachments. The desire for small size to enable continual carrying without bother or fatigue has led to the pocket-fitting shape of most of the new 8-mm cameras. The French EMEL 8-mm camera is of interest as a small 8-mm version of a 16-mm type of camera. The Briskin 8, Keystone K-40, Revere C-16, C-19, 60, and 70 represent cameras of known character in which the

Fig. 2—RCA Type PA142 film recording amplifier.

chief progress is the reduction in cost. The DeMornay Budd Automatic Eight Camera with battery drive was announced and is similar to the wartime 16-mm electric gun cameras. The Franklin President Dual Editor, being interchangeable between 8- and 16-mm with no sacrifice of essential performance, is of interest.

B. SOUND RECORDING

1. 35-Mm

A moderate amount of new recording equipment was introduced last year and broad progress was made in the widespread use in this country and abroad of facilities announced in 1947.[1, 16]

Supersonic-radio playback transmitters with miniature receivers and earphones that can be hidden in the hair or clothing of an actor, were used by several sound departments in Hollywood.[17] These systems proved useful and timesaving in playing back records or for cuing the actors without interfering with recording of dialog.

A new recording amplifier was brought out by the Radio Corporation of America for small studio and mobile installations that incorporates the facilities of voltage amplifier, compressor, high- and low-pass filters, dialog and film-loss equalizer, ground-noise reduction amplifier, and power amplifier in a single unit. (Fig. 2.) This amplifier provides an over-all gain of 69 decibels with an inherent fre-

Fig. 3—RCA MI-10332 35-mm magnetic recorder.

quency range of 20 to 15,000 cycles at the normal operating output level of +24 dbm.*   An electronic mixer-compressor by RCA for deluxe studio installations has been described.[18] A new method for matching variable-gain tubes also has been described[19] and is being used to insure dynamically matched pairs for compressor use.

Some daily prints of picture and sound were made on new safety-base stock by Hollywood laboratories, but its use as negative stock was limited by availability.[20] A large amount of engineering effort was devoted to magnetic coatings on 35-mm, safety-base film and

* Decibels with respect to 0.001 watt.

quantities were made available primarily by Du Pont and Minnesota Mining.[21]

A 35-mm magnetic film recorder was introduced by RCA for either portable or studio use.[22] (Fig. 3.) Operating at 18 inches per second, it provides for wide-range recording and excellent film motion. Recording and playback heads are provided and an erase head can be added if desired. The bias oscillator and playback preamplifier are mounted inside of the recorder.

A similar film recorder was introduced by Western Electric, which may be used for either the magnetic or optical method, thereby facilitating a transition period or re-recording operations in a small studio. High-quality re-recording and review-room machines were introduced on a field-trial basis. A 35-mm magnetic recording and reproducing unit was also demonstrated by Reeves Sound to the Atlantic Coast Section of the Society in November.

An adaptation unit for converting editing machines of the Moviola type was designed for reproducing 35-mm magnetic films with perforations. A method of making a magnetic sound track visible by means of a coating of extremely fine carbonyl iron dispersed in a liquid was also developed to check head alignment and to facilitate editing.[23]

By the end of the year most of the studios in Hollywood were equipped with at least one 35-mm magnetic recording machine capable of operation in synchronism with a camera or projector. Some studios gained additional operating experience through the use of tape recorders in applications where synchronism was not essential. Extensive laboratory tests and limited studio use have established that magnetic recording is of considerable importance for all types of work where re-recording is involved. Excellent frequency response up to 15,000 cycles has been obtained with an inherent ground-noise-to-signal ratio of 50 decibels or better. Ground noise does not appear to increase with film usage and the magnetic sound record is long lived. Other advantages include film re-usage, immediate playback, elimination of lightfast requirements, and simple operation. Important economies can be realized by the reduction of film and processing costs.

Re-recording operations at Warner Brothers[24] studio were simplified and reduced in cost by first combining up to twenty sound-effects tracks into a single reel of magnetic film. In the final re-recording operation, two magnetic sound tracks were made simultaneously: one containing all the speech, music, and sound effects and the other having only the combined music and sound effects. The latter track is then available for making 16-mm versions. as discussed below and for the use of the foreign department in combining the music and sound-effects track with a foreign-speech track.

Fig. 4—Western Electric RA-1417 16-mm re-recording machine.

2. 16-Mm Recording and Processing

There was unusual activity in this field, both in original recording and in re-recording from 35-mm or 16-mm films, which resulted in better sound quality than was obtained two or three years ago. The recording was done by using negative-positive methods and equipment previously announced or by the direct-positive method which eliminates the necessity of making a negative. A direct-positive variable-density recording technique was introduced by Western Electric[25] in which a 24-kilocycle bias was applied to the light valve along with the audio signal to reduce distortion and improve the volume range.

New reproducing equipment for re-recording from 16-mm films was introduced by RCA[26] (Coded PB-176) and by Western Electric.[27] (Fig. 4.) These machines provided the features of high-quality and excellent film motion characteristic of their 35-mm counterparts.[28, 29]

The use of the so-called 35-32-mm process[30] as a step in the release of 16-mm prints increased appreciably during the year. It employs special sound negative film 35 mm in width having 16-mm perforations along each edge. Two re-recorded sound tracks are placed near the center of the film by recording at 36 feet per minute in opposite directions, and then processed in standard 35-mm developing machines of the sprocketless type. Printing is done from the double-track sound negative along with the picture to 32-mm release print stock, is developed by standard positive processing except for rollers 32 mm in width, and then is split. This general method has the advantage of standard 35-mm processing equipment and control as well as locating the sound track in the center of the film where it is protected from rollers and sprockets. At the present time seven Hollywood studios can do this type of recording and at least four laboratories can do the film processing.

Improved printers for optically reducing 35-mm sound negatives to 16-mm prints were announced by RCA[31] and Eastman Kodak[32] and were used to a limited extent. It may be noted, however, that many 16-mm prints have been released which do not come up to present standards.

C. PICTURE AND SOUND REPRODUCTION

1. 35-MM

There was relatively little indoor theater construction and only a minor amount of modernization activity in this country last year. The installation of soundheads, amplifiers, and modern two-way loudspeaker equipment to replace outmoded and wornout equipment continued at a moderate pace. Some increase in modernization was noted toward the end of the year, which may indicate that more exhibitors plan to meet competition by better presentation of the film program.

The outstanding 1948 development in the 35-mm field of theaters and equipment was the enormous increase in the number of drive-in installations, which at the end of the year were variously estimated to number somewhere between 800 and 1000. Part of the popularity of the drive-in is due to the now almost universal use of individual in-car speakers, which eliminate the interference and sound-transit-time problems encountered in the first drive-ins using central-speaker equipment. Screen "presence" and "illusion" are generally satisfactory in spite of the displacement between picture and speakers.

Drive-in theater screens range up to 65 feet in width. To put a reasonably satisfactory picture on such screens, a trend developed toward the use of higher and higher powered light sources, faster lenses, and filters and blowers to cool the film at the projector aperture. The use of high-intensity arc lamps of the condenser type burning 150 to 170 amperes continued and reflector-type lamps were improved and their operating current ranges were increased. Late in the year the Motiograph-Hall reflector-type lamp with rotating positive and thermostatic positioning of the positive crater was announced. Its current range is 85 to 115 amperes and its high lumen output found ready acceptance in the drive-in field. Double-shutter projector mechanisms were widely used in drive-ins, and all manufacturers worked diligently to improve mechanical operation, light output, and aperture cooling.

Sound-equipment designs remained more or less standard except for use of large Class B output amplifiers to deliver the considerable amounts of audio power required by the hundreds of individual speakers of relatively low efficiency in the average installation. Motiograph brought out multiple-amplifier systems, with each amplifier serving only a small group of speakers for improved reliability and greater emergency protection, and also developed a system of lighting for the speaker-junction boxes to reduce collision risks. Toward the end of the year two manufacturers brought out speaker-heater combinations to extend the season in temperate climates and promote comfort in those having twelve-month seasons, but chilly nights. Heaters dissipate approximately 250 watts and have small blowers to distribute the heat and keep surface temperatures to more or less reasonable values.

During 1948 acetate safety-base 35-mm film for release prints came into limited use, principally for certain types of color films. The performance was generally satisfactory, though considerable difficulties with splices were observed, possibly because projectionists do not always recognize the safety-base film and hence do not use the special splicing techniques it requires. No relaxation in projection-room safety requirements was reported, which is to be expected so long as any nitrate film is in common use.

Release prints generally were of excellent quality with respect to both sound and picture during the year, with the exception of some of those made for reissued pictures. Sound on these was substandard in quality, and cases were reported where the original negatives were apparently so badly shrunk that picture frames failed to fill projector apertures completely, causing light streaks at the picture borders.

2. 16-Mm Reproduction

Figures just released by the Government show that 57,409 16-mm projectors were manufactured during 1947;[15] these and others were used last year by churches, clubs, schools, industrial organizations, etc. Numerous projectors having special picture-projection features for television also were produced. Although similar data covering new equipment produced in 1948 are not available, it is understood that quantities were comparable to the preceding year.

It appears that relatively few basic improvements in sound or picture reproduction became available last year. However, some new equipment was introduced which was more sturdy or quieter in operation. In several cases photoconductive tubes[33] were employed and there were numerous simplifications in design or manufacture for reducing cost. There was a slight trend toward arrangements for mounting larger projector lenses having characteristics which will improve picture quality. For example, a new series of projection lenses for professional 16-mm use was announced by Bausch and Lomb.[34] These Super-Cinephor lenses have a speed of $f/1.6$ and focal lengths of two inches to four inches in half-inch steps, will resolve 90 lines per millimeter, and provide practically uniform illumination of the screen.

Several manufacturers were engaged in the development of improved projectors to meet the specifications of the American Standards Association. It is understood that some of this work sponsored by the Government reached a status involving official tests of models and preparations for manufacture. Some progress was made toward standardizing the frequency characteristics of sound projectors. The subject was discussed in a paper[35] at the Washington Convention and an SMPE committee was established to study the problem.

D. TELEVISION

Millions of spoken and written words already have been devoted to the phenomenal rise of television. Its economic, social, and technical aspects were covered under the familiar heading "the impact of television." This section, therefore, is confined to a brief summary of the technical progress in television during 1948 as related to motion pictures.

At the end of the year there were forty-six television stations operating in approximately thirty cities, within reach of about one million receiving sets. Additional coaxial cable and microwave-transmission facilities were undergoing operating tests which would (by January 15, 1949) enable the area bounded by Milwaukee, St. Louis, Boston, and Richmond, Virginia, to receive network programs simultaneously. Transcriptions with picture and sound on film were of at least equal importance in the distribution of program material.

As mentioned earlier (16-mm photography) several new producers and studios concentrated on the production of films for this field. In addition, many film programs were made at the major stations by photographing the picture of a live program as presented on a kinescope tube and recording the sound either by the double- or single-film methods. Special 16-mm cameras as previously described by Eastman[36, 37] and Wall[38] were used extensively for converting the 30-image-per-second television picture to the 24-frame-per-second motion picture speed. Such facilities were used by major stations from ten to forty hours per week, and frequently consumed some 150,000 feet of film for picture and sound during such a period.

An analysis of the over-all photographic process involved in the use of iconoscope and orthicon pickup tubes, kinescope monitor tubes, and film characteristics was presented at the SMPE Convention in May.[39] A nonlinear electrical network was advocated in combination with the iconoscope for gamma correction along with a direct-positive photographic technique from a negative monitor picture. A complete system for 16-mm photography including high speed development was also described by RCA.[40]

Several of the major stations were equipped with special projectors to enable them to reproduce 16-mm and 35-mm films, but the 16-mm equipment was much more widely used. An intermittent, krypton-filled light source known as Synchrolite, which eliminates the need for a projector shutter, was introduced by General Electric.

In the field of theater television there was considerable progress in equipment and field experience with several successful large-screen demonstrations and a few commercial applications in this country during the year.[41] A 15- × 20-foot picture was shown by Warner Brothers in Hollywood for the meeting of the Society of Motion Picture Engineers and the National Association of Broadcasters, and a similar picture was demonstrated by Twentieth Century-Fox and RCA in Philadelphia on the occasion of the Louis-Walcott fight in June.[42] These two demonstrations of direct projection were made with reflective optical systems employing 42-inch spherical mirrors.

RCA demonstrated a 15- × 20-foot picture in September at the St. Louis convention of the Theater Equipment Supply Manufacturers Association using smaller equipment with comparable results. The optics had been reduced from the 42-inch, 500-pound mirror and its 21-inch glass lens to a 20-inch mirror and $15^{1}/_{2}$-inch molded-plastic lens weighing only 50 pounds. Subsequent developments further improved performance and made it possible to remove the optical barrel from the rest of the equipment. The smaller barrel (30 inches in diameter, 36 inches in length) is the only element of the equipment now required in the theater auditorium. It may be mounted 40 to 65 feet from the screen and with proper selection of lenses will project a 15- × 20-foot picture.

A 35-mm film-storage-projection system was installed at the Paramount Theater in New York City and used on several occasions.[41, 43] It consisted of television receiving equipment with a 10-inch cathode-ray tube, a special recording camera with an electronic shutter operating at 24 frames per second, forced processing and drying accomplished in about one minute, and a standard 35-mm projector.

E. STANDARDS

The work on this subject through 1947 was well described in a Report of the Standards Committee.[44] During 1948, the technical committees of the SMPE and the Research Council were active in preparing standards proposals to submit to ASA Sectional Committee Z22 for approval as American Standards. Eight standards in all were given formal approval by Z22 during 1948 and have been published for inclusion in the standards binder.

Five of these were new standards for 35-mm test films developed by the Motion Picture Research Council. Films meeting these specifications are now available from the Society or Research Council.

Two of the eight standards established dimensions for push-pull sound tracks ASA-Z22.69 and Z22.70. Since sound tracks of this type are not used on release prints, there probably is not the industry-wide interest in these standards as has been shown in other film-dimension standards. There was, however, a need for them because of the frequent exchange of original material among the various studios.

The last standard approved in 1948 dealt with the dimensions of theater screens ASA-Z22.29. This work is a revision of the 1946 edition and has been extended to specify not only screen size but recommended masking and placing of grommets in the screen border.

A number of other standards, while not as yet in the hands of Z22, represent considerable work of the various technical committees of the Society.

During the year, the whole problem of apertures for 16- and 8-mm projectors and cameras was reviewed by the 16- and 8-mm Committee. In these proposals, a complete new approach to the aperture problem has been taken and has now received the approval of the Standards Committee. Copies of these proposed standards have been published in the JOURNAL for a period of trial and comment.

Three new cutting and perforating standards for 32-mm film have also been developed by the Film Dimensions Committee. While films of this type have been in use since 1934, they never had been established as formal standards. Consequently, over the intervening years a number of changes have taken place in the dimensions. The values now proposed, however, are believed to be acceptable to all manufacturers and users alike. Copies of these proposals have also been published for trial and criticism.

A start was made also during 1948 to establish standard mounting dimensions for 16- and 8-mm camera lenses. While still in the formative stages, it is expected they will be completed and adopted as American Standards during the coming year. Similarly, a revised standard for both 16- and 8-mm splices has received considerable attention.

In the field of magnetic recording, ASA Sectional Committee Z57, of which the SMPE is cosponsor with The Institute of Radio Engineers, held two meetings. Although no standards were approved, a proposed method of determining flutter content of sound records was drawn up and has received the approval of the Sound Committees of both the SMPE and Research Council.

In conclusion, it is apparent that this is an extremely interesting period in the broad field of motion pictures and that remarkable technical progress was made during the year, which gives promise for the future.

The Chairman wishes to express his appreciation for the co-operation of the members of the Committee and the many other individuals and organizations who provided information and comments.

C. R. SAWYER, *Chairman*

J. E. AIKEN	W. L. TESCH
C. W. HANDLEY	J. W. THATCHER
R. E. LEWIS	R. T. VAN NIMAN
W. A. MUELLER	W. V. WOLFE

REFERENCES

(1) "Report of SMPE Progress Committee," J. Soc. Mot. Pict. Eng., vol. 50, pp. 523–542; June, 1948.

(2) W. M. Harcourt, "British progress in kinematograph engineering," British Kinematography, vol. 14, pp. 33–37; February, 1949.

(3) A. Howard Anstis, "Lens manufacture and design," British Kinematography, vol. 14, pp. 37–42; February, 1949.

(4) H. C. Harsh and J. S. Friedman, "New one-strip color-separation film in motion picture production," J. Soc. Mot. Pict. Eng., vol. 50, pp. 8–13; January, 1948.

(5) Hollis W. Moyse, "Latensification," Amer. Cinematographer, p. 49; February, 1949.

(6) "Report of the Studio Lighting Committee," J. Soc. Mot. Pict. Eng., vol. 51, pp. 656–666; December, 1948.

(7) R. E. Farnham, "Lamps for high-speed photography," J. Soc. Mot. Pict. Eng., vol. 52, March, 1949, Part II, pp. 35–42.

(8) Frederick Foster, "Packaged illumination," Amer. Cinematographer, p. 49; February, 1949.

(9) R. E. Farnham, "Mercury-cadmium lamps for studio set lighting," Amer. Cinematographer, p. 47; February, 1949.

(10) E. W. Beggs, "New developments in cadmium-mercury lamps for motion picture and television studio lighting," presented SMPE 64th Convention, October 25, 1948, to be published.

(11) Bart Sheridan, "Three and a half minute take," Amer. Cinematographer, vol. 50, pp. 304–305; September, 1948.

(12) "Are you one in a million?" Amer. Cinematographer, vol. 29, p. 240; July, 1948.

(13) John H. Waddell, "Foreword," J. Soc. Mot. Pict. Eng., vol. 52, March 1949, Part II, p. 3.

(14) Rudolph Kingslake, "New series of 16-mm camera lenses," J. Soc. Mot. Pict. Eng., this issue, pp. 509–522.

(15) Film Daily, vol. 95, no. 74, p. 6; April 15, 1949.

(16) R. E. Warn, "Recording equipment throughout the world," presented SMPE 64th Convention, October 27, 1948, to be published.

(17) Bruce H. Denney and Robert J. Carr, "Silent playback and public-address system," J. Soc. Mot. Pict. Eng., vol. 52, pp. 313–319; March, 1949.

(18) Kurt Singer, "High-quality recording electronic mixer," presented SMPE 64th Convention, October 27, 1948, to be published.

(19) Kurt Singer, "Preselection of variable-gain tubes for compressors," presented SMPE 64th Convention, October 27, 1948, to be published.

(20) Charles R. Fordyce, "Improved safety motion picture film support," J. Soc. Mot. Pict. Eng., vol. 51, pp. 331–350; October, 1948.

(21) R. Herr, B. F. Murphey, and W. W. Wetzel, "Some distinctive properties of magnetic-recording media," J. Soc. Mot. Pict. Eng., vol. 52, pp. 77–88; January, 1949.

(22) O. B. Gunby, "A portable magnetic recording system," presented SMPE 64th Convention, October 27, 1948, to be published.

596 SMPE Progress Committee

(23) *Sound Tips*, Bulletin 5, December 17, 1948, Minnesota Mining and Manufacturing Company.

(24) William A. Mueller and George R. Groves, "Magnetic recording in the motion picture studio," to be published.

(25) C. R. Keith and Vincent Pagliarulo, "Direct positive variable-density recording with the light valve," presented SMPE 64th Convention, October 26, 1948, to be published.

(26) Carl E. Hittle, "16-mm film phonograph for professional use," J. Soc. Mot. Pict. Eng., vol. 52, pp. 303–308; March, 1949.

(27) G. R. Crane, "Studio 16-mm re-recording machine," presented SMPE 64th Convention, October 26, 1948, to be published.

(28) M. E. Collins, "Lightweight recorders for 35- and 16-mm film," J. Soc. Mot. Pict. Eng., vol. 49, pp. 415–424; November, 1947.

(29) Wesley C. Miller and G. R. Crane, "Modern film re-recording equipment," J. Soc. Mot. Pict. Eng., vol. 51, pp. 399–417; October, 1948.

(30) Frank LaGrande, C. R. Daily, and Bruce H. Denny, "16-mm release printing using 35- and 32-mm film," J. Soc. Mot. Pict. Eng., vol. 52, pp. 211–222; February, 1949.

(31) J. L. Pettus, "Improved optical reduction sound printer," J. Soc. Mot. Pict. Eng., vol. 51, pp. 586–589; December, 1948.

(32) C. W. Clutz, F. E. Altman, and J. G. Streiffert, "35-mm to 16-mm sound reduction printer," presented SMPE 64th Convention, October 28, 1948, to be published.

(33) Norman Anderson and Serge Pakswer, "Comparison of lead-sulfide photoconductive cells with photoemissive tubes," J. Soc. Mot. Pict. Eng., vol. 52, pp. 41–48; January, 1949.

(34) A. E. Neumer, "New series of lenses for professional 16-mm motion picture projection, J. Soc. Mot. Pict. Eng., this issue, pp. 501–509.

(35) John K. Hilliard, "Co-ordination of 35-mm and 16-mm sound-reproducing characteristics," presented SMPE 64th Convention, October 27, 1948.

(36) J. L. Boon, W. Feldman, and J. Stoiber, "Television recording camera," J. Soc. Mot. Pict. Eng., vol. 51, pp. 117–126; August, 1948.

(37) Thomas T. Goldsmith, Jr., and Harry Milholland, "Television transcription by motion picture film," J. Soc. Mot. Pict. Eng., vol. 51, pp. 107–116; August, 1948.

(38) John M. Wall, "Television recording camera intermittent," presented SMPE 64th Convention, October 25, 1948, to be published.

(39) Fred G. Albin, "Sensitometric aspect of television monitor tube photography," J. Soc. Mot. Pict. Eng., vol. 51, pp. 595–612; December, 1948.

(40) Ralph V. Little, Jr., "Equipment for television photography," presented SMPE 64th Convention, October 25, 1948, to be published.

(41) "Theater Television," J. Soc. Mot. Pict. Eng., vol. 52, pp. 243–272; March, 1949.

(42) Roy Wilcox and H. J. Schlafly, "Theater installation, instantaneous large-screen television," J. Soc. Mot. Pict. Eng., this issue, pp. 549–561.

(43) Richard Hodgson, "Paramount Pictures system of theater television," J. Soc. Mot. Pict. Eng., this issue, pp. 540–549.

(44) "Report of SMPE Standards Committee," J. Soc. Mot. Pict. Eng., vol. 51, pp. 230–241; September, 1948.

Sixty-Fifth
Semiannual Convention

DURING THE WEEK of April 4–8, 1949, the 65th Semiannual Convention of the Society was held at the Hotel Statler in New York City. Seven hundred and seventy-five members and guests registered for the nine sessions which featured television, high-speed photography, and a 16-mm sound forum. In addition to the forty-seven technical papers that were presented, two speeches were given. Dr. Allen B. DuMont, who was the guest speaker at the Get-Together Luncheon, spoke on "The Relation of Motion Pictures to Television." On Thursday evening, Dr. Henry B. Hansteen delivered a popular lecture on "Nuclear Energy and Its Application."

The Luncheon on Monday was attended by 265. Mr. Sponable presided, and Jackie Miles gave a humorous talk. At the Cocktail Party and Banquet, there were 259 members and guests. During the Banquet, Past-President Ryder presented a plaque to Mr. Donald E. Hyndman in recognition of the great services he has performed for the Society over the years.

Mrs. Earl I. Sponable, assisted by Mrs. William H. Rivers and a committee of eight, provided an interesting program for the women guests at the Convention.

————————◆————————

Plaque Presented to
Donald E. Hyndman

IN RECOGNITION of his outstanding work as an officer and member of the Society, Donald E. Hyndman was presented with a plaque on April 6, 1949, during the 65th Semiannual Convention of the Society of Motion Picture Engineers. Not only were his efforts as Chairman of the Sustaining Membership Committee publicly acknowledged, but also the fact that unceasingly during the years Mr. Hyndman devoted himself to furthering the interests of the Society and in helping materially to increase its prestige, both technically and in its ability to serve its members and their industry.

Mr. Hyndman was born in Denver, Colorado, on April 14, 1904. He was graduated from the University of Denver in 1926 with a Bachelor of Science Degree in Chemical Engineering; during 1926 and 1927 he attended the University of Rochester.

In the summer of 1926 Mr. Hyndman was employed as assistant to Dr. K. C. D. Hickman in the Research Laboratories of the Eastman Kodak Company. In 1928 he went on a trip around the world

597

Donald E. Hyndman

for Eastman as Manager on one of the Cine Processing Department Cruises. Upon his return, he was a member of the Cine Processing Department and early in 1929 he returned to the Research Laboratories to prepare to become a member of the staff of the Motion Picture Film Department. In July, 1929, he was transferred to New York City as a member of the staff of the East Coast Division of the Motion Picture Film Department; in 1940 he was made Assistant Manager; and in 1946 Manager of the East Coast Division.

Mr. Hyndman is a member of many organizations, and he is a Fellow of the Society of Motion Picture Engineers. He was Engineering Vice-President of the Society from 1939 to 1945; President during 1945 to 1946; and Past-President during 1947 to 1948.

He is coauthor of "The Occurrence and Present Chemical Status of the Female Sex Hormone"; "Plastic Cellulose in Scientific Research"; "Automatic Silver Recovery from Hypo"; and has written numerous articles on motion picture engineering. Past-President Ryder prefaced his presentation of the plaque to Mr. Hyndman with the following remarks:

April 23, 1949, will be the 53rd anniversary of the first exhibition of motion pictures in a theater. On that evening in 1896, Thomas Armat operated a projector of his own design in Koster and Bial's Music Hall in New York City. Twenty years later, the Society of Motion Picture Engineers was founded to fill an important and growing need in the then infant industry.

It is a fitting tribute to the early objectives of our motion picture pioneers that in 1949 the Society is still young in spirit and is able to adapt itself to the changing needs of the motion picture art which now have been extended to include television, just as they were extended to include "talking pictures" twenty-three years ago.

Many of us who remember vividly the early growing pains that accompanied the introduction of sound motion pictures are now in the throes of attempting to orient properly the "equivalent arts of motion pictures and television," so that each will be better able to support the other as they grow, together.

In recent years, the Society has had its own gradual evolution and I am pleased to report that it has never been more adequately equipped to accept new responsibilities than it is today.

During the war years, our engineering committees formed the basis of the American Standards Association's War Committee on Photography, whose work is now a respected part of the history of motion picture standardization.

Toward the end of the war, it became evident that many of the temporary war standards could also serve our industry in peacetime as continuing American Standards, but that a great deal of work and expense on the part of the Society would be required. To get the job done properly, it was necessary to plan a wholesale reorganization of our Society headquarters facilities, increase the staff, and secure additional financing through a dignified but most persistent Sustaining Membership campaign.

Many of our members shared in this work of increasing the stature of our Society, helping it to arrive at its present position of prestige and value to the related arts of motion pictures and television, but the basic plan that was followed throughout this period of recent growth was primarily the work of one man, who also, during his terms as Engineering Vice-President, President, and Past-President, has given the continued guidance and calm mature counsel which made the plan fruitful. All of his associates hold him and his work in high esteem.

Donald E. Hyndman, more than any other single person, has been the guiding hand in our recent history of growth, and in recognition of his unselfish and outstanding personal endeavors, I would like to present this plaque.

Society of Motion Picture Engineers

presents this

Special Award

to

Donald E. Hyndman

in recognition of

his unselfish, outstanding personal endeavors as an Officer of the Society, and his continued enthusiastic guidance in improving the prestige and financial stability of the Society by increased participation of Industry

in

Sustaining Membership

in the

Society of Motion Picture Engineers

April 6, 1949

QUESTIONNAIRE RESPONSE

EARLY IN JANUARY, questionnaire cards were sent to the membership in order to learn what topics are of greatest interest to our readers, which of our readers might prepare papers for publication, and to have our readers suggest the names of prospective authors and possible topics.

On March 31, an analysis was made of the cards returned, and the response was most gratifying. Of the 473 cards received at that time, there were 776 requests for certain types of material; eighty-five authors signified their willingness to prepare manuscripts for consideration; and the names of twenty-one others were suggested.

All of the cards have been forwarded to Dr. Norwood L. Simmons, chairman of the Papers Committee, who will utilize the information given in the solicitation of manuscripts for presentation at Conventions and for publications in the JOURNAL.

There are listed below, in the order of number of requests, the various topics which our readers submitted.

Subject	Requests	Subject	Requests
1. Television	106	15. Studios	17
2. Color	77	16. Production	16
3. Sound	66	17. Cinematography	15
4. 16-mm	63	18. Acoustics	13
5. Processing	55	19. Sound Reproduction	13
6. Equipment	51	20. Theater Television	9
7. Sound Recording	47	21. 35-mm	8
8. Miscellaneous	36	22. Education	6
9. Lighting	34	23. High-Speed Photography	6
10. Magnetic Recording	30	24. Projectors	6
11. Theater	30	25. Animation	3
12. Projection	25	26. Editing	3
13. Films	20	27. Historical	3
14. Optics	18	Total	776

The Editor wishes to thank the members of the Society for their co-operation and to assure them that every effort will be made to give our readers the type of JOURNAL which they desire.

HELEN M. STOTE
Editor

PUBLIC RELATIONS

With the work of the Society extending even further into High-Speed Photography and Television, as well as into the design of motion picture theaters and many other fields, an adequate public information program is a necessity. To present the Society in its proper light to other industries and to the American public, a Public Relations Committee, under the Chairmanship of Irving B. Kahn and including

> Harold Desfor
> R. B. Austrian
> R. T. Van Niman
> Peter Mole

has been appointed to advise the President, Mr. E. I. Sponable. They will review the public relations work of Society headquarters, together with the Executive Secretary, and will implement a conservative but timely program of reports, publicity releases, articles, and so forth.

———————◆———————

Section Meeting

Central

Seventy-four members and guests were present at the March 15, 1949, meeting of the Central Section. This meeting was session nine of the Conference and Production Show sponsored by the Chicago Technical Societies, March 14 to 17.

C. E. Heppberger outlined the plans for the Central Section Regional Meeting to be held in Toledo, Ohio, on June 10, 1949.

The first paper on "High-Speed Photography in Industry" was delivered by its author, Richard O. Painter of General Motors Proving Ground. Mr. Painter reviewed the features of present 16-mm high-speed cameras and described their uses at the Proving Ground. With slides he showed typical arrangements for photographing the various points being studied. Two reels of very interesting film showing some of the studies concluded the presentation.

"New Developments in X-Ray Motion Pictures," by C. M. Slack, L. F. Ehrke, C. T. Zavales, and D. C. Dickson of Westinghouse Electric Corporation, was presented by J. H. Terhorst of the X-Ray Division of Westinghouse in Milwaukee.

Current Literature

THE EDITORS present for convenient reference a list of articles dealing with subjects cognate to motion picture engineering published in a number of selected journals. Photostatic or microfilm copies of articles in magazines that are available may be obtained from The Library of Congress, Washington, D. C., or from the New York Public Library, New York, N. Y., at prevailing rates.

American Cinematographer
30, March, 1949

A New, Vest-pocket Color Temperature Meter (p. 85) D. NORWOOD

A. S. C. Inaugurates Research on Photography for Television (p. 86) V. MILNER

The Cinematographer's Place in Television (p. 87) J. DE MOS

New Lens Testing Method May Improve TV Picture Quality (p. 88) R. B. HARTWELL

Planning the 16-mm Commercial Film (p. 94) C. LORING

International Projectionist
24, 3, March, 1949

The Present Status of Theatre TV (p. 10)

British Kinematography
14, 2, February, 1949

British Progress in Kinematograph Engineering .(p. 33) W. M. HARCOURT

Lens Manufacture and Design (p. 37) A. H. ANSTIS

The Laboratory and 16 mm Colour (p. 43) J. H. COOTE

Ideal Kinema
15, 164, March 10, 1949

What Standardisation Has Done for Kinematography (p. 17) R. H. CRICKS

International Photographer
21, 3, March, 1949

Cameras of Yesteryear. Pt. 3 (p. 16) W. W. CLENDENIN

Meetings of other Societies During 1949

August—
Institute of Radio Engineers West Coast Convention

August 29 through September 1
San Francisco, California

September—
Illuminating Engineering Society National Technical Conference

September 19 through September 23
French Lick, Indiana

October—
Photographic Society of America Convention

October 19 through October 22
St. Louis, Missouri

Journal of the
Society of Motion Picture Engineers

VOLUME 52   JUNE 1949   NUMBER 6

Subscription to nonmembers, $10.00 per annum; to members, $6.25 per annum, included in their annual membership dues; single copies, $1.25. Order from the Society's General Office. A discount of ten per cent is allowed to accredited agencies on orders for subscriptions and single copies. Published monthly at Easton, Pa., by the Society of Motion Picture Engineers, Inc. Publication Office, 20th & Northampton Sts., Easton, Pa. General and Editorial Office, 342 Madison Ave., New York 17, N. Y. Entered as second-class matter January 15, 1930, at the Post Office at Easton, Pa., under the Act of March 3, 1879.

Society of Motion Picture Engineers

342 Madison Avenue—New York 17, N. Y.—Tel. Mu 2-2185
Boyce Nemec . . . Executive Secretary

OFFICERS
1949–1950

PRESIDENT	PAST-PRESIDENT	SECRETARY
Earl I. Sponable	Loren L. Ryder	Robert M. Corbin
460 W. 54 St.	5451 Marathon St.	343 State, St.
New York 19, N. Y.	Hollywood 38, Calif.	Rochester 4, N. Y.

EXECUTIVE VICE-PRESIDENT	EDITORIAL VICE-PRESIDENT	CONVENTION VICE-PRESIDENT
Peter Mole	Clyde R. Keith	William C. Kunzmann
941 N. Sycamore Ave.	120 Broadway	Box 6087
Hollywood 38, Calif.	New York 5, N. Y.	Cleveland 1, Ohio

1948–1949   1949

ENGINEERING VICE-PRESIDENT	TREASURER	FINANCIAL VICE-PRESIDENT
John A. Maurer	Ralph B. Austrian	David B. Joy
37-01—31 St.	1270 Avenue of The Americas	30 E. 42 St.
Long Island City 1, N. Y.	New York 20, N. Y.	New York 17, N. Y.

Governors
1949

1948–1949	James Frank, Jr. 426 Luckie St., N. W. Atlanta, Ga.	1949–1950
Alan W. Cook	William B. Lodge	Herbert Barnett
25 Thorpe St.	485 Madison Ave.	Manville Lane
Binghamton, N. Y.	New York 22, N. Y.	Pleasantville, N. Y.
Lloyd T. Goldsmith	William H. Rivers	Fred T. Bowditch
Warner Brothers	342 Madison Ave.	Box 6087
Burbank, Calif.	New York 17, N. Y.	Cleveland 1, Ohio
Paul J. Larsen	Sidney P. Solow	Kenneth F. Morgan
508 S. Tulane St.	959 Seward St.	6601 Romaine St.
Albuquerque, N. M.	Hollywood 38, Calif.	Los Angeles 38, Calif.
Gordon E. Sawyer	R. T. Van Niman	Norwood L. Simmons
857 N. Martel Ave.	4431 W. Lake St.	6706 Santa Monica Blvd.
Hollywood 46, Calif.	Chicago 24, Ill.	Hollywood 38, Calif.

Magnetic Recording
in the Motion Picture Studio

By WILLIAM A. MUELLER and GEORGE R. GROVES

WARNER BROTHERS PICTURES, BURBANK, CALIFORNIA

Summary—This paper deals with the use of magnetic recording in the production of motion pictures and describes the improvements and economies which magnetic recording effects. Its use is considered in original production recordings, talent testing and coaching, playbacks, re-recordings, foreign versions, publicity recordings, reverberation control, anticipated noise reduction, newsreel single-system recordings, and "electrical printing."

D URING 1948, the art of recording on a magnetic medium has received a great impetus, largely because of the appearance on the market of a number of nonprofessional types of magnetic recording equipment. In other fields of sound recording, such as film and disk recording, it has been the professional equipment that has been the forerunner of popular equipment for the amateur. In the case of magnetic recording, this order of things has been reversed. As a result, the professional recording engineer in the motion picture studio has been obliged to put this new recording medium to work using inadequate equipment while awaiting the development of more precise and standardized equipment.

Whenever absolute synchronism between recorded sound and picture is required, professional-type magnetic recording machines are now being designed and put to use. These are referred to as "synchronous equipments." In some cases, existing 35-mm film recording machines have been converted to record on both 35-mm photographic film and 35-mm nitrate-base film coated with ferrous oxide for magnetic recording. Tests are in progress pointing to future use of 35-mm acetate film base for both magnetic and film recording.

In the sound department of Warner Brothers-First National Studios, many uses have been made to date of both the nonsynchronous and synchronous magnetic recording facilities. Among the uses to which the nonsynchronous machines have been put, we might cite the following as typical examples.

In one production, the unusual technique was employed of shooting each reel as a complete 1000-foot take. This required intensive rehearsing in order to obtain perfection, not only on the part of the actors, but also in the complicated mechanics of camera movement, movement of props, and movement of various parts of the set structure to allow unrestricted action through the four rooms constituting the set. During each rehearsal, a magnetic-tape recorder was used to record the dialog and any comments made by the director to the cast and crew. By playing back such a recording after each rehearsal, many hours of duplicated effort were saved.

Instructions regarding reading of lines, positions, and movements of the actors, instructions on the almost continuous movement of the camera dolly, with dialog cues for the start and stop of each movement, specific cues to the electricians for light changes, instructions and cues to the propmen for shifting position of pieces of furniture and other props, cues for the sliding in and out of movable walls—all such information which could easily be forgotten or misunderstood during a scene ten minutes long—easily could be reviewed and memorized by the crew members and cast. Time was saved and confusion minimized.

Simultaneously with each 1000-foot take recorded on film, a duplicate recording was made on the paper tape which could be played back immediately. Where such complicated scenes of 1000 feet in length were being shot, this provided the director with an immediate and very necessary check on the quality of the performance without having to wait at least twenty-four hours for the film recording to be processed.

In some productions where an actor has to memorize long speeches, say five or six minutes long, the paper tape has been used to excellent advantage in aiding the actor to perfect his delivery before the scene is shot. A particular instance of this occurred in the filming of a picture in which the star had a six-minute speech to deliver in a courtroom. He was able to record his speech as often as he desired in the quiet of his dressing room and rehearse himself to the point where he could unhesitantly give a polished performance before the camera. In the same scene the tape recording was played back to the people in the courtroom for audience-reaction shots, thus eliminating much fatiguing repetition of the speech from the star himself. In this same production, the director made excellent use of the tape recorder in preproduction rehearsing of the whole cast. The ability to record and play back immeditely gave all members of the cast a chance to preview their own performances. This resulted in both production time saved and improved performances.

In the filming of another production, a magnetic-tape recording was made simultaneously with each take on film. It was found that many of the star's best and funniest scenes were the result of spontaneous ad libbing, and no better means could be found than the paper-tape recorder to make a record of these scenes that could be used for immediate reference. The tape recorder was consequently used by the script clerk on this company so that he could transcribe any desired section and have it immediately available for the shooting of different camera angles of the particular scene involved.

As is well known, it is the practice in shooting musical numbers, particularly vocals, to record the music prior to the photographing of the scene. The "prescored" music is played back to the actors, and they mouth in synchronization to the played-back sound while the picture is being photographed. The sound has heretofore been played back from a disk record or from film. In many productions, particularly westerns, where the actor may be traveling along, say on horseback while singing, it is necessary to move the playback equipment along with the camera over considerable stretches of rough terrain. In some cases, film playback has been the only practical means. This is expensive since it involves the making of sound-track prints especially for this purpose which become mutilated with start marks and worn with use. Consequently, they have to be discarded and are of no further use after the scene is shot. By using magnetic film for this purpose, the cost of film stock and laboratory processing is saved. The magnetic film can be marked in the same way as photographic film for start marks and cues and, when the scene has been shot, the film can be erased, cleaned, and put to future use.

In the recording of music, multiple-channel scoring is used a great deal so the balance between the vocalist, the chorus, and the orchestra, or even various sections of the orchestra, can be varied at will when the picture is finally edited and is being re-recorded. This has been done in the past on separate photographic film recording machines. The development of suitable magnetic-recording equipment placing a number of magnetic sound tracks on the same piece of film will greatly simplify this process, reducing substantially the equipment and personnel required and effecting large economies in film consumption.

For some time past, it has been the practice of the publicity department to make radio transcription disk recordings of interviews with the stars appearing in their pictures. These disks receive nation-wide radio playing time. The making of these recordings directly on acetate disks, caused many difficulties and inconveniences which have been overcome by the use of paper-tape recording. For instance, the making of disk records involved the use of a complete disk recording channel which was frequently in use on music-scoring work at the time it was requested for publicity recordings. Frequently all the actors required for the publicity recording were not available at the same time and this made editing necessary. In some cases, changes of picture titles or script changes made between the time of recording and time of release of the disks made editing again necessary. In these cases where the recording had to be made in sections, or where editing was necessary, the publicity recordings were first made on film and later transferred to disk, a costly operation. By using the paper-tape recorder all of these difficulties have been eliminated. The tape recorder, being small, compact, and portable, can be taken to a production stage where an actor might be working on a picture, and the publicity recording is made in a dressing room between production setups. In this way, no production time is lost, and it is no longer necessary to disturb the production recording setup or take the actor away from the stage on which he is working. The recording can be played back for an immediate check and edited. No film-processing charges are incurred and very fast service can be given to the publicity department since the finished disk can be made by re-recording from the tape in minimum time.

Considerable use of the tape recorder has been made in training young acting talent. A special school is maintained at the studio where young artists are taught the various phases of voice culture, delivery, deportment, and such other accomplishments as are necessary to fit them for appearance before the camera. The tape recorder, with its simplicity of operation and immediate playback facilities, has proved ideal for this type of training. Of course, the ability to erase the magnetic tape and use it over and over again makes it most economical for talent training and rehearsal uses.

The nonsynchronous magnetic recording machine has proved very useful in recording many kinds of sound effects. Its small size and portability make it ideally suited for recording in airplanes in flight, in speedboats, automobiles, horse-drawn vehicles, and so forth. In the recording of planes in flight from the ground, the very short starting time of the portable magnetic recorders has made it possible to record the sound of high-speed planes which would have been lost with slower operating equipment. In the recording of animals and birds, it is often necessary to record for long periods of time before the animal or bird makes the desired kind of sound. The recording engineer has no qualms about the cost of allowing his recorder to run for thirty minutes or more when using magnetic tape. On one occasion the tape recorder ran continuously for half a day to obtain a satisfactory recording of the cooing of a couple of doves. The desired part of the recording only lasted thirty seconds but nothing was lost since all the rest of the tape could be erased and used over again. Imagine the cost of doing such a thing with film recording.

In the recording of sound effects having a steep wave front, such as gun shots and explosions, the magnetic tape has proved ideal. Its high signal-to-noise ratio permits excellent volume range for these sounds and, most important, since no noise-reduction equipment is necessary, there is no effect of the initial impact being clipped off by noise-reduction shutters, as occurs in film recording. The result is a very realistic reproduction of these sounds.

Where the more elaborate and expensive synchronous magnetic-tape equipment has been available, it has been the answer to a long-felt need in the recording of sound for motion pictures shot in foreign locations, where film-laboratory facilities are very poor and quite inadequate for sound-track processing and printing. In such cases, film sound track is usually shipped back to the United States for processing, and it might be weeks before a director would receive a report on his product. By the use of magnetic recording, he has an immediate check on his sound, and production can proceed without that feeling of shooting in the dark.

Synchronous magnetic recording has been put to continuous use by the Warner Brothers re-recording department as a time, film, and money saver. Many pictures require the combining of as many as twenty sound-effects tracks into a single reel. This requires the skilled services of several re-recording mixers and considerable rehearsal time. The practice now is to combine any desired number of these effects tracks into a single composite track, magnetically recorded. In this way, the number of tracks to be handled in the final re-recording operation is considerably reduced. This results in more efficient operation of the re-recording personnel, much time saved, and a better product as the end result. The use of magnetic recording enables this to be done without additional cost beyond the initial investment in oxide-coated film stock. Since this stock can be used over and over again, its cost per picture over a year's production is negligible.

When a production is put through the final re-recording operation, two re-recorded sound tracks are made simultaneously; one carries all the speech, music, and sound effects and the other carries only the combined music and sound effects. This music and effects track is then available for the making of 16-mm versions and for the use of the foreign department in making foreign versions.

In making the 16-mm versions it becomes a simple job of combining the original speech track and the one music and effects track in the proper balance and with the appropriate volume range for 16-mm reproduction.

The foreign department likewise has the simple job of combining only two tracks, namely, the music and sound-effects track and the foreign-speech track. All Spanish and Portuguese versions are re-recorded in the Hollywood plant and, consequently, no film is shipped to these countries until the dubbing operation is completed, at which time the shipment of final composite sound-and-picture prints is made.

The current cost of magnetic film is $20.00 to $30.00 per thousand feet cheaper than the cost of 1000 feet of processed sound-track negative, plus 1000 feet of print therefrom. It can be seen, therefore, that large savings can be made by recording the combined music and sound-effects tracks to magnetic film which, after completion of the job, can be erased and used over again.

The fact that magnetic recording has now been developed to a point where it is every bit as good, or even better, than film recording, makes it adaptable to a number of special uses in connection with photographic film recording. Most noteworthy of these is probably its use to obtain anticipated noise reduction on the film sound track. By recording or re-recording initially onto magnetic film and then reproducing the sound therefrom by two reproducing heads spaced a specified distance apart, the output from one head can be made to operate the film-noise-reduction equipment in advance of the signal modulating the recording galvanometer or light valve. No appreciable deterioration of sound quality occurs in this transfer and this is more than offset by the elimination of noise-reduction clipping at the beginning of words.

The use of two reproducing heads, placed a short distance apart, on one magnetic sound track has been used effectively to produce "reverberation" effects. All that is required to accomplish this is a small continuous loop of magnetic film on which the sound is recorded by one head, then reproduced by two staggered reproducing heads, and then erased by a fourth erase head. Such a loop can be run at fairly high speed to permit spacing of the two reproducing heads. By varying the spacing between the two reproducing heads and mixing their outputs together in varying degrees, effects of reverberation and echoes can be obtained.

In the single-film type of system, as used by newsreel companies, in which the sound and picture negative images both appear on the same strip of film, it has been impossible to obtain optimum sound-and-picture quality because both the sound and picture images require a different type of development for best results. A solution to this problem now appears in the production of a film carrying magnetic coating of normal sound track width on one edge. The remainder of the film carries photographic emulsion for picture taking. The magnetic coating is unaffected by immersion in photographic solutions and as a result, the picture can be given any desired processing without affecting the magnetic sound track. There is, therefore, a single system now capable of giving optimum sound-and-picture quality.

In certain foreign countries, mainly Italy and Spain, where American pictures are shown, picture prints are mutually exchanged between countries in the interest of economy, particularly where color prints are concerned. In order to present these pictures in the appropriate language, without reverting to the use of a different picture print for each language, it has been customary to use a process for removing the narrow strip of emulsion from the film that carries the sound track and then recoat this strip with a new emulsion upon which the new language sound track can be printed. It can be seen that if and when the use of magnetic sound track becomes universal for motion picture release prints, the problem of changing the language on any picture print will be an easy matter of erasing one language and replacing it with another by a simple re-recording process.

Since the signal-to-noise ratio of magnetic recording is high, it is quite possible that the magnetic recorder may prove a useful tool in the making of acoustic and noise measurements. It could well serve as a permanent record of such tests as the measurement of camera noise, process projector noise, stage noise level, traffic noise level in specified locations, and many other instances where a ready reference of noise level would be a valuable aid in the comparison of equipments and determination of the suitability of equipments and locations for recording purposes.

There is considerable work being done on "electrical printing" whereby direct positive photographic sound tracks are produced from original magnetic recordings by a simple re-recording process. The advantage is that this eliminates all photographic processes excepting the development of the final print and avoids the flutter inherent in all present sound printing machines. The result is a superior sound record.

"Electrical printing" has also been put to use by the newsreel companies for the high-speed production of the duplicate sound-track negatives required for rapid release printing.

Therefore it can be seen that there are many applications of magnetic recording to motion picture sound recording work at the present time. As magnetic recording devices and materials become better and are standardized, and as more reproducing equipments become available, these uses will be extended until the time will probably arrive when magnetic recording eventually will replace the photographic method entirely.

Portable
Magnetic-Recording System*

By O. B. GUNBY

RCA VICTOR DIVISION, HOLLYWOOD, CALIFORNIA

Summary—This paper reviews the progress in synchronous magnetic recording since it was first demonstrated in May, 1948. Design specifications
for portable applications are discussed.

AT THE 63RD Semiannual Convention of the Society in Santa Monica
on May 18, 1948, there was demonstrated a synchronous magnetic recorder using 35-mm perforated magnetic stock operating at
the standard film speed of 90 feet per minute. So far as is known
this represented the first public demonstration of magnetic recording
providing synchronous operation and over-all performance characteristics comparable to photographic-film-recording equipment.

The enthusiastic and continued interest in magnetic recording that
followed this demonstration demanded immediate investigation as
to how it could best be made available to the industry for a thorough
trial under actual operating conditions. Many helpful suggestions
were received from the Hollywood motion picture studios. As the
result of these suggestions it was decided to provide modification
kits for existing RCA PR-23 recorders and associated equipment that
would permit high-quality magnetic recording without interfering
with the normal function of the equipment for photographic recording. This procedure, it was believed, would permit the exploration
of techniques and provide professional performance in accordance
with Hollywood engineering practice, and at the same time impose
minimum requirements for space and new equipment.

Further conferences with studio sound department personnel suggested the following basic specifications:

1. The recording material should be an iron-oxide coating placed
on 35-mm safety stock having approximately the same mechanical
properties and durability as the base used at present with photographic film.

2. The film speed should be 90 feet per minute.

* Presented October 27, 1948, at the SMPE Convention in Washington.

3. The magnetic heads should be installed in photographic recording machines in such a manner that they would not impair their immediate use for photographic recording.

4. The amplifiers supplied as part of the photographic channel should also be used where possible in magnetic recording and suitable switching provided so that when desired they can be used immediately for photographic recording.

5. The magnetic track should be located in a general position corresponding to that of a photographic sound-track print.

6. The magnetic track should be moved far enough toward the center of the film to minimize disturbances caused by the film perforations.

7. The magnetic record head should produce a track approximately 0.200 inch wide and the playback head should scan a track width of approximately 0.184 inch. These dimensions correspond to those being used for 200-mil push-pull track. The erase head should erase an area approximately 0.208 inch in width.

These basic specifications made it possible for a photographic recording channel to be modified, so that it can be used either for magnetic or photographic recording without doing more than throwing the appropriate switches and threading the desired film stock in the recorder. New items required are the magnetic heads, radio-frequency bias oscillator, playback preamplifier, and installation of mechanical elements in the recorder.

Several recorders were modified in this manner and are being used in Hollywood. Another is being used in a New York studio to determine the potential capabilities of this recording method in sound motion picture production. Current reports from these machines indicate that the over-all quality of the magnetic recordings made is at least equal to that obtained by the photographic process. Additional operating experience with these recorders and others to be installed shortly, should provide studios and producers with the "know how" and techniques necessary to apply magnetic recording to motion picture production purposes.

Discussions with sound department personnel also indicated the need for a truly portable double-film recording system that would permit sound recording at locations inaccessible to the present equipment used for this purpose. Ideas collected from the various studios for this equipment were found to be in close agreement. The general requirements for such an equipment were as follows:

1. Equipment small and light enough to be carried easily in a plane, on shipboard, in a Pullman compartment, or in the trunk of a passenger car.
2. Operation by means of portable storage batteries.
3. Performance equivalent to that obtained by present-day photographic film-recording trucks.
4. Simplified operation.
5. Minimum power drain.

As the study of the best way to meet these requirements continued it became increasingly apparent that the use of magnetic recording in this equipment would give the necessary performance and would also assist materially in meeting each of the other objectives. The advantages of magnetic recording for a portable equipment of this type include the following:

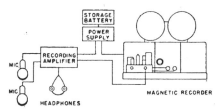

Fig. 1—PM61 portable magnetic-recording system.

1. Recording optical system not required.
2. Exposure lamp, power supply, and associated controls not required.
3. Noise-reduction amplifier and associated power supply not required.
4. Volume compressor and associated power supply not required.
5. Recording machine need not be of lighttight construction.
6. Lighttight film magazines not required.
7. Lighttight film-loading facilities not required.

Fortunately, many of the requirements for a portable magnetic-recording system had been anticipated in recent RCA designs, hence standard equipment could be used with very little modification as component units. The resulting over-all assembly, designated as the PM-61 portable magnetic-recording system is shown in Fig. 1.

The two microphones with their associated suspension mountings, stands, or booms are identical with those supplied on studio-type recording channels. Likewise, the monitor headphones are the standard RCA high-fidelity type.

The recording amplifier is a compact four-stage unit thoroughly field-tested in newsreel work. It provides for low-level mixing of two 250-ohm inputs, 105-decibel gain, and +22 dbm* output at 500/600 ohms. The vibrator-type plate-power-supply unit, storage battery, and battery case have been field-tested in newsreel systems. The storage battery consists of two independent sections, one for the tube heaters and one to operate the plate-power-supply unit.

The magnetic recorder consists of the new RCA 35-mm lightweight recorder, Model PR33, equipped with the necessary facilities for magnetic recording. This recorder has been described previously.[1] Provision was made in the original design for the addition of magnetic-recording facilities, hence no difficulty was experienced in properly placing the recording and play-

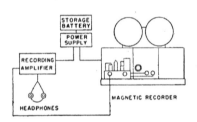

Fig. 2—PM61 connected for playback through monitor headphones.

back heads. After careful deliberation, no erase head was included because (1) studios prefer to use a controlled centralized erasing procedure to prevent accidental losses of recorded material in the field, and (2) film manufacturers will supply the raw stock free of magnetic signal or imperfections.

If desired, however, erasing facilities may be added as an optional feature.

Facilities are included for the use of standard 1000-foot reels with this equipment. This convenient arrangement is possible inasmuch as magnetic film imposes no requirements for lighttight construction or fire-hazard protection. Provision can be made for the use of standard lighttight magazines, if desired, in order to give additional protection from dust and dirt when operating under unfavorable conditions.

The high-frequency bias oscillator is mounted in the enclosure at the left of the film compartment together with a playback preamplifier and its associated equalizer. This preamplifier provides an output level of approximately −27 dbm* and a frequency response suitable for reproducing the playback signal through a voltage amplifier and power amplifier having a flat frequency response.

* Decibels with respect to 0.001 watt.

Rewinding, to permit immediate playback, can be accomplished by threading the film directly from the take-up reel to.the feed reel, and throwing the motor switch to the "rewind" position. The rewind speed is approximately 225 feet per minute, hence a 400-foot film can be rewound in approximately one and three quarter minutes. Playback through the monitor headphones is accomplished by connecting the output of the playback preamplifier to the recording

Fig. 3—M1-10332 portable synchronous magnetic recorder.

amplifier through one of the microphone input circuits as shown in Fig. 2. If playback through a loudspeaker is desired, a suitable power amplifier and loudspeaker should be provided.

Any appropriate RCA recorder motor can be used, but for general applications, a multiduty type to be operated from a 96-volt storage battery will be supplied. This motor can also be operated from a 220-volt, 3-phase, 60-cycle power supply.

The film motion is comparable to that of the corresponding photographic recorders which have less than 0.1 per cent total flutter.

The over-all frequency response is flat ±1 decibel from 80 to 8000 cycles. Signal-to-noise ratio is 57 decibels, weighted in accordance with the 40-decibel equal-loudness curve found in Z24.3-1944 as published by the American Standards Association.

Fig. 3 shows this truly professional, lightweight synchronous magnetic recorder threaded for normal operation. It is believed that it offers many advantages over corresponding photographic recording equipment wherever the utmost in portability and simplicity of operation is required.

REFERENCE

(1) M. E. Collins, "Light weight recorders for 35-mm and 16-mm film," J. Soc. Mot. Pict. Eng., vol. 49, pp. 415–425; November, 1947.

————————◆————————

Television in the SMPE

The interest of Society members and motion picture engineers generally in the new field of television was shown in our reader-preference survey conducted earlier this year by the JOURNAL editor and by the turnouts at the television sessions of the last several conventions. The Society's work in that field has been recognized in other ways and in view of the close relationship developing between these kindred arts, the Society has attempted to keep its members up to date.

The Board of Governors has favorably received a recommendation that this new field of interest be acknowledged through a change in the name of the Society from:

Society of Motion Picture Engineers, to
Society of Motion Picture and Television Engineers

The change would require an amendment to the Society's Constitution and will be discussed during the annual business meeting at the 66th Convention in Hollywood this October. Shortly thereafter all Active, Fellow, and Honorary Members will express their preferences by letter ballot.

All Society members were advised of this proposal by Earl I. Sponable, President of the Society, in his letter of April 6, 1949, in which he said ". . . the Society's Board of Governors, at its April third meeting, received favorably a recommendation that the name of our Society be changed to 'Society of Motion Picture and Television Engineers.' I am, therefore, taking this opportunity of enlisting your support of the change in name."

Factors Affecting Spurious Printing in Magnetic Tapes*

By S. W. JOHNSON

RCA VICTOR DIVISION, CAMDEN, NEW JERSEY

Summary—This paper represents a study to determine the amount of spurious printed-through signal from adjacent layers in a roll of magnetic tape. Effects of time, temperature, and output level of the original recording are taken into consideration, and conclusions drawn.

A GENERAL STUDY of magnetic-tape characteristics should include data on changes that may take place with various kinds of tape in storage, particularly at room temperatures somewhat higher than normal. Although we were interested to find the effects of higher temperatures physically on the tape itself, the primary purpose of this survey was to find if any of the presently available tapes indicated a high level of spurious print-through signal from adjacent layers when exposed to high temperatures. Indications are that the Germans also were sufficiently concerned about this problem that they kept all of their recordings in temperature-controlled rooms.

The equipment used for these measurements consisted of an original experimental model tape recorder, fitted with RCA ring-type heads, and operating at 15 inches per second. The amplifiers used were of RCA design, "broadcast" quality, to keep hum and noise to a minimum. As a further precaution, a 250-cycle high-pass filter was inserted in the output of the reproducing amplifier to make certain no hum components were present in the measured signal. Recordings were made with optimum bias as found from a family of output-versus-distortion curves[1] for the particular tape used. It may be well to note here that these output-versus-distortion curves were used as a basis for establishing all of the output levels referred to in this paper. The reproducing equipment was kept calibrated throughout to have a constant gain of 86 decibels.

The original or "master" recording was then made, the output level being set at a point where the total root-mean-square distortion at 400 cycles per second[1] was 4.5 per cent, as measured on a General

* Presented October 27, 1948, at the SMPE Convention in Washington.

Radio 732A distortion and noise meter. This 4.5 per cent level is, of course, a point where slight overloading occurs, such as is encountered in peaks of a normal recording, in most cases about 6 decibels above the normally recommended recording level of 2 per cent total root-mean-square distortion. The frequency was then raised to 500 cycles per second, and the final master was made at the same level. The 500-cycle-per-second frequency was chosen because the wavelength becomes so short at higher frequencies that it was feared any slight slippage of the two tapes when being removed from the exposing drum would cause partial erasing and thus not give a true reading.

This master was then wrapped around a Bakelite exposing drum built especially for this purpose. Then a piece of freshly erased tape

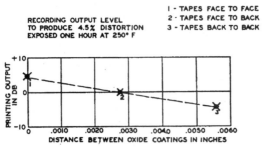

Fig. 1—Effect of oxide-layer spacing to spurious print-through signal.

of the same type was placed over it, with the oxide-coated side in the same direction as the master, as they would be if wound up in a roll. Finally a canvas belt was placed over the outside that would apply a constant two-pound tension on the tapes by means of a spring clamp. The drum was then heated for specified periods in a thermostatically controlled oven. At the end of each period, the tapes were removed and measured on our calibrated setup, the original master being checked also, to make sure the temperatures were not affecting it in any way. All of the tapes tested appeared to be able to withstand temperatures as high as 250 degrees Fahrenheit without any physical signs of deterioration, except perhaps a very slight curling at the edges.

The illustrations are plotted to show the results of the measurements which were made. They show the amount of spurious print-through signal in decibels as compared to exposure time in minutes at temperatures ranging from 80 to 250 degrees Fahrenheit in each case. The original master output and noise levels are also shown. The amount of printing in relation to the output level of the original master for two tapes at two temperatures is shown. Representative distortion-versus-output curves in relation to bias current used for setting output levels are also shown.

A representative group of presently available tapes was chosen which will henceforth be designated by A, B, C, D, and E. The Type C is a high-coercive material, and all of the others are various forms of lower coercive or "red" oxides. The thickness of the various tapes

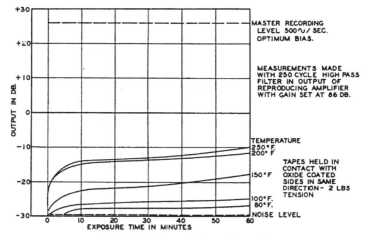

Fig. 2—Spurious printing of Type A.

can be an important factor in tests of this type, so we should note here that the actual samples of tape used for this series of tests have the following average over-all thicknesses:

Type A—0.0028 inch   Type C—0.0028 inch
Type B—0.0020 inch   Type D—0.0023 inch
                    Type E—0.0048 inch

In order to find out just how spurious print-through signal would vary with thickness of the tape, a master was prepared on Type C tape, at 500 cycles per second, recorded to a level producing 4.5 per cent distortion, the same as for all other tests. In this series, however, the freshly erased tape was placed in contact, face to face, with the original master, and heated to a temperature of 250 degrees

Fahrenheit for one hour. Then a second test was made with the oxide-coated sides in the same direction, and finally a third sample was made with the tapes back to back. The results of these measurements will give a curve showing how spurious printing is decreased as the space between the oxide layers is increased. This curve is shown in Fig. 1, illustrating that when the oxide layers are in contact, face to face, with no space in between, the signal is about 5 decibels higher than when the tapes are placed on contact with oxide-coated sides in the same direction, making 0.0028-inch spacing between them. This curve also shows that when the tapes are placed back

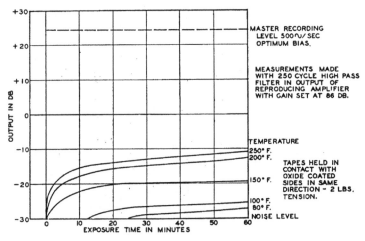

Fig. 3—Spurious printing of Type B.

to back and the spacing is doubled, the spurious-signal level drops another 4 decibels, indicating the importance of the thickness of the tape in relation to spurious print-through signal.

Fig. 2 illustrates the spurious print-through signal as recorded for Type A, which we have arbitrarily adopted as a standard of comparison for all other tests. The curves show that when the temperature is raised to 250 degrees the spurious signal rises to $15^1/_2$ decibels above noise in ten minutes, but only rises about 4 more decibels after an hour of exposure. It may be well to note here that a couple of samples were exposed as long as 380 hours with only a couple of decibels increase in spurious signal from that produced in one hour. These curves show that at ordinary temperatures (80 to 100 degrees Fahrenheit) the spurious printed signal rises only 3 to $4^{1}/_{2}$ decibels above noise after one hour of exposure.

Fig. 3 shows the spurious print-through output of Type B tape. At 250 degrees Fahrenheit the spurious-signal output is about the same as Type A but at lower temperatures, it takes a somewhat longer exposure to produce the same amount of printing.

Fig. 4 is a record of the spurious print-through characteristics of Type C tape. At 250 degrees Fahrenheit the spurious print-through signal is 11 decibels higher than Type A. Temperatures of

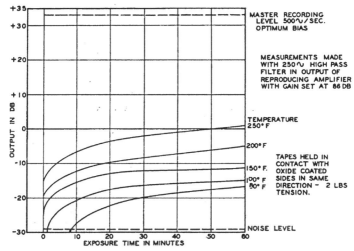

Fig. 4—Spurious printing of Type C.

80 to 100 degrees Fahrenheit also indicate 13 to 14 decibels of printing after one hour's exposure, which is a considerable amount, although this is offset to some extent by the high recording level of this tape.

Fig. 5 represents Type D tape. These curves show it to be just slightly better than Type A or Type B.

Fig. 6 shows the spurious print-through. curves of Type E tape. They show it to be about the same as Type A at high temperatures, but showing practically no spurious printing at room temperatures of about 80 degrees Fahrenheit. This may be due to the thicker base of the Type E tape, therefore making a wider space between layers.

The previous curves were all made with a constant output level in the masters at 4.5 per cent distortion, as high as would be encountered

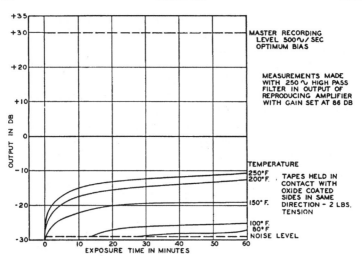

Fig. 5—Spurious printing of Type D.

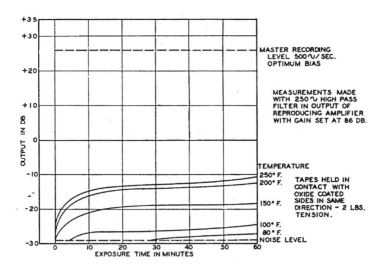

Fig. 6—Spurious printing of Type E.

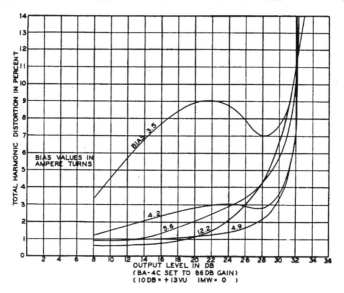

Fig. 7—Distortion versus reproducing level—Type C.

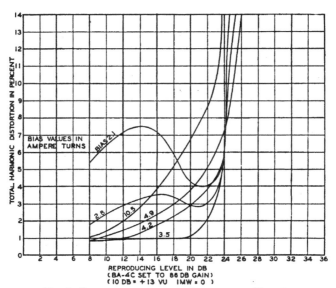

Fig. 8—Distortion versus reproducing level—Type B.

in a number of peaks in any ordinary recording. The effect of keeping all peaks under 2 per cent distortion level, is shown in distortion-versus-output curves of the Type C and B tapes in Figs. 7 and 8. The levels which will give distortion under 2 per cent at optimum bias are about 22 decibels for Type B and $27\frac{1}{2}$ decibels for Type C tape. Fig. 9 shows curves of the amount of spurious printing that occurs as the output level of the master is gradually increased to the overload point. Transferring the point where maximum level at 2 per cent distortion occurs in curves 7 and 8 to the curves in Fig. 9, it is found that if the level is kept within this range there are

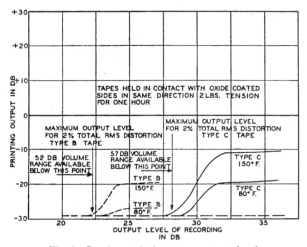

Fig. 9—Spurious printing versus output level.

practically no spurious-printing effects at temperatures up to 150 degrees Fahrenheit, as shown by the arrows. From this it is clear that if a limiter were installed in the recording-amplifier equipment that would keep all peaks under the maximum level allowed for 2 per cent distortion at optimum bias, as determined by distortion-versus-output curves in relation to bias, the spurious print-through effects will be eliminated. In view of the above, there is a workable volume range of between 55 to 60 decibels.

The results of these investigations show that in all of the presently available tapes, except Type C, the spurious print-through signal averages about 4 decibels above noise at room temperatures, rising to about 10 decibels at 150 degrees Fahrenheit when the original reaches a level giving 4.5 per cent distortion.[1] These conditions were observed with a total average volume range of about 55 decibels. The curves also indicate that the spurious print-through effect is largely dependent on temperature, because in all cases the curves do not show very great changes after the first few minutes of exposure. Finally, it was also shown that the output level of the original is an extremely important factor; that is, the higher the level, the more spurious printing effects can be expected, whereas, if the output level is kept down by means of a limiter, so that the distortion is under 2 per cent, there will be practically no spurious print-through signal. As a final audition test, a number of short 1500-cycle-per-second recordings were made with a blank space of several seconds between each two. The level used was that producing 4.5 per cent total root-mean-square distortion. The 1500-cycle frequency was chosen because indications are that this is the most audible frequency at which spuriously printed signals are heard. Portions of this recording were then heated for one hour at 80, 150, and 250 degrees Fahrenheit, respectively. After heating, the recordings were played back to a representative group of listeners, all of whom agreed that "echoes," which were the printed-through signal, could be heard even in the sample heated only to 80 degrees Fahrenheit, and on the samples that were heated to higher temperatures, as many as four "echoes" were heard. This tends to indicate that the spuriously printed signals from the peaks can be heard in quiet passages, if the peaks are allowed to reach a level giving 4.5 per cent distortion.

REFERENCE

(1) G. L. Dimmick and S. W. Johnson, "Optimum high-frequency bias in magnetic recording," J. Soc. Mot. Pict. Eng., vol. 51, pp. 489–501; November, 1948.

DISCUSSION

DR. ROBERT HERR: One thing was not clear to me. Mr. Johnson said that these tapes were erased immediately prior to the time of the test.

MR. S. W. JOHNSON: That is right.

DR. HERR: How long a time?

MR. JOHNSON: Three or four minutes. They were erased and put on the drum.

DR. HERR: In the case of the recording made on the reel on which listening tests were conducted, at the time of recording, was the material also erased during the recording process?

MR. JOHNSON: By the recording process. The tape was erased as the recording was made.

DR. HERR: One other time factor I should like to question is the preparation of the tests on the drum, the removal of the strip which was subjected to the test.

How long a time elapsed between the printing and the playing back of the sample to get the level?

MR. JOHNSON: We took the tape off the drum as quickly as possible, spliced it together, and played it back.

DR. HERR: It would be pretty much the same in all cases?

MR. JOHNSON: That is right.

MR. GEORGE LEWIN: It might be interesting to point out that this echo effect, at least in our opinion in the Signal Corps, is more serious than you might suspect from mere consideration of constant tone. We found with a Brush recorder that the echo is sufficiently distinct so that you can actually understand a given word in a phrase and it appears as an anticipation rather than an echo. While the actual measurements might be down to 30 or 40 decibels from the original signal, it still was very audible; in fact, it still could be heard when the re-recording was made. We should like some comment to suggest ways and means of reducing it. We heard this effect at normal temperature. We did not have to heat the tape up to 100 or 200 degrees Fahrenheit.

MR. JOHNSON: As was noted in the paper, if a limiter was installed in the amplifier so the distortion was under 2 per cent, we found no such effect at normal temperatures.

MR. LEWIN: While we did not measure the actual harmonic distortion, the signal level produced, it was well below the recommended maximum level and yet it was still completely audible and understandable.

———————◆———————

Theater Television

The SMPE Theater Television Committee met twice recently with the Television Committee of the Theater Owners Association. Donald E. Hyndman, Chairman of the Society's Committee, and Mitchell Wolfson, who is Chairman of the Theater Owners group, both report that the exhibitors and the engineers agree television does have a place in the motion picture theater.

After several years of active work on theater television, which was reported on in detail in the March, 1949, issue of the JOURNAL, members of the Society's Committee are pleased to see some activity on the part of the motion picture exhibitors. While the exhibitors are seeking answers to a number of economic problems connected with the introduction of television to the theater, they are also seeking answers to a number of technical problems, and the Society is offering its support through the Committee on Theater Television.

Initial costs of theater television equipment and the lack of program material at the present time are both serious obstacles, but no doubt they can be overcome before long. Present picture quality seems adequate for a beginning, and as soon as the other problems are solved, our theater patrons will be enjoying the timely benefits of theater television in some form.

Theater Loudspeaker Design, Performance, and Measurement*

By JOHN K. HILLIARD

ALTEC LANSING CORPORATION, HOLLYWOOD 38, CALIFORNIA

Summary—Referring to the on-axis measurement out of doors and the correction factor obtained at the Academy Award Theater indoors, the listening tests with Motion Picture Research Council Test Reel ASTR-3 and current studio product verify the electrical-response curve needed for adequate bass response. Additional low-frequency rise was used to determine its effect. It was found that no appreciable equalization could be used. In the case of the high-frequency droop, the rate of attenuation for this particular loudspeaker system was determined to be optimum using a response curve which has been in use for many years with this type of diaphragm and horn construction.

The Motion Picture Research Council Bulletin on Standard Electrical Characteristics dated April 20, 1948, states:

"Although the standard electrical characteristics for the newer-type speaker systems are essentially the same as for the older-type systems, acoustical-response measurements and listening tests have shown that a substantial increase in response is being obtained at the higher frequencies with the newer systems. The increase in efficiency and extension of frequency range is desirable so that the theater may take advantage of the improvements that are continually raising the standard of quality of the release print."

It has been found over the years that attempts to correct major irregularities in response by upward equalization yield sound distinctly inferior to that produced by a system which does not require this equalization. The outdoor free-field measurement technique described has supplied information in such a manner that improvements in design are indicated and then checked for results. Smoothness of response is constantly being sought and this tool definitely indicates the progress that is being obtained.

To our knowledge, this is the first time a standardized method of measuring theater loudspeakers has been conducted jointly by designers and users. The outcome of the tests described here demonstrates the practical value of concentrated action along this line.

THE PRESENT-DAY loudspeaker systems now available for the highest quality sound reproduction have materially approached closer to the goal of "faithful reproduction" that is, reproduction that gives the audience an impression that it is listening to a true likeness of the original. As this gap is narrowed, it becomes increas-

---

* Presented May 18, 1948, at the SMPE Convention in Santa Monica.

ingly necessary and important to evaluate the actual differences which exist between the real and reproduced sound. Technical advances in many fields have contributed to the improvement in the art of design and measurements of performance. In the individual loudspeaker units, the application of improved magnetic materials permits the achievement of higher flux density in a smaller space. This increase in flux density provides higher initial efficiency at the low frequencies with proportionately increased higher frequency response and improved damping. Studies of low-frequency cone materials and their shapes have resulted in a reduction in break-up and standing-wave patterns on the surfaces of the cone, thereby improving the smoothness of the low-frequency response. Improved coupling between low-frequency units and their horns has reduced distortion and increased efficiency. The use of straight open low-frequency horns as compared to the folded horn has reduced phase delay and permitted smoother response through the crossover region. This point is being increasingly recognized as evidenced by the universal application of this type of low-frequency horn in new loudspeakers now being made available. High-frequency units, using diaphragm and voice-coil assemblies of low mass, enable efficient operation throughout all of the useful frequency range.

The use of a theater loudspeaker system is the adaptation of objective equipment to an art. Any method of rating such a tool must include the effects of the subjective factors. With this in mind, there is no standardized method of rating loudspeaker systems or components at the present time. The desirability of such a program is obvious. In rating loudspeaker systems for motion picture use, the following points should be considered and the ratings determined from the limiting factors for each system.

Factors that mainly determine quality considerations affecting both high- and low-frequency units are (1) maximum permissible amplitude without overshooting the uniform field of flux from the magnet; (2) crossover frequency; (3) rate of attenuation of the crossover network; (4) loading and cutoff frequency of the low-frequency horn; (5) outside diaphragm suspension; and (6) resonance and Q of cone assembly on low-frequency unit and resonance and Q of diaphragm and voice coil on high-frequency unit.

Factors that mainly determine reliability in operation are (7) physical dimensions of the voice coil; (8) maximum rise in temperature that can be tolerated without damage to insulation of voice-coil wire; (9) maximum safe voltage on voice coil before arcover to pole piece; and (10) study of component failures by accelerated life tests. It is quite obvious that point (1) affects intermodulation mainly, whereas points (2), (3), (4), (5), and (6) affect smoothness (uniformity of response). Good systems engineering usually indicates that all components should be designed to contribute about equally to distortion where cost considerations are not excessive. Experience to date indicates that, in this instance also, the distortion at the "so-called" operating level should not greatly exceed that which is considered permissible on the film and in reproducing system amplifiers. This, of course, is a lower value than manufacturers would probably assign to loudspeakers for other than motion picture work where higher distortion frequently is tolearted.

For many years, subjective listening tests have been the principal or sole method of evaluating final loudspeaker performance. However, the engineer developed a number of objective tests which materially aided him in this design work. Recently, there has been considerable progress in the development of measuring equipment with which fundamental measurements are obtained which are correlatable with subjective listening tests. These measurements are made outdoors or in any other place where there are no appreciable reflections anywhere within the measured range. Measurements made under these conditions are most suitable for reference use between organizations doing this type of work in various locations. Experience now indicates that these fundamental measurements provide an accurate estimate of sound-pressure level and response.

In the case of theater two-way systems, it has been recognized, for several years, that certain mechanical arrangements and spacing of the low- and high-frequency horns must vary, depending upon (1) the crossover frequency; (2) the character of the dividing network; and (3) the acoustic properties of the individual auditorium in which they are installed. The execution of this adjustment is known in a broad sense as "acoustical phasing."

It is the primary purpose of these outdoor tests to provide information on the individual components of a system such that the method of combining these components into a system may be specified in terms of the measurements. When this information is suitably correlated with listening tests, it should be possible to insure that the proper combination of these components will yield a response curve which is pleasing. It now appears that this pleasingness is largely achieved by having the over-all measured response smooth, that is, free from hills and valleys.

The Theater Sound Standardization Committee of the Research Council has for years conducted subjective listening tests on theater loudspeakers for the purpose of setting optimum-response curves. In the fall of 1946, this committee undertook to make subjective listening tests on a number of loudspeaker systems on which recommended amplifier-frequency characteristics had not been previously published. As a part of this work, it was later decided to make some outdoor measurements on speaker systems in the hope that some correlation could be established between the outdoor measurements and the subjective listening tests. To make these outdoor tests practical, the Altec Lansing Corporation offered its facilities at the RKO Studio Ranch, Encino, California. Various theater speaker systems, which had been previously checked by subjective listening tests in the Academy Award Theater, were measured. These theater systems represented new, current, and older two-way systems. Upon completion of these outdoor tests, some of the speaker systems were reinstalled in the Academy Award Theater for additional subjective listening tests.

The loudspeaker system on which the outdoor tests were made was measured with the following apparatus and routine which constitutes the Altec Lansing standard procedure for loudspeaker measurements on theater systems.

The microphone used for all measurements is the Western Electric 640AA condenser microphone which feeds into a special cathode-follower-type preamplifier.[1] The extremely stable performance over long periods of time, the known response characteristic over the required frequency range, and the low phase shift of this microphone provide the necessary qualities required for reliable measurement purposes. Measurements were made at a height of 40 feet above a flat terrain, and the essentially nondirectional microphone placed 20 feet away from the loudspeaker under test. Calculations indicate that at this height, using a nondirectional source, reflections from the ground will cause a modulation of the response curve 1.9 decibels in amplitude with a frequency interval of about 16 cycles. As the vertical angle of distribution becomes narrower, this modulation effect becomes less in amplitude. The reflected wave will be about 13 decibels less intense than the direct wave. The resulting measurements can be corrected, if necessary, for this reflection. The data obtained from the recorder are referred directly to sound-pressure level in decibels reference, 0.0002 dyne per square centimeter (10^{-16} watt per square centimeter). The combination of the high-speed recorder together with the slow three-minute sweep time, provides information on abrupt discontinuities in response to such a degree that it is felt that there is very little or no smoothing-out occurring in the measuring process.

The procedure for obtaining the necessary information to evaluate loudspeaker performance is as follows:

1. The loudspeaker is located in the proper position and adjusted in accordance with the recommendations of the manufacturer. These adjustments were made under the supervision of a representative of the manufacturer.

2. An impedance measurement is obtained with a continuous sweep frequency and recorded on a chart with the high-speed level recorder. This is obtained by sending a constant current through the loudspeaker and measuring the voltage developed across the loudspeaker. From this curve the lowest value of impedance is selected as the rated impedance of the loudspeaker system. All other measurements involving response and distortion use this value as the generator impedance except those cases where the manufacturer recommends a specific value of generator impedance.

3. Frequency-response curves on high-quality systems are measured from 30 to 16,000 cycles per second at various angles with respect to the loudspeaker axis. These angles in general are 0, 15, 30, 45, 60, 90, and 180 degrees. The input power used usually is in the region of 1 to 10 watts, depending upon the ambient-noise conditions. Before the frequency-response curves are taken, the effect of electrical phase reversal on the response is measured. This is accomplished by making one on-axis measurement of response and then reversing the leads to the high-frequency units and making another response curve.

4. A typical measurement of this type is indicated in Fig. 1. After the correct polarity is determined, the shelving or attenuation to be used in either the low- or high-frequency units for smooth crossover is determined, depending upon their respective efficiencies.

5. There has been considerable discussion in many organizations regarding an appropriate method of measuring loudspeaker distortion. A comprehensive analysis of this distortion should include the phase-frequency characteristic of the system. However, this field has not

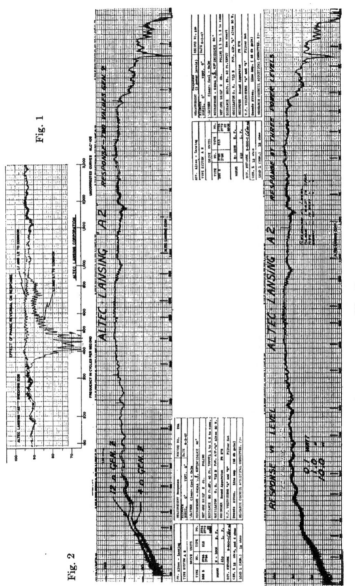

Fig. 1

Fig. 2

Fig. 3

TABLE I

HARMONIC DISTORTION
(20 WATTS TO SPEAKER AT NOMINAL IMPEDANCE)
A2 THEATER SYSTEM

All readings are in per cent of fundamental F1 equals fundamental; F2, F3 are second and third harmonics, etc.

	F1	F2	F3	F4	F5	F6	F7	F8	F9
Frequency	100	200	300	400	500	600	700	800	900
Per cent amplitude	100	0.34	0.65		0.51		0.22		0.08
Frequency	220	440	660	880	1100	1320	1540	1760	1980
Per cent amplitude	100	0.34	0.94		0.18				
Frequency	1000	2000	3000	4000	5000				
Per cent amplitude	100	5.40	0.50	0.18	0.13				

been developed to the point where reliable phase-delay measurements can be obtained, and for the present, distortion is being determined by harmonic and intermodulation methods. Since most large theater two-way loudspeaker systems have a crossover in the region from 400 to 800 cycles, it was considered expedient to use the harmonic method for the low-frequency spectrum up to and including the crossover point and using the two-tone intermodulation technique above the crossover point. For systems having a rating of approximately 50 to 80 watts, harmonic distortion is measured with a sine-wave input of 20 watts using fundamental frequencies of 100, 200 crossover frequency, and 1000 cycles. Intermodulation test tones use paired frequencies of 1000 and 1500 cycles; 2000 and 2500 cycles; 4000 and 4500 cycles. The sine-wave power for each frequency alone is 5 watts.

A typical example of the harmonic distortion of an 80-watt theater system with a power input of 20 watts is shown in Table I. The harmonic distortion was obtained by impressing the fundamental and measuring each separate harmonic. Readings below 0.1 per cent were not included in the data. The same analyzer was used to

TABLE II

HARMONIC DISTORTION

(AT THREE POWER LEVELS)

System	Frequency	F1 100	F2 200	F3 300	F4 400	F5 500	F6 600	F7 700	F8 800	F9 900
At 20 Watts to Nominal Impedance										
A-2	Per cent amplitude	100	0.65	0.91	..	0.53	....	0.29	....	0.09
At 2 Watts to Nominal Impedance										
A-2	Per cent amplitude	100	0.22	0.86	..	0.33	....	0.13	....	..
At 60 Watts to Nominal Impedance										
A-2	Per cent amplitude	100	1.23	1.28	0.22	0.62	....	0.28	....	0.10
At 60 Watts to Nominal Impedance										
Distortion in Amplifier System only with A-2 speaker load	Per cent amplitude	100	1.65	1.48	0.08	0.40	....	..	....	..

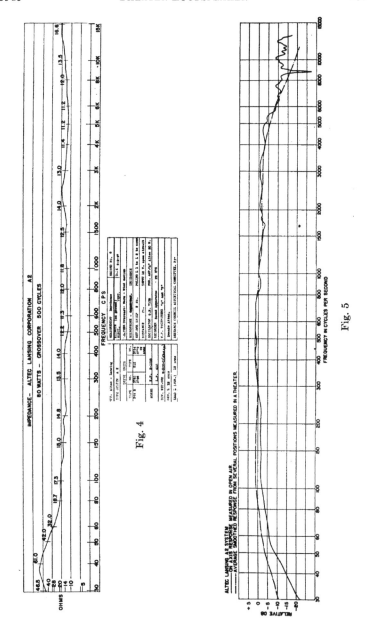

Fig. 4

Fig. 5

measure the sum and difference pairs generated when the intermodulation test tones were applied.

Table II shows the harmonic distortion at 100 cycles for the three following power levels: 2, 20, and 60 watts. There is also shown the harmonic distortion of the amplifier system with the loudspeaker as the load at an output of 60 watts. It should be noted in the columns $F2$ and $F3$ that the nonlinear distortion in the loudspeaker is of such a type as partially to cancel distortion from the amplifier system.

Fig. 2 shows the frequency response obtained using two values of generator impedance. The measurement indicates that appreciable changes in response at very low frequencies are obtained. Therefore, the establishment of a standard response curve for a loudspeaker must specify the generator and the loudspeaker impedance.

Fig. 3 shows the response of the system for each of three inputs covering a range of 20 decibels, namely; 0.1, 1, and 10 watts. It should be noted that the resulting outputs are proportional to the inputs and that the system is therefore linear.

Fig. 4 shows the over-all impedance curve of an 80-watt theater system which uses four low-frequency and two high-frequency units.

Fig. 5 (solid line) is the on-axis free-field sound-pressure level for the same 80-watt system. It shows a droop below 100 cycles and at frequencies lower than 60 cycles the response falls off at a rate of approximately 12 decibels per octave. When this same system was measured indoors at the Academy Award Theater (dotted curve) its average on-axis response was maintained to within 2 decibels of being flat down to 60 cycles. This change in response between indoors and outdoors is attributed to the decreased absorption at the lower frequencies indoors, and the wider distribution at low frequencies. This is verified by the measured reverberation time at the Academy Award Theater.

Table III shows the intermodulation distortion of two loudspeaker systems known as the "800" and $A2$ "Voice of the Theater." The intermodulation values for three pairs of frequencies are shown; namely, 1000 and 1500; 2000 and 2500; and 4000 and 4500.

Some two-way loudspeaker systems have been designed in the past which use separate amplifiers to operate the low- and high-frequency loudspeaker, and there has been an opinion that it was necessary to provide additional power for the low-frequency units over that used on the high-frequency units. In order to evaluate the distribution of energy that occurs in speech, music, and sound effects, a recording

TABLE III
DISTORTION—FREQUENCY DIFFERENCE PAIRS
(20 WATTS TOTAL TO SPEAKER AT NOMINAL IMPEDANCE)

Frequency in Kilocycles

		0.5	1.0	1.5	2.0	2.5	3.0	3.5	4.0	4.5	5.0	5.5	6.0	6.5	7.0	7.5	8.0	8.5	9.0	9.5
$F1 = 1.0$ kilocycle; $F2 = 1.5$ kilocycles																				
A-2	Per cent	1.10	100.0	98.0	3.05	10.20	5.40	0.56	0.33	0.42	0.15	0.12	0.09	..	..	..	..	..	..	..
800	Per cent	0.54	100.00	98.0	2.10	3.50	2.20	0.51	0.75	0.22	0.15	..								
$F1 = 2.0$ kilocycles; $F2 = 2.5$ kilocycles																				
A-2	Per cent	0.79	....	0.71	100.0	97.0	1.10	0.09	6.60	11.20	4.40	..	0.62	0.58	1.30	0.22	..	0.12	0.19	..
800	Per cent	0.25	....	0.69	100.0	78.0	0.32	..	2.60	6.10	4.00	0.10	0.27	0.89	0.90	0.30	..	..	..	
$F1 = 4.0$ kilocycles; $F2 = 4.5$ kilocycles																				
A-2	Per cent	0.74	....					0.68	98.0	100.0	1.00	..	..	..		0.12	6.00	6.20	5.50	0.10
800	Per cent	0.14	....					0.52	78.0	100.0	0.90	..	..	..			4.50	7.70	5.20	0.10

chart was placed across the input to the low-frequency units and the peak voltage was recorded from a Motion Picture Research Council Test Reel (ASTR-3) which had a variety of music, speech, and sound effects recorded. After this measurement, the recording chart was then placed across the terminals of the high-frequency units, and the same film repeated. In order to correlate the frequency range associated with the voltage observed, several notes are made at the top of the chart indicating the subject matter at the time. From this information, it was observed that the high peak energy is associated with frequencies above 500 cycles, as well as below 500 cycles. For this reason, it is felt that equal power should be used to operate the low- and high-frequency units in cases where the crossover is in the neighborhood of 400 to 800 cycles. This also requires that both the low- and high-frequency units be designed so that they both will be capable of radiating a flat power characteristic at least up to 5000 cycles.

REFERENCE

(1)   Paul S. Veneklasen, "Physical measurements of loudspeaker performance," J. Soc. Mot. Pict. Eng., this issue, pp. 641–657.

DISCUSSION

MR. LEWIN: Would you explain briefly the type of distortion measurement, where you show the frequency? That is not the customary intermodulation type.

MR. J. K. HILLIARD: These are closely spaced. As we said, the conventional intermodulation test is at very low frequency, combined with intermediate or high frequency. In this case, we have two frequencies which are 500 cycles apart. We did that because of the difficulty in a two-way system where you have a network and the distortion has to be confined to either one leg or the other of the network.

Physical Measurements of Loudspeaker Performance

By PAUL S. VENEKLASEN

ALTEC LANSING CORPORATION, HOLLYWOOD 38, CALIFORNIA

Summary—Physical measurements of loudspeaker performance are becoming recognized increasingly as a reliable guide for the evaluation of loudspeaker excellence. Facilities for the calibration of loudspeakers and microphones are described. The measurement techniques are illustrated with
data describing a recent model loudspeaker showing the principal performance criteria; namely, frequency response, angular distribution, and distortion. Suggestions are given for a uniform presentation of performance
data and specifications.

DURING THE PAST THREE YEARS the Altec Lansing Corporation has
set up extensive facilities for acoustical measurements as a part
of its program of loudspeaker development. These facilities have
already served to separate many facts from fancy in the loudspeaker
art. Factors, hitherto either unknown or ignored, have been studied
in detail and shown to be of great importance. Accurate measurements are complementing aural judgment in the design process. The
ear is indeed critical but not very analytical. Lacking constant
reference to original sounds, one may quickly become adjusted to, or
even prefer, unnatural reproduction. Physical measurements are
becoming increasingly recognized as one of the reliable indexes of
loudspeaker performance and generally agree with critical listening
tests. It has been found that those loudspeakers which measurements show to be most uniform in response, generally sound most
pleasing and natural.

Through the generous co-operation of the RKO Studios, the measurement facilities are located on a corner of their ranch lot in Encino,
California. The general layout is shown in Fig. 1. Southern California has the climatic conditions which permit the almost constant
use of nature's own free-field room.

The practice of measuring loudspeaker performance in open air is
not new. Recently, several very excellent free-field rooms have been
constructed and described.[1-3] The improvement in the design and
performance of these rooms is of tremendous importance in the field

of acoustical measurement and the advantage of their use in terms of freedom from outdoor noises and weather disturbances is a great blessing to those who may use them.

In the development of large loudspeaker systems, measurements must be made at large distances and at low frequencies. The excellent analysis of performance given in the description of the Harvard

Fig. 1—Measurement site on the RKO ranch in Encino, California.

free-field room indicates that at a distance of 20 feet an error of about 1.5 decibels may be made at 60 cycles. This error increases very sharply at lower frequencies. It has been the practice to measure the performance of loudspeakers an octave below this frequency. A distance of 20 feet is minimal for adequate measurement of a large theater system. Under these circumstances, the known and readily calculated error due to the single reflecting surface of the ground may be a fair exchange for the low-frequency difficulties in a costly enclosed space.

Therefore, the description of the facilities used by the author is presented to show how the older techniques, when assembled with the aid of recent practices in instrument design, may be used to secure reliable results without extravagant cost.

Fig. 2—Loudspeaker mounted on 20-foot steel tower.

Fig. 3—Use of a pole for the measurement of large loudspeaker systems.

For the final calibration of smaller loudspeakers, including small theater systems, the steel tower shown in Fig. 2 is used. The platform of this tower is 20 feet above the ground. The rigging is such that systems weighing several hundred pounds can be hoisted single-handed. The microphone boom pivots around a centerline in the front face of the tower. The boom motion is motor-controlled from the shack and is synchronized with the recorder. The microphone may be extended to 10 feet from the loudspeaker and rotates through plus and minus 90 degrees from the center.

Larger theater-type loudspeaker systems which must be measured at greater microphone distances are calibrated on a pole rig as shown in Fig. 3. Such a system is hoisted until its base is 40 feet above the ground and lashed to the pole. The microphone boom may be set at various angles.

Significant departure from free-field conditions is due only to the proximity of the ground. The effect of ground reflections is shown in Fig. 4. The maximum fluctuations caused by reflection from hard

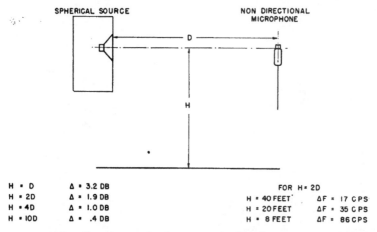

H = D	Δ = 3.2 DB		FOR H = 2D	
H = 2D	Δ = 1.9 DB	H = 40 FEET	ΔF = 17 CPS
H = 4D	Δ = 1.0 DB	H = 20 FEET	ΔF = 35 CPS
H = 10D	Δ = .4 DB	H = 8 FEET	ΔF = 86 CPS

Fig. 4—The effect of ground reflections on loudspeaker measurements. Δ is the maximum error caused by the ground reflection. Δf is the frequency interval of the modulation due to the reflected wave.

ground are given for several ratios of source to ground distance. The frequency interval of these fluctuations is also shown. Such fluctuations are easily recognized in the charts and removed when the curves are corrected. Fig. 5 shows the instrumentation arranged within the shack. Any novelty in the equipment is due to an insistence on convenience, reliability, and accuracy.

The accuracy and the absolute basis of our calibrations depend upon the use of the Western Electric 640AA condenser microphone with an open-circuit calibration from the Bell Telephone Laboratories. This microphone is used with a cathode-follower-type preamplifier shown in Fig. 6. This preamplifier and its complementary voltage amplifier and power supply were described by the writer in an earlier paper.[4]

Conversion of records from preamplifier output to sound-pressure level is accomplished by the calibration and correction curves shown in Fig. 7. At the top is the open-circuit pressure calibration of the 640AA microphone. Next is the correction relating pressure calibration to free-field calibration at parallel incidence. Next is the preamplifier insertion loss relative to the open-circuit input voltage. These curves combined produce the over-all correction curve at the

Fig. 5—Instruments in the shack.

bottom of the figure which permits conversion from preamplifier output in decibels re one microvolt, as plotted by the recorder, to sound-pressure level in decibels re 0.0002 dyne per square centimeter. Hence, the traces from the recorder may be corrected to give an absolute calibration of the loudspeaker under test.

Filters are used in the microphone circuit to decrease the annoyance of wind noise. As frequency sweeps upward, successive filters cut off below 25 cycles per second, then 200, then 500 cycles per second.

Wind shields around the microphone alone are not successful, because at higher frequencies considerable modulation of the signal is caused by refraction of the sound.

The output of the microphone preamplifier is recorded on a Sound Apparatus Company Type FR level recorder. This instrument has been completely remodeled. Its speed is increased three times. Complete revision of the electronic circuit increases the clutch currents by six times and assures positive movement of the pen for 0.2 decibel off-balance signal. Separate motors drive the chart and the clutch mechanisms. The chart drive is synchronized with the oscillator drive. Accuracy and dependability have been greatly improved. The oscillator sweeps from 30 to 16,000 cycles per second in three minutes. The speeding up of the recorder, together with the slow sweep, are very essential to avoid smoothing out abrupt discontinuities in response.[5, 6]

A special power amplifier is designed for extremely low distortion and low output impedance. The internal resistance is reduced by feedback to about 0.3 ohm. Thus, in loudspeaker testing, a known open-circuit voltage may be inserted in series with any desired source impedance, and reaction of the load back on the amplifier is eliminated. The remaining equipment in the shack is more or less standard.

Fig. 6—Western Electric 640AA condenser microphone mounted on the preamplifier.

These then are the facilities used for acoustical measurements. They are in almost constant use in connection with development programs. They were selected and used to good advantage in connection with the standardization program of the Research Council of the motion picture industry. The characteristics of the principal theater loudspeaker systems were measured during this program.

The same equipment is being used increasingly for making comparison calibrations of the various types of microphones used in the broadcasting and recording fields, as well as in the acoustical standardization program of the aircraft industry on the West Coast.

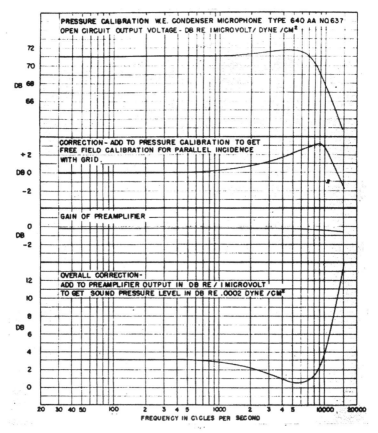

Fig. 7—Calibration curves and correction curves for use with the 640AA microphone.

There are many properties of loudspeakers which may conceivably be measured and more are being suggested daily. To be useful a given measurement must be readily interpreted in terms which can be translated into design changes which will result in recognizable improvements in performance.

There remain several simple categories of loudspeaker performance which, when accurately measured, show great variation between speakers and probably account for most of the audible performance variations. These include (1) the input impedance of the system;

(2) frequency response on axis; (3) angular distribution; and (4) distortion.

I should like to show how these simple measurements are made, show the results of these measurements on a sample loudspeaker, and suggest a simple form for describing these performance criteria in technical specifications and in advertising.

The input impedance of a loudspeaker as a function of frequency is measured as shown in Fig. 8. The quantity measured is simply the magnitude of the vector.

A constant open-circuit voltage of 10 volts is maintained while the frequency range is swept. Assuming that the magnitude of the measured impedance remains small compared to 1000 ohms, the voltage e, appearing across the loudspeaker terminals, will be related

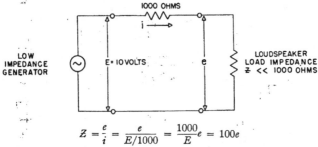

$$Z = \frac{e}{i} = \frac{e}{E/1000} = \frac{1000}{E}e = 100e$$

Fig. 8—Circuit and equations for the measurement of loudspeaker impedance. E = open-circuit voltage supplied to the test circuit; i = current flowing in-circuit; e = voltage measured across the loudspeaker terminals.

to its impedance as shown by the equation on the figure. The voltage e is recorded. The result of this measurement on the sample loudspeaker is shown in Fig. 9. In this case the nominal impedance is called 12 ohms.

Having chosen a value of nominal impedance the frequency response on the axis can be determined and a useful measure of loudspeaker efficiency can be obtained for this microphone position. This is done as follows: For a given loudspeaker, the manufacturer will state a recommended amplifier output impedance. In this case the recommended value is 10 ohms. For the frequency-response measurement therefore the loudspeaker is fed through a 10-ohm series resistor with an open-circuit voltage which will deliver 1 watt to a 12-ohm resistance load. This is 6.4 volts for the sample. The frequency response is run with this input condition.

Fig. 9—Impedance curve for the sample loudspeaker.

Fig. 11—On-axis frequency response of sample loudspeaker, showing the recorded trace, the smoothed curve, and the corrected sound-pressure level.

The matter of microphone position needs clarification. Depending upon the type of loudspeaker being measured, the frequency response may vary considerably with the vertical position of the microphone. In the case of small loudspeakers for home use, a uniform practice of measuring the "on-axis" response with the microphone in a position

Fig. 10—Standard microphone position for the measurement of on-axis response and efficiency.

corresponding to the ears of a person seated at a distance of 10 feet from the loudspeaker is used. This places the microphone about 45 inches above the base of the loudspeaker cabinet as shown in Fig. 10. The results of the measurement of frequency response on-axis for the sample loudspeaker are shown in Fig. 11.

This curve will serve to illustrate the process of correcting the

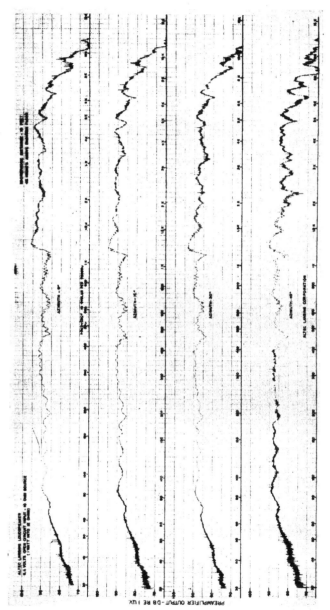

Fig. 12—Recordings of loudspeaker response at various azimuth positions.

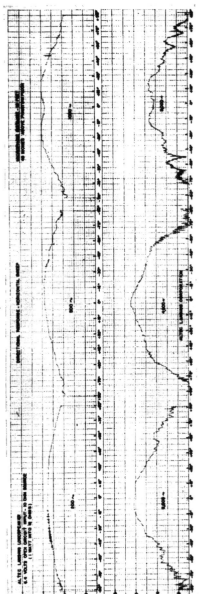

Fig. 13—Recording angular distribution by sweeping the microphone at a constant input frequency.

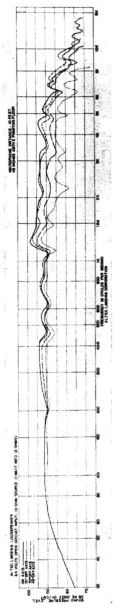

Fig. 14—Superimposed, corrected curves of loudspeaker response.

recorded curve to eliminate ground reflections as shown by the dotted line drawn through the recorded trace. The conversion of this curve to sound-pressure level has also been carried out in accordance with the correction curve of Fig. 7 so that the corrected calibration curve is as shown on the graph. The process of correcting these curves should not be confused with the common practice of publishing "expected," "calculated," or "idealized" curves. The smoothed curve in no place departs from the recorded curve by a greater amount than that which can be accounted for by ground reflections, which for this particular setup is ± 1.8 decibels.

Angular distribution can be measured in two different ways. The microphone may be turned to specific angular positions with respect to the axis and the frequency-response record repeated. The results of this type of measurement are shown for the sample loudspeaker in Fig. 12. Additional detail is furnished by another method; namely, to maintain a constant input frequency and sweep the microphone around the loudspeaker. The results of this measurement are shown in Fig. 13. The response curves may be superimposed after correction to furnish a fairly complete description of the distribution characteristics as shown in Fig. 14. Vertical distribution measurements may be made in similar fashion with the loudspeaker turned on its side.

Many methods have been suggested for measuring loudspeaker distortion. In the present state of the art there is ample evidence to indicate that any but fundamental measurements are misleading. Accordingly, one or more tones are inserted into the loudspeaker and the components which appear in the microphone output are measured with a wave analyzer. The results of such measurements on the sample loudspeaker are shown in Fig. 15 for an input of 4 watts. The use of pairs of tones and the search for sum and difference components permit the examination of distortion at high frequencies. This could not be accomplished by simple harmonic analysis. The search for intermodulation between tones at 100 and 5000 cycles per second for this loudspeaker was fruitless, this distortion bearing no obvious relation to the harmonic distortion at 100 cycles per second.

Interpretation of distortion measurements should be approached with great skepticism. The recorded values do represent distortion, defined simply as the difference between what comes out and what goes in. Nominal values for permissible distortion as usually recited are probably a superficial answer to a complicated subject. The

determination of distortion tolerance by inserting known distortion
and defining a distortion threshold is probably a tedious approach,
lacking in quantitative accuracy and general value.   A more general

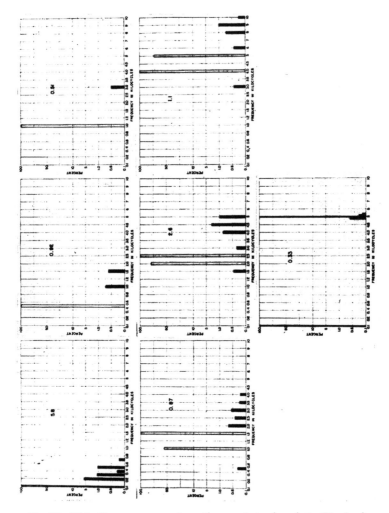

Fig. 15—Distortion measurements on the sample loudspeaker. The fundamental tones are shown by the open bars, and the distortion products by solid
bars. The figure on each graph gives the total distortion.   Input level is 4
watts.

answer may perhaps be derived by a study of masking. While the existing data are very limited a few indications of the complexity of the problem and hints as to order of magnitude may be obtained from the data of Wegel and Lane.[7] The shapes of masking areas indicate that the following may be expected:

1. Distortion tolerance will increase with loudness.

2. The tolerance for components above the fundamentals will be much greater than for components below the fundamentals.

3. The tolerance will be less for components more widely separated from the fundamentals.

4. At a loudness level of about 80 decibels, the tolerance for a second harmonic is about 3 per cent, for a third harmonic, about 1 per cent, and for a subtone about an octave below the fundamentals, about 0.03 per cent.

This writer believes that a study of masking will present a reliable guide for establishing distortion limits. Further investigations of masking thresholds would be a great contribution to this field. The above indications should be a sufficient guide to stress that it is essential for a system to be capable of producing sufficient loudness with nominal power input. Hence the importance of electroacoustic efficiency.

The extravagance of claims for loudspeaker performance as evidenced in advertising literature is deplorable. The measurement techniques which have been described suggest a simple and honest system for loudspeaker performance specification.

The on-axis response as shown in Fig. 14 for a standard microphone position may be stated as ±4.5 decibels from 100 to 5000 cycles per second; down 10 decibels at 45 and 7600 cycles per second; down 20 decibels at less than 30 and at 11,000 cycles per second.

This type of rating is based upon this conviction, derived from complete measurements on, and considerable listening to, many loudspeakers of various manufacturers. Entirely too much stress is being placed on wide-range reproduction to the neglect of smooth and clean reproduction over a limited range of frequencies. Many so-called "hi-fidelity" loudspeakers have response curves which vary over 30 decibels in mid-range! They do not sound well with even limited range input. *Smoothness of response in a limited range should be achieved and stressed before wide-range reproduction will be worth while.*

On-axis efficiency may be described for the sample by stating the average sound-pressure level created at the standard microphone position with 1 watt input. This is 94 decibels (relative to 0.0002 dyne per square centimeter).

Angular distribution may be described as within 5 decibels of on-axis response for a horizontal spread of 60 degrees to 10,000 cycles per second.

Distortion ratings should place due emphasis on efficiency by giving the distortion for a standard sound-pressure level at the standard microphone position, i.e., 10 feet distance, seated ear height in open air. Thus the sample may be rated: less than 4 per cent distortion for 94 decibels sound-pressure level at 10 feet in the frequency range from 100 to 5000 cycles per second.

A rating system such as has been discussed herein is based on objective and meaningful measurements and correlates well with subjective listening tests. It should be a reliable guide to the public and a stimulus to the manufacture of better loudspeakers.

ACKNOWLEDGMENT

The writer wishes to express his sincere appreciation to G. L. Carrington, president of the Altec Lansing Corporation, for far-sighted judgment in establishing and supervising this measurement program; to J. K. Hilliard for sponsorship and continuing encouragement of the program; to the many colleagues who have contributed to its success; and to the RKO Studios for the laboratory site.

REFERENCES

(1) Office of Scientific Research and Development Report No. 4190, "The Design and Construction of Anechoic Sound Chambers," Electro-Acoustic Laboratory, Harvard University, October 15, 1945.

(2) E. H. Bedell, "Some data on a room designed for free-field measurements," J. Acous. Soc. Amer., vol. 8, pp. 118–125; October, 1936.

(3) H. F. Olson, "Acoustic laboratory in the new RCA laboratories," J. Acous. Soc. Amer., vol. 15, pp. 96–102; October, 1943.

(4) P. S. Veneklasen, "Instrumentation for the measurement of sound pressure level with the Western Electric 640AA microphone," J. Acous. Soc. Amer., vol. 20, pp. 807–817; November, 1948.

(5) E. C. Wente, "Principles of measurement of room acoustics," J. Soc. Mot. Pict. Eng., vol. 26, pp. 145–154; February, 1936.

(6) C. P. Boner, H. W. Jones, and W. J. Cunningham, "Indoor and outdoor response of an exponential horn," J. Acous. Soc. Amer., vol. 10, pp. 180–183; January, 1939.

(7) R. L. Wegel and C. E. Lane, "Auditory masking," Phys. Rev., vol. 23, pp. 265–285; February, 1924.

Theater Reproducer for Double-Width Push-Pull Operation*

By G. R. CRANE

WESTERN ELECTRIC COMPANY, HOLLYWOOD 38, CALIFORNIA

Summary—This paper describes a new soundhead for use in studio review rooms which provides a new optical system similar to that used in new re-recording equipment for double-width tracks. A single control permits operation with standard single-track, 100-mil push-pull or 200 mil push-pull. The film motion is of the new double-arm type equipped with hydraulic damping and controlled compliance. It is also readily adapted for magnetic reproduction.

THE INCREASING USE of double-width push-pull sound tracks in the studio together with the various advances in recording techniques, has accentuated the need for a high-quality theater-type sound reproducer with adequate optical facilities for both wide and standard sound tracks as well as adaptability to magnetic recording.

The Western Electric RA-1435 reproducer was developed for this purpose and provides an optical system similar to that recently developed for studio re-recording equipment. Good film motion is provided by a double-arm, tight-loop film path having controlled compliance and hydraulic damping. The design minimizes the number of parts to provide the utmost simplicity in operation and maintenance. It may be equipped with a magnetic reproducing head which is retractable to clear the film when optical equipment is used.

DESCRIPTION

A front view of the machine is shown by Fig. 1 and the general construction is somewhat similar to most commercial sound units with the motor mounted in the rear. It is adaptable to most of the commercial projection heads, pedestals, and related equipment for use in studio review rooms. It is shown equipped with a so-called dummy head for re-recording use. This reproducer is normally supplied with a double-V belt drive from its motor for review-room purposes with which double film attachments are frequently used. However, a silent chain drive is available for use with interlock, synchronous, or a special combination motor to provide either synchronous or interlock operation.

* Presented October 27, 1948, at the SMPE Convention in Washington.

As shown by Fig. 1, the film path consists of a tight loop between the top sprocket, which is that of the projection head or dummy head, and the single lower sprocket contained in the sound unit. Two movable rollers are located in the path on each side of the scanning drum with the lower roller connected to a fluid dashpot for damping. A threading indicator is provided just above the scanning drum so that the operator may observe the relative position of the filter arms when the proper length of loop is threaded. Since this type of

equipment ordinarily uses relatively new film having a minimum of shrinkage, special film sprockets are employed having relatively large teeth so that there is a minimum of clearance between the sprocket hole and the sprocket tooth in the direction of film travel. This insures satisfactory flutter performance with the single lower sprocket·being used as both feed and holdback sprocket. The analysis of this filter system has appeared previously.[1]

Fig. 2 shows the film path with the magnetic adaptation. This places a magnetic reproducing head in the path between the scanning drum and the lower fil-

Fig. 1—Front view of the reproducer equipped with a "dummy" head for use in re-recording.

ter roller. The head is on a retractable mounting controlled by a rotatable knob at the lower left so that it is removed from contact with the film when photographic tracks are used.

OPTICAL SYSTEM

Years of practical experience in optical systems have demonstrated the safety of so-called front-scanning systems, but these systems are not readily observed for scanning performance. The convenience of rear-scanning systems for visually determining scanning is well known, but many such systems have contained fire hazards because of the relatively large amount of light placed upon the film by the condenser-lens system. In order to retain the advantage of each of these systems, the one used in this machine combines the convenience of rear scanning with the safety of front scanning and is similar to that used in recent studio re-recording equipment.[2] As shown by Fig. 3, a cylindrical-lens system places a scanning line upon the film 1 mil high, and of sufficient length to more than cover the area oc-

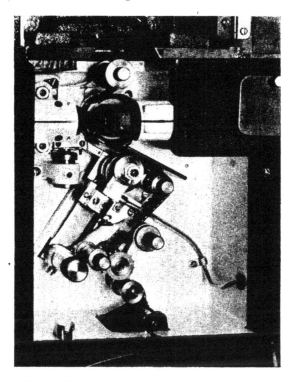

Fig. 2—Close-up view of film compartment showing the addition of a magnetic reproducing head, which is retractable from the film path by 180 degrees rotation of the knob at the lower left-hand corner.

cupied by all sound tracks in current use. The light source is a 10-volt, 5-ampere, curved-filament, prefocused lamp and the optical constants are such that its vertical position is not critical, thereby facilitating interchangeability and essentially eliminating microphonic noise generated by vibration of the lamp or optical parts.

A rear-scanning type of system is employed to produce an enlarged image of the film and this scanning line on a mask, so that the limits

of scanning are readily observable and adjustable at any time.
Fig. 3 shows the system schematically, omitting a prism and mirror
which merely turn the optical axis by 90 degrees for mechanical
convenience. This system produces a vertical enlargement of the
scanning line of the order of 100:1 to permit more convenient observation of the scanning limits relative to the mask, and a lateral magnification of approximately 3:1. A scanning mask contains three sets
of openings, one of which is always registered with the projected light

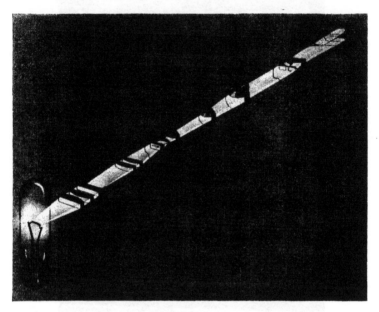

Fig. 3—Pictorial schematic of the optical system, omitting a prism and a
mirror which merely turn the optical axis for convenience.

beam to limit the scanning. This mask is controlled by a knob at
the upper right of the machine front surface as shown in Fig. 1. The
openings provide for 200-mil push-pull in either the standard or
offset positions and for 100-mil single, and 100-mil push-pull. In the
vertical direction all of the projected light beam being scanned passes
through each opening in the mask and is directed to each of the dual
cathodes of an RCA 929 phototube. These images are essentially
variable in intensity only, regardless of the manner in which the light
is cut at the film plane.

ELECTRICAL SYSTEM

The dual phototube is mounted by means of a phototube mesh assembly which is removable as a unit with the portion containing the phototube and its electronic components flexibly mounted to eliminate microphonic noise. Its circuit provides for switching from push-pull to parallel operation and a balancing potentiometer is provided for attaining balance in push-pull operation.

The phototube mesh circuit is intended for operation with several feet of concentric cable to an RA-1277 phototube amplifier which is contained in an auxiliary control cabinet shown in Fig. 4. This RA-1420 control cabinet also contains the exciter-lamp controls and space is provided for mounting a variable attenuator and film-loss equalizer as optional equipment for re-recording purposes.

CONCLUSION

This equipment has been found to give optical performance equivalent to that of high-quality re-recording equipment and is readily adaptable either to studio double-width or the standard single sound tracks. The scanning conditions are readily visible and adjustable during operation. Other essential performance characteristics such as flutter, signal-to-noise ratio, and frequency response are achieved

Fig. 4—Control unit, containing the phototube amplifier, attenuator, and variable equalizer as well as the exciter-lamp controls.

with a minimum of maintenance. The simplicity of design provides for rapidity in operation which is vital in reducing costs in re-recording operations. Although the magnetic-sound program is at this date still in the preliminary stage this equipment is readily adaptable for magnetic reproduction.

REFERENCES

(1) C. C. Davis, "An improved film-drive filter mechanism," *J. Soc. Mot. Pict. Eng.*, vol. 46, pp. 454–465; June, 1946.

(2) Wesley C. Miller and G. R. Crane, "Modern film re-recording equipment," *J. Soc. Mot. Pict. Eng.*, vol. 51, pp. 399–418; October, 1948.

Studio 16-Mm Re-Recording Machine*

By G. R. CRANE

WESTERN ELECTRIC COMPANY, HOLLYWOOD 38, CALIFORNIA

Summary—This paper describes a completely new 16-mm reproducer for studio re-recording operations which is comparable in basic performance to recently developed 35-mm equipment and contains similar facilities for convenience in rapid operation.

The film-drive mechanism gives excellent flutter performance with a maximum of simplicity and requires a minimum of maintenance. Various motor speeds may be accommodated.

The optical system provides for a maximum signal-to-noise ratio and contains an operating adjustment for optimum focus with emulsion on either side of the film. An adjustable speed rewind is provided with automatic shutoff.

The machine is contained in a steel-rack-type cabinet which stands from the floor and permits machines to be mounted side by side. Rack-mounting space is available for associated apparatus on standard panels.

THE INCREASING USE of 16-mm film by the motion picture industry has made apparent the need for improved high-quality reproducing equipment, particularly for use in the re-recording process. To meet this need a new 16-mm re-recorder has been developed, the over-all design of which follows that of new 35-mm re-recording machines recently described in the JOURNAL.[1] In addition to providing good basic performance, the facilities for rapid and economical operation and a minimum of maintenance, every effort has been made to simplify the equipment and to keep its cost commensurate with a 16-mm medium.

ARRANGEMENT OF EQUIPMENT

The RA-1417 re-recorder is shown by Fig. 1. The machine is housed in a relay-rack type of cabinet of conventional design which places the film-moving equipment in the upper section of the cabinet at a conveneint level for operation, and leaves the lower section available for the mounting of equipment such as lamp and high-voltage power supplies. This type of cabinet permits placing machines side by side in a row which facilitates operation and presents a

* Presented October 26, 1948, at the SMPE Convention in Washington.

pleasing appearance in the re-recording room. Ample space is provided for the use of 1600-foot reels with space for manipulation and threading without congestion. The rear of the cabinet has a hinged door for access to equipment and openings are provided at the top of each cabinet for exit of the film to a suitable overhead loop rack. The cabinets are finished in blue-gray and the large mechanism panel is heavy aluminum. All of the machine mechanism, amplifiers, wiring, and terminal strips are attached to this one member which may be removed as a complete unit.

Film-Pulling Mechanism

The film-pulling mechanism used in this machine is very similar to that used in a recording machine previously described in the Journal.[2] The mechanical filtering is provided by the tight-loop type of film path having controlled compliance with fluid damping. The theoretical considerations for this type of film motion have also been described[3] and therefore will not be repeated here. Experience with this film propulsion in recording equipment has demonstrated its ability to produce and maintain excellent film motion with both 16- and 35-mm film with a minimum of maintenance. Two flanged sprockets are used, the teeth of which are designed to fill the sprocket

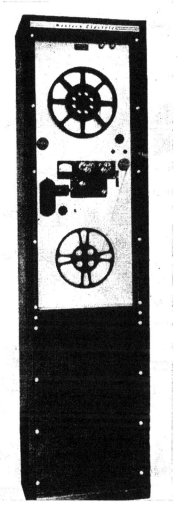

Fig. 1—Front view of re-recorder with film threaded for operation. The lower compartment is shown with a blank cover panel.

holes as nearly as possible consistent with reasonable shrinkage values. This reduces the so-called "crossover" effect which results from take-up and feed disturbances occurring when the film tensions approximate that in the filtered path between the two sprockets. As shown by Fig. 2, the film passes over two rollers, one in each path between each sprocket and the scanning drum. These rollers are mounted on rigid, pivoting arms, with the roller shafts coming through slots in the front surface to permit approximately vertical motion of each

Fig. 2—Close-up showing film path and optical system with focus control for emulsion in front or rear.

roller. The arms are provided with compliance by the use of a single, common spring which determines film tension in the path between the sprockets. The geometrical arrangement is such that the gravitational forces on the parts are compensated, and the system operates to keep the filter rollers in a mean position without requiring any periodic adjustments. In threading the machine there are no free loops and the proper threading position is indicated by a small target which causes a black line to be centered in a circular opening when the right sprocket hole is engaged. The line cannot be seen for one sprocket error in either direction. The handwheel for moving the drive mechanism manually consists of a disk, a section of which protrudes through the panel to the right of the mechanism. The flutter generated by this machine does not exceed ±0.08 per cent total or ±0.05 per cent at any given rate.

Rewind

To save time in the re-recording routine the upper-feed reel shaft is provided with a motor drive for rewinding the film. This drive consists of a 110-volt alternating-current, series motor driving the reel shaft through a friction drive consisting of a small steel-drive roller on the motor shaft bearing against a large pulley on the reel shaft having a special rubber surface, which gives a quiet and relatively efficient drive. This efficiency is important since the film in normal operation must drive the motor as a load and the maximum pull on the film at the smallest hub diameter of a 400-ft reel must not be excessive from the standpoint of sprocket-hole damage. A rheostat is provided as shown in Fig. 3 to alter the speed of this rewind and it becomes useful for slowing it down to inspect the track or select portions thereof. As shown by Fig. 3, the film is threaded for the rewind operation around an idler at the left-hand side of the machine. This idler is movable horizontally through a small distance and actuates a microswitch located in the rear to provide for automatic shutoff of the rewind as the film runs out. The rewind motor is controlled by a lock-in-type relay which is started by a push button labeled "start." The motor is normally stopped by the roller switch but it may be stopped at any time by a red push button labeled "stop."

Optical System

The optical system chosen for this re-recorder is similar to that which has long been used in theater-type 35-mm reproducing equipment but is, of course, modified to produce a scanning-light beam which is 72 mils long and $1/_2$ mil high. It is a projection type of optical system and a 5-ampere curved-filament lamp is used. All lens elements are contained in a tube with a working distance from the film to the first lens surface of approximately $1/_4$ inch, giving ample space for threading rapidly and cleaning the lens front element. The optical-scanning losses correspond closely to those of an ideal 0.5-mil slit, and amount to approximately 2.5 and 4.5 decibels at 5000 and 7000 cycles per second, respectively. Carlson has described both the optical system and the advantage of the curved-filament lamp in an earlier paper in the JOURNAL.[4]

Fig. 3—Close-up showing film path for rewinding, with rewind controls in upper right-hand corner.

The mounting of the lens tube provides for completely independent azimuth and focus controls as shown by Figs. 2 and 3. Azimuth is adjustable by means of a screw shown on the front of the assembly and focus is controlled by extremely fine thread incorporated in the lens-mounting sleeve. Both focus and azimuth controls are spring-loaded so that there is no backlash or looseness in either case. When focus has been determined with the emulsion either toward or away from the lens an index ring is then locked to the mounting sleeve in such a manner that it operates between stops with index lines to indicate optimum focus for the emulsion in either position. As an additional safety precaution a latch is provided to prevent the ring from being accidentally moved after being set for the emulsion position required.

The phototube is located within the scanning drum and is flexibly mounted to eliminate microphonic noise caused by mechanical vibration. It is coupled by concentric cable to the phototube amplifier and the lightproof hood over the phototube is readily removable for inspection or cleaning. The over-all performance is such that the illuminated area on the phototube closely approximates variable intensity as the scanning beam is cut in any manner at the film plane. An RCA No. 927 gas-filled, red-sensitive phototube normally is used, but the RCA No. 5583 blue-sensitive tube may be used interchangeably.

AMPLIFIER

The phototube amplifier contains a special feedback circuit to permit several feet of shielded cable to be used between the phototube and the amplifier. This allows the amplifier to be mounted in a convenient location removed from conflict with mechanical parts. The feedback reduces the effective input impedance so that the cable losses normally encountered are eliminated to give an over-all electrical frequency response which is substantially flat. The signal-to-noise ratio is better than that of conventional methods and the noise caused by the amplifier components is considerably below phototube hiss and film noise. The amplifier is readily removable from its four rubber mounts through the rear of the machine.

CONCLUSION

This machine has, of course, dispensed with a few of the more de luxe features provided for 35-mm equipment but no compromise has been made with the basic performance. High-quality reproduction is obtained with a minimum of maintenance and the equipment layout provides facilities for rapidity in operation which is vital in reducing costs in re-recording operations. The cabinet-style assembly offers convenience in operation, affords space for mounting associated equipment, eliminates the necessity of supplying auxiliary support structures, and greatly improves the appearance of the re-recording installation.

REFERENCES

(1) Wesley C. Miller and G. R. Crane, "Modern film re-recording equipment," *J. Soc. Mot. Pict. Eng.*, vol. 51, pp. 399–418; October, 1948.

(2) G. R. Crane and H. A. Manley, "A simplified all-purpose film recording machine," *J. Soc. Mot. Pict. Eng.*, vol. 46, pp. 465–475; June, 1946.

(3) C. C. Davis, "An improved film-drive filter mechanism," *J. Soc. Mot. Pict. Eng.*, vol. 46, pp. 454–465; June, 1946.

(4) F. E. Carlson, "Properties of lamps and optical systems for sound reproduction," *J. Soc. Mot. Pict. Eng.*, vol. 33, pp. 80–97; July, 1939.

------◆------

High-Speed Photography

Society members were surprised and pleased to receive the 130-page supplement on high-speed photography which was part of the March, 1949, JOURNAL. Past work of the Society in this field and plans for the future were outlined in the Foreword written by John H. Waddell of the Bell Telephone Laboratories, who is Chairman of the High-Speed Photography Committee.

As the first publication of its kind in this rapidly expanding field of scientific photography, the supplement has been well received by branches of industry that are not normally interested in the work of the Society of Motion Picture Engineers. For example, quantity orders for additional copies of the supplement have come from aircraft manufacturers, steel companies, the National Military Establishment, and a multitude of research laboratories.

Additional copies of the supplement are available to schools or individuals who would like to refer to the material contained in this "first standard text."

Members are asked to call this publication to the attention of their friends and business associates so that it may be used to advantage by engineers or researchers everywhere.

Copies are $1.50 each and should be ordered from

THE SOCIETY OF MOTION PICTURE ENGINEERS
342 Madison Avenue
New York 17, N. Y.

35-Mm-to-16-Mm
Sound Reduction Printer*

By C. W. CLUTZ, F. E. ALTMAN, and J. G. STREIFFERT

EASTMAN KODAK COMPANY, ROCHESTER 4, NEW YORK

Summary—While retaining the basic principles of the earlier Eastman sound reduction printers, this new model incorporates significant improvements such as: a newly designed apochromatic objective system, a condenser system of increased efficiency, reduction of film wear by elimination of all stationary members in the film path, increased film capacity, torque motor take-ups, lifetime lubrication of most bearings, pedestal mounting for greater convenience of operation, and complete operating controls conveniently located.

CONTINUOUS OPTICAL REDUCTION PRINTING has been one of the principal methods of producing 16-mm sound prints since the first commercial introduction of sound reduction printers in 1934[1, 2] and 1935.[3] Because of the inherent differences in size and number of perforations between 35-mm and 16-mm film, it is logical that, with comparable effort and care, a procedure which uses the 35-mm medium prior to release printing by reduction must, in general, produce results of higher quality than one which uses the 16-mm medium throughout. The continually increasing demand for higher-quality 16-mm sound films, particularly in connection with the rapidly expanding television industry, emphasizes the necessity for using those procedures most productive of high quality.

In order to transfer the sound-track image from 35-mm film to 16-mm film with maximum fidelity, the reduction printer must have several attributes. First, an optical system of the highest quality must be used to achieve maximum high-frequency response with the various black-and-white and color materials being used. Second, the relative motion between the optical image projected onto the 16-mm film and the film itself must be held to a minimum to prevent the introduction of perceptible wows and flutter. Third, for long troublefree service, such a printer must be built sturdily, simple and convenient to operate, and free from the necessity of making or checking delicate adjustments by the user.

* Presented October 28, 1948, at the SMPE Convention in Washington. Communication No. 1231 from the Kodak Research Laboratories.

The same basic design is used in the Eastman Sound Reduction Printer, Model D, as in the earlier models. Briefly, this is an arrangement for projecting an image of the 35-mm track, properly reduced and oriented, onto the 16-mm film while the two films are engaged with appropriate sprockets on opposite ends of a common shaft. The use of improved optical and mechanical materials, designs, and components, and the provision of greater convenience and ease of operation have been the principal objectives in designing the new printer.

Fig. 1, illustrating the 16-mm side of the printer, shows the pedestal mounting which permits easy access to both sides for threading. Fig. 2 is a close-up of the 35-mm film path. Cushion-tired casters in the base facilitate moving the printer from place to place. The motor and lamp controls are located at the front where they are convenient for the operator when viewing both film paths.

The various mechanical and optical components are mounted on one rigid main casting made of a special nickel-iron alloy. This casting is annealed for dimensional stability both before and after rough machining, and, in addition, is made impervious to oil by vacuum-pressure impregnation with Bakelite varnish.

Fig. 1—Eastman Sound Reduction Printer, Model D, 16-mm side.

The main driving mechanism consists of a high-quality integral worm-gear reduction motor which has the printing sprockets mounted on opposite ends of its output shaft. The supply and take-up sprocket shafts are coupled to this same output shaft by means of a silent chain running in oil. The take-up spindles are driven by individual torque motors. By this means good take-up characteristics for all roll sizes are obtained without the use of friction drives and their inherent maintenance problems.

The simplicity of this mechanical arrangement, together with the use of standard, commercially available components wherever pos-sible, should greatly ameliorate any mechani-cal servicing problem that might arise. Very little lubrication service is required because of the use of oil baths and of lifetime lubricated and sealed ball bearings throughout. The printer can be ·equipped with motors for any standard voltage or frequency.

Speeds of travel of the 35-mm and 16-mm films are 150 and 60 feet per minute, respectively. Roll capacities are 3000 feet of 35-mm film and 1600 feet of 16-mm film.

At no point in their travel through the printer do the film surfaces come into contact with sta-tionary members. This not only reduces to a minimum surface wear of the films, but also im-

Fig. 2—Eastman Sound Reduction Printer, Model D, 35-mm side.

proves the film motion by eliminating variations in film tensions above and below the printing sprockets. Such variations in ten-sion are caused by differences in the coefficient of friction from one part of a roll to another and by various types and conditions of film base, backing, and so forth.

The films are fed on and taken off the printing sprockets by means of rollers set at the thickness of the film from the sprockets. With the path of the film as it engages and disengages the printing sprocket thus defined, the base diameter of the sprocket and the best tooth shape are determined by the shrinkage range to be accommodated, as outlined by Chandler.[4, 5] The 35-mm printing sprocket is thus designed to accommodate film of from 0 to 0.5 per cent shrinkage, and the 16-mm sprocket for film of 0 to 0.3 per cent shrinkage. Using Chandler's method of computing average theoretical flutter, for various values of swelling or shrinkage, we obtain the values indicated in Table I.

TABLE I

16-Mm Film Pitch	Flutter, %	35-Mm Film Pitch	Flutter, %
0.300 in. +0.0%	0.0	0.1875 in. +0.0%	0.0
−0.1	0.13	−0.1	0.08
−0.2	0.19	−0.2	0.14
−0.3	0.22	−0.3	0.17
		−0 4	0.19
		−0.5	0.20

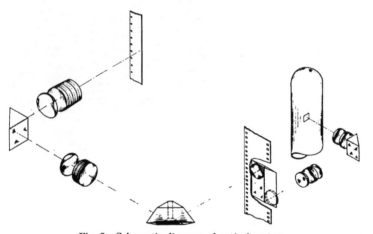

Fig. 3—Schematic diagram of optical system.

Experience with these printers to date indicates that flutter values of this order are being achieved.

Run-out of the base circle of the printing sprockets and indexing of the teeth are kept to very close tolerances to avoid the introduction of long-period flutter or wows.

Fig. 3 is a schematic drawing of the complete optical layout. Since an anamorphote objective system is required to obtain the proper difference in image reduction between the vertical and horizontal dimensions, a cylindrical lens is introduced into the condenser system to fill completely the exit pupil of the objective system with light in both meridians. The maximum illumination available is approximately ten times that currently used for printing Kodachrome or black-and-white films.

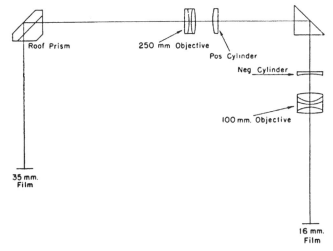

Roof Prism   250 mm Objective
Pos Cylinder
Neg Cylinder
100 mm. Objective
35 mm. Film
16 mm. Film

Fig. 4—Detail of objective system.

Fig. 4 shows a plan view of the components of the objective system. Basically, it consists of a 250-mm and a 100-mm telescopic objective, each located at its focal length from the 35-mm and 16-mm films, respectively, so that there is collimated or parallel light between them. This provides the proper vertical reduction in image size. A positive cylindrical lens and a negative cylindrical lens constituting an afocal system, i.e., one which does not change the convergence or divergence of the rays but merely changes the magnification, is introduced between the spherical objectives to achieve the proper horizontal magnification. Fig. 5 illustrates the high order of correction for spherical and chromatic aberration which has been achieved in the telescopic objectives over an effective aperture of $f/5.0$ and for the wavelength range of 410 to 650 millimicrons through judicious design and selection of glasses with special regard to their effect on secondary spectra.

The fact that the three curves are parallel and practically super-imposed indicates that marginal and zonal spherical aberration have been corrected over the entire wavelength range, while the straight-

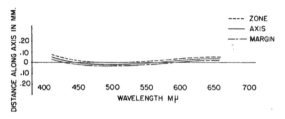

Fig. 5—Graph showing the high degree of spherochromatic correction of the objective system.

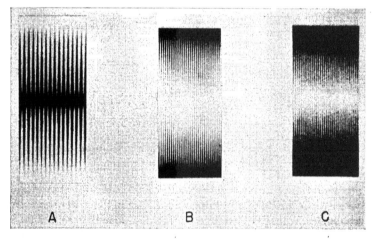

Fig. 6—Photomicrographs of: A, 9500-cycle, 35-mm negative; B, aerial image at 16-mm printing point; and C, 16-mm reduction print on fine-grain release positive film.

ness of the lines indicates the excellent correction for secondary spectra.

Photographic tests using tricolor filters and panchromatic film have verified these computed values by indicating a change of less than ± 0.02 mm in the position of optimum focus for the various filters.

The theoretical resolving power of an optical system can be computed by taking the reciprocal of the radius of the first dark ring in the diffraction pattern surrounding the Airy disk image of an artificial star, namely,

$$\text{resolving power} = 1/r = d/1.22 f \lambda,$$

where d is the diameter of the last lens, f is its focal length, and λ is the wavelength of light. Using this formula, we arrive at a value of 160 lines per millimeter at 500 millimicrons. Both visual and photographic observations indicate that this theoretical value is being reached or exceeded. A 6000-cycle, 16-mm track has approximately 30 cycles or "lines" per millimeter so that the system is capable of resolving a pattern at least five times as fine as that of a 6000-cycle track.

Further evidence of this high resolution is shown in Fig. 6, in which A is a photomicrograph of a 9500-cycle-per-second, 35-mm negative, B is a photomicrograph of the aerial image of this negative at the 16-mm printing plane, and C is a photomicrograph of an actual 16-mm No. 7302 print of the same negative. These photographs indicate that the imaging qualities of the system are well in excess of what the film is able to record.

In conclusion, we should like to emphasize the importance of insuring that the films in any printer travel in planes which are parallel within a high order of accuracy to prevent the introduction of an azimuth error in the print. By using flanged rollers above and below the printing points, the azimuth error introduced by the printer is kept within approximately ±6 minutes.

REFERENCES

(1) A. F. Victor, "Continuous optical reduction printing," J. Soc. Mot. Pict. Eng., vol. 23, pp. 96–100; August, 1934.

(2) G. L. Dimmick, C. N. Batsel, and L. T. Sachtleben, "Optical reduction sound printing," J. Soc. Mot. Pict. Eng., vol. 23, pp. 108–117; August, 1934.

(3) O. Sandvik and J. G. Streiffert, "A continuous optical reduction sound printer," J. Soc. Mot. Pict. Eng., vol. 25, pp. 117–127; August, 1935.

(4) J. S. Chandler, "Some theoretical considerations in the design of sprockets for continuous film movement," J. Soc. Mot. Pict. Eng., vol. 37, pp. 164–177; August, 1941.

(5) J. S. Chandler, D. F. Lyman, and L. R. Martin, "Proposals for 16-mm and 8-mm sprocket standards," J. Soc. Mot. Pict. Eng., vol. 48, pp. 483–521; June, 1947.

High-Quality Recording Electronic Mixer*

By KURT SINGER

RCA Victor Division, Hollywood 28, California

Summary—This paper describes an electronic mixer having performance characteristics suitable for high-quality sound recordings. Rapid acting time and comparatively fast restoring time are some of the features enumerated. Complete performance characteristics and mechanical arrangements are shown.

THE ELECTRONIC MIXER described in this paper is another item in RCA's new line of recording equipment. Other components of this line have been described in three papers which have been given previously.[1-3] Considering the high performance of these equipment items, it was only natural that in the quest for continuous refinement of recording quality, the need for an improved electronic mixer or, as it is also called, compressor, made itself felt. The main objectives in the design of this device which had to be met were as follows:

1. Sufficient gain to provide all the necessary amplification between the mixer and the bridging bus.

2. Input impedance 600 or 250 ohms; output impedance 500 to 600 ohms.

3. Freedom from distortion and fast acting time so as not to impair the high quality of the remaining recording channel components.

4. Flat frequency characteristic within 0.5 decibel from 20 to 10,000 cycles.

5. At least 50 decibels signal-to-thump ratio.

6. Mechanical construction to provide for 100 per cent front service. This was judged particularly important for installations where the rear of the amplifier rack is not accessible.

7. Alternating- or direct-current operation so as to permit use in studio or truck channels.

The result of the development of a device which meets the above objectives is shown in the following illustrations. Fig. 1 shows the amplifier mounted, partially pulled out for access to tubes. The mechanical construction is identical with the arrangement shown in

---

* Presented October 27, 1948, at the SMPE Convention in Washington.

two previously presented papers.[1, 2] This construction permits pulling out the chassis from the rack and tilting it in such a manner that the underside of the chassis becomes instantly accessible. By means of plug-in cable connections which can be made or broken rapidly, it is possible to remove the entire amplifier chassis from the rack without use of tools for bench test or service. No attempt will be made to go into the details of the mechanical construction since they have been adequately covered previously.

The controls on the front panel and their purpose are as follows: An input gain control, which provides twenty steps of 1 decibel each

Fig. 1

is located at the left. A selector switch at the right permits selection of "Electronic" or "Manual" operation of the electronic mixer, which means selection of compressed or uncompressed operation. It also provides for a position which is used for balancing the variable gain stage. In addition, another selector switch in the center permits an indication of correct vacuum-tube operation and power-supply voltages. A balancing potentiometer facilitates obtaining tube balance of the variable-gain stage, by means of an external balancing voltage easily introduced through a test jack pair.

Fig. 2 shows the circuit schematic. It will be noticed that this electronic mixer or compressor is of the backward-acting type: which means that a voltage derived from the output of the electronic mixer is amplified, rectified, filtered, and reapplied as a gain-

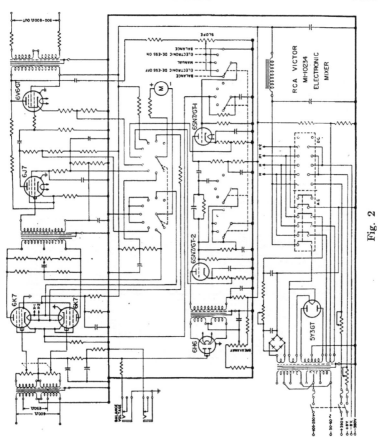

Fig. 2

control voltage to the variable-gain stage.   This circuit arrangement is by no means new; however, it contains a few refinements.   The input to the electronic mixer is applied to an input transformer, the secondary of which is terminated in a dual potentiometer-type gain control.   The swingers of this gain control connect to the grids of a pair of variable-mu tubes, 6K7's in this case, which are connected in a push-pull arrangement and transformer coupled to a two-stage output amplifier which employs negative feedback.   A simple balancing arrangement located between the plates of the variable-gain stage and the plate-coupling transformer permits adjustment for minimum longitudinal output from these two tubes.   A subsequent paper will elaborate on the manner in which longitudinal balance is obtained over the dynamic operating range of these tubes. The transformer which couples the variable-gain stage to the fixed-gain output amplifier limits any unbalance which may occur in the variable-gain stage to this stage itself. Consequently the characteristic of the tubes following the variable-gain stage is relatively unimportant so far as longitudinal balance of the variable-gain stage is concerned. In some previous compressors which employed resistance coupling between the variable and the fixed-gain stages it was necessary to balance also the fixed-gain stage since cancellation of longitudinal voltages did not occur until they reached the output transformer. To continue with the description of the circuit, voltage derived from the output amplifier is amplified in two resistance-coupled triode stages and fed to a full-wave rectifier through a coupling transformer. A frequency-selective network, located between the two resistance-coupled triode stages, permits the introduction of a frequency characteristic in the control amplifier conforming to the requirements set up for minimum spectral-energy distortion. This

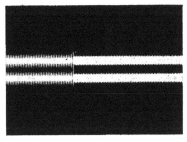

Fig. 3

phenomenon was first discussed by Miller[4] and subsequently has been covered in a paper by Rettinger and Singer.[5] The rectified control voltage is filtered by means of a simple resistance-capacitance network and then applied to the control grids of the variable-gain stage as a variable bias. The acting time, which is a function of the charging time of the capacitors in the timing filter, has been kept as short as practical without having to go to an unsuitably long restoring time. The acting time of this electronic mixer is less than one-half millisecond. Fig. 3 shows a recording of a 5000-cycle tone whose amplitude increases suddenly by 20 decibels. The restoring time is adjustable by selection of various values of discharge timing resistors; however, this change cannot be made rapidly since it entails the use of a soldering iron and replacement of the discharge timing resistor. From experience in the studios, a restoring time of 100 milliseconds for 99 per cent gain restoration has been found the most universally accepted. Longer restoring times tend to introduce echo effects which are definitely objectionable. In re-recording, sometimes a restoring time of half a second for 99 per cent gain restoration is used. In most studios, however, electronic mixers used for re-recording purposes are not normally used for original recordings and are permanently adjusted for the longer restoration time.

A rotary selector switch in conjunction with a 500-microampere meter permits metering of all tubes and of the A and B supply. The metering resistors are of such values that, when a meter indication in the center of the dial is obtained, correct operation is assured.

A simple switching arrangement permits selection of either alternating- or direct-current supply, the alternating-current supply

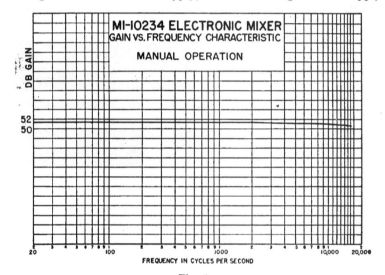

MI-10234 ELECTRONIC MIXER
GAIN VS. FREQUENCY CHARACTERISTIC

MANUAL OPERATION

DB GAIN

52
50

20   100   1000   10,000   20,000
FREQUENCY IN CYCLES PER SECOND

Fig. 4

being self-contained. In the studios the electronic mixer is intended to be operated from the alternating-current mains in conjunction with a line-voltage regulator. It will be noticed that a small selenium-bridge-type rectifier supplies direct current to the heaters of the 6K7 tubes in the variable-gain stage. This expedient was found necessary in order to fulfill the signal-to-noise ratio requirements that had been set up. For direct-current operation, direct current has to be supplied from external sources such as batteries, regulated power supplies, or dynamotors.

Figs. 4 and 5 show the frequency characteristics of the device. Fig. 4 shows the uncompressed characteristic which is, for all practical purposes, flat from 20 to 14,000 cycles. Fig. 5 illustrates the frequency characteristics that can be obtained in electronic operation. The upper curve shows electronic operation without the use of the spectral-energy-distortion eliminator which is usually called "de-esser." The lower curve illustrates the frequency characteristic which is obtained when the "de-esser" is in the circuit. It will be seen that both of these curves show a secondary portion in dashed lines. The solid curves represent the frequency characteristic that is obtained if measured with a conventional gain set. Due to the

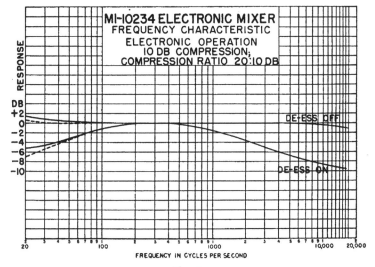

Fig. 5

extremely fast attack time the filtering of the control voltage is reduced at the very low frequencies (below speech frequencies), hence, some of the unfiltered ripple components contribute to the apparent low-frequency characteristic as measured on a gain set. However, if the frequency characteristic is measured by means of a frequency-discriminating device as, for instance, a wave analyzer, then the true frequency characteristic will be obtained, which is shown in dashed lines. The distortion as obtained from this electronic mixer in the uncompressed position is less than 0.5 per cent from 50 to 7500 cycles at an output level of +28 dbm.* The distortion occurring during compressed operation is a function of filtering

* Decibels with respect to 0.001 watt.

of the control voltage which in turn is dependent upon the restoring time. With a restoring time of 100 milliseconds and an output level of +18 dbm,* a 400-cycle distortion on the order of 0.8 per cent will be measured. However, it must be understood that this 0.8 per cent does not represent nonlinear distortions which occur in the variable-gain or fixed-gain stages of the electronic mixer. The nonlinear distortion is on the order of about 0.2 per cent; however, the ripple components of the control voltage consist mainly of the second

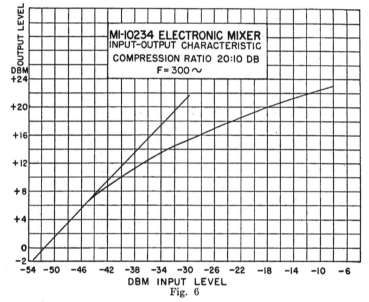

Fig. 6

harmonic of the fundamental and consequently will be measured and expressed in terms of per cent distortion of the fundamental. It is therefore not surprising that under high compressions, distortions on the order of 3 per cent can be measured at 60 cycles with a restoring time of 100 milliseconds. This distortion goes down to around 1 per cent as the restoring time is increased to half a second.

The input-output characteristic compressed and uncompressed is shown in Fig. 6. The electronic mixer was adjusted for a compression ratio of 20 into 10 decibels. The breakaway point occurs at an output level of around +8 dbm.* Full compression, that is 10-decibel gain reduction, is obtained at an output level of +18 dbm.* This input-output characteristic covers the most widely used operating range of this electronic mixer. Another favorite with some of the studios is a compression ratio of 30 into 15 decibels. Other compression ratios are not illustrated because the subject has been well covered in the past. This device can also be adjusted for limiting and then a compression ratio of 20 into 4 can be obtained. Slope and breakaway controls for compression-ratio adjustments are located on the underside of the chassis and are readily accessible by pulling the electronic mixer from the rack and tilting the chassis by means of the tilting arrangement. The signal-to-noise ratio is 75 decibels referred to an output level of +18 dbm.* The signal-to-thump ratio can be held consistently better than 50 decibels. This is accomplished by preselection of 6K7 tubes which are to be used in the variable-gain stage.

REFERENCES

(1) Kurt Singer, "A high-quality recording power amplifier," J. Soc. Mot. Pict. Eng., vol. 48, pp. 560–569; June, 1947.

(2) Kurt Singer, "Versatile noise-reduction amplifier," J. Soc. Mot. Pict. Eng., vol. 50, pp. 562–571; June, 1948.

(3) M. E. Collins, "A de luxe film recording machine," J. Soc. Mot. Pict. Eng., vol. 48, pp. 148–156; February, 1947.

(4) B. F. Miller, "Elimination of relative spectral energy distortion in electronic compressors," J. Soc. Mot. Pict. Eng., vol. 39, pp. 317–324; November, 1942.

(5) M. Rettinger and K. Singer, "Factors governing the frequency response of a variable-area film recording channel," J. Soc. Mot. Pict. Eng., vol. 47, pp. 299–327; October, 1946.

DISCUSSION

CHAIRMAN F. L. HOPPER: In the electronic mixer, have you observed in past operation in the mixer part any delay in attenuation of the signal when the gain is rapidly cranked down or off, and whether there are any low-frequency transients? In mixing as you get a fadeout or rapid change in the mixing attenuator, do you notice any delay in gain change?

MR. HOLLIS D. BRADBURY:* The operating time is so very fast that you could not hear it. In fact, you must have at least a 5000-cycle frequency to measure it. Half a millisecond is pretty short. It is difficult even in making the test to key the signal so cleanly that you get an accurate indication of the amplifier performance; otherwise, you read key clicks on the film. Anything you do in the way of nominal recording is much slower than in the test condition.

I do not feel competent to answer the other question. Perhaps there is someone here who would answer it from RCA. Would you restate the second half of the question?

CHAIRMAN HOPPER: If in rapid gain changes made as suggested is any low-frequency transient introduced or filter added to eliminate it?

MR. BRADBURY: If the tubes are sufficiently well balanced (and these days that is not a problem), there is no unbalance developed, 50 decibels or better below the signal.

* NOTE: Mr. Bradbury read the Electronic Mixer paper.

Preselection of Variable-Gain Tubes for Compressors*

By KURT SINGER

RCA VICTOR DIVISION, HOLLYWOOD 28, CALIFORNIA

Summary—A novel method of obtaining balanced pairs of variable-gain tubes for compressors and limiters is described. Virtually thumpless operation is obtained through the use of these balanced variable-gain tube pairs.

IN THE PAST, selection and balance of the tubes in the variable-gain stage of any compressor or limiter has always been a problem. Various means of obtaining tubes with matched characteristics over their dynamic range have been used; however, they all lacked a 100 per cent safety factor. It has been known that when 6K7 tubes are paired by matching their static grid-voltage–plate-current characteristics, usable pairs after tedious selection could be obtained; however, some pairs, although carefully matched did not prove free from thump in actual operation. To remedy this, various means of obtaining dynamically matched, variable-mu, tube pairs were investigated. From the methods suggested, a test procedure was worked out which has proved very successful. It has been possible to obtain matched pairs of 6K7 tubes which can be sent to the field with certainty that they will work properly in various electronic mixers. If the balancing control in these electronic mixers is adjusted properly, freedom from thump and ease of maintenance will be the reward.

The following method is employed to obtain dynamically matched pairs of 6K7 tubes. First, all 6K7 tubes are aged for 100 hours using the circuit shown in Fig. 1. The tubes are triode-connected and a plate potential of 100 volts and proper cathode bias are applied. Any number of tubes can be aged simultaneously provided individual cathode resistors are employed. The heater voltages are kept constant by means of voltage regulators and are never allowed to fall below 6.3 volts. Experience has shown that a tube in which the

---

* Presented October 27, 1948, at the SMPE Convention in Washington.

heater voltage has fallen below 6.3 volts during aging will not maintain its characteristic and will drift. It is desirable to work these
tubes at a plate current somewhat higher than in actual operation to
stabilize them. After the tubes have aged for 100 hours, they are
removed from the aging racks and put in the new tube-matching
curcuit which is illustrated in Fig. 2. This tube-matching circuit
provides a rapid method of selecting and determining the dynamic
characteristic of 6K7 tubes over a wide range of grid bias and also
provides a static test by exploring only a short section of the grid-
voltage–plate-current characteristics. The circuit used in this tube-

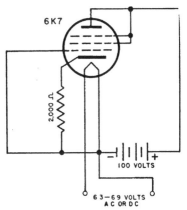

Fig. 1—Tube-aging circuit schematic.

matching circuit is essentially a duplication of the circuits surrounding the 6K7 tubes in actual operation. An apparent difference in the
circuit employed for matching as compared to the circuit used in
actual operation lies in the fact that the two grids of the tubes being
matched are connected in phase, that is, in parallel, whereas in actual
operation the two grids are connected 180 degrees out of phase, that
is, in push-pull relation. It must be borne in mind however that, so
far as longitudinal voltages are concerned, the two grids are in parallel
also in actual operation. The plates of the tubes are connected in
push-pull in the matching circuit as well as during operation.

Briefly, the following takes place when the dynamic characteristic
of two tubes is investigated and compared in this circuit. By means
of a clamp circuit consisting of a rectifier tube in conjunction with
suitable capacitor and resistor, a certain negative bias is impressed

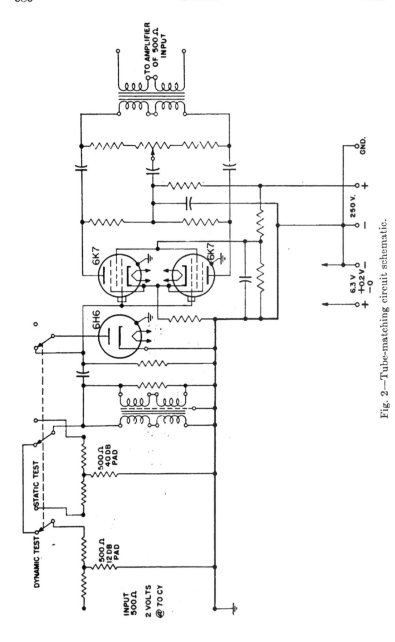

Fig. 2—Tube-matching circuit schematic.

upon the control grids of the tubes under investigation. An alternating-current potential is also applied which swings these grids negatively to a voltage corresponding to 20-decibel gain reduction and positively to a voltage representing zero control voltage. Since the plates of the two tubes under test are connected in push-pull, no differential output will be obtained if the dynamic characteristics of these two tubes are identical over the entire range of grid swing. Since this is an ideal which is not usually attained in practice, certain compromises have to be made. First, it may happen that two tubes have dynamic characteristics of practically identical slope which are, however, displaced from each other. In order to make such tubes usable, a balancing potentiometer is incorporated in the electronic mixer and a corresponding balancing potentiometer of lesser range is also used in this tube-matching circuit. By an adjustment of this potentiometer for minimum output, the dynamic-tube characteristics which are alike in slope but displaced can be made to coincide and produce the same minimum output as two tubes whose dynamic characteristics are identical. By means of this test, tube pairs are selected which will perform satisfactorily over a gain-reduction range of 20 decibels. However, to insure that the tube pairs will match without readjustment of the potentiometer at close to zero gain reduction or, say 2 or 3 decibels compression, another similar test is performed during which the grids of the tubes under test are biased to zero gain-reduction voltage and swung over a very small range by means of a reduced alternating-current potential. The zero gain-reduction region is important since in actual operation the tube balance is checked at zero gain reduction and a readjustment of the balance potentiometer for a match at high values of gain reduction cannot be effected.

For simplicity and ease of understanding, the tube-matching test circuit is shown containing only two tubes under investigation. Actually four multideck selector switches permit the comparison of 21 tubes. Consequently the comparison between tubes is very much simplified since they do not have to be removed from their sockets nor is there any delay in waiting for stabilization of heater temperature.

In order to facilitate understanding of the function of this tube matcher, consider the circuit in the light of the preliminary information given during discussions of the matching procedure itself. An input signal of 2 volts at approximately 70 cycles is applied to a

12-decibel pad. This pad isolates the oscillator from any impedance variation which might occur due to loading effects of the rectifier tube across the secondary of the transformer. This 12-decibel pad is followed by a 40-decibel pad. By means of a switch the 40-decibel pad can be connected or disconnected from the circuit. The position labeled "Dynamic Test" disconnects this 40-decibel pad. In the dynamic test position, a signal is fed to the primary of a transformer, the secondary of which is loaded to present a 500-ohm impedance to the input pad. The loaded secondary is connected to a series capacitor and a shunt resistor in parallel with which is a half-wave rectifier, a 6H6 tube specifically. This circuit combination of series capacitor, shunt resistor, and shunt rectifier forms what is known as a clamp circuit. The function of this clamp circuit is as follows. Let us visualize the current which occurs at the secondary of the transformer as an alternating-current sine wave having a zero axis which charges the series capacitor to substantially the peak amplitude of the first positive cycle of the wave. However since the rectifier tube shunts the secondary of the transformer in one direction only, the actual wave impressed on the grids of the tubes under test is shifted in such a manner that substantially all of the wave is negative, e.g., below the zero axis of the wave at the secondary of the transformer. There is only a very small flat portion at the crests of this wave caused by the discharge of the series capacitor through the resistor in shunt with the rectifier. However, for all practical purposes, the entire wave impressed on the grids is negative and its peak value is essentially the same as that of the wave across the secondary of the transformer. This negative displacement of the zero axis is desired since the gain-control voltage impressed on the tubes under actual operation is all in the negative direction. The input voltage and constants of the series capacitor and shunt resistor are dimensioned in such a manner that the grids of the tubes under investigation are swung negatively to a value corresponding to a 20-decibel gain reduction and positively corresponding to about zero gain reduction. The circuit constants surrounding the two 6K7 tubes in the test are identical to the ones used in the electronic mixer, with the exception that the balancing potentiometer located in the plate circuit has less range than in the production units. This added safety feature was found desirable since component tolerances of production units have to be considered. Upon switching to the "Static Test" position, the grid swing of the tubes under investigation is reduced by 40 decibels and the clamp circuit is disconnected by opening the connection to the 6H6 rectifier plate. The grids are now kept at zero potential with respect to ground which corresponds to zero gain reduction and the characteristics are investigated only over a very small range since they receive now 40 decibels less grid swing than in the dynamic test position.

This test circuit has been used in actual production for the purpose of producing an adequate stock of dynamically matched pairs of 6K7 tubes and has been found extremely satisfactory both from a consideration of ease of operation and of higher yield than can be obtained by plotting grid-voltage versus plate-current characteristics. In addition, once two tubes have met the criteria set up for satisfactory operation, it is certain that these tubes will perform satisfactorily in production units. Several of the Hollywood studios have used this circuit and all report uniformly satisfactory results.

ACKNOWLEDGMENT

The tube-matching arrangement was worked out by J. J. DeMuth of RCA's Hollywood Engineering Products Division from a general suggestion of W. N. Masters also of the Hollywood office.

Direct-Positive Variable-Density Recording with the Light Valve*

By C. R. KEITH and V. PAGLIARULO

WESTERN ELECTRIC COMPANY, NEW YORK 5, NEW YORK, AND
LOS ANGELES 38, CALIFORNIA

Summary—A method is described for making "direct-positive" variable-density sound records. Improvements over previous records are high output level, high signal-to-noise ratio, low distortion, and elimination of printer loss and distortion. Standard light valves with a superposed high-frequency bias are used in making the record.

A NEW METHOD of photographic sound recording has recently been developed which combines the advantages of a variable-density record with those of a direct positive. As is well known, a direct-positive record is one which does not require printing; that is, the film which runs through the recorder is, after development, suitable for high-quality reproduction. Elimination of printing improves frequency response, reduces flutter, and avoids the cost of negative film and development. The advantages of variable-density records in minimizing the effects of improper adjustment of the reproducer scanning beam are well known. In addition to these advantages, the direct-positive record here described has about 8 decibels higher output and 6 decibels higher signal-to-noise ratio than a standard variable-density record without noise reduction, so that in many cases noise reduction, with its attendant distortions, may be eliminated.

Direct reproduction of sound from an original variable-density record requires either special nonlinear electrical circuits to compensate for the nonlinearity of the negative[1] or accurate linearity in the record itself. While the former has been shown to be feasible, it is, of course, preferable to have a linear record which does not require special reproducing circuits. The so-called "toe" records[2] are also a form of direct positive, but their use is limited due to insufficient linearity to give good quality at full modulation.

---

* Presented October 26, 1948, at the SMPE Convention in Washington.

Fig. 1 is a typical curve showing the relation between transmission and exposure of a "toe" record. As would be expected the intermodulation is rather high for signal modulation of 50 per cent or higher. This has been a serious limitation to the use of toe records for high-quality sound reproduction.

In the present method, photographic distortion has been very considerably reduced by the superposition of an alternating-current bias on the light-valve ribbons along with the signal currents. The alternating-current bias is applied at constant amplitude and at a frequency high enough so that it is not recorded on the film to any

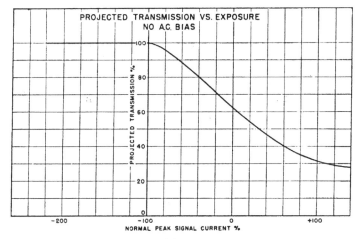

Fig. 1

great extent and is not reproduced by normal reproducing equipment. A bias frequency of 24 kilocycles has been found satisfactory. Its purpose is to convert the light transmitted by the valve to a wave shape which, when combined with the photographic properties of the film, will give a linear relation between light-valve signal current and film transmission (projected). It has been found that this is accomplished quite well when the peak amplitude of the bias current is sufficient to open and close the ribbons by approximately twice the amount of the unbiased, unmodulated ribbon spacing. When speech or music currents are applied, they have the effect of shifting the midpoint of the bias oscillation. With the above bias amplitude (called 200 per cent bias), the valve is more than closed for at least a portion of each bias cycle for usual values of signal amplitude. However, the ribbons do not strike each other since they are strung in different planes. The light is, of course, entirely cut off during the portion of the bias cycle for which the ribbons overlap, giving several pulsations of light for each signal cycle.

The effect of the alternating-current bias is shown in a general way in Fig. 2. While this figure is drawn on the basis of a single-ribbon valve, the same result is obtained, with a two-ribbon valve. When there is no alternating-current bias, the light through the valve

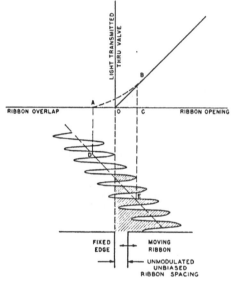

Fig. 2.—Change in exposure produced during recording with alternating-current bias.

is directly proportional to the opening between the ribbons and is shown as the line OB. Also with no alternating-current bias there is no light transmitted for any closed positions of the ribbons such as OA. When a high-frequency current is superposed on the valve strings the spacing for each instantaneous value of signal current goes through a sine-wave variation corresponding to the peak amplitude of the alternating-current bias. When the signal is greater (in the direction opening the valve) than the bias amplitude, as at E, the light varies throughout the bias cycle, but the presence of the bias does not change the average amount of light during a bias cycle.

However, when the signal has lower values, as between D and E, the valve is closed during part of a bias cycle and the transmitted light takes the form of pulses as shown in the shaded portions of Fig. 2. For signal currents of sufficient negative polarity, as at D, the valve is not open for even the peaks of the bias wave so that the average light is zero. Therefore signal currents varying from D to E vary the average transmitted light in the presence of an alternating-current bias along the curve AB. This curve may be readily calculated by Fourier analysis, in which the points on the curve AB are the average or direct-current components of the portions of the

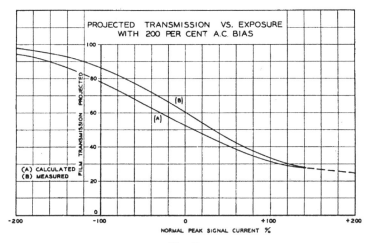

Fig. 3

bias cycle during which the valve is open. Fig. 3 (A) shows the calculated effect of the exposure as modified by the alternating-current bias on the exposure-transmission curve of Fig. 1. The fact that the curve extends into the region of valve closure (negative valve opening) simply means that this portion of the curve represents a signal amplitude which, without any alternating-current bias, would more than close the valve.

However, the above assumption that the effective exposure is proportional to the curve AB of Fig. 2 is not quite correct. Actually the exposure consists of very short pulses reaching considerably higher peak illumination than when no alternating-current bias is used. Consequently the effect of short pulses of light must be considered as compared to a continuous exposure of the same product of intensity by time.

According to the well-known recipiocity law, the effective exposure is proportional to the product of intensity by time, or more exactly $\int Idt$ when the intensity varies during a given interval of time. If this law held exactly, the effective exposure would be the average exposure. However, it is also well known that there is an appreciable deviation from this relation particularly at the high intensities and short exposure time used in sound recording. In general, the effective exposure is decreased with respect to the average exposure when the intensity is increased. Since the alternating-current bias has the effect of increasing the peak intensity for a given average illumination, the effective exposure is decreased and the film transmission as actually measured is shown in curve (B) of Fig. 3.

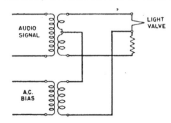

Fig. 4—Method of applying alternating-current bias to light valve.

It will be noted that the curve showing the relation between signal current in the light valve and film transmission is quite linear over a large portion of its length and then gradually approaches the limiting values. This is a type of characteristic noted for "graceful overload" and a minimum of volume or nonlinear distortion. Confirmation of these qualities is shown in the results of recording tests described below.

There are several ways in which the alternating-current bias may be applied to the light valve. In the present experiments the circuit was that shown in Fig. 4. It will be noted that the alternating-current bias is introduced in essentially the same type of circuit as is commonly used for direct-current noise-reduction bias. Standard permanent-magnet two-ribbon light valves were used in all cases, with unbiased ribbon spacing of 0.5 to 1.0 mil. The strain on the ribbon is, of course, reduced for the smaller spacing, and the power requirements for both signal and alternating-current bias are also reduced. Since the valve strings resonate at 8 to 9 kilocycles, the bias power at 24 kilocycles is considerably greater than the maximum speech power. However, it is not enough to injure the ribbons or to require a large power supply.

Positive-type film developed to a IIB sensitometer gamma of

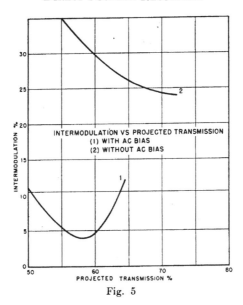

Fig. 5

approximately 2.3 has been found satisfactory, the value of gamma not being critical. Fig. 5 shows the great reduction in distortion obtained by the use of alternating-current bias. It will be noted that

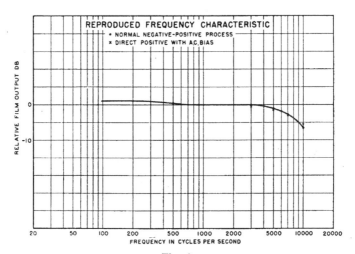

Fig. 6

the minimum distortion occurs near the transmission which gives maximum output. Table I gives other data on the new method as compared to previous methods (100-mil standard tracks, 90 feet per minute in all cases).

TABLE I

	Direct-Positive Process with Alternating-Current Bias				Toe Record		Straight-Line Process*
Light-valve modulation, per cent	100	126	159	200	100	200	100
Film modulation, per cent	56	66.5	74.0	96.5	...	...	56.5
Film output, decibels	66.3	67.8	69.7	71.0	68.7	72.0	57.0
Film noise, decibels	15.0	15.0	15.0	15.0	15.0	15.0	12.5
Signal-to-noise, decibels	51.3	52.8	54.7	56.0	53.7	57.0	44.5
Film projection transmission, per cent	56.9	...	...	...	56.8	...	17.8
Second harmonic, per cent	0.8	2.5	4.0	4.0	8.5	1.5	0.5
Third harmonic, per cent	0.5	1.0	2.0	3.0	0.5	12.0	0.3

* No noise reduction.

Particularly notable in the above table are the figures of 51-decibel signal-to-noise ratio, weighted for 40-decibel level, at low distortion for the new process and 56-decibel signal-to-noise ratio at comparatively low distortion for 100 per cent "overload." Since film noise depends to a large extent on laboratory processing, the above figures on signal-to-noise ratio may vary appreciably depending on the quality of laboratory work.

The use of alternating-current bias has little effect on frequency response, the slightly larger loss at high frequencies due to wider average valve spacing being compensated by the fact that it is not necessary to print a negative to obtain a positive. Fig. 6 shows a frequency-response curve for this method as compared to the normal negative-positive process (35 mm, 18 inches per second). Fig. 7 shows that the input-output characteristic is quite linear over a range of signal inputs up to 200 per cent of normal levels. Records of speech and music made according to the new method fully confirm the expectations indicated by intermodulation and other tests. The cleanness of reproduction and absence of noise reduction effects are particularly noticeable. Present indications are that the new method will find extensive use, particularly in 16-mm and television fields.

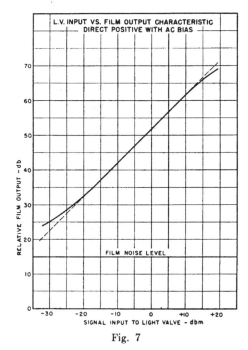

Fig. 7

REFERENCES

(1) W. J. Albersheim, "Device for direct reproduction from variable density sound negatives," J. Soc. Mot. Pict. Eng., vol. 29, pp. 274–281; September, 1937.

(2) Donald MacKenzie, "Straight-line and toe records with the light-valve," J. Soc. Mot. Pict. Eng., vol. 17, pp. 172–203; August, 1931.

DISCUSSION

MR. GEORGE W. SOMES: I understand in this method there is no direct-current noise reduction applied.

MR. C. R. KEITH: No, sir, there is not.

MR. GEORGE LEWIN: Is it possible to apply direct current as well, so as to reduce the noise.

MR. KEITH: I refer that to Dr. Frayne.

DR. J. G. FRAYNE: I believe it is possible, although we have not done so. There is nothing in the theory that does not permit it.

MR. LEWIN: Is it true to say with this system you have a 50 per cent transmission when it is unmodulated and, therefore, it is bound to be noisier than a variable area, which has a direct-current bias in the unmodulated sections?

DR. FRAYNE: In plotting the noise curve, noise on film versus density on film, the noise increases with increasing density, which reaches a density of 0.3 and then the curve flattens over and comes down again. Then you get the best signal-to-noise ratio at transmission of 50 per cent. At variable area you have a density of zero or 1.4. The noise on variable area comes from clear area mostly, so you cannot compare the two.

CHAIRMAN BARTON KREUZER: As I understand this system, the electrical transmission characteristic is curved to offset the curve of the photographic characteristics. If that is so, considering the variations in photographic processing and perhaps in light valves, does it not narrow the photographic processing tolerances and perhaps make them rather critical?

MR. KEITH: The system is not critical with respect to either light-valve or photographic characteristics. The only variations of any consequence between light valves (of the same type) are small differences in sensitivity due to slight variations in ribbon spacing. The alternating-current bias is set with reference to the spacing of a particular valve and this spacing normally remains constant for long periods of time.

As for variations in photographic processing, we have found that the "toe" of the positive development curve is held remarkably constant by film laboratories. The reason for this is that the densities in this region are ones which are important in positive picture development. Any changes in this part of the H-and-D curve would be immediately apparent in picture high lights and flesh tones.

CHAIRMAN KREUZER: Is the system in commercial use?

MR. KEITH: The system is not in commercial use at the present time, but we expect it to be in the near future.

MR. CYRIL J. STAUD: Suppose you made the "toe" sharper for example, what happens to the bias effect? Could you compensate for it?

MR. KEITH: No doubt, by use of a different value of bias current.

MR. HARRY J. REED, JR. Am I correct in assuming that the bias in the wave form is sinusoidal?

MR. KEITH: Yes. There are no special precautions taken with regard to the harmonic content of the frequency. It is essentially sinusoidal.

CHAIRMAN KREUZER: When this system is used as a direct positive, can the film be printed on Kodachrome? Can you use this as a master direct positive for Kodachrome film?

DR. FRAYNE: It can be done; we have done so. I would not say the results were 100 per cent, but you can use Kodachrome for that.

Section Meeting

Central

Approximately 100 members and guests were present at the April 19, 1949, meeting of the Central Section. J. R. Montgomery of the Revere Camera Company delivered the first paper on, "A New General Purpose 16-Mm Sound Projector." The machine described and demonstrated is intended for the medium-priced market composed of home, school, and commercial users of 16-mm sound films. Its single case is readily portable because of its relatively small size and low weight; the loudspeaker is mounted in the removable top cover, which becomes a resonated baffle for increased low-frequency response when placed upon a flat surface. Numerous mechanical design features resulting in both good operating characteristics and significant economies in manufacturing were outlined. Optical and electrical designs are based upon the use of the lead-sulfide phototube because of its inherent high sensitivity, freedom from microphonic noise pickup, excellent signal-to-noise ratio, and electromechanical sturdiness. The exciter lamp is operated at conventional temperatures on supersonic alternating current delivered by an oscillator. An extremely wide volume-control range is afforded by varying both the lamp temperature and the amplifier gain; very little field trouble with dye-type sound tracks has been reported. With so much volume-control range available, tracks which do not fully modulate the light beam can be reproduced at usable level, although with impaired signal-to-noise ratio. The latter is almost inevitably blamed upon the recording, since films with more opaque tracks reproduce normally. The extended control range, plus the relative insensitivity of the lead-sulfide phototube to voltage-supply variations, make the projector usable under the widely varying line-voltage conditions encountered in field use.

The second presentation of the evening was given by Eugene W. Beggs of the Westinghouse Electric Corporation's Lamp Division, under the title of "New Developments in Cadmium-Mercury Lamps and Other Vapor and Gas-Discharge Lamps for Motion Picture and Television Studio Lighting." This was originally given at the fall 1948 Washington convention of the Society of Motion Picture Engineers and was re-presented at this section meeting for the benefit of the local studio people.

Mr. Beggs gave a brief history of the development of gas-discharge lamps and outlined their salient characteristics. Mercury-vapor lamps of high efficiency and large lumen output have been unsuitable for color photography because of their lack of output in the red end of the spectrum. This deficiency has been corrected in lamps now under development by adding cadmium to the mercury. Experimental lamps of this type were demonstrated in direct comparison with plain mercury lamps using various colored objects to show the improved color composition of the light. Auxiliary power-supply equipment was described and illustrated in slides. Mr. Beggs stated that while the new lamps are not yet commercially available, it is probable that eventually they will find wide use in motion picture and television studio lighting because of their efficiency, excellent color quality, and convenience in operation.

699

66th Semiannual Convention

—PAPERS PROGRAM

Authors who plan to prepare papers for presentation at the 66th Convention should write at once for Authors' Forms and important instructions to the Papers Committee member listed below who is nearest. Authors' Forms, titles, and abstracts must be in the hands of Mr. Grignon by August 15 to be included in the Tentative Program, which will be mailed to members thirty days before the Convention.

N. L. SIMMONS, *Chairman*
6706 Santa Monica Blvd.
Hollywood 38, California

J. E. AIKEN, *Vice-Chairman*
116 N. Galveston St.
Arlington, Virginia

LORIN GRIGNON, *Vice-Chairman*
20th Century-Fox Films Corp.
Beverly Hills, California

E. S. SEELEY, *Vice-Chairman*
Altec Service Corp.
161 Sixth Ave.
New York 13, New York

R. T. VAN NIMAN, *Vice-Chairman*
4501 Washington Blvd.
Chicago 24, Illinois

H. S. WALKER, *Vice-Chairman*
1620 Notre Dame St., W.
Montreal, Que., Canada

————————— ♦ —————————

Meetings of other Societies During 1949

August—
Institute of Radio Engineers West
Coast Convention

August 29 through September 1
San Francisco, California

September—
Illuminating Engineering Society
National Technical Conference

September 19 through September 23
French Lick, Indiana

October—
Photographic Society of America
Convention
Optical Society of America
Annual Meeting

October 19 through October 22
St. Louis, Missouri
October 27 through October 29, 1949
Buffalo, New York

October-November—
Radio Fall Meeting
Joint IRE-RMA

October 31 through November 1
Syracuse, New York

March, 1950—
Optical Society of America
Winter Meeting

March 9 through March 11, 1950
New York, New York

Current Literature

THE EDITORS present for convenient reference a list of articles dealing with subjects cognate to motion picture engineering published in a number of selected journals. Photostatic or microfilm copies of articles in magazines that are available may be obtained from The Library of Congress, Washington, D. C., or from the New York Public Library, New York, N. Y., at prevailing rates.

American Cinematographer
30, 4, April, 1949
Technicolor Photography Under Water (p. 122) J. HOUSLER
Directors of Photography Report on Television Research (p. 124) J. FORBES
Films for Television (p. 125) N. KEANE

Audio Engineering
33, 5, May, 1949
Conversion Method to Increase Film Recording Studio Facilities (p. 15) N. T. PRISAMENT

Communications
29, 4, April, 1949
Variable-Density Recording on 16-Mm Film for TV (p. 31) L. W. MARTIN

Electronics
22, 5, May, 1949
High-Speed Production of Metal Kinescopes (p. 81) H. P. STEIER and R. D. FAULKNER

Tele-Tech
8, 5, May, 1949
Lighting Requirements of Television Studios. Pt. 1 (p. 24) R. E. BLOUNT

International Photographer
21, 4, April, 1949
History and Development of the Animated Film (p. 5) J. V. NOBLE
The New Light (p. 14) A. WYCKOFF
Just What Is Television. Pt. III (p. 18) J. H. WILLOUGHBY
Care of Lenses (p. 20) A. E. MURRAY
21, 5, May, 1949
History and Development of the Animated Film (p. 13) J. V. NOBLE
Recording Television Images on Film (p. 18)

International Projectionist
24, 4, April, 1949
Projection Preparation for the "Seasonal" Theatre (p. 7) R. A. MITCHELL
The Use of Films in Television (p. 12)

RCA Review
10, 1, March, 1949
Development of a Large Metal Kinescope for Television (p. 43) H. P. STEIER, J. KELAR, C. T. LATTIMER, and R. D. FAULKNER
The Graphecon—A Picture Storage Tube (p. 59) L. PENSAK

701

～ New Products ～

Further information concerning the material described below can be obtained by writing direct to the manufacturers. As in the case of technical papers, publication of these news items does not constitute endorsement of the manufacturer's statements nor of his products.

Variable-Speed Motor

A new variable-speed motor with tachometer for the 16-mm camera field to fit the Cine Special and Maurer cameras has been developed by **National Cine Equipment, Inc.,** 20 W. 22 Street, New York 10, N. Y.

It has a professional-type motor, designed for complete versatility, compactness, economy, and interchangeability, speed range of 8 to 50 frames per second, variable speed, determined by the mechanical governor and read on a bold-faced tachometer graduated in frames per second, facing the cameraman for easy reading and operation.

The basic unit of the motor, consisting of a separate base for Cine Special camera, with interchangeable motor, is as follows: 115-volt alternating- or direct-current, universal; variable speed with tachometer; 12-volt direct-current variable speed with tachometer; 115-volt, 60-cycle, alternating-current, single-phase, synchronous; and 220-volt, 60-cycle, alternating-current, three-phase, synchronous.

For use on the Maurer camera a special adapter plate is attached to the camera and any of the above motors can be used.

Book Review

Hochstromkohlebogen, by Wolfgang Finkelnburg

Published (1948) by Springer-Verlag, Berlin, Germany. Paper covered. 214 pages + 4-page bibliography + 3-page index + viii pages. 132 illustrations. $6^1/_2$ x $9^1/_2$ inches. Price, $22.50.

This book, in German, is substantially identical with the English work, "The High-Current Carbon Arc," by the same author, which was reviewed on page 112 of the January, 1949, issue of the JOURNAL. A number of minor additions and corrections have been made in the text of the subject edition, and the illustrations are much superior to those in the English version. Particular attention is called to Figs. 122 and 126, and the associated texts, which refer to the 1000-ampere arc stream and to the very fine 450-ampere searchlight arc lamp mechanism, respectively.

urnal of the
ciety of Motion Picture Engine

INDEX

Volume 52

January–June, 1949

INDEX TO AUTHORS

Volume 52

January, 1949, through June, 1949

3

INDEX TO SUBJECTS

Volume 52

January, 1949, through June, 1949

6

---◆---

JOURNAL OF THE SMPE AND INDEXES

Nonmembers and libraries may subscribe to the JOURNAL at the rate of $12.50 per year.

A ten-year Index, covering the years 1936 through 1945 is available at $2.00 per copy. Previous Indexes for 1916–1930 and 1930–1935 are $1.25 each.

In the June and December issues of the JOURNAL for each year there will be found Indexes for the preceding six-month period.

A limited number of back issues of the JOURNAL and the TRANSACTIONS are also available.

Please address all inquiries to

The Society of Motion Picture Engineers
342 Madison Avenue, New York 17, N. Y.

INDEX TO NONTECHNICAL SUBJECTS

Volume 52

January, 1949, through June, 1949

Lightning Source UK Ltd.
Milton Keynes UK
UKHW040311060119
335017UK00009B/352/P